ASTROCYTES AND EPILEPSY

science & technology books

Companion Web Site:

http://booksite.elsevier.com/9780128024010

ASTROCYTES AND EPILEPSY

Jacqueline A. Hubbard and Devin K. Binder

Please access the following content on the Companion Website:

- About the Authors
- About the Book
- Volume Table of Contents
- Chapter Abstracts
- Volume Tables
- Figures in .jpg
- Figures in Powerpoint
- References

ASTROCYTES AND EPILEPSY

JACQUELINE A. HUBBARD and DEVIN K. BINDER
Center for Glial–Neuronal Interactions,
Division of Biomedical Sciences,
School of Medicine, University of California,
Riverside, CA, USA

AMSTERDAM • BOSTON • HEIDELBERG • LONDON
NEW YORK • OXFORD • PARIS • SAN DIEGO
SAN FRANCISCO • SINGAPORE • SYDNEY • TOKYO
Academic Press is an imprint of Elsevier

Academic Press is an imprint of Elsevier
125 London Wall, London EC2Y 5AS, United Kingdom
525 B Street, Suite 1800, San Diego, CA 92101-4495, United States
50 Hampshire Street, 5th Floor, Cambridge, MA 02139, United States
The Boulevard, Langford Lane, Kidlington, Oxford OX5 1GB, United Kingdom

British Library Cataloguing-in-Publication Data
A catalogue record for this book is available from the British Library

Library of Congress Cataloging-in-Publication Data
A catalog record for this book is available from the Library of Congress

ISBN: 978-0-12-802401-0

For Information on all Academic Press publications
visit our website at https://www.elsevier.com/

Publisher: Mara Conner
Acquisitions Editor: Melanie Tucker
Editorial Project Manager: Kristi Anderson
Production Project Manager: Chris Wortley
Designer: Mark Rogers

Typeset by MPS Limited, Chennai, India

Image on the cover: Marked histological changes occurring during epileptogenesis in the mouse hippocampus. *Top:* healthy hippocampus. *Bottom:* 21 days after intrahippocampal kainic acid-induced status epilepticus (corresponding to the epileptic state with spontaneous seizures). Cell nuclei are stained dark blue (DAPI marker). Neurons (red, NeuN marker) occupy organized cell layers (dentate gyrus, CA1–CA3) in the healthy hippocampus, but marked neuronal loss (eg loss of CA1 pyramidal cells) and cell layer disorganization (eg dentate granule cell dispersion) is observed in the epileptic hippocampus. Astrocytes (green, GFAP marker) and microglia (cyan, Iba1 marker) display normal morphology in the healthy hippocampus (*top*) but develop markedly abnormal "reactive" morphology in epilepsy (*bottom*). These changes together account for "hippocampal sclerosis", a hallmark of temporal lobe epilepsy.

Dedication

Jacqueline A. Hubbard

To Oleg for taking care of me when I was too busy to take care of myself

Devin K. Binder

For my wife Kellie
And my children Alexis and Grant
Whose love and dedication
Sustain and nourish me daily

Contents

Preface

Epilepsy is a devastating group of neurological disorders characterized by periodic and unpredictable seizure activity in the brain. There is a critical need for new drugs and approaches given that at least 1/3 of all epilepsy patients do not become seizure-free with existing medications and become "medically refractory". Much of epilepsy research has focused on neuronal targets, but current drugs cause severe cognitive, developmental, and behavioral side effects. Recent discoveries regarding the role of astrocytes in neuronal and network excitability and seizure activity together with the changes in astrocytes that occur in epileptic brain tissue will be reviewed and discussed in this book. These new insights indicate that astrocytes may represent an important therapeutic target in the control of epilepsy. There are several good books on astrocytes but there is no current book on astrocytes and epilepsy that brings together all of the exciting recent studies on astrocytes and epilepsy.

This book will uniquely provide both a resource on recent developments in the field of astrocyte biology and neuroscience related to astrocytes as well as providing an overview of the field of epilepsy and how animal models and human studies in epilepsy tissue have shown a significant role of astrocytes. Following an introduction to the history and modern conception of astrocytes, we summarize types of epilepsy and varieties of human neuropathology, then explore the role of specific astrocyte systems in epilepsy. In particular, individual chapters cover astrocyte calcium signaling, potassium channels, water channels, glutamate metabolism, adenosine metabolism, gap junctions, blood-brain barrier disruption, and inflammation. Finally, we synthesize information about astrocytes and epilepsy and suggest new potential new therapeutic targets.

This book is intended for both scientists and clinicians. General neuroscientists, as well as investigators in the field of astrocytes, will find an overview of basic scientific findings and translational studies relevant to astrocytes and epilepsy. Epilepsy researchers who have not yet fully considered the interaction of astrocytes and epilepsy will hopefully be stimulated by this book to pursue glial cell research. Students of all levels, in particular graduate students and postdoctoral students, will learn about astrocyte biology in health and disease. Clinicians and clinician-scientists in the field of epilepsy including neurologists and other specialists involved in the care of patients with epilepsy will find plenty herein of translational interest. Last, translational drug development opportunities outlined in the chapter on potential new therapeutic targets will be of interest to pharmaceutical companies involved in development of new antiepileptic drugs. We hope that this book will serve as a frequently referenced learning tool for all levels of interested scientists, healthcare, and industry professionals. We will have succeeded in our goals if we have fired the interest of our readers in this exciting and rapidly emerging topic and ultimately if development in this field contributes to patient care.

Jacqueline A. Hubbard and
Devin K. Binder

Acknowledgments

I would like to acknowledge three eminent scientists who form the "glutamate, water, and potassium" of my own individual synaptic gliostasis. To Helen Scharfman (GLUTAMATE), seminal explorer of hippocampal electrophysiology, for unflagging support of a developing physician-scientist and gentle guidance and endless ideas. To Alan Verkman (WATER), seminal explorer of aquaporin biology, for enabling me to join his laboratory for a time and initiate studies of aquaporins, seizures, and extracellular space physiology. To Christian Steinhäuser (POTASSIUM), seminal explorer of astrocyte physiology and pathology, for enabling me to join his laboratory and teaching me a great deal about the cells that now have a STARRING role in the brain. To all three, for friendship and ongoing collaboration. To have been connected with them in intellectual syncytium has been glorious. Unlike epileptic astrocytes, may we never become uncoupled!

Helen Scharfman

Alan Verkman

Christian Steinhäuser

I would like to acknowledge my colleagues at UC Riverside, including Todd Fiacco, Emma Wilson, Seema Tiwari-Woodruff, Monica Carson, Khaleel Razak, Iryna Ethell, Hyle Park, Victor Rodgers, Guillermo Aguilar, Ameae Walker, Jonny Lovelace, Jenny Szu and Jennifer Yonan.

Last, I would like to acknowledge my mother and father, both synthetic organic chemists, for instilling a love of science, creativity and discovery.

Devin K. Binder

I would like to acknowledge my family. To my father who got me interested in science and has served as my role model for my entire life. To my mother who taught me to be strong and to never give up. To my siblings who are all exceptional human beings—I worked hard to keep up with all of you!

I would also like to acknowledge my lab mates, Jen and Jenny, for their continuous support.

Jacqueline A. Hubbard

CHAPTER

1

History of Astrocytes

OUTLINE

Astrocytes and Epilepsy
DOI: http://dx.doi.org/10.1016/B978-0-12-802401-0.00001-6

OVERVIEW

In this introduction to the history of astrocytes, we wish to accomplish the following goals: (1) contextualize the evolution of the concept of neuroglia within the development of cell theory and the "neuron doctrine"; (2) explain how the concept of neuroglia arose and evolved; (3) provide an interesting overview of some of the investigators involved in defining the cell types in the central nervous system (CNS); (4) select the interaction of Wilder Penfield and Pío del Río-Hortega for a more in-depth historical vignette portraying a critical period during which glial cell types were being identified, described, and separated; and (5) briefly summarize further developments that presaged the modern era of neurogliosci-ence. In an endeavor of this kind, one must be sure to acknowledge those authors who have previously made strong attempts to synthesize information regarding the history of the concept of neuroglia. In particular, the efforts and publications of Helmut Kettenmann, Bruce Ransom, George Somjen, Alexei Verkhratsky, Arthur Butt, and Vladimir Parpura should be commended highly [1–7]. In contradistinction to the historical aspect of this chapter, see *Chapter 2: Astrocytes in the Mammalian Brain* will describe a more modern understanding of astrocytes in the mammalian brain.

NEURON DOCTRINE

Neuron theory or neuron doctrine asserts that nerve tissue is composed of individual cells which are anatomical and functional units called neurons. While generally attributed to Wilhelm Waldeyer who coined the term "neuron" in 1891 [8] and the neurohistological work of Santiago Ramón y Cajal [9], there were many scientific antecedents enabling the development of the neuron doctrine [10–13]. Microscopic imaging of the nerve cell in 1836 by Gabriel Valentin, visualization of the axon in 1837 by Robert Remak, development of "cell theory" by Schleiden and Schwann (1838–39), and discovery of the "Purkinje" cell by Jan Purkinje in 1839 were all seminal. The early proponents of the neuron doctrine, Wilhelm His, Fridtjof Nansen, and Auguste Forel, based partly on direct observations of Golgi-stained nerve cells and partly on credible generalizations, independently concluded that neurons are likely to be separate units [14–16]. In this section, we will explore the scientific antecedents to the neuron doctrine and the fascinating investigators involved.

Purkinje and Valentin's *Kugeln*

Jan Evangelista Purkinje (1787–1869) (Fig. 1.1) served as the Chair of Physiology in Breslau (the modern-day Polish city, Wrocław) and made many valuable contributions to cell biology and neurobiology. Gabriel Valentin (1810–83) was one of his students assigned to study the nervous system. In 1836, Valentin described the nerve cell with remarkable accuracy using water-prepared tissue. He noticed small *Kugeln* (globules) which were clearly nerve cells. He described that the globule had a sharp outline (cell membrane) and an interior substance called *parenchymasse* (parenchyma) filled with viscous fluid containing numerous granules. Inside this was the nucleus and within it a small corpuscle (nucleolus).

FIGURE 1.1 **Jan Evangelista Purkinje (1787–1869).**

The very first description of the nerve cell body was made. Valentin and his mentor Purkinje had generally implied the idea of a "cell" but it was Schwann who proposed the unified "cell theory" and extolled its significance.

Scheiden, Schwann, and Cell Theory

> The development of the proposition, that there exists one general principle for the formation of all organic productions, and that this principle is the formation of cells, as well as the conclusions which may be drawn from this proposition, may be comprised under the term cell-theory, using it in its more extended signification, whilst in a more limited sense, by theory of the cells we understand whatever may be inferred from this proposition with respect to the powers from which these phenomena result [17,18].

Matthias Jakob Schleiden (1804–81) (Fig. 1.2) was a lawyer who went back to school to study medicine and eventually became a botanist. In 1838, Schleiden published a paper on the origins of the plant cell; despite some flawed conclusions, this paper contained the essential idea that plants are composed of cells [19]. Schleiden's friend, the zoologist Theodor Schwann (1810–82) (Fig. 1.3), heard one of his ideas over dinner one October evening in 1838 [20]. Together, they went to the laboratory and examined sections of spinal cord that Schwann studied. Schleiden recognized the nuclei and their similarity to those of plant cells. That night, the first notion of a unified cellular theory was born. A year later, in 1839, Schwann published the cell theory for all living tissues (plants and animals) [17]. Interestingly, no mention of Schleiden was made! (Incidentally, it was Schleiden who influenced the young Carl Zeiss to form his subsequently very important optical firm, the fruits of which were much improved microscopes to the present day.)

FIGURE 1.2 **Matthias Jakob Schleiden (1804–81).**

FIGURE 1.3 **Theodor Schwann (1810–82).**

Remak's Remarkable Observation

In his 1838 thesis, Robert Remak (1815–65) (Fig. 1.4) described peripheral nerves as a "primitive band" (axon) within a thin walled "primitive tube" (myelin sheath). These bands were later referred to as the "band of Remak" and synonymous with axon or axis cylinder. He also described sympathetic nerves which lacked the tubular covering and called them primitive fibers. They were later referred to as the "fiber of Remak." A third contribution was relationship between the nerve and the nerve fiber: he stated that "[t]he organic fibers

FIGURE 1.4 **Robert Remak (1815–65).**

arise from the very substance of the nucleated globules." Remak never conclusively backed his statement with evidence. It was a "remarkable observation" and in many ways years ahead of his time. The water preparation of these specimens and the associated swelling or shrinkage cast doubt over the accuracy of these findings. However, Remak was vociferous in his opinions and publicly acknowledged that his findings were contrary to Gabriel Valentin's findings. In the pinnacle of his career, Remak was the first to suggest in 1855 that

1: Each multipolar anterior horn cell connects with one motor root fibre.
2: The other processes of the cells are (physically and chemically) different from that fibre.

This is the first recorded, written statement alluding to the principle that a nerve cell gives rise to one process, different from the rest, which extends out to become an axon [13].

The Neurohistology of Robert Bentley Todd

Robert Bentley Todd (1809–60) (Fig. 1.5) was an anatomist, neurologist, and physiologist best known for the eponym "Todd's paralysis" [21,22]. However, his neurohistological contributions are not well known [23–25].

Todd was aware of Schwann's cell theory; however, he was not sure of the nature of the connection between "nerve vesicles" (cell bodies) and "nerve tubes":

> It is in vain, in the present state of our knowledge, to speculate upon the use of these caudate processes. Do they constitute a bond of union between the nerve-vesicles and certain nerve-tubes? or are they commissural fibers serving to connect the grey substance of different portions of the nervous centres? Until a more extended research has made us better acquainted with peculiarities of these vesicles in various localities, it would be premature to offer any conjecture concerning their precise relation to the other elements of the nervous centres. They exist, with different degrees development, in the locus niger of the crus cerebri, in the laminae of the cerebellum, in the gray matter of the spinal cord and medulla oblongata, and in the ganglions, and in the grey substance of the cerebral convolutions, in which latter situation they are generally of small size [24].

FIGURE 1.5 **Robert Bentley Todd (1809–60).**

Todd appeared to maintain a neutral position and supported neither the reticular theory nor the neuron theory, holding that more study was required to draw a scientifically sound conclusion to this debate. Despite the subsequent excitement over the cell theory [11], it was not widely accepted at the time that cell theory could be applied to the nervous system [12]. Todd and Bowman also described the environment in which the nerve-vesicles exist in greater detail. While the nerve-vesicles are not physically touching, "they are either imbedded in a soft, granular matrix, as in the brain, or enveloped in a capsule of nucleated cells, as in the ganglia" [26]. Perhaps here they are describing supportive cells, such as astrocytes and oligodendroglia (not well distinguished until the work of Pío del Río-Hortega, one of Cajal's disciples).

Todd and Bowman describe the cross-section of a myelinated axon or dendrite as a white substance of Schwann (not termed myelin until Virchow in 1858) forming a tube around the "axis cylinder" (not termed "axon" until Kölliker in 1896) and further suggested that a difference in function exists between the "white substance" and the "nerve tube":

> It is a conjecture by no means devoid of probability, that the processes of the caudate vesicle may, after passing some way, become invested by the tubular membrane and by the white substance of Schwann [26].

While Todd and Bowman do describe the cytoarchitecture of the nerve-vesicle, they do not accurately describe its connections. Instead of suggesting that the axis cylinder is contiguous with the cell body, Todd and Bowman do not distinguish between the myelin sheath and the neurite processes, naming the two components the "fibrous nervous matter." They do state that "the nerve-vesicles do not lie in immediate contact with each other" [26].

The exact nature of this "juxtaposition" was necessarily vague during Todd's era, but this statement certainly seems closer to neuron doctrine than to reticularism (with no "admixture" of material). Todd was acutely aware of their inability to visualize actual nerve connections:

> How the fibres comport themselves with respect to the elements of the vesicular matter is not exactly known. It is certain, however, that the nerve-tubes frequently adhere to the sheaths of nerve-vesicles, and that many of them pass between the nerve-vesicles, probably to form a connection with more distant ones [26].

He recognized that they did not yet possess the proper histological techniques to study "the intimate connection of this granular sheath to the vesicle, and to its processes when they exist" [26]. It would not be until Ramón y Cajal fully utilized and perfected Camillo Golgi's *reazione nera* ("black reaction" or silver stain, developed in 1873) that clear morphological evidence supporting the neuron doctrine could be adduced. Even so, as is well known, Golgi still defended reticularism in their joint (and contentious) Nobel ceremony of 1906.

While the overwhelmingly important work of Ramón y Cajal in solidifying the acceptance of the neuron doctrine has been well understood and described [10], the earlier events leading up to the acceptance of cell theory in the nervous system are not as often recognized [27]. In this, Robert Bentley Todd's contributions have for the most part been overlooked. He proposed that the nerve-vesicle was the primary center for electrical activity in the brain and outlined its histology. At the time, Todd (and Bowman) did not have the appropriate techniques to elucidate further anatomic details of cell–cell interactions in the CNS; and furthermore the functional concept of the synapse was not introduced until Sherrington in 1897. Nevertheless, just as Todd's electrical theory of epilepsy predated that of Hughlings-Jackson by 20 years [22]. Todd identified the primacy of the "nerve-vesicle" more than 40 years before Waldeyer's 1891 formulation of the neuron doctrine [8].

Wilhelm His' Seminal Contributions

Wilhelm His Sr. (1831–1904) (Fig. 1.6) was a pupil of Robert Remak and a noted embryologist. He trained under several renowned developmental scientists such as Rudolf Virchow and Albert von Kölliker. As the professor of Anatomy and Physiology at University of Leipzig, His was the first to observe the development of the nerve cell. In 1886, he published that the nerve cell and its prolongations formed an independent unit [12]. Furthermore, he elucidated that the function of a nerve cell was based on proximity rather than continuity. In his studies on the peripheral nervous system, he observed that the nerve fiber was in close proximity to the muscle but not continuous [28]. He reasoned that the same underlying principle applied to the nerve cells in the CNS. In 1890, His coined the term "dendrites" to describe the branching, ramifying processes of the nerve cells [29]. Thus, he made a profound contribution to the neuron theory although his subsequent work focused exclusively on development and embryology. His son Wilhelm His Jr. (1863–1934) became a cardiologist and anatomist and discovered the eponymous "bundle of His."

FIGURE 1.6 **Wilhelm His Sr. (1831–1904).**

Fridtjof Nansen: The Renaissance Man

Fridtjof Nansen (1861–1930) (Fig. 1.7) was a Professor of Zoology and Oceanography at the Royal Frederick University of Oslo. He made several remarkable contributions to neuroscience through his study of marine animals. Nansen embarked on an Arctic exploration to Greenland and after his return in 1882 he became the curator at the Bergen museum. Here, he interacted with experts studying the peripheral nervous system in marine animals. In 1886, he left to Pavia to train under the master neurohistologist Camille Golgi. He returned to Bergen museum and published his findings that nerve cells were bound by membranes and touched each other but were *not fused to each other*, an important step in the subsequent "neuron doctrine" [30]. He coupled this publication with unique illustrations with hand-etched figures [31]. Interestingly, he also postulated that neuroglia might be the "seat of intelligence, as it increases in size from the lower to the higher forms of animal" [30,32]. Eventually in 1887, he defended his dissertation and left for his second expedition to the Arctic where he was marooned for over a year [33]. After his rescue from the North Pole, his mind drifted to diplomacy and state affairs. He worked in the League of Nations, was its first president, developed the refugee passport, and helped establish monarchy in the newly independent Norway. He received the Nobel Peace prize in 1922 for his humanitarian work.

Auguste Forel: Neurohistology, Myrmecology, and Sexology

Auguste-Henri Forel (1848–1931) (Fig. 1.8) was a Swiss psychiatrist, trained under Von Gudden in Munich in neuroanatomy and psychiatry. He thus became a proponent of the

FIGURE 1.7 **Fridtjof Nansen (1861–1930).**

FIGURE 1.8 **Auguste-Henri Forel (1848–1931).**

Gudden method, which applied the technique of secondary degeneration to study the interrelationship between cortex and subcortical structures. After his medical training in Munich, Forel took the position of Professor of Psychiatry at the Burghölzli asylum in Zurich. Here in solitude, he pursued his neuroanatomical work and developed an interest in ant colonies and their social structure. Using the Gudden method of controlled retrograde degeneration, he published in 1887 that all nerve fibers were processes of cells [10].

> I think all the systems of fibres and the so-called networks of fibres of the nervous system are no more than nervous processes of cells that ramify to a greater or lesser distance in the form of ramifications within other ramifications, but not anastomosized [34].

Interestingly, Forel and Wilhelm His both began writing their independent assertions, unaware of each other's work, that later formed the tenets of the neuron doctrine in 1886. But, due to a publishing delay, Forel's work was printed and disseminated later than Wilhelm His' work.

During his service at Burghölzli asylum, Forel was fascinated by his studies on myrmecology and even named his home as La Fourmilière (The Ant Colony) and gradually shifted his focus away from neurohistology. In his later years, he suffered a stroke on his right side, learned to use his left hand and authored several controversial publications in sexology, eugenics, and penal law related to mental illnesses [35].

Rudolph Albert Von Kölliker: Neurohistologist and Cajal Champion

Rudolph Albert von Kölliker (1817–1905) (Fig. 1.9) was a prolific neurohistologist and embryologist and among other things named the "axon." He is well known for his role in the dissemination of Cajal's observations, but Kölliker was initially a firm proponent of the

FIGURE 1.9 **Rudolph Albert von Kölliker (1817–1905).**

reticular theory. In 1889, the German Anatomical Society meeting was held at the University of Berlin and both Cajal and Kölliker were present. At that time, Kölliker was already a respected scientist and a giant in the field of histology. In this meeting, Cajal was assigned a table for demonstration of his neurohistology slides. Cajal's phenomenal staining of the nerve cells made a lasting impression on Kölliker's mind, despite Cajal's broken German and halting French [36]. Kölliker invited the young Spaniard to a private dinner and introduced him to several renowned German histologists. The extraordinary contribution to the neuron doctrine by Kölliker was his vocal support and enthusiastic promulgation. Despite being in his seventies, Kölliker wrote to several histologists extolling Cajal's observations, publicly abandoned the erroneous reticular theory and adopted the neuron theory [9]. Even more extraordinary is the fact that Kölliker himself translated some of Cajal's work from Spanish to German, accelerating the acceptance of Cajal's work.

Waldeyer and the *Neuronlehre* (Neuron Doctrine)

Heinrich Wilhelm Gottfried von Waldeyer-Hartz (1836–1921) (Fig. 1.10) published over 269 papers spanning a range of topics from gross anatomy, histology, physiology, pathology, anthropology, education, history to the arts [37]. In a series of review articles in 1891 titled "Über einige Neuere Forschungen im Gebiete der Anatomie des Centralnervensystems" (On some recent research in areas of the anatomy of the CNS) published in *Deutsche Medizinische Wochenschrift* (German Medical Weekly), Waldeyer founded the word "neuron" from the Greek word "sinew." The doctrine of separate, distinct entities as the building blocks of the CNS had been formulated by others before him based on histological works of Ramón y Cajal and Nansen [38]. However, Waldeyer's new term "neuron" bestowed this doctrine and its functional entity awareness and consequent acceptance of the concept.

FIGURE 1.10 **Wilhelm Waldeyer (1836–1921).**

Waldeyer was probably directly influenced by having read Kölliker's recent German translation of Cajal's works. He had a specific talent for coining new scientific terms. He had also named "chromosome," "plasma cells," and "gastric canal" before coining the term "neuron" [37,39,40]. Waldeyer was the Chair of Anatomy and Professor at the prestigious research institution of Berlin University, a military surgeon and a consulting physician for Emperor of Prussia Friedrich III, which indubitably helped the new term "neuron" to gain prominence. Waldeyer unified diverse and controversial data into a single coherent theory and published in a popular, widely read journal (notably without contributing a single original observation) and yet obtained worldwide and long-standing recognition for the discovery of the "neuron" [41].

Conclusion

The neuron doctrine was based on a series of observations that culminated in the concept of the neuron as the single independent unit in the nervous system. It is clear from the work described above that several important investigators laid important groundwork for the neuron doctrine. Jan Purkinje, Gabriel Valentin, Robert Remak, and Robert Bentley Todd made important contributions from 1836–45. However, none of these scientists directly formulated the neuron doctrine. Nearly half a century later, Wilhelm His, Fridtjof Nansen, and Auguste Forel independently published documents anticipating the neuron doctrine within a span of 1 year (1886–87). Interestingly, these three investigators inferred similar conclusions despite different methodologies such as retrograde degeneration, development of nerve cells, and marine biology. Thus, while Kölliker gets credit for championing Cajal's work after 1889 and Waldeyer published the neuron doctrine in 1891, it is clear that in the development of neurohistology in the 19th century, these previous scientists played a crucial part.

DEVELOPMENT OF THE CONCEPT OF NEUROGLIA

Rudolf Virchow

> Ich habe bis jetzt, meine Herren, bei der Betrachtung des Nervenapparatus immer nur der eigentlich nervösen Theile gedacht. Wenn man aber das Nervensystem in seinem wirklichen Verhalten im Körper studieren will, so ist es ausserordentlich wichtig, auch diejenige Masse zu kennen, welche zwischen den eigentlichen Nerventheilen vorhanden ist, welche sie zusammenhäund dem Ganzen mehr oder weinger seine Form gibt.
>
> [Hitherto, gentlemen, in considering the nervous system, I have only spoken of the really nervous parts of it. But if we would study the nervous system in its real relations in the body, it is extremely important to have a knowledge of that substance also which lies between the proper nervous parts, holds them together and gives the whole its form in a greater or lesser degree] [42].

Rudolph Virchow (1821–1902) (Fig. 1.11) was primarily a pathologist who studied diseased tissue. The above quote comes from a lecture that Rudolf Virchow delivered to medical students at the Charité Hospital in Berlin on April 3, 1858. He had previously introduced the actual term *Nervenkitt* (neuroglia) 2 years earlier as a "connective substance, which

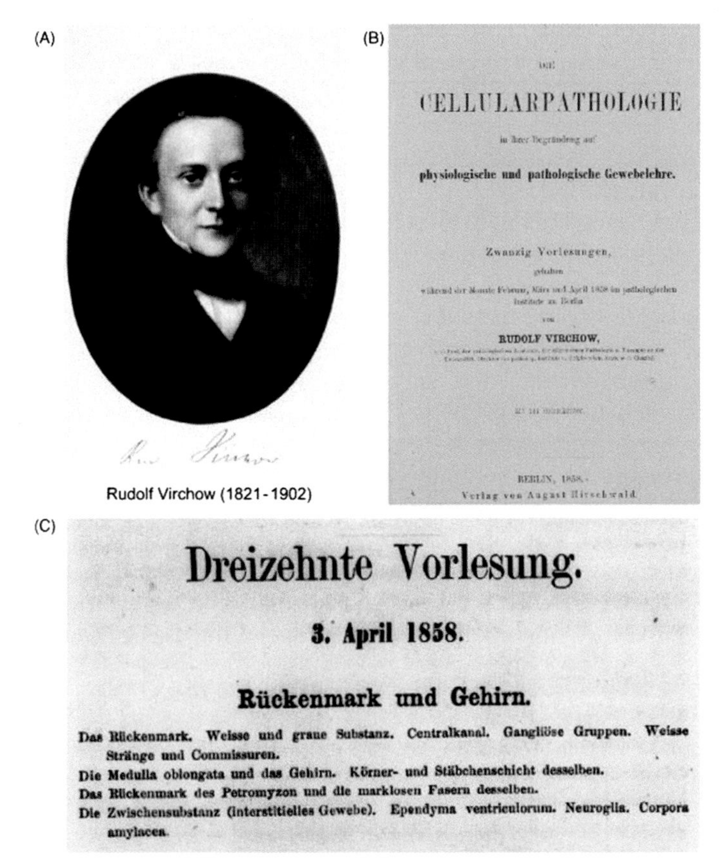

Dreizehnte Vorlesung.

3. April 1858.

Rückenmark und Gehirn.

Das Rückenmark. Weisse und graue Substanz. Centralkanal. Gangliöse Gruppen. Weisse Stränge und Commissuren.
Die Medulla oblongata und das Gehirn. Körner- und Stäbchenschicht desselben.
Das Rückenmark des Petromyzon und die marklosen Fasern desselben.
Die Zwischensubstanz (interstitielles Gewebe). Ependyma ventriculorum. Neuroglia. Corpora amylacea.

FIGURE 1.11 **Rudolf Virchow (1821–1902) and the concept of neuroglia.** (A) Portrait of Rudolf Virchow in the 1850s. Rudolf Virchow was born on October 13, 1821, in Schivelbein, which was then under the rule of the Prussian kingdom and now is a city of Swidwin in Poland. He studied medicine in Berlin and worked as a pathologist at the Charité. After the failure of the 1848 revolution, in which he had actively participated, he was forced to leave Berlin and to move to Würzburg, where he became Professor of Pathology. He returned to Berlin in 1856 and occupied the Chair in Pathology for the rest of his life. He had a broad interest in science, ranging from cancer research and neuroscience to anthropology and, as the editor-in-chief of Virchow's Archiv, he could oversee it all. Rudolf Virchow was not only a highly influential scientist, but was actively engaged in different aspects of political and cultural life. He initiated laws for meat inspection at slaughterhouses and, as a member of the German parliament, he was instrumental in installing a modern sewage system in the city of Berlin, borne out of the recognition that there is a relationship between infections and hygienic conditions. His interest in anthropology let him to participate in excavations carried out by his friend Heinrich Schliemann in Troy, and it was Virchow who convinced Schliemann to donate the treasures of Priam to the city of Berlin. Virchow assembled a tremendous collection of pathologic specimens and, at the end of his life, he opened a Pathologic Museum, not only for students and medical practitioners, but also for the public. He died in Berlin on September 5, 1902. (B) The frontispiece of *Cellular Pathology*, published in 1858. (C) Lecture 13 ("Spinal cord and the brain") from *Cellular Pathology*, where the name neuroglia was first coined. *Source: Reproduced with permission from Verkhratsky A, Butt AM. Glial physiology and pathophysiology. Oxford, UK: Wiley-Blackwell, 2013 (Figure 1.13).*

forms in the brain, in the spinal cord, and in the higher sensory nerves a sort of Nervenkitt (neuroglia), in which the nervous system elements are embedded" [43]. Thus, in his invention of the term neuroglia in 1856, Virchow thought of the neuroglia (nerve glue) as a true connective tissue (*Zwischenmasse* or "in between tissue") rather than containing individual cell types. Note however, from above that cell theory had already been introduced by Schleiden and Schwann in 1838–39.

Twenty of Virchow's lectures delivered between February and April 1858 were published as *Cellular pathology* [42]. This contained Virchow's first illustrations of neuroglia (Fig. 1.12). In this volume he also explains that "the real cement, which binds the nervous elements together, and that in all its properties it constitutes a tissue different from the other forms of connective tissue, has induced me to give it a new name, that of *neuroglia* (nerve cement)" [42]. *Neuroglia* is variously translated as "nerve cement," "nerve glue," and "nerve putty" [4–6]. This would seem to indicate merely an acellular matrix, however, he goes on to add elsewhere that "where neuroglia is met with, it also contains a certain number of cellular elements" [42]. Ultimately, as Kettenmann and Ransom indicate [1], the term *neuroglia* came to refer to *neuroglial cells* as its cellular constituents became more and more apparent.

At the same time (1858), Virchow was the one to introduce the term *myelin* to refer to the fatty sheath surrounding "axis cylinders" (not termed "axon" until Kölliker): "It is this substance, for which I have proposed the name of medullary matter (Markstoff), or *myeline*, that in extremely large quantity fills up the interval between the axis-cylinder and the sheath in primitive nerve fibers" [42].

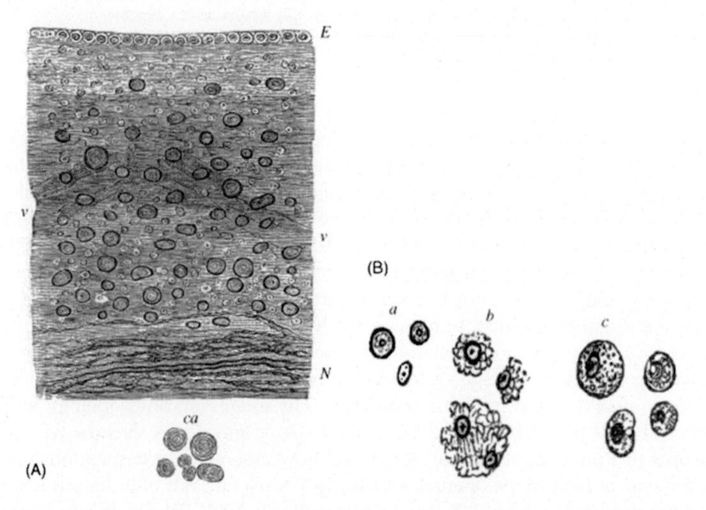

FIGURE 1.12 **Virchow's illustrations of neuroglia.** (A) Ependyma and neuroglia in the floor of the fourth ventricle. Between the ependyma and the nerve fibers is "the free portion of the neuroglia with numerous connective tissue corpuscles and nuclei." Numerous corpora amylacea are also visible, shown enlarged below the main illustration (ca). *E*, ependymal epithelium; *N*, nerve fibers; *v*, blood vessels. (B) Elements of neuroglia from white matter of the human cerebral hemispheres. *a*, free nuclei with nucleoli; *b*, nuclei with partially destroyed cell bodies; *c*, complete cells. *Source: Reproduced with permission from Verkhratsky A, Butt AM. Glial physiology and pathophysiology. Oxford, UK: Wiley-Blackwell; 2013 (Figure 1.14).*

Other Investigators Develop More Detailed Images of Neuroglial Cells

Heinrich Müller (1820–64) was a pathologist who performed a thorough analysis of retinal histology in Kölliker's department. He noted that the retina contains radial fibers [44], which Kölliker called Müller fibers in 1852 [45]. He provided detailed histological images of retinal glial cells (thereafter called Müller cells) [44,46].

Several years later, Max Schulze (1825–74) provided even more detailed and precise images of retinal glial cells (Fig. 1.13) [47].

Otto Deiters (1834–63) was the first to provide histological images of stellate cells which resemble astrocytes [48]. *Deiter's cell* was synonymous with *astrocyte* for many years. Upon Deiters' early death at the age of 29 from typhoid fever, his friend Max Schulze published Deiters' findings 2 years later [48]. In it he clearly identifies stellate cells from both white and gray matter (see Kettenmann and Ransom for details) [1,48].

Karl Bergmann (1814–65) discovered fibers in the molecular layer of the cerebellum and described them as processes of glial cells, subsequently termed "Bergmann" glia [49].

In 1869, the anatomist Jakob Henle (1809–85) (yes he of the loop of Henle in the kidney) and his son-in-law Friedrich Merkel (1845–1919) described networks of stellate cells (presumably astrocytes) in the gray and white matter of the spinal cord [50].

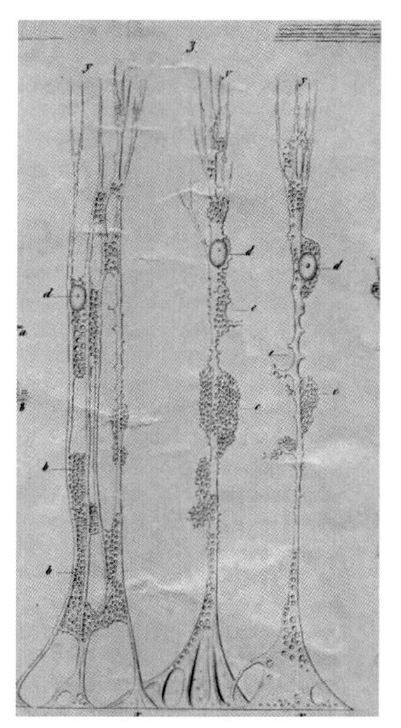

FIGURE 1.13 **Retinal "Müller" glial cells.** Müller fiber of the sheep retina, inspected by Max Schulze with a microscope from Amici. *yyy*, brush-like fibrils extending from the outer Müller fiber in the outer granular layer; *xx*, internal limiting membrane; *a*, opening in the limiting membrane; *b*, very delicate network of fenestrated membranes similar in the ganglion cell layer; *cc*, network in the so-called molecular layer; *ddd*, nuclei as part of the Müller fibers; *ee*, cavity in which the nuclei or the cells of the internal granular layer is located. *Source: From Schulze, 1859. Image kindly provided by Prof. Helmut Kettenmann, Max Delbruck Center for Molecular Medicine, Berlin. Reproduced with permission from Verkhratsky A, Butt AM. Glial physiology and pathophysiology. Oxford, UK: Wiley-Blackwell; 2013 (Figure 1.15).*

Camillo Golgi

Camillo Golgi (1843–1926) (Fig. 1.14), born in Brescia, went to Pavia for medical school and then assumed the title of Extraordinary Professor of Histology and then in 1881 Chair of General Pathology at the same institution. He is associated with supporting the reticular theory of brain organization (*contra* the neuron doctrine). Golgi performed the first in-depth histological observations of neuroglia in the 1870s [51,52]. Golgi described cells with the features of astrocytes and oligodendrocytes, and was the first to describe glial–vascular contacts or "end feet":

> Studying thin slices of treated cortical substance (osmium treated for 4 to 6h) with addition of glycerin with the microscope, revealed numerous, long, fine and never arborized prolongations originate...Many (prolongations) extend to the vessel walls, including capillaries, vessels of medium size (particularly to those) and directly attach to the walls of the capillaries or to the lymphatic border of vessels with larger diameter...The surround of the vessels is demarcated by a dense, regular network of fibers [53].

Based on these findings, which clearly suggest astrocyte endfeet, Golgi theorized that glial cells provide the bridge between parenchyma and vasculature (Fig. 1.15). This led to his concept of metabolic support and substance exchange, which remains quite valid today.

Of course, Golgi is best known for his development of the *reazione nera* ("black reaction" or silver nitrate stain, developed in 1873) that allowed visualization of entire neurons and glial cells. Others such as Cajal and Río-Hortega (see below) would subsequently use and modify Golgi's techniques to develop specific stains for glial cells [54,55].

FIGURE 1.14 **Camillo Golgi (1843–1926).**

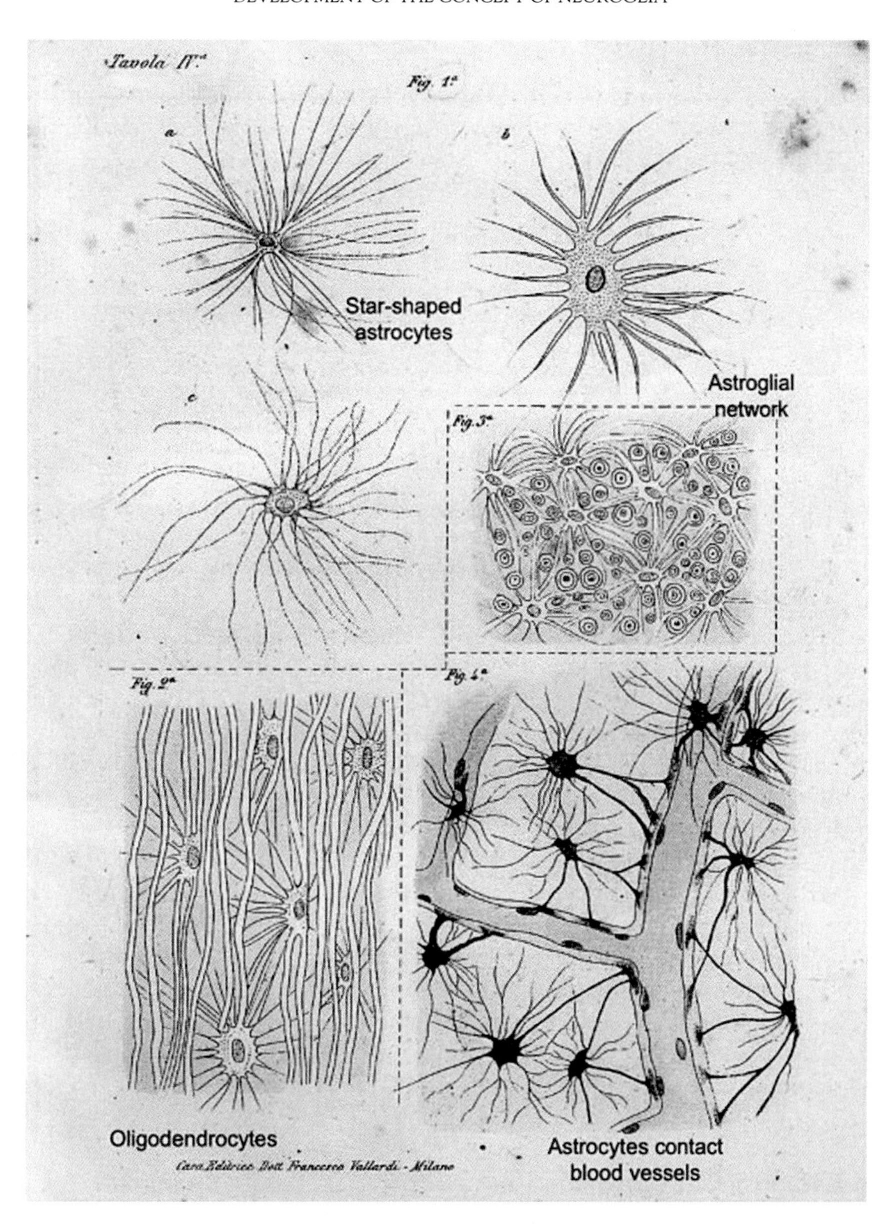

FIGURE 1.15 **Neuroglial cells stained by the silver-chromate technique and drawn by Camillo Golgi (Golgi, 1903).** Top panels show individual star-shaped astrocytes and astroglial networks. At the bottom right, astrocytes forming numerous contacts (the end feet) with brain capillaries are demonstrated. The bottom left panel shows the drawing of white matter with numerous cellular processes oriented parallel to axons, which most likely represent oligodendrocytes. *Source: Reproduced with permission from Verkhratsky A, Butt AM. Glial physiology and pathophysiology. Oxford, UK: Wiley-Blackwell; 2013 (Figure 1.16).*

Naming of the "Astrocyte"

Stellate CNS glial cells were named "astrocytes" (from the Greek *astron—star* and *kutos*—container or vessel, later cell, thus literally "star cells") by Michael von Lenhossék from Würzburg (1863–1937) in 1893:

> I would suggest that all supportive cells be named spongiocytes, and the most common form in vertebrates be named spider cells or **astrocytes**, and use the term neuroglia only cum grano salis [with a grain of salt], at least until we have a clearer view…Astrocytes are the small elements, which form the supportive system of the spinal cord. They are star shaped and indeed no other comparison describes their form so clearly [56].

Andriezen further classified them into fibrous glia (*Langstrahler*, "long projectors," found in white matter) and protoplasmic glia (*Kurzstrahler*, "short projectors," found in gray matter) [57].

It is noteworthy that around this time, in the last decade of the 19th century, just 3 years after the term "neuron" was coined by Waldeyer in 1891 and "astrocyte" by Lenhossék in 1893, that in 1894 Carl Ludwig Schleich (1859–1922) hypothesized for the first time that neuronal–glial interactions are important for brain function. In his book *Schmerzlose Operationen* (literally "Pain-free Operations"), which was mainly a treatise on the principles of local anesthesia, he also hypothesized that interactions among glial cells and neurons determine the status of excitation in the CNS (Fig. 1.16) [58]. In particular, according to Schleich, excitation from neuron to neuron via intercellular gaps is checked by generally inhibitory glial cells filling these interneuronal gaps. In particular, glial swelling was postulated to inhibit and glial shrinkage to promote neuronal excitation, a prescient idea but exactly converse to what is now known about the excitability-promoting effects of astrocyte swelling such as release of glutamate! Nevertheless, Schleich was the first to (correctly) promote the role of glial cells in controlling neuronal excitation, and recent findings in this field clearly indicate profound effects of glia on excitability. A related prescient hypothesis was that of Ernesto Lugaro, who was the first to propose (in 1907) that astrocytes function to take up and metabolize chemical transmitters at "neuronal articulations" (synapses) [59]; this was experimentally confirmed by Mennerick and Zorumski in 1994 [60], and we now know that astrocyte glutamate transporters are responsible for approximately 90% of glutamate reuptake from the extracellular space [61].

Santiago Ramón y Cajal and Glial Functions

The great Spanish neurohistologist Santiago Ramón y Cajal (1852–1934) (Fig. 1.17) was born in a small village in the north of Spain and studied medicine in the Faculty of Medicine in Zaragoza. In 1883, Cajal was appointed Professor of Descriptive and General Anatomy at the University of Valencia. In 1887, he moved to the University of Barcelona, where he was appointed to the Chair of Histology and Pathological Anatomy. In 1892, he was appointment to the same Chair of Histology and Pathological Anatomy at the University of Madrid, where he remained until retirement.

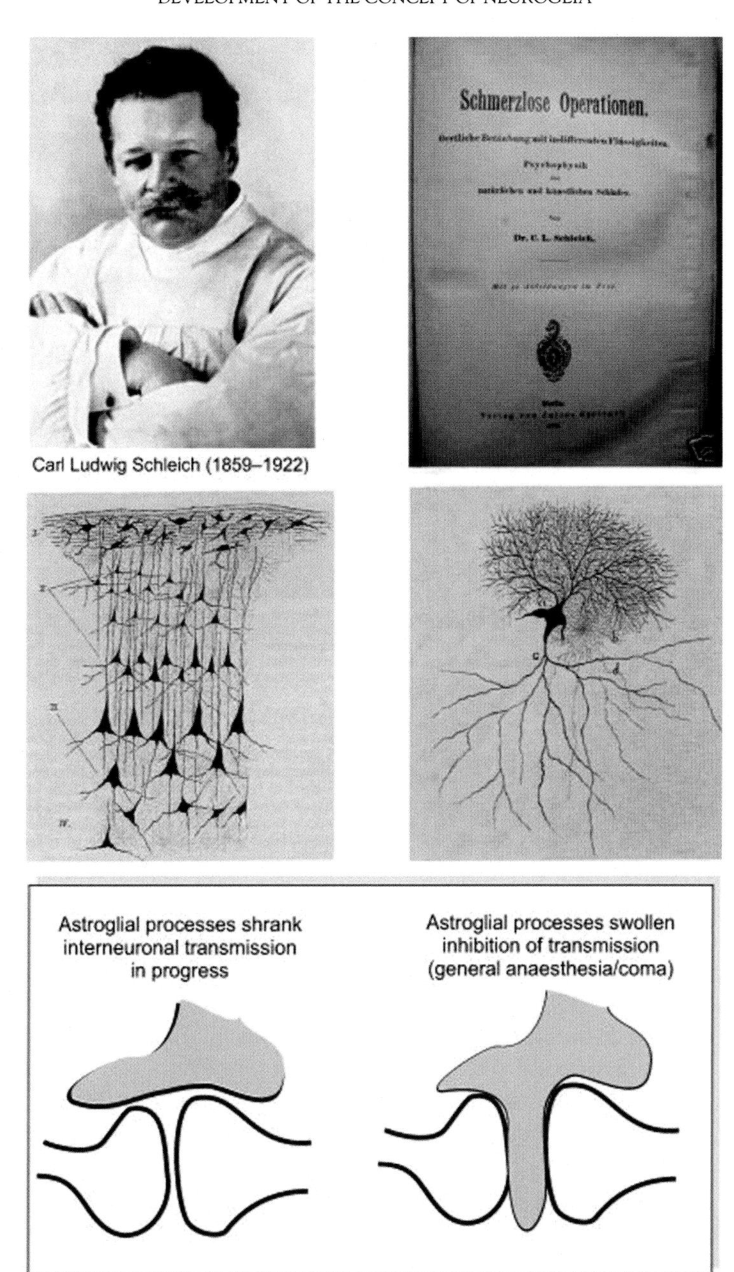

FIGURE 1.16 **Carl Ludwig Schleich (1859–1922) and the neuronal–glial interactions hypothesis.** Schleich was a pupil of Virchow and surgeon who introduced local anesthesia into clinical practice. In 1894, he published a book *Schmerzlose Operationen*, the frontispiece of which is shown in the right upper panel. Apart from describing the principles of local anesthesia, this book also contained the first detailed essay on interactions in neuronal–glial networks as a substrate for brain function. Mid panels show original drawings from this book, depicting intimate contacts between glial cells and neurons, and the lower panel shows, in a schematic manner, Schleich's theory of neuroglia controlling information flow in neuronal networks. *Source: Reproduced with permission from Verkhratsky A, Butt AM. Glial physiology and pathophysiology. Oxford, UK: Wiley-Blackwell; 2013 (Figure 1.24).*

FIGURE 1.17 **Santiago Ramón y Cajal (1852–1934).**

Between 1887 and 1903, Cajal was able to apply the Golgi method to describe in detail almost every part of the CNS (Fig. 1.18). In his pioneering summary treatise *Histology of the Nervous System*, published in Spanish in 1899–1904 and translated into French in 1909 and 1911 and into English only in 1995 [62], Cajal queried: "What is the function of glial cells in neural centers? The answer is still not known, and the problem is even more serious because it may remain unsolved for many years to come until physiologists find direct methods to attack it" [62]. At the time, microglia and oligodendrocytes, the mysterious "third element" that his disciple Pío del Río-Hortega (1882–1945) (Fig. 1.19) would be instrumental in deciphering, were not yet described. Cajal dismissed Golgi's hypothesis that glial cells carry important nutritive functions by virtue of the proximity to capillaries on the one hand and neuronal cell bodies on the other. Similarly he did not subscribe to Weigert's (1895) "filling theory" in which glia are entirely passive, just filling spaces between neurons [63]. Held (1904) suggested that glial fibers form a syncytial network (prefiguring the modern knowledge that astrocytes are interconnected by gap junction networks) [64]. Cajal rejected the notion of a syncytium (as he did with neurons in support of the neuron doctrine against reticularism), and instead he subscribed to the "isolation theory" (developed by his brother Pedro) in which astrocytes act as physical insulation [62]. Ironically of course, Cajal's disciple Pío del Río-Hortega was ultimately to describe oligodendrocytes (see below) and proposed that they were the source of CNS myelin. Of course, Cajal's studies provided overwhelming evidence in favor of the neuron doctrine, which contrasted with the reticular theory [13]. Nageotte (1910) described secretory granules in glial cells and suggested an endocrine function for glia [65]; Cajal (1913) agreed that protoplasmic astrocytes may be secretory [66]; and Cajal's disciple Achúcarro (1915) hypothesized that secretions from glial cells might enter the bloodstream to influence distant organs [67].

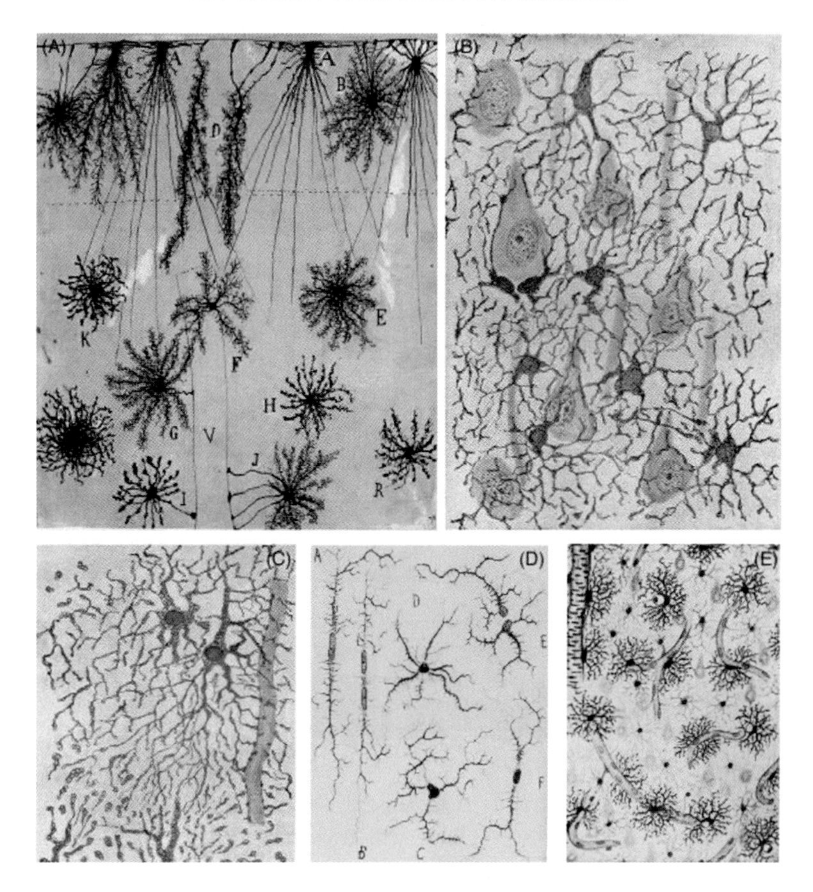

FIGURE 1.18 **Drawings of glial cells by Santiago Ramón y Cajal and Pío del Río-Hortega.** (A) Cajal's drawing of Golgi impregnated glia, showing human cortical neuroglial cells of the plexiform layer (a–d), cells of the second and third layers (e–h and k, r) and perivascular glia (i, j). (B), (C) Astrocytes in the stratum lucidum of the human CA1 area of the hippocampus, with particular emphasis on the anatomy of perivascular astrocytes in the CA1 stratum radiatum. (D), (E) Drawings of Pío del Río-Hortega, showing the different morphological types of microglial cells in the rabbit Ammon's horn and cortical perivascular neuroglia. *Source: Reproduced with permission from Verkhratsky A, Butt AM. Glial physiology and pathophysiology. Oxford, UK: Wiley-Blackwell; 2013 (Figure 1.18).*

Glial Alterations in Neurological Disease: Early Concepts

Following developments in the concepts of neuroglia, neuropathologists started to recognize glial alterations in tissues from patients with neurological diseases. Carl Frommann (1831–92) found changes in glial cell morphology in the vicinity of demyelinating plaques [68]. Andriezen indicated that astrocytes "exhibit a morbid hypertrophy in pathological conditions" [57]. Franz Nissl (1860–1919) described two pathological cell types: *Stäbchenzellen* (rod cells), which appeared in demented patients' brains; and *Körnerzellen* (granule cells) or *Gitterzellen* (lattice cells) which were associated with disruption of the blood–brain barrier [69,70]. Alois Alzheimer (1864–1915) described ameboid change in glial cells in response to acute and chronic neurologic diseases, such as epilepsy, syphilis, and the disease that

FIGURE 1.19 **Pío del Río-Hortega (1882–1945).** As seen with a time exposure photograph taken from Wilder Penfield's simple box camera. Río-Hortega's inscription reads, "Al gran artista de la oligodendroglia y de la fotografía y excelente amigo—Wilder G. Penfield." ("To the great artist of oligodendroglia and of photography and excellent friend—Wilder G. Penfield.") *Source: Reproduced with permission from Gill AS, Binder DK. Wilder Penfield, Pío del Río-Hortega, and the discovery of oligodendroglia. Neurosurgery 2007;60(5):940–8.*

bears his name [71]. Nicolás Achúcarro, a Basque-born psychiatrist and neurohistologist who was a colleague and alumnus of Cajal, advanced the idea that dysfunction of neuroglia may itself produce brain diseases [72]. Critically for epilepsy and brain injury, the role of astrocytes in glial scar formation was recognized by Cajal, Río-Hortega, and Penfield in the 1920s (see below) [73]. Cajal's later career (from 1903 through his death in 1934) was devoted largely to the investigation of pathological processes within the nervous system. This work was summarized in another classic of neuroscience, *Degeneration and Regeneration of the Nervous System* [74].

WILDER PENFIELD, PÍO DEL RÍO-HORTEGA, AND DELINEATION OF THE "THIRD ELEMENT"

To the modern neuroscientist, Wilder Penfield (1891–1976) (Fig. 1.20) needs little introduction. Many are familiar with the basic outline of his legendary career (Table 1.1). Penfield is justifiably famous for two main achievements: his tremendous contributions to the functional mapping of the human brain in the surgical treatment of epilepsy and the founding of the Montreal Neurological Institute. However, lesser known but seminal as well was his initial neuropathological work examining and characterizing glial cells. As the function of glial cells increasingly is seen as complimentary rather than subsidiary to neurons [75–77], it seems appropriate to look back at Penfield's important contributions to

FIGURE 1.20 **Wilder G. Penfield (1891–1976) c.1924.** *Source: Reproduced with permission from Gill AS, Binder DK. Wilder Penfield, Pío del Río-Hortega, and the discovery of oligodendroglia. Neurosurgery 2007;60(5):940–8.*

TABLE 1.1 Some Significant Events in the Life of Wilder Penfield

Year(s)	Event
1904	At 13 years, Penfield discovers existence of Rhodes Scholarship
1913	Graduation from Princeton
1914	Awarded Rhodes Scholarship
1915–16	Medical student at Oxford, exposed to William Osler and Charles Sherrington
1916–18	Medical student at Johns Hopkins Medical School
1918–19	Surgical internship at Peter Bent Brigham Hospital, Boston
1919–20	Returned to Oxford for postgraduate study, works with Sherrington assistants, Cuthbert Bazett on decerebrate rigidity and Harry M. Carleton on neurocytology
1921	Joins staff as neurosurgeon at Presbyterian Hospital, New York (surgical chief: Allen Whipple)
1924	Leaves Presbyterian Hospital to go to Madrid
September 1924	Returns to New York and Presbyterian Hospital, meets William Vernon Cone and inaugurates Laboratory of Neurocytology
1928	Works with Otfrid Foerster in Breslau, Germany for 6 months
1928	Arrives in Montreal (Royal Victoria Hospital and McGill University)
December 11, 1928	Operates on sister Ruth for oligodendroglioma
1932	Publishes *Cytology and Cellular Pathology of the Nervous System*, a 3-volume text with contributions from world-leading authorities, many of whom Penfield knew personally
September 27, 1934	Opening of the Montreal Neurological Institute

the early characterization of glial cells. The modern reader may be surprised that Penfield was, together with his Spanish mentor Pío Del Río-Hortega, the first to properly describe oligodendroglial cells. In this section, we review Penfield's trip to Spain and the contributions arising from his fruitful albeit short time in the *Residencia des Estudiantes* in Madrid, 1924 [78].

Penfield's Idea to Go to Spain

In 1923, a young Wilder Penfield was attempting to stain brain scars in the laboratories of the New York Presbyterian Hospital in the hope of identifying the mysterious cause of posttraumatic epilepsy in both animal models and humans. Penfield had begun this project 2 years earlier, when he began teaching medical students at the Columbia College of Physician and Surgeons under the authority of William C. Clarke, professor of surgical pathology. Clarke challenged Penfield immediately, asking "Wouldn't you like to see how the nerve cells, and all the other cells that surround them and nourish them, behave when… you make an incision in the brain? What is the cause of epilepsy?" [79].

Penfield was fascinated by epilepsy from his very first exposure to the CNS, mentioning that as early as his undergraduate years, he "had filled [his] index cards with notes from the writings of Hughlings Jackson" [79]. Clarke gave Penfield his first opportunity to begin studying the disease. He recognized immediately, however, "that the methods we use show only half the picture" [79], and indeed 2 years later, Penfield met a "dead end" in his work due to the inability to stain the nonneuronal cells of the CNS. Penfield believed these cells were crucial to demonstrating the healing process of the brain and in helping elucidate why a "healing scar so often leads to epilepsy" [79].

It was then that Penfield recalled the trouble he had staining neurons in the laboratory of another famous mentor he had worked with while on a Rhodes Scholarship, Sir Charles S. Sherrington of Oxford University (1857–1952) (Fig. 1.21), who had admonished him "Don't give up until you have tried the methods of Ramón y Cajal" [79]. (Sherrington had invited Cajal to stay at his home during Cajal's Croonian Lectureship in 1894 and the awarding of an honorary degree from Oxford; Cajal was rumored to have turned the guest bedroom into a histology laboratory! [80]) Remembering the brilliant success he had met with using Cajal's staining techniques, Penfield immediately went to the New York Academy of Medicine to read Cajal's articles in the hopes of adopting his staining techniques once again, only this time for glia rather than neurons. Though the techniques proved fruitful, Penfield was not able at first to emulate the beautiful stains of Cajal's greatest disciple, Pío Del Río-Hortega, nor was he able to completely interpret the results of the stains he used.

In January of 1924, Penfield decided to approach the man who had recruited him to the Presbyterian Hospital, surgical chief and professor of surgery Allen O. Whipple (Fig. 1.22), in the hopes of securing funds to travel to Spain and study under Pío Del Río-Hortega. Río-Hortega had published detailed drawings of the nonneuronal cells Penfield stated were "no more than ghosts" [79] in his preparations. Though Penfield was unsure how Whipple would respond, Whipple was quite supportive and decided to call upon Mrs. Percy Rockefeller. He had operated on her daughter free of charge and was able to secure a generous grant from Mrs. Rockefeller. With the help of a few other benefactors, Whipple

FIGURE 1.21 **Sir Charles S. Sherrington (1857–1952).** *Source: Reproduced with permission from Gill AS, Binder DK. Wilder Penfield, Pío del Río-Hortega, and the discovery of oligodendroglia. Neurosurgery 2007;60(5):940–8.*

FIGURE 1.22 **Allen O. Whipple (center) and his surgical team, including Wilder Penfield (far left), c.1924 at the Presbyterian Hospital.** William V. Cone is standing behind Penfield. Whipple was known for developing the pancreaticoduodenectomy ("Whipple procedure") and describing Whipple's triad (diagnostic triad for insulinoma). *Source: Reproduced with permission from Gill AS, Binder DK. Wilder Penfield, Pío del Río-Hortega, and the discovery of oligodendroglia. Neurosurgery 2007;60(5):940–8.*

secured enough funds to allow for Penfield, his wife, and two children to spend 6 months in Madrid. Here, Penfield hoped to work closely with Río-Hortega to "study the brain of man, and then move on to the effects of disease on the brain" [79]. Interestingly, Penfield and family set sail for Spain before receiving any word or invitation from Río-Hortega. Penfield wryly describes finally receiving Río-Hortega's one-word imperative, "*venga*" ("come"), while halfway across the Atlantic [79].

Types of Neuroglia

It was not until the publication of Golgi's method in 1886 [81,82] that a means for the more exact study of nerve cells and neuroglia was made possible. Cajal's pioneering modifications of Golgi's technique made possible the detailed anatomical study of the CNS. In fact, Cajal's modifications and studies were so fundamental that many, including Sherrington and Penfield, regarded Cajal (Fig. 1.23) as the true father of neuroanatomy [79]. His studies on glial cells came after his celebrated studies of neurons. Using his new gold chloride-sublimate method, Cajal demonstrated the morphology of the protoplasmic neuroglia as well as the fibrous neuroglia of the white matter in 1913 [66,83]. Cajal also recognized that the "satellites" and "interfascicular cells" were of a different class from neuroglia, which he termed "*the third element*" [66].

However, the remainder of the "adventitial" or nonnervous cells remained unstained until Cajal's disciple Pío Del Río-Hortega (1882–1945) (Fig. 1.19) described a method of using silver carbonate to stain neuroglia and connective tissue in 1918 [84,85]. For the first time, this clearly distinguished two cell types with distinct cytoplasmic expansions, which Río-Hortega termed microglia and oligodendroglia. Thus the nonneural interstitial cells

FIGURE 1.23 **Santiago Ramón y Cajal (1852–1934) in his laboratory.**

FIGURE 1.24 **Penfield (lower row, second from left) and Río-Hortega (lower row, center) in Spain with colleagues, 1924.** *Source: Reproduced with permission from Gill AS, Binder DK. Wilder Penfield, Pío del Río-Hortega, and the discovery of oligodendroglia. Neurosurgery 2007;60(5):940–8.*

could be divided into four classes: (1) fibrous neuroglia; (2) protoplasmic neuroglia; (3) microglia; and (4) oligodendroglia. Oligodendroglia were so named because they exhibited fewer (Greek *oligo*—few) and smaller branches (Greek *dendro*—branch) than astrocytes, and were later called *oligodendrocytes* (as astroglia are *astrocytes*). Río-Hortega would go on to focus his research efforts on microglia, elucidating their genesis, function, and pathologies in remarkably precise fashion [86]. He indicated that microglia are histiocytes of mesodermal origin as opposed to the epithelial origin of "classical" glia. Regarding microglia he stated: "Since it is of different ancestry and its characteristics differ from those of the nerve cells (first element) and the neuroglial astrocytes (second element) (containing astrocytes and oligodendrocytes), the microglia constitutes the *true* third element of the central nervous system" [87].

However, there was still much debate on the existence of oligodendroglia as a distinct CNS cell type, due in part to the difficulty involved in staining these cells. In fact, Cajal himself was unable to produce an effective and reproducible stain, leading him to dismiss these cells altogether as a true class of glial cells and to declare that the "third element" was made up exclusively of microglia. Given Cajal's enormous influence, his dismissal was perhaps just as important as any staining difficulty with respect to the further characterization of oligodendroglia. This put a strain on the relationship between Cajal and Río-Hortega. On Penfield's arrival in Spain, there had been no resolution to this debate (Fig. 1.24).

Penfield's Description of Oligodendroglia

Penfield noted that prior to the method of Río-Hortega, oligodendroglia were very difficult to stain completely. The first to stain oligodendroglial cells was the Scottish investigator William Ford Robertson (1867–1923), who employed a platinum method to describe cells that he termed *mesoglia* [88,89]. However, in his term *mesoglia*, he had just described a group of cells he believed to be of mesodermal origin. Penfield went back to examine an original preparation of Robertson and compared it with sections stained by Río-Hortega's method. He was able to verify that the "mesoglia" of Robertson were indeed identical to the "oligodendroglia" of Río-Hortega. Based on this, Penfield suggested that the term "mesoglia" be abandoned.

Penfield then learned the method of Río-Hortega for staining oligodendroglial cells and added his own modifications. Using the "ammoniacal silver carbonate" method of Río-Hortega, originally developed by Achúcarro [90], the results were variable. In 1924, Penfield reported on his modifications to the method, and used it to stain oligodendroglia in the CNS of rabbits.

The remainder of the staining procedure consisted of washing, toning, fixing, dehydrating, and clearing the specimen. Toning, originally described by Cajal, consisted of substituting gold for silver. With his modifications, Penfield had finally succeeded in developing a reliable stain for oligodendroglia. Don Pío, having been shown Penfield's exceptional slides, remarked, "*Casi mejor que yo*" (almost better than I could do). Penfield later reminisced that he "might have laughed at his use of the word 'almost,'" but that "he could expect no higher praise" [79].

Río-Hortega then asked Penfield to publish his results confirming oligodendroglia as the remaining cell type of the "third element." Penfield studied and drew many of the oligodendroglia that he stained (Fig. 1.25). While noting that neuroglia and oligodendroglia both possessed "the asteroid body with expansions, centrosome and Golgi apparatus of similar appearance" [91], he nevertheless was able to distinguish many characteristics of oligodendroglia. He noted that oligodendroglial nuclei are larger than those of microglia but smaller

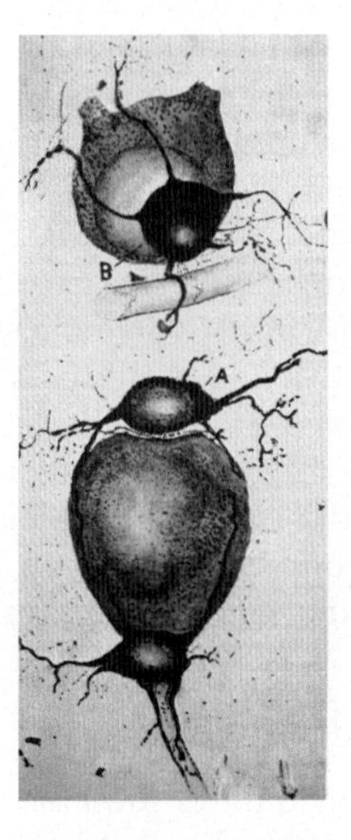

FIGURE 1.25 **Penfield's original drawing of perineuronal and perivascular oligodendroglia.** Published in his 1924 *Brain* paper, *Oligodendroglia and its relation to classical neuroglia. Source: Reproduced with permission from Gill AS, Binder DK. Wilder Penfield, Pío del Río-Hortega, and the discovery of oligodendroglia. Neurosurgery 2007;60(5):940–8.*

than those of neuroglia. He also commented on the ability to distinguish neuroglia from oligodendroglia by the presence of "sucker feet" (modern-day "vascular endfeet") on the former group of cells [91]. Penfield also stated that with his improved methods he could get a better view of oligodendroglial cell cytoplasm, showing that the "expansions" of cytoplasm were directed along the length of the "neuron cable system" (ie, along white matter tracts) (Fig. 1.26). In addition to studying white matter oligodendroglia (termed "interfascicular glia" by Río-Hortega), Penfield carefully described oligodendroglia in gray matter as well. Penfield clearly showed that "perineuronal satellites" included *both* oligodendroglia and microglia. Similarly, he showed that "perivascular satellites" could also be either oligodendroglia or microglia. Critically, he noted that while the cell bodies of these two types of perivascular satellites were "applied closely to the blood-vessel," neither one were like neuroglia in this respect: with neuroglia it was the "neuroglia expansions that are applied to the neuron and vessel." Interestingly, he may have simultaneously underestimated neuroglia and overestimated neurons in claiming that "oligodendroglia forms by far the most numerous group of cells in the central nervous system, after nerve cells" [91].

In 1924, Wilder Penfield published his work from *La Residencia* in a seminal article in the journal *Brain*, "Oligodendroglia and its relation to classical neuroglia" [91]. In this article, he paid homage to his mentors Cajal and Río-Hortega:

> In spite of untiring study of the central nervous system which has demonstrated the intricate morphology of neurones and neuroglia, a very numerous body of small cells (the third element of Cajal) continued to be refractory to staining methods…The brilliant studies of Del Río-Hortega show that these cells possess complicated expansions. By demonstrating their detailed structure, he was able to show that they fall into two groups, differing in form and function. One group, which he chose to denominate microglia, is of mesodermal origin, and the other, oligodendroglia, composing the more numerous portion of the cells, he believes to be of ectodermal origin [91].

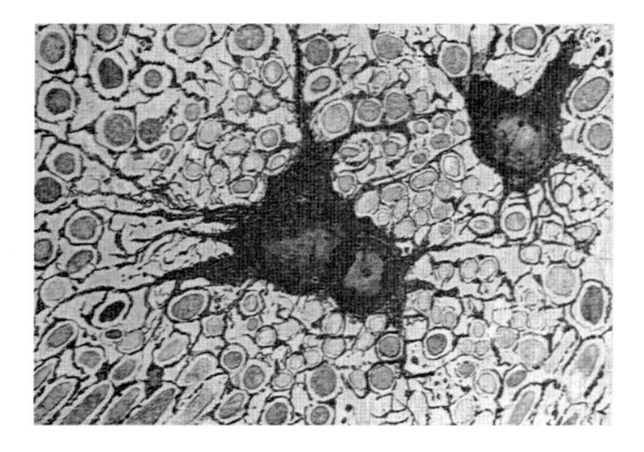

FIGURE 1.26 **Penfield's drawing depicting oligodendroglial expansions encircling myelin sheaths in the CNS of a rabbit.** Published in his 1924 *Brain* paper, *Oligodendroglia and its relation to classical neuroglia. Source: Reproduced with permission from Gill AS, Binder DK. Wilder Penfield, Pío del Río-Hortega, and the discovery of oligodendroglia. Neurosurgery 2007;60(5):940–8.*

TABLE 1.2 Penfield's Formulation of the Interstitial Cells of the Central Nervous System (1924)

Interstitial Cells	In White Matter	In Gray Matter
Neuroglia, classical (ectodermal)	Fibrous	Protoplasmic
Oligodendroglia (probably ectodermal)	Interfascicular	Perineuronal (satellite)
Microglia (probably mesodermal)	Present (no subdivision)	Present and numerous (no subdivision)

In this historic paper, Penfield summarized his work along with the work of Cajal and Río-Hortega in formulating an overall classification of the interstitial cells of the nervous system (Table 1.2). This gross classification has changed remarkably little since.

Coming Together: The Fruit of Penfield's Spanish Expedition

When he initially used Río-Hortega's stains, before leaving for Madrid, Penfield explained the results were, "very exciting, but also very confusing. What I saw was difficult to interpret" [79]. In a short 6 months in Madrid, not only did Penfield perfect a specific stain for oligodendroglia and describe this "third element" in the seminal 1924 Brain paper, but with Río-Hortega also moved from "pure" neuroscience to neuropathology in studying the reaction of glial cells to injury. In an article entitled "Cerebral cicatrix: the reaction of neuroglia and microglia to brain wounds" [73], they provide several observations regarding cellular changes following simple stab wounds:

> The formation of a simple cicatrix in the brain presents the following stages: The first cellular change is observed in microglia cells which begin their phagocytic activity early and continue it for a long period. Later, the neuroglia astrocytes about the wound become swollen and those closest to the area of destruction or to obliterated vessels undergo clasmatodendrosis. There follows rapid amitotic division of the other astrocytes and the cells then become fibrous and arrange themselves typically in a radial fashion about the wound. Most of their expansions, and particularly the robust ones, are arranged like the spokes of a wheel with the site of the former stab as the hub.... A connective tissue core forms at the center, connective-tissue collagen fibrils are laid down and the wound contracts [73].

It is noteworthy that their basic idea of these injury stages and formation of the "cicatrix" (contracting scar), is essentially correct. Compare, for example, a modern neuropathology textbook:

> In regions of tissue damage hematogenous monocytes infiltrate the CNS and phagocytose dead cells and necrotic debris. Swelling of astrocytes is a relatively rapid response. With time, reactive astrocytes proliferate and insinuate long cytoplasmic processes into the adjacent brain parenchyma, which appear as fibrils in appropriately stained preparations [92].

Others have recently verified "clasmatodendrosis," their term for the loss of distal astrocytic processes, in degenerative disorders such as Alzheimer's disease [93–95]. Though the details of gliosis and microglial response were not complete, the numerous observations and conjectures put forth by Río-Hortega and Penfield in this single publication are astonishing. "Cerebral cicatrix" [73] was actually intended by Penfield to be a combined publication with his 1924 Brain paper. The latter paper was to be a combined publication with

FIGURE 1.27 **Penfield and William V. Cone (1897–1959) outside the Royal Victoria Hospital, Montreal (c.1932).** *Source: Reproduced with permission from Gill AS, Binder DK. Wilder Penfield, Pío del Río-Hortega, and the discovery of oligodendroglia. Neurosurgery 2007;60(5):940–8.*

the former. However, "Pío procrastinated. This was his old-time enemy. I could not get our results into print for 3 years, not until 1927" [79].

In 1927, back at the New York Presbyterian Hospital, along with his research associate William V. Cone (Fig. 1.27), Penfield decided to write a "textbook on the general principles of neuropathology without describing specific diseases" [79]. (Cone was actually a remarkably astute innovator and scholar whom, if not for his reluctance to write papers, would undoubtedly be widely known to every modern neurosurgeon [96].) Penfield thought that moving beyond a simple description of various diseases to a mechanistic description of disease, pathophysiology was an essential step in the eventual treatment of various neurological diseases. However, Penfield "realized, far too often, that someone else, somewhere in the world, could write a better chapter. I wrote to several to see if they would do a chapter for us. I was surprised when the invitations were readily accepted, since my name carried no prestige as editor" [79]. Numerous sources [97–100] belie this self-effacing claim: Penfield was widely known and respected, even in 1927, by many of the most eminent scientists of the day. In any event, all of his requests to contribute to Cytology and Cellular Pathology of the Nervous System were accepted, save for Cajal who "alone refused, saying he had advancing arteriosclerosis, the histologist's way of describing old age" [79]. Río-Hortega also gave Penfield pause, for he had a habit of not responding to letters and telegrams. Penfield describes bombarding Río-Hortega "with letters and finally, received a telegram from him followed by a letter. 'Of course I will write for your book. How could you think otherwise'" [79].

Cytology, dedicated to Cajal, was finally published in 1932 with 26 eminent contributors. It proved to be an instant and influential success. It was the first tome written on neuropathology from a basic science perspective, a common staple of many pathology texts written today both for graduate and medical study. Penfield describes the reaction he received after the first edition went out of print:

> When, eventually, the first edition of this reference book went out of print, I received letters of inquiry from all over the world. But I was too busy making clinical use of what I had learned to undertake a second edition. At long last, in 1965, Hafner, New York, reprinted it without change [79].

For such a text to be reprinted in original form more than 30 years after its first publication is a testament to its lasting influence.

Penfield credits the time he spent in Europe, specifically in Madrid, providing him the "keys to understanding" [79]. At a time when glial cells were just being described and differentiated by Río-Hortega and himself, Penfield immediately studied their reaction to injury, and their potential role in epilepsy. In his autobiography *No Man Alone*, he stated, "if one desired to throw new light on the effect of disease, or injury, and on the process of healing in the brain, the best hope lay in the study of the nonnervous cells, using Hortega's little-tried methods" [79]. More than 80 years after this statement was made, recent evidence is accumulating for a critical glial contribution to epilepsy [75,101].

Penfield was humble in acknowledging his mentors. He credited Sherrington with influencing his scientific thinking "more than anyone else" [102], saying "I looked through his eyes and came to realize that here in the nervous system was the great unexplored field—the undiscovered country in which the mystery of the mind of man might someday be explained" [79]. In the obituary he wrote for Río-Hortega in 1945, Penfield makes no mention of his own role in the authentication of oligodendroglia or the elucidation of microglia, giving full credit to Río-Hortega [103].

In deep irony, in the late 1920s Penfield's older sister Ruth Inglis developed an oligodendroglioma, a tumor of the same cell type that Penfield had substantiated in Madrid. Interestingly, on describing a tumor resection in a letter to his mother in 1921, Penfield had stated, "Brain surgery is a terrible profession. If I did not feel it would become very different in my lifetime, I should hate it" [79]. Despite craniotomies for resection of her tumor by Penfield on December 11, 1928 and for recurrence by Harvey Cushing on November 6, 1930, Ruth ultimately died in September 1931.

This vignette outlining Penfield's contributions to the elucidation of Cajal's "third element" provides a wonderful example of critical neuroscience research performed by a neurosurgeon. Penfield later made seminal contributions to epilepsy surgery, the mapping of the intact human brain, and the founding of the Montreal Neurological Institute [97–100,104,105]. As an Oslerian "medico-chirurgical neurologist" [98], Penfield indeed embodied the ultimate combination of neuroscientist and neurosurgeon. When comparing physician investigators to pure scientists, Penfield offers advice, that is, as true today as it was then: "We have our practical purposes. We must select our weapons and plan our researches with the patient and his unique problems in mind" [79].

BEGINNING OF THE MODERN ERA

Could the "something else" needed to pull together disparate facts, harmonize apparent contradictions, and put an end to our journeys down blind alleys just be the physiological properties of that other cell population of the brain, the glia? [106]

Despite Río-Hortega's worldwide recognition for describing oligodendrocytes and microglia, his dispute with Cajal undermined his reputation in Spain. Ultimately a reconciliation between Río-Hortega and Cajal took place in 1928, when Cajal was 77 years [107]. Cajal passed away in 1934, and Río-Hortega continued his research until his death in 1945 [85].

In 1952, Alan Hodgkin and Andrew Huxley published a series of classic papers in which they described, analyzed, and modeled the ionic conductances occurring during the nerve action potential, a seminal event for modern neurophysiology [108–114]. The modern era in glial physiology began with several key discoveries in the late 1950s and 1960s. Walter Hild and colleagues obtained microelectrode recordings from cultured astrocytes for the first time in 1958 [115], and demonstrated electrophysiological properties distinct from neurons in 1962 [116]. Already in 1961, Robert Galambos advanced a "glial-neural theory of brain function" [106]. In 1965, Leif Hertz was the first to emphasize the importance of glia in uptake of extracellular potassium released by active neurons [117]. The Swedish neurophysiologist Holger Hydén described "a two-cell collaboration responsible for brain activity" in 1960 [118], and together with his colleagues demonstrated alterations in gene expression in the neuron–glia unit during learning [119], degenerative brain disease [120], and exposure to psychoactive drugs [121,122]. The great Stephen Kuffler (Fig. 1.28), working with David Potter, showed that glial cells have lower resting membrane potential than neurons and did not generate action potentials in 1964 [123]. Later, working with Richard Orkand

FIGURE 1.28 **Stephen W. Kuffler (1913–80).**

and John Nicholls, Kuffler demonstrated electrical coupling between glial cells in 1966 [124]. These experiments also led to formulation of the potassium spatial buffering hypothesis [125]. Milton Brightman and Tom Reese identified structures connecting astrocytes (now known to be gap junctions) in 1969 [126].

Nevertheless, glial cells were still thought to be passive participants in CNS function, playing a supportive role. However, in 1984 the research teams of Helmut Kettenmann and Harold Kimelberg identified glutamate and GABA receptors in astrocytes and oligodendrocytes [127–129]. In the 1980s, Glenn Hatton and colleagues showed that pituicytes (pituitary astrocytes) actively modulate hormone secretion, a direct glial–neuronal interaction in the neurohypophysis [130]. In 1990, Ann Cornell-Bell and colleagues found that networks of astrocytes can communicate via propagating calcium waves [131]. In 1994, Maiken Nedergaard, Philip Haydon, and Vladimir Parpura discovered that astrocytes can stimulate calcium elevations in adjacent neurons when grown in co-culture [132,133]. Thus, bidirectional communication between astrocytes and neurons was already clearly documented in the mid-1990s. Further work, some summarized later in this book, has indicated that astrocytes express a multitude of neurotransmitter receptors and ion and water channels; that there is marked astrocyte heterogeneity within and between distinct brain areas; and that there are major new roles for astrocytes in both physiology and disease.

References

[1] Kettenmann H, Ransom BR. The concept of neuroglia: a historical perspective. In: Kettenmann H, Ransom BR, editors. Neuroglia (*2nd ed.*). New York, NY: Oxford University Press; 2005. p. 1–16.

[2] Parpura V, Verkhratsky A. Astrocytes revisited: concise historic outlook on glutamate homeostasis and signaling. Croat Med J 2012;53:518–28.

[3] Parpura V, Verkhratsky A. Neuroglia at the crossroads of homeostasis, metabolism and signalling: evolution of the concept. ASN Neuro 2012;4:201–5.

[4] Somjen GG. Nervenkitt: notes on the history of the concept of neuroglia. Glia 1988;1:2–9.

[5] Verkhratsky A, Butt A. Introduction to glia. In: Verkhratsky A, Butt A, editors. Glial neurobiology: a textbook. Hoboken, NJ: John Wiley & Sons, Ltd.; 2007. p. 3–12.

[6] Verkhratsky A, Butt A. History of neuroscience and the dawn of research in neuroglia. In: Verkhratsky A, Butt A, editors. Glial physiology and pathophysiology. Oxford, UK: Wiley-Blackwell; 2013. p. 21–73.

[7] Verkhratsky A, Parpura V. General pathophysiology of neuroglia: neurological and psychiatric disorders as gliopathies. In: Parpura V, Verkhratsky A, editors. Pathological potential of neuroglia. New York, NY: Springer; 2014. p. 1–12.

[8] Waldeyer-Hartz HWG. Uber einige neuere Forschungen im Gebiete der Anatomie des Centralnervensystems. Deutsche Med Wochenschr 1891;17:1213–8. 44–46, 67–69, 87–89, 331–332, 352–356.

[9] Rapport RL. Nerve endings: the discovery of the synapse. New York: W.W. Norton; 2005.

[10] López-Muñoz F, Boya J, Alamo C. Neuron theory, the cornerstone of neuroscience, on the centenary of the Nobel prize award to Santiago Ramon y Cajal. Brain Res Bull 2006;70(4–6):391–405.

[11] Mayr E. The growth of biological thought: diversity, evolution, and inheritance. Cambridge, MA: Harvard University Press; 1982.

[12] Mazzarello P. A unifying concept: the history of cell theory. Nat Cell Biol 1999;1(1):E13–5.

[13] Shepherd GM. Foundations of the neuron doctrine. New York: Oxford University Press; 1991.

[14] Gjerstad L, Gilhus NE, Storstein A. A retrospective view on research in neuroscience in Norway. Acta Neurol Scand Suppl 2008;188:3–5.

[15] Akert K. August Forel—cofounder of the neuron theory (1848–1931). Brain Pathol 1993;3(4):425–30.

[16] Probst A, Langui D. The neuron theory: one of the main scientific achievements of the 19th century. Schweiz Rundsch Med Prax 1994;83(16):462–9.

[17] Schwann T. Mikroskopische Untersuchungen uber die Ubereinstimmung in der Strukturund dem Wachstum der Tiere and Pflanzen. Berline: Sander'schen Buchhandlung; 1839.
[18] Schwann T. Microscopical researches into the accordance in the structure and growth of animals and plants. London: The Sydenham society; 1847.
[19] Schleiden M. Beitrage für Phytogenesis. Arch. Anat. Physiol. Wiss Med. 1838;5:137–76.
[20] Locy WA. Biology and its makers. New York: H. Holt and Company; 1908.
[21] Binder DK. A history of Todd and his paralysis. Neurosurgery 2004;54(2):480–7.
[22] Reynolds EH. Todd, Hughlings Jackson, and the electrical basis of epilepsy. Lancet 2001;358:575–7.
[23] Todd RB. The cyclopaedia of anatomy and physiology. London: Sherwood, Gilbert, and Piper; 1847.
[24] Todd RB. The descriptive and physiological anatomy of the brain, spinal cord, and ganglions, and of their coverings, adapted for the use of students. London: Sherwood; 1845.
[25] Binder DK, Rajneesh KF, Lee DJ, Reynolds EH. Robert Bentley Todd's contribution to cell theory and the neuron doctrine. J Hist Neurosci 2011;20(2):123–34.
[26] Todd RB, Bowman W. The physiological anatomy and physiology of man. London: Parker; 1856.
[27] Guillery RW. Relating the neuron doctrine to the cell theory. Should contemporary knowledge change our view of the neuron doctrine? Brain Res Rev 2007;55:411–21.
[28] Finger S. Origins of neuroscience: a history of explorations into brain function. New York: Oxford University Press; 1994.
[29] Stuart G, Spruston N, Häusser M. Dendrites. Oxford; New York: Oxford University Press; 1999.
[30] Nansen F. The structure and combination of the histological elements of the central nervous system. Bergen: Bergens Museum Aarbs; 1886.
[31] Fodstad H, Kondziolka D, de Lotbiniere A. The neuron doctrine, the mind, and the arctic. Neurosurgery 2000;47(6):1381–8. discussion 8–9.
[32] Fields RD. The other brain. New York: Simon & Schuster; 2011.
[33] Huntford R. Nansen: the explorer as hero. London: Duckworth; 1997.
[34] Forel A. Einige hirnanatomische Betrachtungen und Ergebnisse. Arch. Psychiat. Nervenkr 1887;18(1):162–98.
[35] Kuechenhoff B. The psychiatrist Auguste Forel and his attitude to eugenics. Hist Psychiatry 2008;19(74 Pt 2):215–23.
[36] Finger S. Minds behind the brain: a history of the pioneers and their discoveries. New York: Oxford University Press; 2000.
[37] Winkelmann A. Wilhelm von Waldeyer-Hartz (1836–1921): an anatomist who left his mark. Clin Anat 2007;20(3):231–4.
[38] Fodstad H. The neuron theory. Stereotact Funct Neurosurg 2001;77(1–4):20–4.
[39] Zacharias H. Key word: chromosome. Chromosome Res 2001;9(5):345–55.
[40] Cremer T, Cremer C. Centennial of Wilhelm Waldeyer's introduction of the term "chromosome" in 1888. Cytogenet Cell Genet 1988;48(2):65–7.
[41] Parent A. Auguste Forel on ants and neurology. Can J Neurol Sci 2003;30(3):284–91.
[42] Virchow R. Die Cellularpathologie in ihrer Begründung auf physiologische und pathologische Gewebelehre. Zwanzig Vorlesungen gehalten während der Monate Februar, März und April 1959 im pathologischen Institut zu Berlin. Berlin: August Hirschwald; 1858.
[43] Virchow R. Gesammelte Abhandlungen zur wissenschaftlischen Medizin. Frankfurt a.M.: Verlag von Meidinger Sohn & Comp; 1856.
[44] Müller H. Zur histologie der Netzhaut. Z Wissenschaft Zool 1851;3:234–7.
[45] Kölliker A. Zur Anatomie und Physiologie der Retina. Verh Physikal-Med Ges Würzburg 1852;3:316–36.
[46] Müller H. Observations sur la structure de la rétine de certains animaux. Comptes Rendus de l'Académie der Sciences de Paris 1856;43:743–5.
[47] Schultz M. Observationes de retinae structura penitiori. Bonn, Germany: Published lecture at the University of Bonn; 1859.
[48] Deiters O. Untersuchungen über Gehirn und Rückenmark des Menschen und der Säugethiere. Braunschweig: Vieweg; 1865.
[49] Bergmann K. Notiz über einige Strukturverhältnisse des Cerebellums und Rückenmarks. Z Med 1857;8:360–3.
[50] Henle J, Merkel F. Über die sogenannte Bindesubstanz der Centralorgane des Nervensystems. Z Med 1869;34:49–82.
[51] Golgi C. Suella Struttura della sostanza grigia del cervello (comunicazione preventiva). Gazzetta Medica Italiana, Lombardia. 1873;33:244–246.

[52] Golgi C. Opera omnia. Milano: Hoepli; 1903.
[53] Golgi C. Contribuzione alla fina Anatomia degli organi centrali del sistema nervosos. Rivista clinica di Bologna; 1872;2:38–46.
[54] Ramón y Cajal S. Contribución al conocimiento de la neuroglia del cerebro humano. Trab Lab Invest Biol 1913;11:255–315.
[55] Del Río-Hortega P. El tercer elemento de los centros nerviosos. Bol Soc Esp Biol 1919;9:69–120.
[56] Lenhossék Mv. Der feinere Bau des Nervensystems im Lichte neuester Forschung. Berlin: Fisher's Medicinische Buchhandlung H. Kornfield; 1893.
[57] Andriezen WL. The neuroglia elements of the brain. Br Med J 1893;2:227–30.
[58] Schleich CL. Schmerzlose Operationen. Oertliche Betäubung mit indiffrenten Flüssigkeiten. Berlin: Julius Springer; 1894.
[59] Lugaro E. Sulle funzioni della nevroglia. Riv Patol Nerv Ment 1907;12:225–33.
[60] Mennerick S, Zorumski CF. Glial contributions to excitatory neurotransmission in cultured hippocampal cells. Nature 1994 Mar 3;368(6466):59–62.
[61] Danbolt NC. Glutamate uptake. Prog Neurobiol 2001;65(1):1–105.
[62] Ramón y Cajal S. Histology of the nervous system. New York: Oxford University Press; 1995.
[63] Weigert C. Beiträge zur Kenntnis der Normalen Menschlichen Neuroglia. Frankfurt, Germany: Ärztlicher Verein; 1895.
[64] Held H. Über den Bau der Neuroglia und über die Wand der Lymphgefasse in Haut und Schleimhaut. Abhandl Math Physikal Klasse Kgl Sachs Ges Wiss 1904;28:199–318.
[65] Nageotte J. Phénomènes de sécrétion dans le protoplasma des cellules névrogliques de la substance grise. C R Soc Biol (Paris) 1910;68:1068–9.
[66] Deleted in Review.
[67] Achúcarro N. De l'évolution de la nevroglie et specialment de ses relations avec l'appareil vasculaire. Trav Lab Rech Biol Univ Madrid 1915;13:169–212.
[68] Frommann C. Untersuchung über die Gewebsveränderungen bei der Multiplen Sklerose des Gehirns. Jena: Gustav Fischer; 1878.
[69] Nissl F. Über einigen Beziehungen zur Nervenzellenerkrankungen und gliosen Erscheinungen bei verschiedene Psychosen. Arch Psychiatrie 1899;32:656–76.
[70] Nissl F. Die Histopathologie der paralytischen Rindenerkrankung. Jena: Gustav Fischer; 1904.
[71] Alzheimer A. Beiträge zur Kenntnis der pathologischen Neuroglia und ihrer Beziehungen zu den Abbauvorgängen im Nervengewebe. In: Nissl F, Alzheimer A, editors. Histologische und histopathologische Arbeiten über die Grosshirnrinde mit besonderer Berücksichtigung der pathologischen Anatomie der Geisteskrankheiten. Jena: Gustav Fischer; 1910. p. 401–562.
[72] Achúcarro N. Notas sobra la estructura y funciones de la neuroglía y en particular de la neuroglía de la corteza cerebral humana. Trab Lab Invest Biol 1913;3:1–31.
[73] Del Río-Hortega P, Penfield WG. Cerebral cicatrix: the reaction of neuroglia and microglia to brain wounds. Bull Johns Hopkins Hosp 1927;41:278–303.
[74] Ramón y Cajal S. Degeneration and regeneration of the nervous system. London: Oxford University Press; 1928.
[75] Binder DK, Steinhäuser C. Functional changes in astroglial cells in epilepsy. Glia 2006;54(5):358–68.
[76] Fields RD, Stevens-Graham B. New insights into neuron-glia communication. Science 2002;298(5593):556–62.
[77] Ransom BR, Behar T, Nedergaard M. New roles for astrocytes (stars at last). Trends Neurosci 2003;26(10):520–2.
[78] Gill AS, Binder DK. Wilder Penfield, Pío del Río-Hortega, and the discovery of oligodendroglia. Neurosurgery 2007;60(5):940–8.
[79] Penfield W. No man alone: a neurosurgeon's life. Boston: Little, Brown and Company; 1977.
[80] Penfield W. The career of Ramón y Cajal. Arch Neurol Psychiatry 1926;16:213–20.
[81] Golgi C. Sulla fina anatomia degli organi centrali del sistema nervoso. Milano: Ulrico Hoepli; 1885.215.
[82] Golgi C. Sur l' anatomia microscopique des organes centraux su Systéme nerveux. Arch Ital Biol 1886;7:15–47.
[83] Ramón y Cajal S. Sobre un nuevo proceder de impregnacion de la neuroglia y sus resultados en los centros nerviosis del hombre y animales. Trab Lab Inv Biol 1913;11:219–37.
[84] Del Río-Hortega P. Noticia de un nuevo y facial metodo para la coloracion de la neuroglia y del tejido conjunctivo. Trab Lab Inv Biol 1918;15:367–78.
[85] Tremblay ME, Lecours C, Samson L, Sánchez-Zafra V, Sierra A. From the Cajal alumni Achúcarro and Río-Hortega to the rediscovery of never-resting microglia. Front Neuroanat 2015;9:45.

[86] Rezaie P, Male D. Mesoglia & microglia—a historical review of the concept of mononuclear phagocytes within the central nervous system. J Hist Neurosci 2002;11:325–74.
[87] Del Río-Hortega P. Microglia Penfield W, editor. Cytology and cellular pathology of the nervous system. New York: Hafner; 1932. p. 483–534.
[88] Robertson WF. The normal histology and pathology of the neuroglia (in relation specially to mental diseases). J Ment Sci 1897;43:733–52.
[89] Robertson WF. A microscopic demonstration of the normal and pathological history of mesoglia cells. J Ment Sci 1900;46:724.
[90] Achúcarro N. Nuevo método para el estudio de la neuroglia y del tejido conjuntivo. Bol Soc Esp Biol 1911;1:139–41.
[91] Penfield W. Oligodendroglia and its relation to classical neuroglia. Brain 1924;47:430–52.
[92] Ellison D, Love S. Neuropathology. London: Mosby International Ltd.; 1998.
[93] Sahlas DJ, Bilbao JM, Swartz RH, Black SE. Clasmatodendrosis correlating with periventricular hyperintensity in mixed dementia. Ann Neurol 2002;52(3):378–81.
[94] Hulse RE, Winterfield J, Kunkler PE, Kraig RP. Astrocytic clasmatodendrosis in hippocampal organ culture. Glia 2001;33(2):169–79.
[95] Tomimoto H, Akiguchi I, Suenaga T, Nishimura M, Wakita H, Nakamura S, et al. Alterations of the blood-brain barrier and glial cells in white-matter lesions in cerebrovascular and Alzheimer's disease patients. Stroke 1996;27(11):2069–74.
[96] Preul MC, Stratford J, Bertrand G, Feindel W. Neurosurgeon as innovator: William V. Cone (1897–1959). J Neurosurg 1993;79:619–31.
[97] Feindel W. Harvey Cushing's Canadian connections. Neurosurgery 2003;52:198–208.
[98] Feindel W. Osler and the "medico-chirurgical neurologists": Horsley, Cushing, and Penfield. J Neurosurg 2003;99:188–99.
[99] Preul MC, Feindel W. "The art is long and the life short": the letters of Wilder Penfield and Harvey Cushing. J Neurosurg 2001;95:148–61.
[100] Preul MC, Feindel W. Origins of Wilder Penfield's surgical technique: the role of the "Cushing ritual" and influences from the European experience. J Neurosurg 1991;75:812–20.
[101] Tian G-F, Azmi H, Takano T, Xu Q, Peng W, Lin J, et al. An astrocytic basis of epilepsy. Nat Med 2005;11(9):973–81.
[102] Scolding N. Myelin biology and its disorders: two-volume set. Brain 2004;127(9):2144–7.
[103] Penfield W. Obituaries: Pío Del Río-Hortega, M.D. (1882–1945). Arch Neurol Psychiatry 1945;54:413–6.
[104] Almeida AN, Martinez V, Feindel W. The first case of invasive EEG monitoring for the surgical treatment of epilepsy: historical significance and context. Epilepsia 2005;46:1082–5.
[105] Feindel W. The Montreal Neurological Institute. J Neurosurg 1991;75:821–2.
[106] Galambos R. A glia-neural theory of brain function. Proc Natl Acad Sci USA 1961;47:129–36.
[107] Cano Diaz P. Una contribución a la ciencia histológica: la obra de Don Pío del Río-Hortega. Madrid: Consejo Superior de Investigaciones Científicas; 1985.
[108] Hodgkin AL, Huxley AF. Propagation of electrical signals along giant nerve fibers. Proc R Soc Lond B Biol Sci 1952;140(899):177–83.
[109] Hodgkin AL, Huxley AF. A quantitative description of membrane current and its application to conduction and excitation in nerve. J Physiol 1952;117(4):500–44.
[110] Hodgkin AL, Huxley AF. The dual effect of membrane potential on sodium conductance in the giant axon of Loligo. J Physiol 1952;116(4):497–506.
[111] Hodgkin AL, Huxley AF. The components of membrane conductance in the giant axon of Loligo. J Physiol 1952;116(4):473–96.
[112] Hodgkin AL, Huxley AF. Currents carried by sodium and potassium ions through the membrane of the giant axon of Loligo. J Physiol 1952;116(4):449–72.
[113] Hodgkin AL, Huxley AF. Movement of sodium and potassium ions during nervous activity. Cold Spring Harb Symp Quant Biol 1952;17:43–52.
[114] Hodgkin AL, Huxley AF, Katz B. Measurement of current-voltage relations in the membrane of the giant axon of Loligo. J Physiol 1952;116(4):424–48.
[115] Hild W, Chang JJ, Tasaki I. Electrical responses of astrocytic glia from the mammalian central nervous system cultivated in vitro. Experientia 1958;14(6):220–1.

[116] Hild W, Tasaki I. Morphological and physiological properties of neurons and glial cells in tissue culture. J Neurophysiol 1962;25:277–304.
[117] Hertz L. Possible role of neuroglia: a potassium-mediated neuronal-neuroglial-neuronal impulse transmission system. Nature 1965;206(989):1091–4.
[118] Hydén H. A two-cell collaboration responsible for brain activity. Acta Univ Gothoburgensis 1960;66:1–26.
[119] Hydén H, Egyhazi E. Glial RNA changes during a learning experiment in rats. Proc Natl Acad Sci USA 1963;49:618–24.
[120] Hydén H. Production of RNA in neurones and glia in Parkinson's disease indicating genic stimulation. In: Costa E, Cote LJ, Yahr MD, editors. Biochemistry and pharmacology of the basal ganglia. New York: Raven Press; 1966. p. 195–204.
[121] Hertz L, Hansson E, Rönnbäck L. Signaling and gene expression in the neuron-glia unit during brain function and dysfunction: Holger Hydén in memoriam. Neurochem Int 2001;39(3):227–52.
[122] Hydén H. The effect of tranylcypromine on synthesis of macromolecules and enzyme activities in neurons and glia. Neurology 1968;18:732–6.
[123] Kuffler WS, Potter DD. Glia in the leech central nervous system: physiological properties and neuron-glia relationship. J Neurophysiol 1964;27:290–320.
[124] Orkand RK, Nicholls JG, Kuffler SW. Effect of nerve impulses on the membrane potential of glial cells in the central nervous system of amphibia. J Neurophysiol 1966;29(4):788–806.
[125] Kuffler WS. Neuroglial cells: physiological properties and a potassium mediated effect of neuronal activity on the glial membrane potential. Proc R Soc Lond B Biol Sci 1967;168:1–21.
[126] Brightman MW, Reese TS. Junctions between intimately apposed cell membranes in the vertebrate brain. J Cell Biol 1969;40(3):648–77.
[127] Bowman CL, Kimelberg HK. Excitatory amino acids directly depolarize rat brain astrocytes in primary culture. Nature 1984;311(5987):656–9.
[128] Kettenmann H, Backus KH, Schachner M. Aspartate, glutamate and gamma-aminobutyric acid depolarize cultured astrocytes. Neurosci Lett 1984;52(1–2):25–9.
[129] Kettenmann H, Gilbert P, Schachner M. Depolarization of cultured oligodendrocytes by glutamate and GABA. Neurosci Lett 1984;47(3):271–6.
[130] Hatton GI. Pituicytes, glia and control of terminal secretion. J Exp Biol 1988;139:67–79.
[131] Cornell-Bell AH, Finkbeiner SM, Cooper MS, Smith SJ. Glutamate induces calcium waves in cultured astrocytes: long-range glial signaling. Science 1990;247(4941):470–3.
[132] Nedergaard M. Direct signaling from astrocytes to neurons in cultures of mammalian brain cells. Science 1994;263(5154):1768–71.
[133] Parpura V, Basarsky TA, Liu F, Jeftinija K, Jeftinija S, Haydon PG. Glutamate-mediated astrocyte-neuron signalling. Nature 1994;369(6483):744–7.

CHAPTER

2

Astrocytes in the Mammalian Brain

OUTLINE

OVERVIEW

In contrast to *Chapter 1: History of Astrocytes,* in which a comprehensive history of the concept of glia was developed, in this chapter we will provide a brief overview of the modern concept of glia. We will summarize the classification of neuroglia, astrocyte types in the CNS, phylogeny of astrocytes, astrocyte heterogeneity, and then mention many of the currently known functions of astrocytes. Each of these subsections is not meant to be comprehensive as there are many excellent review articles and textbooks [1] that address each of these issues in more detail. Finally, the emerging concept of neurological diseases as "gliopathies" is considered. Subsequent chapters will more specifically enumerate glial changes in the context of epilepsy, in both human tissue and animal models.

Astrocytes and Epilepsy
DOI: http://dx.doi.org/10.1016/B978-0-12-802401-0.00002-8

CLASSIFICATION OF NEUROGLIA AND ASTROCYTE TYPES IN THE CNS

Neuroglia are classified into peripheral nervous system (PNS) and central nervous system (CNS) glia (Fig. 2.1). There are various forms of specialized PNS glia including Schwann cells (responsible for myelinating and surrounding peripheral nerve axons), olfactory ensheathing cells (OECs, found in the olfactory system), satellite glia (found in peripheral ganglia), and enteric glia (in the enteric nervous system). CNS glia includes macroglia (astrocytes, NG2-glia, and oligodendrocytes, derived from ectoderm) and microglia (derived from mesoderm).

Astrocytes ("star-like cells") are quite diverse, and in fact the term astrocyte or astroglia encompasses multiple types of structurally and functionally distinct cells. The more that is being learned about astrocytes, the more it is clear that there is great structural and functional heterogeneity among astrocytes, both within and between distinct brain regions [2–4]. Nevertheless, it is useful to briefly review the main features of astrocytes. The largest morphological subdivision has been between *protoplasmic astrocytes* and *fibrous astrocytes*. Protoplasmic astrocytes, found in the gray matter of the brain and spinal cord, have several (5–10 in rodent) primary processes which then expand into a complex arborization. Interestingly, this arborization delineates a "domain" and in the normal state there is little overlap between astrocyte domains [5,6] (Fig. 2.2). Each protoplasmic astrocyte extends processes to invest synapses (perisynaptic processes) [7], blood vessels (forming a specialized structure called a perivascular endfoot), and sometimes the pial surface (subpial endfeet, collectively forming the *glia limitans*) (Fig. 2.3). The processes from a single rodent protoplasmic astrocyte contact approximately 100,000 synapses [5,8].

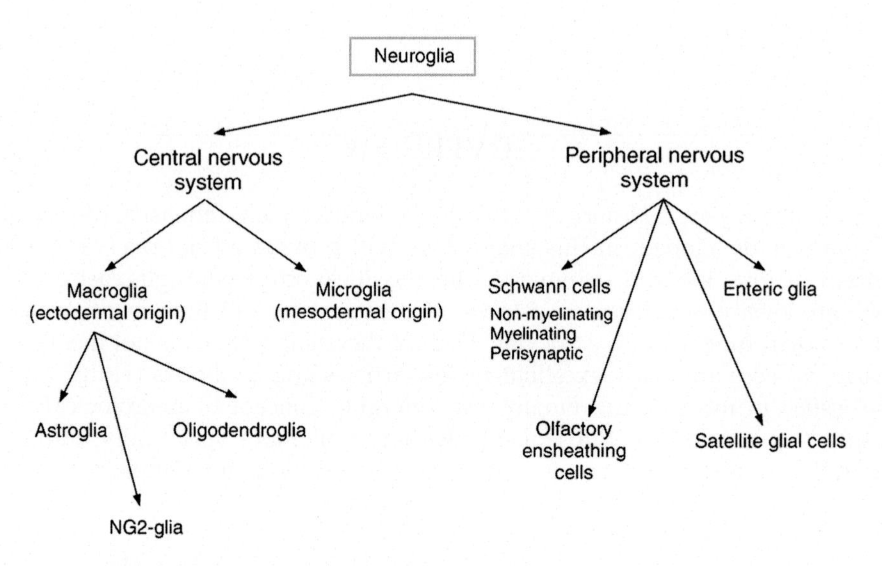

FIGURE 2.1 **Classification of neuroglia.** *Source: Reproduced with permission from Verkhratsky A, Butt AM. Glial physiology and pathophysiology. Oxford, UK: Wiley-Blackwell; 2013 (Figure 3.1).*

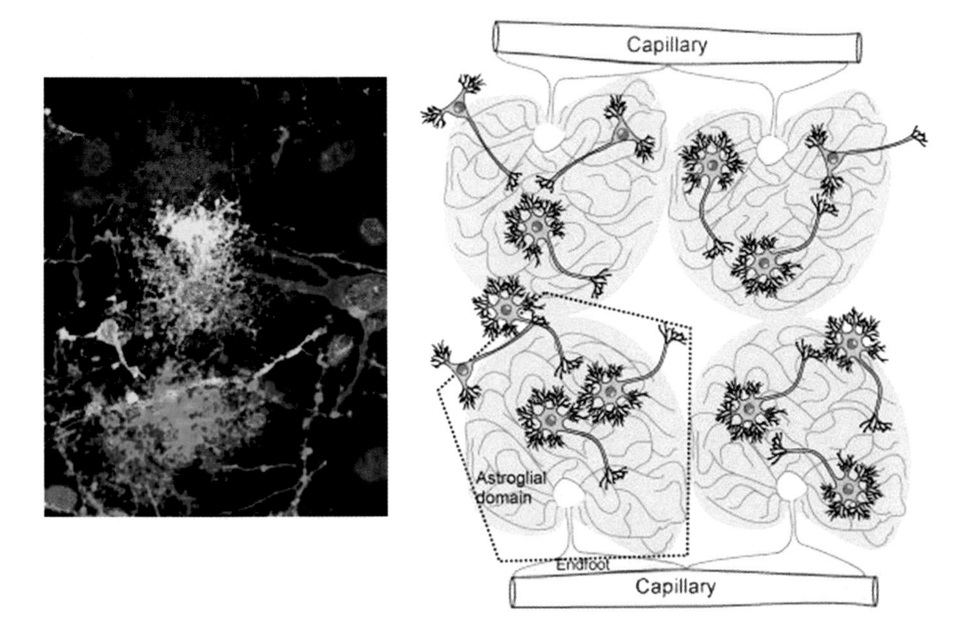

FIGURE 2.2 **Astrocyte domains form the microarchitecture of gray matter.** Left: Each single astrocyte occupies a well-defined territory; astroglial contacts occur only through distal processes and, overall, the overlap between astrocyte territories does not exceed 3–5%. Images show neighboring hippocampal astrocytes stained with different fluorescent dyes. Right: Schematic representation of astrocytic domains, which are organized in rows along the vessels, the latter typically being positioned in the narrow interface between astrocytes. *Source: Reproduced with permission from Verkhratsky A, Butt AM. Glial physiology and pathophysiology. Oxford, UK: Wiley-Blackwell; 2013 (Figure 4.33).*

Fibrous astrocytes, found in the white matter of brain and spinal cord and in the optic nerve and nerve fiber layer of the retina, are less arborized than protoplasmic astrocytes, and their cell bodies are located in a regular pattern between the axon bundles. In addition to perivascular and/or subpial endfeet, they also extend perinodal processes to contact the axons at nodes of Ranvier [9]. While fibrous astrocytes exhibit diverse morphology, in general they lack domain organization [10].

Originally, it was thought that astrocytes could be segregated by molecular markers, notably by expression of the intermediate filament protein glial fibrillary acidic protein (GFAP). However, GFAP expression is heterogeneous among astrocyte subtypes. For example, fibrous astrocytes express GFAP to a greater extent than protoplasmic astrocytes, and there is heterogeneity of GFAP expression among protoplasmic astrocytes of distinct brain regions. Astrocyte heterogeneity of form and function in the rule rather than the exception (Fig. 2.4).

A specialized group of astroglia is called *radial glia*. These cells are elongated astroglia that are prominent during development. They are bipolar cells with ovoid cell bodies and processes that span from the ventricular to the pial surface. Radial glia serve as a scaffold for neuronal migration (Fig. 2.5) [11]. Radial glia-like cells may persist into adulthood, such as in the retina (*Müller glia*) and cerebellum (*Bergmann glia*). Radial glia also have a neurogenic role [12,13].

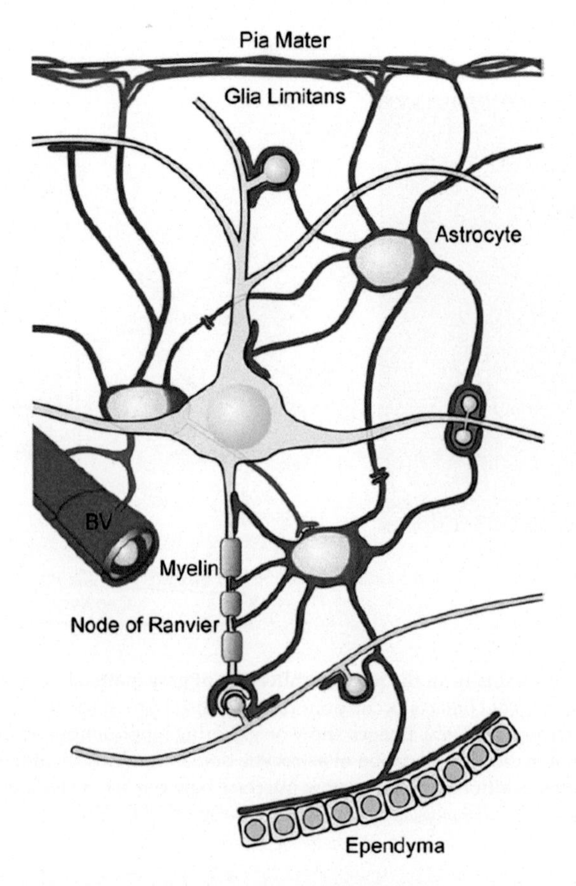

FIGURE 2.3 **Astrocytes.** In the adult central nervous system (CNS), astrocytes interact with multiple cell types. As depicted in this schematic, they form endfeet on capillaries that induce the blood–brain barrier properties of the cerebral microvasculature. They interact with the cells along the pial surface, as well as at the ventricular lumen, and their processes interdigitate among the neurons, synapses, nodes of Ranvier, and oligodendrocytes. Additionally, they form gap junctions with other astrocytes. *Source: Reproduced with permission from Levinson et al. Astrocyte development. In: Rao MS, Jacobson M, editors. Developmental neurobiology. New York: Kluwer Academic/Plenum Publishers; 2005. p. 197–222.*

Given the morphological and functional heterogeneity of astrocytes, identification of "classical" astrocytes requires careful consideration. Kimelberg put forth the following criteria [14,15]:

1. Absence of electrical excitability
2. Very negative membrane potential (−80 to −90 mV)
3. Expression of neurotransmitter transporters (GABA, glutamate)
4. Expression of intermediate filaments (particularly GFAP)

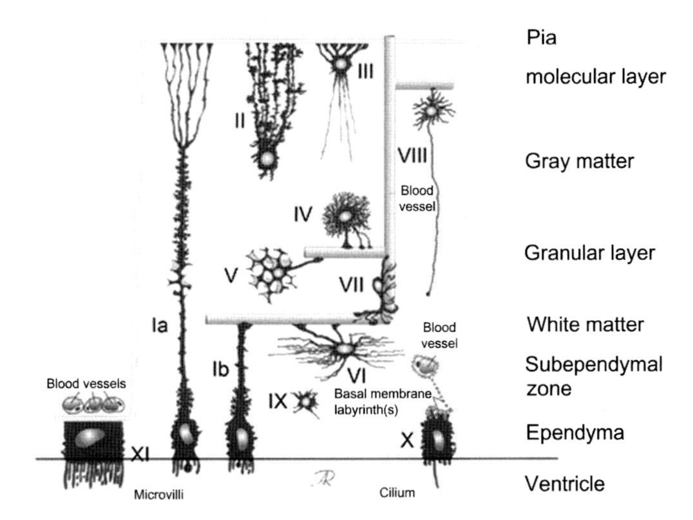

FIGURE 2.4 **Morphological diversity of astrocytes.** Ia, pial tanycyte; Ib, vascular tanycyte; II, radial astrocyte (Bergmann glial cell); III, marginal astrocyte; IV, protoplasmic astrocyte; V, velate astrocyte; VI, fibrous astrocyte; VII, perivascular astrocyte; VIII, interlaminar astrocyte; IX, immature astrocyte; X, ependymocyte; XI, choroid plexus cell. *Source: Reproduced with permission from Reichenbach A, Wolburg H. Astrocytes and ependymal glia, In: Kettenmann H, Ransom BR, editors. Neuroglia. New York, NY: Oxford University Press; 2005. p. 20.*

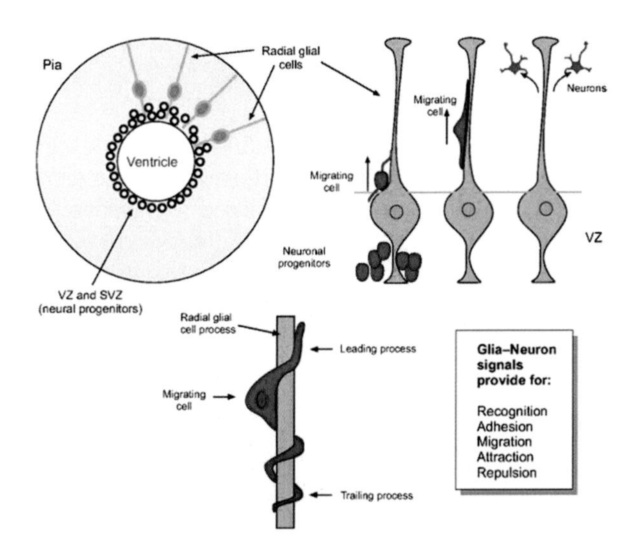

FIGURE 2.5 **Radial glial cells form a scaffold that assists neuronal migration in the developing nervous system.** Radial glial cells extend their processes from the ventricular zone (VZ) and subventricular zone (SVZ), where neural progenitors reside, toward the pia. Neuronal precursors attach to the radial glial cells and migrate along their processes toward their final destination. Numerous reciprocal factors released by both neurons and glia regulate the processes of mutual recognition, attraction, adhesion, migration, and final repulsion. *Source: Reproduced with permission from Verkhratsky A, Butt AM. Glial physiology and pathophysiology. Oxford, UK: Wiley-Blackwell; 2013 (Figure 4.31).*

5. Glycogen granules
6. Processes contacting blood vessels
7. Perisynaptic processes
8. Connection to other astrocytes by gap junctions.

However, while these criteria have heuristic value in categorization, there are many exceptions to these criteria leading to heterogeneity, such as the existence of: (1) GFAP-negative protoplasmic astrocytes; (2) GFAP-positive cells that do not form gap junctions; (3) "GluR" cells that express neurotransmitter receptors but not neurotransmitter transporters [16,17]; and (4) cells that may or may not extend processes to contact distinct boundaries/surfaces (blood vessels, pia, ependyma). Astrocyte heterogeneity is being increasingly recognized [2,3,18].

PHYLOGENY OF ASTROCYTES

How has the morphology and function of astrocytes evolved in phylogenetic terms? While the evolutionary origins of glia are obscure, in general terms glia have evolved to subserve by-product removal, structural support, phagocytic needs, developmental programming, and circuit modulation [19]. Studies of simpler nervous systems and comparison among mammalian astrocytes (eg, rodent vs human) have led to important clues to the diversity of astrocyte specializations. As we will see, the roles of mammalian glia have become even more complex and integral to nervous system function.

Caenorhabditis elegans

The nervous system of the nematode *Caenorhabditis elegans* contains exactly 302 neurons and 56 supportive/glial cells, 50 of which are ectodermal and 6 mesodermal in origin. Most of the epithelial-derived glial cells (46/50) are associated with sensory organs called sensilla. The six mesodermal glia are located around a central "nerve ring" and make gap junctions with neurons and muscle cells [1]. Functional roles of *C. elegans* glia are thought to include: sensilla function, ion homeostasis, development and morphogenesis, phagocytosis, and modulation of neuronal activity and behavior [20–23].

Drosophila

Insects developed diversification of glia [24]. In *Drosophila*, neuroglia comprises approximately 10% of the cells in the CNS (which has in total about 90,000 cells) but are divided into many types, including *surface glia* (make the hemolymph–brain barrier), *cortex glia* (located in the neuropil around axons and synapses), and *tract glial cells* (covering axonal tracts). Distinct areas of the *Drosophila* CNS, such as the optic lobe and optic lamina, have further specialization of glia. In insects, the hemolymph–brain barrier is formed by septate junctions between glia (rather than by tight junctions between endothelial cells which comprise the blood–brain barrier of mammals) [1]. Functions of glial cells in insects include ion (especially K^+) homeostasis, redistribution of ions via gap junction-connected networks, neurotransmitter homeostasis (histamine in the retina, and *Drosophila* CNS glia express glutamate transporters dEAAT1 and dEAAT2 and glutamine synthetase), control of circadian rhythms, and trophic support for neurons [1,25].

Mammals

The remarkable increase in complexity and size of the mammalian brain is paralleled by similarly remarkable increases in glial number and complexity [26,27]. In addition, comparison among mammalian species has led to interesting insights. Glia:neuron ratio has generally increased in phylogeny as brain size has increased [27,28], although this is controversial [29]. Much more impressive than glia:neuron ratio is the marked difference in morphology of individual astrocytes by species. Notably, human astrocytes are much larger and more complex than rodent astrocytes [26,30] (Fig. 2.6). The mean diameter of a human protoplasmic astrocyte (142 μm) is about 2.5 times larger than rat (56 μm) and the astrocyte domain volume is about 16.5 times larger. Human fibrous astrocytes also are about 2.2 times larger in humans than rodents [30]. Perhaps most remarkably, a single human protoplasmic astrocyte has about 10 times more primary processes and manifold more fine processes than a rodent astrocyte [30]. Therefore a single human astrocyte may contact about 2 million synapses within its domain volume compared to about 100,000 in the rodent [5,26,30]. By contrast, volumetric neuronal synaptic density and the number of synapses per neuron is quite similar between rodents and primates. In addition, primates appear to have distinct populations of specialized astrocytes not found in other mammals: "interlaminar astrocytes" whose processes penetrate through the cortex and end in layers III and IV [31–34]; and "polarized astrocytes" dwelling in layers V and VI. Terminal processes of interlaminar astrocytes were reported to be particularly large in Albert Einstein's brain [35] but the significance is unclear. To assess the cell-autonomous properties of human astrocytes, Han and colleagues engrafted human glial progenitor cells into neonatal immunodeficient mice. Engrafted human glia integrated with host astrocytes, propagated Ca^{2+} signals threefold faster than native mouse astrocytes, and most strikingly supported enhancement of long-term potentiation of synaptic transmission (LTP) and improved learning in multiple paradigms [36]. This remarkable result suggests a relationship between cognitive performance and astrocyte complexity for the first time.

MULTIFACETED PHYSIOLOGICAL ROLES OF MAMMALIAN ASTROCYTES

Rather than "just brain glue" and also beyond basic homeostatic functions, recent "astroscientific" work has clearly identified many key physiological roles for CNS astrocytes [15,37,38]. Astrocytes are critical contributors to CNS development, structural support, barrier function, homeostatic function, metabolic support, synaptic transmission, regulation of blood flow, higher/integrative brain functions, brain defense, neuroprotection, and response to injury (Table 2.1). Many of these roles have been discovered recently, and a flowering of astrocyte and glial–neuronal–vascular interactions research is currently underway. For each role, many associated references and review articles are available, which are

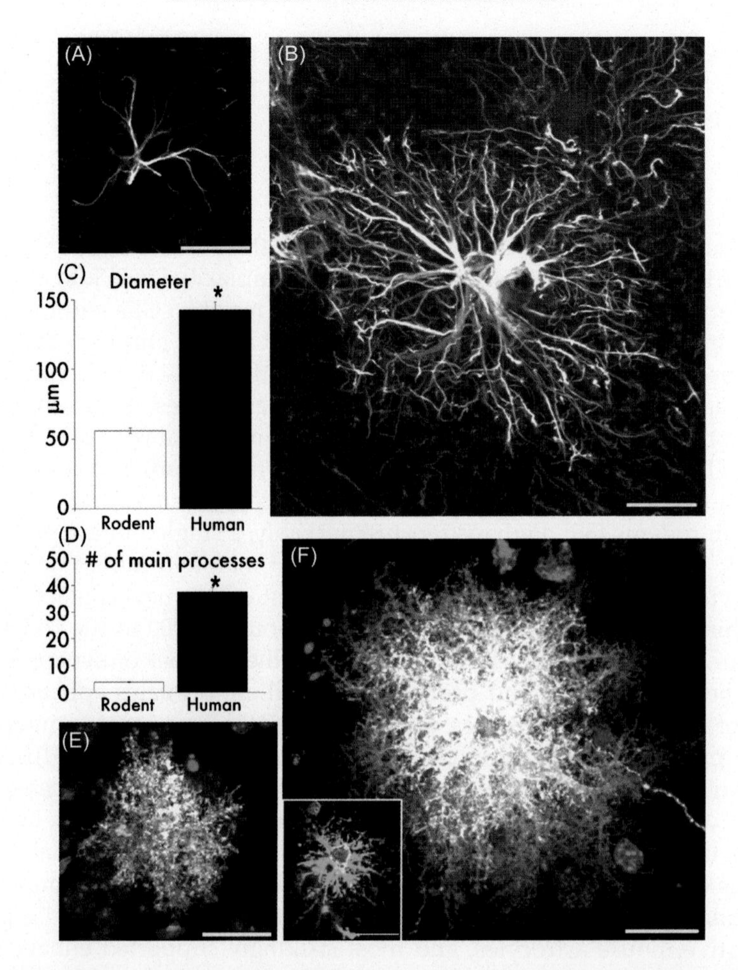

FIGURE 2.6 **Protoplasmic astrocytes are larger and more complicated than the rodent counterpart.** (A) Typical mouse protoplasmic astrocyte. GFAP, White. Scale bar, 20 μm. (B) Typical human protoplasmic astrocyte in the same scale. Scale bar, 20 μm. (C) and (D) Human protoplasmic astrocytes are 2.55-fold larger and have 10-fold more main GFAP processes than mouse astrocytes (human, $n = 50$ cells from 7 patients; mouse, $n = 65$ cells from 6 mice; mean ± SEM; $^*P = 0.005$, t test). (E) Mouse protoplasmic astrocyte diolistically labeled with DiI (white) and sytox (blue) revealing the full structure of the astrocyte including its numerous fine processes. Scale bar, 20 μm. (F) Human astrocyte diolistically labeled demonstrates the highly complicated network of fine process that defines the human protoplasmic astrocyte. Scale bar, 20 μm. Inset, Human protoplasmic astrocyte diolistically labeled as well as immunolabeled for GFAP (green) demonstrating colocalization. Scale bar, 20 μm. *Source: Reproduced with permission from Oberheim et al. Uniquely hominid features of adult human astrocytes. J Neurosci 2009;29:3276–87.*

outside the scope of this chapter. Also, given the complexity and diversity of astrocytes, this is not a comprehensive list. In the context of epilepsy, as will be detailed clearly in the succeeding chapters, these "novel" roles of astrocytes are critical and ultimately will help in identifying truly new ways of approaching epilepsy as a disease and with new therapeutic targets.

TABLE 2.1 Functions of Astrocytes

Development of the CNS	Neurogenesis Neural cell migration and formation of layered gray matter Synaptogenesis
Structural support	Parcellation of the gray matter through "tiling" Delineation of *pia mater* and the vessels by perivascular glia Formation of the neurovascular unit
Barrier function	Regulation of formation and permeability of blood–brain and CSF–brain barriers Formation of glial–vascular interface (astrocyte endfeet)
Homeostatic function	Control over extracellular K^+ homeostasis through local and spatial buffering Control over extracellular pH Regulation of water transport Removal of neurotransmitters from extracellular space Homeostasis of pH, CO_2, Na^+, and glucose
Metabolic support	Uptake of glucose; deposition of glycogen Provision of lactate to neurons in activity-dependent manner
Synaptic transmission	Regulation of synapse maintenance and assisting in synaptic pruning Provision of glutamate for glutamatergic transmission (through de novo synthesis and glutamate–glutamine shuttle) Regulation of synaptic plasticity Integration of synaptic fields Humoral regulation of neuronal networks through secretion of neurotransmitters and neuromodulators
Regulation of blood flow	Regulation of local blood supply (functional hyperemia) through secretion of vasoconstrictors or vasodilators
Higher brain functions	Respiration Sleep and circadian rhythms Learning and memory
Brain defense, neuroprotection, and response to injury	Reactive astrogliosis and scar formation Modulation of immune response and secretion of cytokines and chemokines

Modified with permission from Verkhratsky A, Butt AM. Glial physiology and pathophysiology. Oxford, UK: Wiley-Blackwell; 2013 (Table 4.3).

TRIPARTITE SYNAPSE

One key concept critical for subsequent discussions of astrocytes and epilepsy is the evolving concept of the "tripartite" synapse [7,39–42] (Fig. 2.7). The goal of the tripartite synapse concept is to convey how closely astrocytes are associated with synaptic transmission, both structurally and functionally. It has been shown that astrocyte processes are very closely associated with synaptic regions, in many cases partially or completely "wrapping" or "cradling" pre- and postsynaptic terminals (Fig. 2.8) [7,43–45]. The distance between perisynaptic astrocyte membrane and synaptic structures in the hippocampus and cerebellum is about 1 μm. What is the function of these perisynaptic glial processes? Studies have

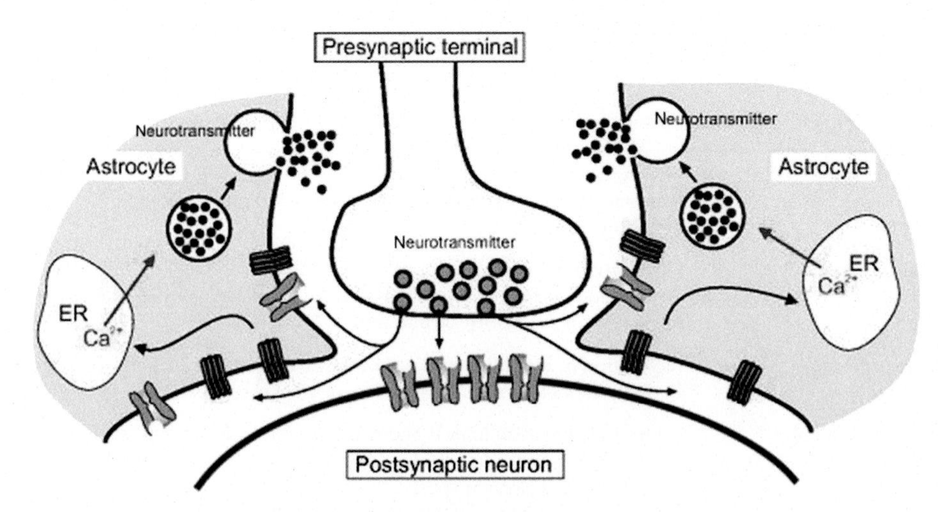

FIGURE 2.7 **The tripartite synapse.** The concept of the tripartite synapse assumes that it is constructed from a presynaptic terminal, the postsynaptic membrane and surrounding astrocyte processes. The neurotransmitter released from the presynaptic terminal interacts with specific receptors located in both the postsynaptic neuronal membrane and in the astroglial membrane. This triggers astroglial Ca^{2+} signals, which induce the release of neurotransmitters from astrocytes that signal back to neuronal compartment. *Source: Reproduced with permission from Verkhratsky A, Butt AM. Glial physiology and pathophysiology. Oxford, UK: Wiley-Blackwell; 2013 (Figure 4.51).*

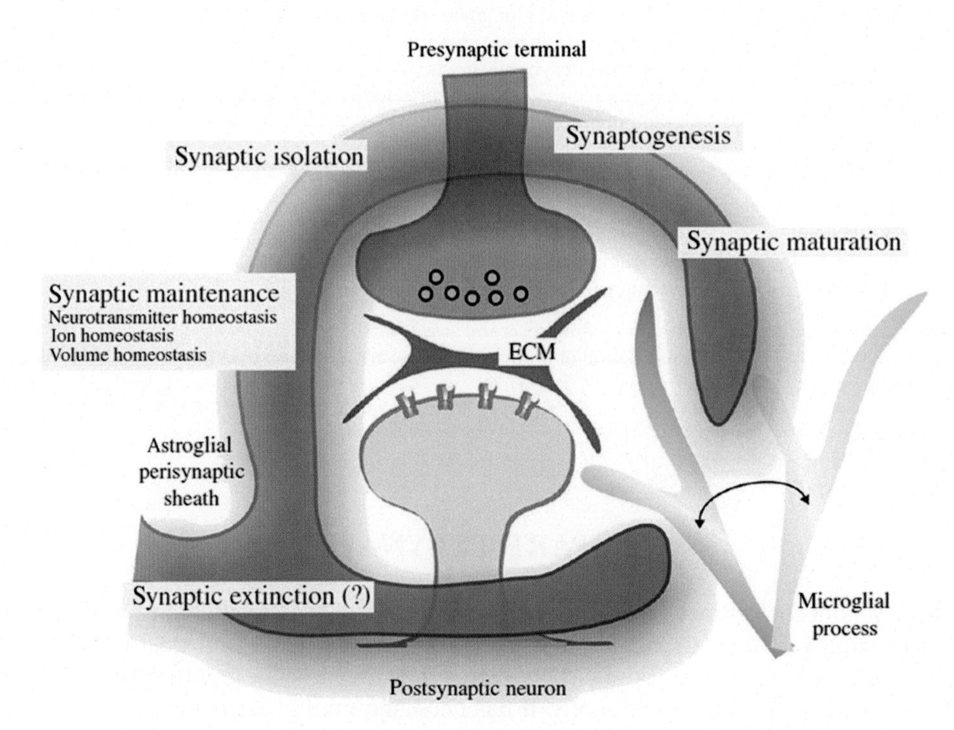

FIGURE 2.8 **The astroglial cradle.** Astroglial cradle embraces and fosters multipartite synapse in the CNS. The majority of synapses in the brain and spinal cord are composed of several components that include: the presynaptic terminal, the postsynaptic part, the perisynaptic process of the astrocyte, the process of a neighboring microglial cell that periodically contacts the synaptic structure, and the ECM present in the synaptic cleft, and also extending extra-synaptically. Astroglial perisynaptic sheaths enwrap synaptic structures and regulate, influence and assist synaptogenesis, synaptic maturation, synaptic maintenance, and synaptic extinction. *Source: Reproduced with permission from Verkhratsky A, Nedergaard AM. Astroglial cradle in the life of the synapse. Phil Trans R Soc B 2014;369:20130595.*

demonstrated that reduction in glial coverage of synapses leads to (1) prolonged EPSPs in the cerebellum [46]; (2) increased activation of mGlu receptors in the supraoptic nucleus [47,48]; and (3) synaptic spillover of glutamate during stimulation in the hippocampus [49]. Thus, the primary function of perisynaptic glial processes, at least at excitatory glutamatergic synapses, appears to be glutamate uptake. Recent evidence also indicates activity-dependent perisynaptic membrane diffusion of the major astrocyte glutamate transporter, GLT1, further shaping synaptic transmission through a glial mechanism [50].

NEUROLOGICAL DISEASES AS "GLIOPATHIES"

Based on the wide variety of important homeostatic roles for astrocytes, their role in modulating synaptic transmission and their key changes in response to injury, more modern conceptions of neurological diseases hold that they may have a "gliopathic" component [51–54]. In the case of epilepsy, it has been demonstrated that there is a loss in astrocyte domain organization in the epileptic brain [55]. The precise functional consequences of this are as yet unclear; however, this and other major changes in astrocyte form, function, and expression of gap junctions, ion channels, receptors, and transporters in epilepsy will be reviewed and considered in detail later in this book. From a translational standpoint, to the extent that *astrocytes* (instead of neurons) are the elements of the nervous system responsible for homeostatic function, and disease represents loss of homeostasis, then therapeutic targets for disease treatment should be directed at restoring homeostasis in astrocytes. That concept leads to many new therapeutic targets that may be putatively more astrocyte-specific. Such a strategy may not have the deleterious side effects of inhibiting normal neuronal function and synaptic transmission, as is the case for many current antiepileptic drugs [56].

References

[1] Verkhratsky A, Butt AM. Glial physiology and pathophysiology. Oxford, UK: Wiley-Blackwell; 2013.

[2] Zhang Y, Barres BA. Astrocyte heterogeneity: an underappreciated topic in neurobiology. Curr Opin Neurobiol 2010;20(5):588–94.

[3] Hoft S, Griemsmann S, Seifert G, Steinhäuser C. Heterogeneity in expression of functional ionotropic glutamate and GABA receptors in astrocytes across brain regions: insights from the thalamus. Philos Trans R Soc Lond B Biol Sci 2014;369(1654):20130602.

[4] Steinhäuser C, Berger T, Frotscher M, Kettenmann H. Heterogeneity in the membrane current pattern of identified glial cells in the hippocampal slice. Eur J Neurosci 1992;4(6):472–84.

[5] Bushong EA, Martone ME, Jones YZ, Ellisman MH. Protoplasmic astrocytes in CA1 stratum radiatum occupy separate anatomical domains. J Neurosci 2002;22(1):183–92.

[6] Wilhelmsson U, Bushong EA, Price DL, Smarr BL, Phung V, Terada M, et al. Redefining the concept of reactive astrocytes as cells that remain within their unique domains upon reaction to injury. Proc Natl Acad Sci USA 2006;103(46):17513–8.

[7] Ventura R, Harris KM. Three-dimensional relationships between hippocampal synapses and astrocytes. J Neurosci 1999;19(16):6897–906.

[8] Halassa MM, Fellin T, Takano H, Dong JH, Haydon PG. Synaptic islands defined by the territory of a single astrocyte. J Neurosci 2007;27(24):6473–7.

[9] Butt AM, Duncan A, Berry M. Astrocyte associations with nodes of Ranvier: ultrastructural analysis of HRP-filled astrocytes in the mouse optic nerve. J Neurocytol 1994;23(8):486–99.

[10] Butt AM, Colquhoun K, Tutton M, Berry M. Three-dimensional morphology of astrocytes and oligodendrocytes in the intact mouse optic nerve. J Neurocytol 1994;23(8):469–85.

[11] Levison SW, de Vellis J, Goldman JE. Astrocyte development. In: Rao MS, Jacobson M, editors. Developmental neurobiology. New York: Kluwer Academic/Plenum Publishers; 2005. p. 197–222.
[12] Kriegstein AR, Gotz M. Radial glia diversity: a matter of cell fate. Glia 2003;43(1):37–43.
[13] Rakic P. Elusive radial glial cells: historical and evolutionary perspective. Glia 2003;43(1):19–32.
[14] Kimelberg HK. The problem of astrocyte identity. Neurochem Int 2004;45(2–3):191–202.
[15] Kimelberg HK. Functions of mature mammalian astrocytes: a current view. Neuroscientist 2010;16(1):79–106.
[16] Wallraff A, Odermatt B, Willecke K, Steinhäuser C. Distinct types of astroglial cells in the hippocampus differ in gap junction coupling. Glia 2004;48(1):36–43.
[17] Matthias K, Kirchhoff F, Seifert G, Hüttmann K, Matyash M, Kettenmann H, et al. Segregated expression of AMPA-type glutamate receptors and glutamate transporters defines distinct astrocyte populations in the mouse hippocampus. J Neurosci 2003;23(5):1750–8.
[18] Griemsmann S, Hoft SP, Bedner P, Zhang J, von Staden E, Beinhauer A, et al. Characterization of panglial gap junction networks in the thalamus, neocortex, and hippocampus reveals a unique population of glial cells. Cereb Cortex 2015;25(10):3420–33.
[19] Hartline DK. The evolutionary origins of glia. Glia 2011;59(9):1215–36.
[20] Bacaj T, Tevlin M, Lu Y, Shaham S. Glia are essential for sensory organ function in *C. elegans*. Science 2008;322(5902):744–7.
[21] Oikonomou G, Shaham S. The glia of *Caenorhabditis elegans*. Glia 2011;59(9):1253–63.
[22] Shaham S. Glial development and function in the nervous system of *Caenorhabditis elegans*. Cold Spring Harb Perspect Biol 2015;7(4):a020578.
[23] Stout Jr RF, Verkhratsky A, Parpura V. *Caenorhabditis elegans* glia modulate neuronal activity and behavior. Front Cell Neurosci 2014;8:67.
[24] Hartenstein V. Morphological diversity and development of glia in *Drosophila*. Glia 2011;59(9):1237–52.
[25] Freeman MR. *Drosophila* central nervous system glia. Cold Spring Harb Perspect Biol 2015;7:11.
[26] Oberheim NA, Wang X, Goldman S, Nedergaard M. Astrocytic complexity distinguishes the human brain. Trends Neurosci 2006;29(10):547–53.
[27] Reichenbach A. Glia:neuron index: review and hypothesis to account for different values in various mammals. Glia 1989;2(2):71–7.
[28] Sherwood CC, Stimpson CD, Raghanti MA, Wildman DE, Uddin M, Grossman LI, et al. Evolution of increased glia-neuron ratios in the human frontal cortex. Proc Natl Acad Sci USA 2006;103(37):13606–11.
[29] Herculano-Houzel S. The glia/neuron ratio: how it varies uniformly across brain structures and species and what that means for brain physiology and evolution. Glia 2014;62(9):1377–91.
[30] Oberheim NA, Takano T, Han X, He W, Lin JH, Wang F, et al. Uniquely hominid features of adult human astrocytes. J Neurosci 2009;29(10):3276–87.
[31] Colombo JA, Reisin HD. Interlaminar astroglia of the cerebral cortex: a marker of the primate brain. Brain Res 2004;1006(1):126–31.
[32] Colombo JA, Sherwood CC, Hof PR. Interlaminar astroglial processes in the cerebral cortex of great apes. Anat Embryol (Berl) 2004;208(3):215–8.
[33] Colombo JA, Yanez A, Puissant V, Lipina S. Long, interlaminar astroglial cell processes in the cortex of adult monkeys. J Neurosci Res 1995;40(4):551–6.
[34] Sosunov AA, Wu X, Tsankova NM, Guilfoyle E, McKhann 2nd GM, Goldman JE. Phenotypic heterogeneity and plasticity of isocortical and hippocampal astrocytes in the human brain. J Neurosci 2014;34(6):2285–98.
[35] Colombo JA, Reisin HD, Miguel-Hidalgo JJ, Rajkowska G. Cerebral cortex astroglia and the brain of a genius: a propos of A. Einstein's. Brain Res Rev 2006;52(2):257–63.
[36] Han X, Chen M, Wang F, Windrem M, Wang S, Shanz S, et al. Forebrain engraftment by human glial progenitor cells enhances synaptic plasticity and learning in adult mice. Cell Stem Cell 2013;12(3):342–53.
[37] Allen NJ, Barres BA. Glia: more than just brain glue. Nature 2009;457(7230):675–7.
[38] Volterra A, Meldolesi J. Astrocytes, from brain glue to communication elements: the revolution continues. Nat Rev Neurosci 2005;6(8):626–40.
[39] Araque A, Parpura V, Sanzgiri RP, Haydon PG. Tripartite synapses: glia, the unacknowledged partner. Trends Neurosci 1999;22(5):208–15.
[40] Halassa MM, Fellin T, Haydon PG. The tripartite synapse: roles for gliotransmission in health and disease. Trends Mol Med 2007;13(2):54–63.

[41] Santello M, Calì C, Bezzi P. Gliotransmission and the tripartite synapse. Adv Exp Med Biol 2012;970:307–31.
[42] Perea G, Navarrete M, Araque A. Tripartite synapses: astrocytes process and control synaptic information. Trends Neurosci 2009;32(8):421–31.
[43] Verkhratsky A, Nedergaard M. Astroglial cradle in the life of the synapse. Philos Trans R Soc Lond B Biol Sci 2014;369(1654):20130595.
[44] Verkhratsky A, Nedergaard M, Hertz L. Why are astrocytes important? Neurochem Res 2015;40(2):389–401.
[45] Nedergaard M, Verkhratsky A. Artifact versus reality–how astrocytes contribute to synaptic events. Glia 2012;60(7):1013–23.
[46] Iino M, Goto K, Kakegawa W, Okado H, Sudo M, Ishiuchi S, et al. Glia-synapse interaction through Ca^{2+}-permeable AMPA receptors in Bergmann glia. Science 2001;292(5518):926–9.
[47] Oliet SH, Piet R, Poulain DA. Control of glutamate clearance and synaptic efficacy by glial coverage of neurons. Science 2001;292(5518):923–6.
[48] Oliet SH, Piet R, Poulain DA, Theodosis DT. Glial modulation of synaptic transmission: insights from the supraoptic nucleus of the hypothalamus. Glia 2004;47(3):258–67.
[49] Rusakov DA. The role of perisynaptic glial sheaths in glutamate spillover and extracellular Ca(2+) depletion. Biophys J 2001;81(4):1947–59.
[50] Murphy-Royal C, Dupuis JP, Varela JA, Panatier A, Pinson B, Baufreton J, et al. Surface diffusion of astrocytic glutamate transporters shapes synaptic transmission. Nat Neurosci 2015;18(2):219–26.
[51] Parpura V, Heneka MT, Montana V, Oliet SH, Schousboe A, Haydon PG, et al. Glial cells in (patho)physiology. J Neurochem 2012;121(1):4–27.
[52] Verkhratsky A, Sofroniew MV, Messing A, deLanerolle NC, Rempe D, Rodriguez JJ, et al. Neurological diseases as primary gliopathies: a reassessment of neurocentrism. ASN Neuro 2012;4(3).
[53] Chung WS, Welsh CA, Barres BA, Stevens B. Do glia drive synaptic and cognitive impairment in disease? Nat Neurosci 2015;18(11):1539–45.
[54] Barres BA. The mystery and magic of glia: a perspective on their roles in health and disease. Neuron 2008;60(3):430–40.
[55] Oberheim NA, Tian GF, Han X, Peng W, Takano T, Ransom B, et al. Loss of astrocytic domain organization in the epileptic brain. J Neurosci 2008;28(13):3264–76.
[56] Crunelli V, Carmignoto G, Steinhäuser C. Novel astrocyte targets: new avenues for the therapeutic treatment of epilepsy. Neuroscientist 2015;21(1):62–83.

CHAPTER

3

Gliotransmitters

OUTLINE

OVERVIEW

Historically, neurons were considered the primary cells responsible for brain function. This view has been challenged, however, by recent research proving the existence of astrocyte–neuron communication resulting in both the modulation of neurotransmission and the maintenance of brain homeostasis. Chemical signaling underlies this bidirectional communication and strong evidence has demonstrated that neurotransmitters can be released from astrocytes in a process known as gliotransmission. We will focus on glutamate release, as it is the most well-studied gliotransmitter. At least six mechanisms of glutamate release from astrocytes have been identified: (1) calcium-dependent exocytosis; (2) reversal of membrane transporters; (3) glutamate exchange through the cystine/glutamate antiporter; (4) release through ionotropic purinergic receptors; (5) cell swelling-induced opening of

Astrocytes and Epilepsy
DOI: http://dx.doi.org/10.1016/B978-0-12-802401-0.00003-X

anion channels; and (6) release through hemichannels. This chapter will discuss the evidence, controversy, and drawbacks for each mechanism of gliotransmission.

A NEUROCENTRIC VIEW

The classical view of glial cells was that they were primarily passive (serving as the "glue" to hold neurons together) whereas neurons were thought to be solely responsible for brain information processing [1]. Glia were largely ignored in terms of brain function and synaptic transmission. This was likely due to the fundamental difference between neurons and astrocytes: neurons are electrically excitable (can produce "all-or-none" action potentials that propagate through neuronal networks) whereas astrocytes are not. Because of this, astrocytes were often thought of as "silent" partners in the brain. This view has been challenged, however, with the use of new techniques, most notably patch-clamp electrophysiology and the use of fluorescent dyes, which has led to the discovery that astrocytes can modulate synaptic transmission.

A number of astrocytic properties have been discovered that led to the dismissal of the "passive" astrocyte theory and to our current understanding of the "active" astrocyte that plays a role in neurotransmission and neuronal physiology. Astrocytes in culture were found to exhibit depolarization in response to certain neurotransmitters, suggesting they possessed ionotropic receptors [2,3]. These cells respond to sensory input [4–8] and exhibit glutamate-induced calcium oscillations [9]. In addition, astrocytes play a role in metabolic support, uptake of potassium (K^+), and clearance of neurotransmitters [10]. Most importantly, astrocytes engage in bidirectional communication with neurons [11,12], suggesting they are intimately involved in synaptic transmission.

The chemical synapse was traditionally thought to be composed of a presynaptic neuron and a postsynaptic neuron. The concept that a third component (the astrocyte) was involved gained popularity in the 1990s, although evidence for a communication system between neurons and astrocytes involving chemical transmitters was hypothesized well before then [13]. This led to the term "tripartite synapse" to highlight the direct role of astrocytes in synaptic function (Fig. 3.1) [14] (also see *Chapter 2: Astrocytes in the Mammalian Brain*).

ASTROCYTES AT THE SYNAPSE

Astrocytes are essential for proper synaptic development. Without them, few synapses would form and the ones that do would be functionally immature [15]. Imaging studies confirmed that astrocytes are organized into exclusive territories or "microdomains" with fine structures of appendages that are found either near or contacting synaptic clefts [16–19]. Through the use of serial electron microscopy and 3D analysis, it was estimated that about 57% of synapses in hippocampal area CA1 had astrocytic processes juxtaposed to them [20]. Overall, there was a large variation in processes surrounding the axon–spine interface, suggesting that astrocytes nonuniformly regulate synapses.

Astrocytes are crucial in maintaining the homeostasis of surrounding synapses. They supply energy metabolites, including citric acid cycle constituents [21–23], participate in the

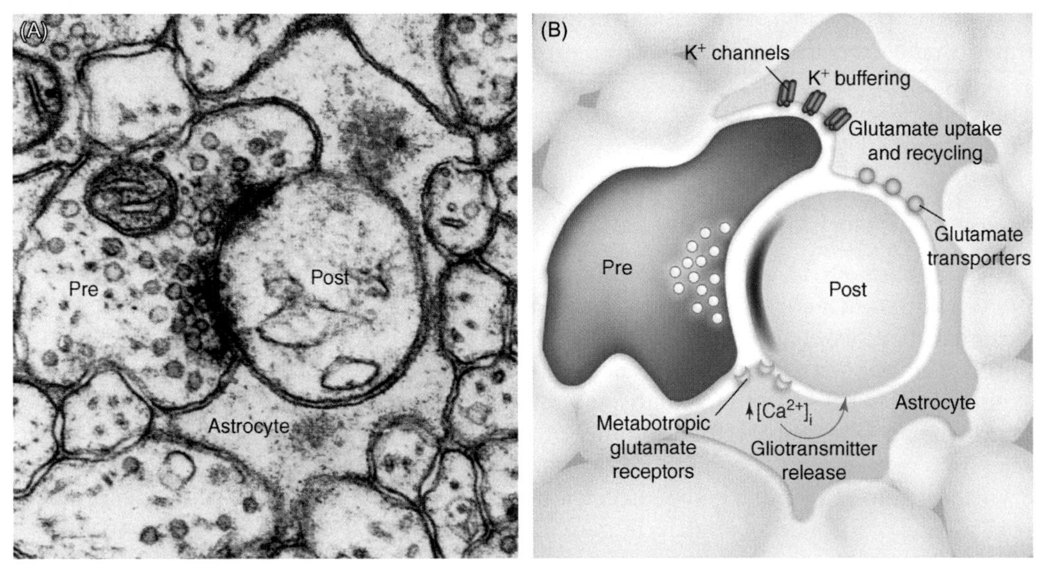

FIGURE 3.1 **The astrocytic process is the third active element forming the tripartite synapse.** (A) Electron micrograph showing a presynaptic (Pre) and postsynaptic (Post) terminal enwrapped by the astrocytic process (green) forming the tripartite synapse. (B) The close association of the astrocytic process with the presynaptic and postsynaptic terminals exerts crucial roles in clearing K^+ ions that accumulate following neuronal activity, and in the uptake of the synaptic transmitter glutamate by the activity of plasma membrane glutamate transporters. Additionally, neurotransmitter release from presynaptic terminals can activate astrocytic metabotropic receptors, which induce the inositol (1,4,5)-trisphosphate (IP_3)-dependent release of Ca^{2+} from internal stores, which in turn triggers the release of several neuroactive compounds (gliotransmitters) from these cells. Locations of astrocytic transporters and receptors in this figure do not necessarily represent their exact spatial distribution. *Source: Reproduced with permission from Halassa MM, Fellin T, Haydon PG. The tripartite synapse: roles for gliotransmission in health and disease. Trends Mol Med 2007;13(2):54–63.*

clearance and spatial buffering of extracellular K^+ [24–27], remove glutamate from the extracellular space [28–31], and control extracellular levels and diffusion of neurotransmitters through astrocyte protrusion into the synaptic cleft [32–34]. Through these functions, astrocytes can modulate neuronal output. In fact, computational modeling has predicted that due to the dynamic relationship between neurons and astrocytes that ensheathe synapses, astrocytes may modulate the probability of vesicle release [35], optimize neural network performance [35,36], control the threshold value of transition from synchronous to asynchronous behavior among neurons [37], and lead to spike-timing-dependent plasticity at remote synaptic sites [38].

BIDIRECTIONAL COMMUNICATION BETWEEN NEURONS AND ASTROCYTES

The morphology and location of astrocytes allows them to both listen (through neurotransmitter receptors) and respond (G-protein-coupled receptors (GPCRs) and other intracellular pathways) to neuronal activity. Indisputable evidence has already demonstrated that astrocytes can respond to neuronal input, often with an increase in intracellular calcium. Emerging evidence, however, now suggests that astrocytes can discriminate between different inputs and that astrocytes can reciprocally respond to neurons with feedback signals.

Astrocytes Discriminate Between Different Inputs

Several studies have demonstrated that astrocytes have the ability to discriminate between different neurotransmitters and synaptic inputs. For example, astrocytes in hippocampal slices responded with calcium elevations to acetylcholine (ACh), but not to alveus stimulation-induced glutamate release, despite the presence of both cholinergic and glutamatergic axons [39,40]. Interestingly, astrocytes are able to integrate these inputs and can react to synaptic activity with a nonlinear input–output response [1]. Evidence for this has come from experiments involving visual, olfactory, and somatosensory stimulation.

Astrocytes in the primary visual cortex have various receptive-field characteristics similar to those seen in neurons, including spatially restrictive receptive fields, orientation tuning, and spatial-frequency tuning [41]. Astrocytes, however, had sharper orientation and spatial-frequency tuning than neurons and responded to visual stimuli with a delayed spike in intracellular calcium [41]. Photostimulation of astrocytes enhanced both excitatory and inhibitory synaptic transmission; this was mediated by metabotropic glutamate receptor (mGluR) activation [42]. Electrical stimulation of the nucleus basalis paired with visual stimulation-induced potentiation of visual responses in neurons found in the primary visual cortex [43]. The mechanism behind these changes was thought to involve muscarinic ACh receptors evoking *N*-methyl-D-aspartate (NMDA) receptor-mediated slow inward currents (SICs) in neurons. This potentiation was absent, however, in mice lacking the primarily astrocytic IP_3 receptor (type 2), suggesting that intracellular calcium elevations in astrocytes were required [43]. Similar to what was seen in the visual cortex, odor stimulation in mice led to calcium transients in astrocytic endfeet and was highly correlated with the dilation of upstream arterioles and the release of glutamate [44]. In addition, odor-evoked functional hyperemia in glomerular capillaries was mediated, in part, by astrocytic mGluR5 and cyclooxygenase (COX) activation.

Activation of the somatosensory cortex through either mechanical limb stimulation [8] or whisker stimulation [7] led to delayed increases in astrocytic calcium in the somatosensory cortex. Whisker stimulation also led to delayed onset local field potential (LFP) responses [45]. In layer 2/3 of the barrel cortex, astrocytic intracellular calcium increases were elicited in response to glutamatergic inputs from layer 4 of the same column, but not to glutamatergic projections from layer 2/3 of adjacent columns [46]. Interestingly, either somatosensory stimulation via tail pinch [47] or whisker stimulation combined with electrical stimulation of the nucleus basalis of Meynert (NBM) [45] induced intracellular calcium spikes in astrocytes and cortical long-term potentiation (LTP). Locomotor performance also led to calcium rises in networks of Bergmann glia, which were abolished by the blockade of either neuronal activity or glutamatergic transmission [6]. Astrocytes in the ventrobasal thalamus preferentially responded to corticothalamic inputs over sensory pathways [48]. Taken together, it is clear that astrocytes respond to a variety of neuronal inputs in a discriminatory manner.

Astrocytes Send Feedback Signals to Neurons

After astrocytes sense synaptic activity, they have the ability to respond with an output signal back to neurons. This has been demonstrated in co-cultures of astrocytes and neurons, where astrocytes were able to trigger calcium increases in adjacent neurons [49,50]. Astrocytes in slices of neonatal rat ventrobasal thalamus displayed spontaneous intracellular calcium oscillations that propagated as waves to neighboring astrocytes and resulted in NMDA receptor-mediated SICs in neurons found along the wave path [51]. In hippocampal slices of rats, direct stimulation of astrocytes increased miniature inhibitory postsynaptic currents (mIPSCs) in pyramidal neurons, an effect that was blocked by inhibition of astrocytic calcium signaling [52]. Similarly, stimulation of either the Schaffer collateral pathway or individual astrocytes resulted in NMDA receptor-mediated SICs in adjacent neurons [53]. Calcium elevations in astrocytic processes have also been proposed to participate in local tuning of neurotransmitter release at excitatory synapses [54].

RELEASE OF "GLIOTRANSMITTERS" FROM ASTROCYTES

Chemical signaling is now known to underlie bidirectional communication between astrocytes and neurons (Fig. 3.2) [11]. Gliotransmitters are neuroactive chemicals released from any glial cell, including oligodendrocytes, astrocytes, and microglia. Several compounds may be released from glia, including excitatory amino acids, nucleotides and nucleosides, eicosanoids, prostaglandins, neurotrophins, cytokines [55], and taurine [56]. The release of gliotransmitters, a process known as gliotransmission, primarily occurs in astrocytes. Evidence for gliotransmission began as early as the 1970s [13,57,58], but this concept was not widely accepted. Now, however, there is a growing body of evidence in support of gliotransmission [1,4,12,14,55,59–72].

MECHANISMS OF GLIOTRANSMISSION

Astrocytes are known to release gliotransmitters through a variety of mechanisms. Early studies used in vitro techniques to demonstrate the ability of astrocytes to release gliotransmitters while more recent studies have observed these phenomena in vivo. The gliotransmitter that has been most well studied is glutamate. At least six potential mechanisms of astrocyte glutamate release have been hypothesized so far: (1) calcium-dependent exocytosis; (2) reversal of membrane transporters; (3) glutamate exchange through the cystine/glutamate antiporter; (4) release through ionotropic purinergic receptors; (5) cell swelling-induced opening of anion channels; and (6) release through hemichannels (Fig. 3.3) [73]. Each of these are discussed in more detail below.

Calcium-Dependent Exocytosis

Several lines of evidence for calcium-dependent exocytosis of glutamate from astrocytes have been accumulating since the 1990s. Initial studies found intracellular calcium rises in

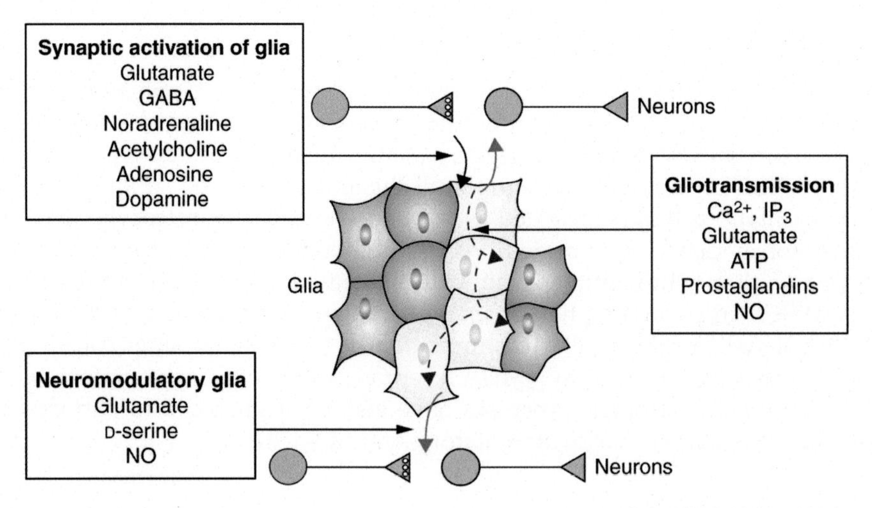

FIGURE 3.2 **Chemical signaling underlying neuron–glia bidirectional communication.** Neurotransmitters spilling over from active synapses can stimulate receptors on glial cells, resulting in internal calcium elevation and activation of the glial cell (*black arrow*, synaptic activation of glia). Several different transmitters have been shown to start neuron-to-glia signaling at various synapses of either the central or peripheral nervous system. $[Ca^{2+}]_i$ rise in the active glial cell (depicted in pale yellow; pink indicates inactive cells) can produce a neuromodulatory response by triggering transmitter release (*blue arrows*, neuromodulatory glia). This glia-to-neuron signaling can act locally, feeding back a response to the original stimulatory synaptic domain, or distally, to modulate a different synaptic domain, if the activation signal is spatially propagated in the glial network. Various transmitters may sustain glia-to-neuron signaling, although the best collected evidence so far is for calcium-dependent "exocytosis" of glutamate from astrocytes. A third form of communication, here called "gliotransmission" (red dashed lines and arrows), allows signals to spread in the glial network with spatial expansion of the original input. Mostly studied in cultured astrocytes, gliotransmission has been classically monitored as intercellular waves of calcium elevation (pale yellow cells) and thought to depend on intracellular signal (eg, inositol triphosphate [IP_3]) diffusion through gap junctions. But recent work indicates that calcium waves are accompanied by waves of extracellular mediators, such as ATP or glutamate. Therefore, gliotransmission must also proceed through autocrine/paracrine glia-to-glia signaling, possibly involving additional transmitters such as prostaglandins and nitric oxide. *Source: Reproduced with permission from Bezzi P, Volterra A. A neuron-glia signalling network in the active brain. Curr Opin Neurobiol 2001;11(3):387–94.*

astrocytes that were oscillatory and propagating [9]. Next, researchers discovered that astrocytes can release gliotransmitters and that glutamate can induce a response in neighboring neurons [50]. Finally, various vesicular release proteins were discovered to be expressed by astrocytes [74]. Collectively, these data support the notion that astrocytes release glutamate through calcium-dependent exocytosis. Although this mechanism has gained a great deal of support, it still remains controversial [75]. All lines of evidence (and controversy) will be discussed in detail in *Chapter 6: Astrocyte Calcium Signaling*.

Reversal of Glutamate Transporters

Astrocytes play an important role in removing glutamate from the extracellular space through specific sodium-dependent glutamate transporters found near synapses. The two

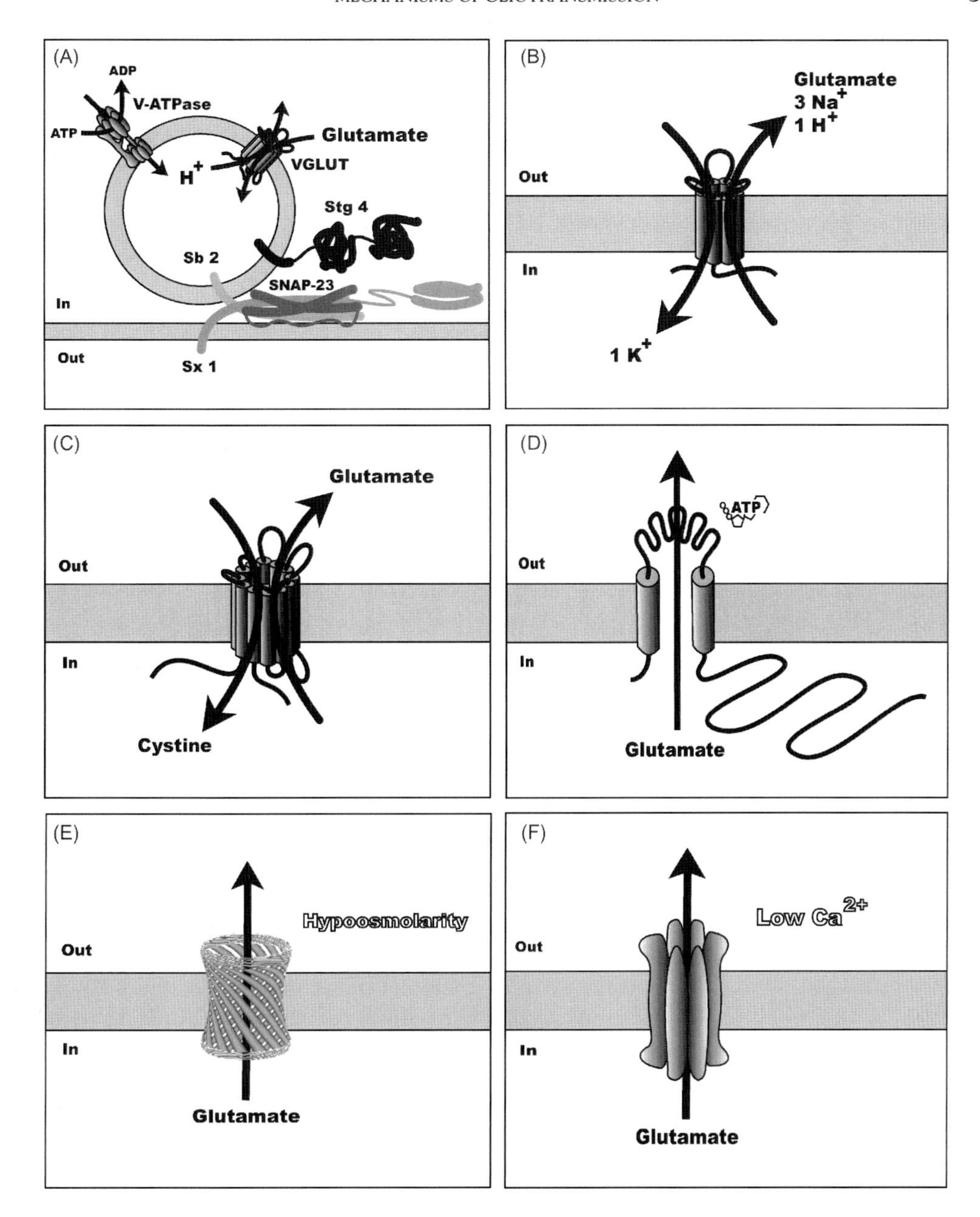

FIGURE 3.3 **Mechanisms of glutamate release from astrocytes.** (A) Ca^{2+}-dependent exocytosis; (B) reversal of uptake by plasma membrane glutamate transporters, (C) glutamate exchange via the cystine/glutamate antiporter, (D) release through ionotropic purinergic receptors (only one subunit shown, but exchanger is believed to be multimeric), (E) anion channel opening induced by cell swelling, (F) release through functional unpaired connexins, "hemichannels." *Sb 2*, synaptobrevin 2; *SNAP-23*, synaptosome-associated protein of 23kDa; *Stg 4*, synaptotagmin 4; *Sx 1*, syntaxin 1; *V-ATPase*, vacuolar type H^+-ATPase; *VGLUT*, vesicular glutamate transporter. *Source: Reproduced with permission from Malarkey EB, Parpura V. Mechanisms of glutamate release from astrocytes. Neurochem Int 2008;52(1–2):142–54.*

major astrocytic transporters are glutamate transporter-1 (GLT1) and glutamate/aspartate transporter (GLAST). Under homeostatic conditions, glutamate is driven into astrocytes through its concentration gradient. During pathophysiological events, however, perturbed extracellular ionic concentrations may favor the reversal of these transporters. This was initially demonstrated with whole-cell clamp of Müller cells from salamander retina by Szatkowski et al. [76]. They found that raising extracellular K^+ concentrations around glial cells evoked an outward current produced by the reversal of glutamate uptake [76]. This current was inhibited by extracellular glutamate and sodium and was increased by membrane depolarization.

Many studies used primary rat or mouse astrocyte cultures to study gliotransmitter release through glutamate transporter reversal. Massive excitatory amino acid efflux was seen under conditions of energy failure, including blockade of glycolytic and oxidative metabolism and ATP production [77,78]. Prolonged exposure to high extracellular K^+ concentrations (60 mM for 15–20 min), however, did not increase extracellular glutamate concentrations [78]. Similar results were seen in pure cultures of retinal neurons and glia preloaded with [^{3}H]-acetate (preferentially metabolized by astrocytes) and [U-^{14}C]-glucose (preferentially metabolized by neurons). After total metabolic blockade with iodoacetate and KCN, release of [^{3}H]-glutamate was significantly increased within 15 minutes whereas [U-^{14}C]-glutamate remained unchanged [79]. Pretreatment with a glutamate transporter blocker inhibited about 57% of the glutamate release after 30 minutes of metabolic block, suggesting that sodium-dependent glutamate transporter reversal is a major contributor, but not the sole component, to increased extracellular glutamate levels during energy deprivation [79]. Similarly, inhibition of sodium-dependent glutamate transport with the astrocytic GLT1 blocker dihydrokainate (DHK) was protective against anoxia-induced glutamate release [80].

Ischemia has been associated with increased extracellular glutamate levels. Forebrain ischemia was induced in anesthetized rats and microdialysate concentrations of glutamate were measured in the presence or absence of various blockers [81]. Inhibition of the astrocytic GLT1 with DHK or anion channel blockers diminished ischemia-induced glutamate release in rat striatum, suggesting that both cell swelling and the reversal of glutamate transporters might contribute to elevated extracellular glutamate levels [81]. A separate study induced ischemia in hippocampal slices from 12-day-old rats and bathed slices with various blockers of different glutamate release mechanisms. They found that reversal of neuronal glutamate transporters was the primary contributor to ischemia-induced glutamate release [82].

Reversal of glutamate transporters has excitotoxic consequences including lipoxygenase pathway-mediated cell death [83]. Exposure of co-cultures of rat cortical neurons and glia to the excitatory amino acid transporter inhibitor L-*trans*-pyrrolidine-2,4-dicarboxylate (PDC) led to a rapid elevation in extracellular glutamate and NMDA receptor-mediated excitotoxicity [84]. The glutamate increase was insensitive to tetrodotoxin, independent of extracellular calcium levels, present in astrocyte-pure cultures, and was suppressed in sodium-free medium. To confirm these findings, Volterra et al. [84] used glutamate transporters functionally reconstituted in liposomes to show that PDC activated carrier-mediated release of glutamate via transporter reversal. Taken together, these data suggest that the net efflux of glutamate may occur in the unhealthy brain, but the contribution to the healthy brain has yet to be determined.

Cystine/Glutamate Antiporter

Transport of cystine into the cell is essential for the synthesis of glutathione, an important antioxidant. In addition, cystine transport contributes to the maintenance of extracellular basal glutamate levels. Uptake of cystine into cells can occur through the cystine/glutamate antiporter (system x_c^-) or the sodium-dependent glutamate transporters (system X_{AG}^-); both of these are present on astrocytes. Once cystine is driven into the cell, glutamate is transported out. The addition of cystine to astrocytes in vitro or in vivo led to increased extracellular glutamate levels [85–90].

Warr et al. [90] used whole-cell patch-clamp analysis of neurons in rat brain slices to measure glutamate activation of NMDAR receptors after the application of cystine. They found that cystine generated a current, which was abolished by glutamate receptor blockers [90]. This current, however, was not due to the direct activation of NMDA receptors or by potentiation of the action of background levels of glutamate receptors. Moreover, experiments using isolated salamander Müller cells demonstrated that cystine did not block glutamate transporters and cystine-induced currents were not affected by the removal of extracellular chloride or the addition of the (Na^+–K^+–$2Cl^-$) cotransporter inhibitor furosemide. Their results suggested that the cystine-generated glutamate current was the result of cystine–glutamate exchange in cells [90].

The use of microdialysis has been used to measure glutamate release in response to cystine application in rat brain slices [88] and in vivo [85]. Blocking the cystine/glutamate antiporter in vivo decreased extracellular glutamate levels in the rat striatum by 60% whereas blocking sodium and calcium channels had minimal effects [85]. Similarly, cystine increased tonic glutamate release in CA1 hippocampal slices, measured by glutamate receptor-mediated currents evoked in pyramidal cells [86]. This release was inhibited by the anion exchange inhibitor 4,4′-diisothiocyano-2,2′-stilbenedisulfonic acid (DIDS) but remained unaffected by blocking gap junctions, ATP receptors, specific anion channels, and calcium-dependent release. It is possible, however, that the release of glutamate only occurred with the addition of cystine, but not at physiological levels [86].

The cystine/glutamate transporter may be negatively regulated by mGluR2/3 via a cyclic adenosine monophosphate (cAMP)-dependent protein kinase mechanism [85]. Interestingly, Tang et al. (2003) found that this regulation can occur bidirectionally; low concentrations of mGluR2/3 agonist increased cystine uptake whereas low concentrations decreased cystine uptake [89]. A similar bell-shaped dose response curve was seen with a PKA inhibitor. Blocking calcium/calmodulin-dependent kinase II, however, stimulated system x_c^-, suggesting it is involved in system x_c^- modulation [89].

Regulation of synaptic activity through type II mGluRs may be a result of the cystine/glutamate antiporter [87,88]. Specifically, bath application of physiological levels of cystine to acute tissue slices from the nucleus accumbens or prefrontal cortex led to elevated glutamate levels but decreased miniature excitatory postsynaptic currents (mEPSCs), spontaneous EPSC (sEPSC) frequency, and evoked EPSC amplitude [87]. The effects were thought to be presynaptic and were blocked by system x_c^- or mGluR2/3 antagonists [87]. This mechanism may play a role in drug-seeking behavior because rats who were trained to self-administer cocaine and were also treated with mGluR2/3 blockers resisted the reinstatement of this behavior [87].

The reduction in extracellular glutamate after cocaine withdrawal has been attributed to a compromised system x_c^- [91]. Both the application of glutamate exchange channel blockers and cocaine diminished extracellular glutamate levels [92]. System x_c^- was also found to be highly expressed in gliomas and inhibition of system x_c^- reduced the levels of intracellular glutathione and consequently caused oxidative cell death of tumor cells [93]. In addition, the use of system x_c^- inhibitors have been used to demonstrate that cystine uptake may be part, but not the sole contributor, to an astrocyte cell death mechanism [83].

Purinergic Receptors

Purinergic signaling may occur through either P1 receptors (ligand is adenosine) or P2 receptors (ligands are ATP, uridine triphosphate (UTP), and their metabolites). P1 receptors are G-protein-coupled whereas P2 receptors can be either GPCRs (P2Y receptors) or ligand-gated ion channels (P2X receptors, Table 3.1). Glial cells express both ionotropic and metabotropic purinergic receptors. Purinergic P2X receptors express several subunits that can assemble into homomeric or heteromeric channels. $P2X_7$ receptors are homomeric, have a large pore, and are found on astrocytes.

Duan et al. [94] demonstrated that $P2X_7$ receptors release excitatory amino acids from astrocytes. The addition of ATP to cortical astrocyte cultures produced inward currents that were characteristic of $P2X_7$ receptors and were amplified in low divalent cation medium. Specifically, the currents were blocked by the P2X receptor antagonist pyridoxal phosphate-6-azophenyl-2,4-disulfonic acid (PPADS), activated by ATP, and more potently activated by the unselective $P2X_7$ agonist benzoyl ATP (BzATP). Using radiolabeled tracers Duan et al. [94] confirmed permeability to different substrates, including glutamate and aspartate, through this channel. Finally, the channel was permeable to Lucifer yellow, indicating a large channel opening [94].

Release of glutamate through P2X receptors was confirmed in situ by Fellin et al. [95]. In hippocampal slices, BzATP triggered SICs and tonic currents in pyramidal neurons after NMDA receptor activation. The source of glutamate was most likely astrocytes because this effect occurred in the presence of tetrodotoxin. In addition, tonic currents were blocked by P2X receptor antagonists, amplified in low, extracellular calcium, and were unaffected by glutamate transporter and hemichannel blockers [95]. Fellin et al. also confirmed a large depolarization in astrocytes and Lucifer yellow uptake in a subpopulation of cells induced by BzATP.

The release of ATP from P2Y receptors has also been demonstrated [96–98]. In co-cultures of hippocampal neurons and astrocytes, the application of reactive blue 2 (RB-2), a P2Y

TABLE 3.1 Classification of Purinergic Receptors

Receptor Type	Receptor Name	Main Ligand(s)
G-protein-coupled receptors (GPCRs)	P1	Adenosine
	P2Y	ATP, ADP, UTP, UDP
Ligand-gated ion channels	P2X	ATP

antagonist, increased amplitude of EPSCs whereas PPADS had no effect [97]. Moreover, the application of ATP on hippocampal cultures reduced the amplitude of EPSCs and inhibitory postsynaptic currents (IPSCs). ATP-induced synaptic suppression was shown to be mediated through presynaptic modulation; inward currents induced by glutamate or GABA were not affected by ATP but both tetrodotoxin and bicuculline decreased the frequency but not the amplitude of spontaneous EPSCs. The use of various antagonists confirmed that glutamatergic activity led to heterosynaptic suppression and was dependent on the activation of P2Y and non-NMDA receptors. In addition, heterosynaptic suppression was also dependent on the presence of astrocytes. Finally, it was shown that glutamate released from neurons induced ATP release from astrocytes [97]. Interestingly, stimulation-induced ATP release from astrocytes activated P2Y receptors and consequently suppressed transmitter release from hippocampal synapses, resulting in heterosynaptic long-term depression (LTD) [96].

A thorough study by Domercq et al. [99] provided convincing data regarding astrocytic release of glutamate through calcium-dependent exocytosis and the P2Y1 receptor subtype. The addition of ATP to astrocyte cultures led to a rapid release of glutamate that was stimulated by P2Y1R agonists, unaffected by P2X receptor-preferring agonists, and blocked by both generic P2 purinergic receptor antagonist and a selective P2Y1R antagonist [99]. Evidence that ATP-induced release of glutamate occurs through calcium-dependent exocytosis came from both total internal fluorescence reflection imaging of fluorescence-labeled glutamatergic vesicles and through pharmacological manipulation. Glutamate release was abolished with calcium chelation, bafilomycin A1 (blocker of vesicular uptake of transmitter), or tetanus neurotoxin which blocks vesicular release through cleavage of the vesicle soluble *N*-ethylmaleimide-sensitive factor attachment protein receptor (v-SNARE) protein synaptobrevin. Both TNF-α and prostaglandins were released through P2Y1 receptor after ATP stimulation of cultured astrocytes (confirmed pharmacologically). Glutamate release was inhibited with prostaglandin inhibitors and, interestingly, TNF-α knockout mice had reduced glutamate responses. Finally, results were confirmed in hippocampal slices. In agreement, a later study confirmed activity-dependent astrocyte release of glutamate occurs through P2Y1 receptors in hemibrain horizontal slices [98]. Taken together, Domercq et al. [99] identified an astrocyte specific, P2Y1 receptor-induced glutamate exocytosis that was dependent on both TNF-α and prostaglandins.

Anion Channel Opening

Studies using 2-photon microscopy demonstrated that astrocytes are more prone to swelling than neurons [100,101]. Astrocytes can compensate for swelling through a number of methods, including the opening of the astrocyte-specific channel aquaporin-4 (discussed in *Chapter 8: Water Channels*) and volume-regulated anion channels (VRACs). The release of water through VRACs is accompanied by a number of small organic anions and amino acids. Kimelberg et al. [102] demonstrated that hypotonic media caused the release of glutamate, aspartate, and taurine in primary astrocyte cultures. The swelling-induced release was inhibited by anion transport inhibitors. These findings have been confirmed in more recent studies [103–105].

Takano et al. [106] demonstrated that astrocytes in culture exhibited transient calcium-dependent cell volume increases that resulted in taurine, aspartate, and glutamine, and glutamate release. This release was ATP-dependent [104,106] and was blocked by anion channel inhibitors. Release was the same in connexin 43 knockout mice, suggesting that it occurred independent of hemichannels. Liu et al. [103] demonstrated that both hypotonic and ischemic stimuli caused glutamate release through channels with properties different from VRACs: maxi-anion channels. The use of cell-attached patches revealed these channels were more permeable to glutamate, insensitive to phloretin (a VRAC inhibitor) and were sensitive to gadolinium (Gd^{3+}, maxi-anion channel blocker) [103]. Together, both channels represent a major potential conductive pathway for swelling-induced glutamate release.

Spreading depression (SD), induced with bath application of ouabain in hippocampal slices, was associated with cell swelling and glutamate release [107]. SD still occurred in calcium-free medium but was depressed by a VRAC blocker. In vivo evidence of amino acid release was provided by Haskew-Layton et al. [108]. They perfused the rodent cortex with a hypoosmotic medium to induce cell swelling. This resulted in a sevenfold increase in glutamate and aspartate, which was inhibited by an anion channel blocker but not by a glutamate transporter or exocytosis blocker [108].

The lack of specific inhibitors makes anion channel gliotransmission difficult to study. For example, certain chloride channel blockers may also inhibit gap junctions [109]. In fact, one study used Gd^{3+} as a maxi-anion channel blocker [103] whereas another study used it as a connexin hemichannel blocker [110]. Better pharmacological tools will be required to further elucidate the role of anion channels in gliotransmitter release. Future studies should focus on observing VRAC release of glutamate in vivo in the healthy, unaltered brain.

Hemichannels

Connexins are transmembrane proteins that assemble into groups to form pores, called hemichannels. They allow the exchange of ions and small molecules between the inside of the cell and the extracellular environment. When two separate hemichannel pores come together, they form a gap junction that allows for cell-to-cell communication. A number of connexin proteins exist and are expressed on a variety of cell types. Astrocytes predominately express connexin 30 (Cx30) and connexin 43 (Cx43). Functional hemichannels that release glutamate, aspartate, prostaglandins, and ATP have been found on astrocytes [110–115].

Mobilization of intracellular calcium may lead to the release of ATP through hemichannels [110,114,116]. Cotrina et al. [112] demonstrated that ATP release required connexin expression and purinergic receptor-activated intracellular calcium mobilization. Furthermore, in nominal divalent cation-free solution, glutamate transporter reversal did not significantly mediate glutamate release [115]. Other groups have proposed that mechanical stimulation could open hemichannels and evoke ATP release [110,117,118]. Glutamate and aspartate released from astrocytes in vitro was blocked by gap junction inhibitors but not by blocking calcium mobilization [115]. Similarly, glutamate release in cultured hippocampal astrocytes after exposure to a hypotonic solution was nearly blocked by gap junction blockers [115]. In contrast to Cotrina et al. [112], Ye et al. [115] found that purinergic antagonists, inhibitors of $P2X_7$ receptors, and blockers of P2 receptors had no effect on glutamate release, therefore, precluding the involvement of purinergic receptor activation.

The release of gliotransmitters through hemichannels may serve a variety of roles in both physiological and pathophysiological conditions, including central nervous control of breathing, brain glucose sensing, and synaptic transmission and memory [119]. Furthermore, hemichannel activity was found to increase in response to amyloid-β peptide fragment [119]. ATP release through mutated connexin 37 hemichannels has been implicated in atherosclerosis [120]. An in vivo study demonstrated that hemichannels in the cochlea can release ATP, which may contribute to hearing sensitivity [116]. This release was increased by a reduction in extracellular calcium and was blocked by gap junction inhibitors. Future studies should focus on in vivo hemichannel gliotransmission in both the healthy and diseased brain.

EFFECTS OF GLIOTRANSMISSION

Astrocytes can sense synaptic activity and respond with intracellular calcium elevations. This can lead to the release of gliotransmitters, which can act on pre- and postsynaptic sites on neurons, thus regulating neuronal activity [1]. This can be accomplished through the modulation of synaptic transmission [121], control of synaptic efficiency [122], or regulation of synaptic scaling [123]. Glutamate released from astrocytes can strengthen synaptic transmission or induce SICs by activation of NMDA receptors [55,98,124,125]. Astrocyte-evoked glutamate release may also lead to synchronized activity among neurons [124]. Conversely, astrocytes may also release GABA and cause synchronous inhibition of neurons, which has been demonstrated in the rat olfactory bulb [125].

D-serine and Synaptic Plasticity

Astrocytic processes contain storage vesicles capable of both storing and releasing D-serine in a calcium- and SNARE protein-dependent manner [126–128]. Evidence for this includes the presence of vesicular release machinery on astrocytes and co-localization of this machinery with D-serine [126–128]. Furthermore, the release of D-serine was inhibited by the removal of extracellular calcium or depletion of intracellular calcium stores and was amplified by increasing extracellular calcium [128]. Finally, glial vesicular uptake of D-serine was dependent on both ATP and chloride [127]. D-Serine may be released along with glutamate [127] and exhibit powerful effects on synaptic plasticity.

Synaptic plasticity, divided into LTP and LTD, involves rapid adjustments in the strength of individual synapses in response to synaptic activity [123]. Many forms of LTP rely on NMDA receptors and D-serine, the principal endogenous coagonist of the glycine site of NMDARs at functional excitatory synapses [129,130]. D-Serine immunoreactivity was highly localized to astrocytes in culture and astrocytic release was stimulated by the activation of non-NMDA glutamate receptors [131]. D-Serine-induced LTP was blocked by clamping internal calcium in individual CA1 astrocytes, which was reversed by the exogenous application of D-serine or glycine [132]. Moreover, depletion of D-serine in astrocytes [132,133] or blocking astrocyte activity with a metabolic inhibitor [129] impaired NMDA receptor-mediated synaptic transmission and blocked LTP, suggesting that the astrocytic calcium-dependent release of D-serine may be responsible for NMDA receptor-induced LTP. Heterosynaptic LTD, on the other hand, was caused by stimulation-induced adenosine triphosphate (ATP) release from astrocytes that led to suppressed transmitter release from hippocampal synapses [96]. Like LTP, astrocytic calcium elevations were required for heterosynaptic depression [134].

Astrocytic Release of ATP

Abundant ATP storage in astrocytic lysosomes has been reported [135], and ATP has been found in secretory granules within astrocytes and purines were released after elevated calcium levels [136]. Furthermore, stimulation of an astrocyte island in mouse astrocyte cultures resulted in the propagation of calcium waves and simultaneous release of ATP [137]. Interestingly, ATP in turn mediated calcium wave propagation [137]. Once ATP is released from astrocytes, it may directly act on purinergic receptors or it may be converted to adenosine first.

ATP may act as a potent modulator of neuronal activity. The release of ATP from Müller glial cells was converted to adenosine and inhibited neuronal firing through the activation of A1 adenosine receptors [138]. Panatier et al. [139], however, studied the CA1 region of the hippocampus and found that adenosine derived from astrocytic ATP activated presynaptic A2A receptors; this led to increased excitatory synaptic transmission.

ATP may play a role in synaptic plasticity. Gordon et al. [140] demonstrated that norepinephrine could trigger astrocytic ATP release and subsequent slow, inward $P2X_7$-mediated currents. ATP also increased synaptic strength at glutamatergic synapses by promoting the postsynaptic insertion of α-amino-3-hydroxy-5-methyl-4-isoxazolepropionic acid receptors [140]. In a preparation of hypothalamic brain slices, the same group also demonstrated a mGluR-mediated rise in calcium and subsequent release of ATP in response to afferent activity [141]. Taken together, the above results suggest that glial cells play an active role in modulating neuronal activity and information processing in an ATP-dependent manner.

GLIOTRANSMISSION CONTROVERSY

Despite the growing support for gliotransmission, at least four main problems with current data have been identified. (1) The majority of experiments have been conducted in vitro, meaning very little in vivo data exists [142–144]. The in vitro data may not be representative of the true brain state. For example, Cotrina et al. demonstrated that ATP secretion occurred through hemichannels in HeLa and glioma cell lines [112]; these may not accurately exemplify the healthy brain. (2) Many experiments have used nonphysiological conditions, including excess application of neuroactive substances [143,144]. (3) Agonists and antagonists used in slice preparations are often nonspecific and have other, unintended effects [143,144]. (4) There has been a lack of necessary experimental tools and ambiguity in experimental design [143]. For example, many slice work preparations are conducted in magnesium-free solution. Removing magnesium from the bath relieves voltage-dependent block on NMDARs, but also has other effects on astrocytes including initiation of spontaneous calcium signaling [145].

A common tool used to study gliotransmission is transgenic mice. This, however, can also lead to conflicting or misleading information as well. A study using transgenic mice with inhibited vesicular release via overexpression of a dominant-negative domain of vesicular SNARE (dnSNARE) in astrocytes demonstrated gliotransmission of ATP [146]. Additional studies with dnSNARE mice found that astrocytic activation (ATP released and converted into adenosine) of A1 receptors resulted in regulation of NMDA receptor surface expression

[147] and modulation of cortical slow oscillations [148]. Astrocyte specificity, however, was called into question when Fujita et al. [149] found widespread expression of the dnSNARE transgene in cortical neurons.

Once glutamate enters astrocytes, it is quickly converted to glutamine by glutamine synthetase. Therefore, astrocytic glutamate levels may be insufficient to be released [142]. In addition, it is difficult to link astrocyte calcium spikes to glutamate release because the calcium increases evoked SICs in neighboring neurons inconsistently with a delay ranging from 1 to 20 seconds [143]. One alternative explanation for gliotransmission that has been proposed is nonspecific leakage through the astrocytic membrane [86]. Glutamate release from astrocytes has been documented to occur with other amino acids, including aspartate, glycine, and D-serine. The relative ratio of amino acids released from cultured astrocytes matched the cytosolic ratio [47,106,150], supporting the notion of leakage through the membrane. Therefore, it is possible that astrocytes do not participate in or rapidly modulate synaptic transmission, but instead serve a more long-lasting, generalized function.

CONCLUSIONS

Growing evidence supports a role of astrocytes in modulation of neuronal signaling and synaptic transmission. Synaptic activity activates ion channels and neurotransmitter receptors that lead to calcium oscillations within astrocytes. Through one of at least six mechanisms, astrocytes may release glutamate, although this topic is hotly debated. Gliotransmitter release from astrocytes may have a variety of neuromodulatory effects and can target pre- or postsynaptic transmission through the activation of different receptors, including NMDA receptors. The molecular mechanisms and physiological conditions underlying neuron–glia communication and their relevance for proper synaptic activity, however, still remain unclear. More in vivo data under physiological and pathological conditions will need to be collected to fully grasp the extent to which astrocytes shape neuronal activity. Newer techniques such as in vivo optogenetic astrocyte photostimulation [42] will help to overcome the in vitro limitations that other studies have faced. In turn, this growing understanding will clarify how gliotransmission may be altered in disease states such as epilepsy.

References

[1] Perea G, Sur M, Araque A. Neuron-glia networks: integral gear of brain function. Front Cell Neurosci 2014;8:378.
[2] Bowman CL, Kimelberg HK. Excitatory amino acids directly depolarize rat brain astrocytes in primary culture. Nature 1984;311(5987):656–9.
[3] Kettenmann H, Backus KH, Schachner M. Aspartate, glutamate and gamma-aminobutyric acid depolarize cultured astrocytes. Neurosci Lett 1984;52(1–2):25–9.
[4] Croft W, Dobson KL, Bellamy TC. Plasticity of neuron-glial transmission: equipping glia for long-term integration of network activity. Neural Plast 2015;2015:765792.
[5] Hirase H, Qian L, Bartho P, Buzsaki G. Calcium dynamics of cortical astrocytic networks in vivo. PLoS Biol 2004;2(4):E96.
[6] Nimmerjahn A, Mukamel EA, Schnitzer MJ. Motor behavior activates Bergmann glial networks. Neuron 2009;62(3):400–12.
[7] Wang X, Lou N, Xu Q, Tian GF, Peng WG, Han X, et al. Astrocytic Ca^{2+} signaling evoked by sensory stimulation in vivo. Nat Neurosci 2006;9(6):816–23.

[8] Winship IR, Plaa N, Murphy TH. Rapid astrocyte calcium signals correlate with neuronal activity and onset of the hemodynamic response in vivo. J Neurosci 2007;27(23):6268–72.
[9] Cornell-Bell AH, Finkbeiner SM, Cooper MS, Smith SJ. Glutamate induces calcium waves in cultured astrocytes: long-range glial signaling. Science 1990;247(4941):470–3.
[10] Zhang Q, Haydon PG. Roles for gliotransmission in the nervous system. J Neural Transm (Vienna) 2005;112(1):121–5.
[11] Bezzi P, Volterra A. A neuron-glia signalling network in the active brain. Curr Opin Neurobiol 2001;11(3):387–94.
[12] Perea G, Navarrete M, Araque A. Tripartite synapses: astrocytes process and control synaptic information. Trends Neurosci 2009;32(8):421–31.
[13] Martin DL. Synthesis and release of neuroactive substances by glial cells. Glia 1992;5(2):81–94.
[14] Halassa MM, Fellin T, Haydon PG. The tripartite synapse: roles for gliotransmission in health and disease. Trends Mol Med 2007;13(2):54–63.
[15] Ullian EM, Sapperstein SK, Christopherson KS, Barres BA. Control of synapse number by glia. Science 2001;291(5504):657–61.
[16] Grosche J, Matyash V, Möller T, Verkhratsky A, Reichenbach A, Kettenmann H. Microdomains for neuron-glia interaction: parallel fiber signaling to Bergmann glial cells. Nat Neurosci 1999;2(2):139–43.
[17] Bushong EA, Martone ME, Jones YZ, Ellisman MH. Protoplasmic astrocytes in CA1 stratum radiatum occupy separate anatomical domains. J Neurosci 2002;22(1):183–92.
[18] Halassa MM, Fellin T, Takano H, Dong JH, Haydon PG. Synaptic islands defined by the territory of a single astrocyte. J Neurosci 2007;27(24):6473–7.
[19] Volterra A, Meldolesi J. Astrocytes, from brain glue to communication elements: the revolution continues. Nat Rev Neurosci 2005;6(8):626–40.
[20] Ventura R, Harris KM. Three-dimensional relationships between hippocampal synapses and astrocytes. J Neurosci 1999;19(16):6897–906.
[21] Allaman I, Belanger M, Magistretti PJ. Astrocyte-neuron metabolic relationships: for better and for worse. Trends Neurosci 2011;34(2):76–87.
[22] Hassel B, Westergaard N, Schousboe A, Fonnum F. Metabolic differences between primary cultures of astrocytes and neurons from cerebellum and cerebral cortex: effects of fluorocitrate. Neurochem Res 1995;20(4):413–20.
[23] Westergaard N, Sonnewald U, Schousboe A. Metabolic trafficking between neurons and astrocytes: the glutamate/glutamine cycle revisited. Dev Neurosci 1995;17(4):203–11.
[24] Butt AM, Kalsi A. Inwardly rectifying potassium channels (K_{ir}) in central nervous system glia: a special role for $K_{ir}4.1$ in glial functions. J Cell Mol Med 2006;10(1):33–44.
[25] Kressin K, Kuprijanova E, Jabs R, Seifert G, Steinhäuser C. Developmental regulation of Na^+ and K^+ conductances in glial cells of mouse hippocampal brain slices. Glia 1995;15(2):173–87.
[26] Orkand RK. Extracellular potassium accumulation in the nervous system. Fed Proc 1980;39(5):1515–8.
[27] Sontheimer H. Voltage-dependent ion channels in glial cells. Glia 1994;11(2):156–72.
[28] Bergles DE, Dzubay JA, Jahr CE. Glutamate transporter currents in Bergmann glial cells follow the time course of extrasynaptic glutamate. Proc Natl Acad Sci USA 1997;94(26):14821–5.
[29] Bergles DE, Jahr CE. Synaptic activation of glutamate transporters in hippocampal astrocytes. Neuron 1997;19(6):1297–308.
[30] Coulter DA, Eid T. Astrocytic regulation of glutamate homeostasis in epilepsy. Glia 2012;60(8):1215–26.
[31] Hansson E, Eriksson P, Nilsson M. Amino acid and monoamine transport in primary astroglial cultures from defined brain regions. Neurochem Res 1985;10(10):1335–41.
[32] Clasadonte J, Haydon PG. Connexin 30 controls the extension of astrocytic processes into the synaptic cleft through an unconventional non-channel function. Neurosci Bull 2014;30(6):1045–8.
[33] Nagelhus EA, Ottersen OP. Physiological roles of aquaporin-4 in brain. Physiol Rev 2013;93(4):1543–62.
[34] Pannasch U, Freche D, Dallérac G, Ghézali G, Escartin C, Ezan P, et al. Connexin 30 sets synaptic strength by controlling astroglial synapse invasion. Nat Neurosci 2014;17(4):549–58.
[35] Nadkarni S, Jung P, Levine H. Astrocytes optimize the synaptic transmission of information. PLoS Comput Biol 2008;4(5):e1000088.
[36] Porto-Pazos AB, Veiguela N, Mesejo P, Navarrete M, Alvarellos A, Ibañez O, et al. Artificial astrocytes improve neural network performance. PloS one 2011;6(4):e19109.
[37] Amiri M, Hosseinmardi N, Bahrami F, Janahmadi M. Astrocyte-neuron interaction as a mechanism responsible for generation of neural synchrony: a study based on modeling and experiments. J Comput Neurosci 2013;34(3):489–504.

[38] Wade JJ, McDaid LJ, Harkin J, Crunelli V, Kelso JA. Bidirectional coupling between astrocytes and neurons mediates learning and dynamic coordination in the brain: a multiple modeling approach. PloS one 2011;6(12):e29445.
[39] Perea G, Araque A. Properties of synaptically evoked astrocyte calcium signal reveal synaptic information processing by astrocytes. J Neurosci 2005;25(9):2192–203.
[40] Araque A, Martín ED, Perea G, Arellano JI, Buño W. Synaptically released acetylcholine evokes Ca^{2+} elevations in astrocytes in hippocampal slices. J Neurosci 2002;22(7):2443–50.
[41] Schummers J, Yu H, Sur M. Tuned responses of astrocytes and their influence on hemodynamic signals in the visual cortex. Science 2008;320(5883):1638–43.
[42] Perea G, Yang A, Boyden ES, Sur M. Optogenetic astrocyte activation modulates response selectivity of visual cortex neurons in vivo. Nat Commun 2014;5:3262.
[43] Chen N, Sugihara H, Sharma J, Perea G, Petravicz J, Le C, et al. Nucleus basalis-enabled stimulus-specific plasticity in the visual cortex is mediated by astrocytes. Proc Natl Acad Sci USA 2012;109(41):E2832–41.
[44] Petzold GC, Albeanu DF, Sato TF, Murthy VN. Coupling of neural activity to blood flow in olfactory glomeruli is mediated by astrocytic pathways. Neuron 2008;58(6):897–910.
[45] Takata N, Mishima T, Hisatsune C, Nagai T, Ebisui E, Mikoshiba K, et al. Astrocyte calcium signaling transforms cholinergic modulation to cortical plasticity in vivo. J Neurosci 2011;31(49):18155–65.
[46] Schipke CG, Haas B, Kettenmann H. Astrocytes discriminate and selectively respond to the activity of a subpopulation of neurons within the barrel cortex. Cereb Cortex 2008;18(10):2450–9.
[47] Navarrete M, Perea G, Fernandez de Sevilla D, Gómez-Gonzalo M, Núñez A, Martín ED, et al. Astrocytes mediate in vivo cholinergic-induced synaptic plasticity. PLoS Biol 2012;10(2):e1001259.
[48] Parri HR, Gould TM, Crunelli V. Sensory and cortical activation of distinct glial cell subtypes in the somatosensory thalamus of young rats. Eur J Neurosci 2010;32(1):29–40.
[49] Nedergaard M. Direct signaling from astrocytes to neurons in cultures of mammalian brain cells. Science 1994;263(5154):1768–71.
[50] Parpura V, Basarsky TA, Liu F, Jeftinija K, Jeftinija S, Haydon PG. Glutamate-mediated astrocyte-neuron signalling. Nature 1994;369(6483):744–7.
[51] Parri HR, Gould TM, Crunelli V. Spontaneous astrocytic Ca^{2+} oscillations in situ drive NMDAR-mediated neuronal excitation. Nat Neurosci 2001;4(8):803–12.
[52] Kang J, Jiang L, Goldman SA, Nedergaard M. Astrocyte-mediated potentiation of inhibitory synaptic transmission. Nat Neurosci 1998;1(8):683–92.
[53] Fellin T, Pascual O, Gobbo S, Pozzan T, Haydon PG, Carmignoto G. Neuronal synchrony mediated by astrocytic glutamate through activation of extrasynaptic NMDA receptors. Neuron 2004;43(5):729–43.
[54] Di Castro MA, Chuquet J, Liaudet N, Bhaukaurally K, Santello M, Bouvier D, et al. Local Ca^{2+} detection and modulation of synaptic release by astrocytes. Nat Neurosci 2011;14(10):1276–84.
[55] Santello M, Calì C, Bezzi P. Gliotransmission and the tripartite synapse. In: Kreutz MR, Sala C, editors. Synaptic Plasticity: Vienna, Austria, Springer; 2012.
[56] Shain WG, Martin DL. Activation of β-adrenergic receptors stimulates taurine release from glial cells. Cell Mol Neurobiol 1984;4(2):191–6.
[57] Martin DL, Shain W. High affinity transport of taurine and β-alanine and low affinity transport of gamma-aminobutyric acid by a single transport system in cultured glioma cells. J Biol Chem 1979;254(15):7076–84.
[58] Bowery NG, Brown DA, Collins GG, Galvan M, Marsh S, Yamini G. Indirect effects of amino-acids on sympathetic ganglion cells mediated through the release of gamma-aminobutyric acid from glial cells. Br J Pharmacol 1976;57(1):73–91.
[59] Agulhon C, Petravicz J, McMullen AB, Sweger EJ, Minton SK, Taves SR, et al. What is the role of astrocyte calcium in neurophysiology? Neuron 2008;59(6):932–46.
[60] Araque A, Carmignoto G, Haydon PG. Dynamic signaling between astrocytes and neurons. Annu Rev Physiol 2001;63:795–813.
[61] Araque A, Carmignoto G, Haydon PG, Oliet SH, Robitaille R, Volterra A. Gliotransmitters travel in time and space. Neuron 2014;81(4):728–39.
[62] Araque A, Parpura V, Sanzgiri RP, Haydon PG. Tripartite synapses: glia, the unacknowledged partner. Trends Neurosci 1999;22(5):208–15.
[63] Calì C, Marchaland J, Spagnuolo P, Gremion J, Bezzi P. Regulated exocytosis from astrocytes: physiological and pathological related aspects. Int Rev Neurobiol 2009;85:261–93.

[64] Chevaleyre V, Takahashi KA, Castillo PE. Endocannabinoid-mediated synaptic plasticity in the CNS. Annu Rev Neurosci 2006;29:37–76.

[65] Guček A, Vardjan N, Zorec R. Exocytosis in astrocytes: transmitter release and membrane signal regulation. Neurochem Res 2012;37(11):2351–63.

[66] Halassa MM, Fellin T, Haydon PG. Tripartite synapses: roles for astrocytic purines in the control of synaptic physiology and behavior. Neuropharmacology 2009;57(4):343–6.

[67] Halassa MM, Haydon PG. Integrated brain circuits: astrocytic networks modulate neuronal activity and behavior. Annu Rev Physiol 2010;72:335–55.

[68] Parpura V, Heneka MT, Montana V, Oliet SH, Schousboe A, Haydon PG, et al. Glial cells in (patho) physiology. J Neurochem 2012;121(1):4–27.

[69] Parpura V, Zorec R. Gliotransmission: exocytotic release from astrocytes. Brain Res Rev 2010;63(1–2):83–92.

[70] Perea G, Araque A. Glial calcium signaling and neuron-glia communication. Cell Calcium 2005;38(3–4): 375–82.

[71] Perea G, Araque A. Synaptic regulation of the astrocyte calcium signal. J Neural Transm 2005;112(1):127–35.

[72] Zorec R, Araque A, Carmignoto G, Haydon PG, Verkhratsky A, Parpura V. Astroglial excitability and gliotransmission: an appraisal of Ca^{2+} as a signalling route. ASN Neuro 2012;4(2).

[73] Malarkey EB, Parpura V. Mechanisms of glutamate release from astrocytes. Neurochem Int 2008;52(1–2):142–54.

[74] Parpura V, Fang Y, Basarsky T, Jahn R, Haydon PG. Expression of synaptobrevin II, cellubrevin and syntaxin but not SNAP-25 in cultured astrocytes. FEBS Lett 1995;377(3):489–92.

[75] Fiacco TA, Agulhon C, Taves SR, Petravicz J, Casper KB, Dong X, et al. Selective stimulation of astrocyte calcium in situ does not affect neuronal excitatory synaptic activity. Neuron 2007;54(4):611–26.

[76] Szatkowski M, Barbour B, Attwell D. Non-vesicular release of glutamate from glial cells by reversed electrogenic glutamate uptake. Nature 1990;348(6300):443–6.

[77] Longuemare MC, Swanson RA. Excitatory amino acid release from astrocytes during energy failure by reversal of sodium-dependent uptake. J Neurosci Res 1995;40(3):379–86.

[78] Longuemare MC, Swanson RA. Net glutamate release from astrocytes is not induced by extracellular potassium concentrations attainable in brain. J Neurochem 1997;69(2):879–82.

[79] Zeevalk GD, Davis N, Hyndman AG, Nicklas WJ. Origins of the extracellular glutamate released during total metabolic blockade in the immature retina. J Neurochem 1998;71(6):2373–81.

[80] Li S, Mealing GA, Morley P, Stys PK. Novel injury mechanism in anoxia and trauma of spinal cord white matter: glutamate release via reverse Na^+-dependent glutamate transport. J Neurosci 1999;19(14):RC16.

[81] Seki Y, Feustel PJ, Keller Jr. RW, Tranmer BI, Kimelberg HK. Inhibition of ischemia-induced glutamate release in rat striatum by dihydrokainate and an anion channel blocker. Stroke 1999;30(2):433–40.

[82] Rossi DJ, Oshima T, Attwell D. Glutamate release in severe brain ischaemia is mainly by reversed uptake. Nature 2000;403(6767):316–21.

[83] Re DB, Nafia I, Melon C, Shimamoto K, Kerkerian-Le Goff L, Had-Aissouni L. Glutamate leakage from a compartmentalized intracellular metabolic pool and activation of the lipoxygenase pathway mediate oxidative astrocyte death by reversed glutamate transport. Glia 2006;54(1):47–57.

[84] Volterra A, Bezzi P, Rizzini BL, Trotti D, Ullensvang K, Danbolt NC, et al. The competitive transport inhibitor L-trans-pyrrolidine-2, 4-dicarboxylate triggers excitotoxicity in rat cortical neuron-astrocyte co-cultures via glutamate release rather than uptake inhibition. Eur J Neurosci 1996;8(9):2019–28.

[85] Baker DA, Xi ZX, Shen H, Swanson CJ, Kalivas PW. The origin and neuronal function of in vivo nonsynaptic glutamate. J Neurosci 2002;22(20):9134–41.

[86] Cavelier P, Attwell D. Tonic release of glutamate by a DIDS-sensitive mechanism in rat hippocampal slices. J Physiol 2005;564(Pt 2):397–410.

[87] Moran MM, McFarland K, Melendez RI, Kalivas PW, Seamans JK. Cystine/glutamate exchange regulates metabotropic glutamate receptor presynaptic inhibition of excitatory transmission and vulnerability to cocaine seeking. J Neurosci 2005;25(27):6389–93.

[88] Moran MM, Melendez R, Baker D, Kalivas PW, Seamans JK. Cystine/glutamate antiporter regulation of vesicular glutamate release. Ann N Y Acad Sci 2003;1003:445–7.

[89] Tang XC, Kalivas PW. Bidirectional modulation of cystine/glutamate exchanger activity in cultured cortical astrocytes. Ann N Y Acad Sci 2003;1003:472–5.

[90] Warr O, Takahashi M, Attwell D. Modulation of extracellular glutamate concentration in rat brain slices by cystine/glutamate exchange. J Physiol 1999;514(Pt 3):783–93.

[91] Baker DA, McFarland K, Lake RW, Shen H, Tang XC, Toda S, et al. Neuroadaptations in cystine/glutamate exchange underlie cocaine relapse. Nat Neurosci 2003;6(7):743–9.
[92] Baker DA, Shen H, Kalivas PW. Cystine/glutamate exchange serves as the source for extracellular glutamate: modifications by repeated cocaine administration. Amino Acids 2002;23(1–3):161–2.
[93] Chung WJ, Lyons SA, Nelson GM, Hamza H, Gladson CL, Gillespie GY, et al. Inhibition of cystine uptake disrupts the growth of primary brain tumors. J Neurosci 2005;25(31):7101–10.
[94] Duan S, Anderson CM, Keung EC, Chen Y, Swanson RA. P2X7 receptor-mediated release of excitatory amino acids from astrocytes. J Neurosci 2003;23(4):1320–8.
[95] Fellin T, Pozzan T, Carmignoto G. Purinergic receptors mediate two distinct glutamate release pathways in hippocampal astrocytes. J Biol Chem 2006;281(7):4274–84.
[96] Chen J, Tan Z, Zeng L, Zhang X, He Y, Gao W, et al. Heterosynaptic long-term depression mediated by ATP released from astrocytes. Glia 2013;61(2):178–91.
[97] Zhang JM, Wang HK, Ye CQ, Ge W, Chen Y, Jiang ZL, et al. ATP released by astrocytes mediates glutamatergic activity-dependent heterosynaptic suppression. Neuron 2003;40(5):971–82.
[98] Jourdain P, Bergersen LH, Bhaukaurally K, Bezzi P, Santello M, Domercq M, et al. Glutamate exocytosis from astrocytes controls synaptic strength. Nat Neurosci 2007;10(3):331–9.
[99] Domercq M, Brambilla L, Pilati E, Marchaland J, Volterra A, Bezzi P. P2Y1 receptor-evoked glutamate exocytosis from astrocytes: control by tumor necrosis factor-α and prostaglandins. J Biol Chem 2006;281(41):30684–96.
[100] Andrew RD, Labron MW, Boehnke SE, Carnduff L, Kirov SA. Physiological evidence that pyramidal neurons lack functional water channels. Cereb Cortex 2007;17(4):787–802.
[101] Risher WC, Andrew RD, Kirov SA. Real-time passive volume responses of astrocytes to acute osmotic and ischemic stress in cortical slices and in vivo revealed by two-photon microscopy. Glia 2009;57(2):207–21.
[102] Kimelberg HK, Goderie SK, Higman S, Pang S, Waniewski RA. Swelling-induced release of glutamate, aspartate, and taurine from astrocyte cultures. J Neurosci 1990;10(5):1583–91.
[103] Liu HT, Tashmukhamedov BA, Inoue H, Okada Y, Sabirov RZ. Roles of two types of anion channels in glutamate release from mouse astrocytes under ischemic or osmotic stress. Glia 2006;54(5):343–57.
[104] Mongin AA, Kimelberg HK. ATP potently modulates anion channel-mediated excitatory amino acid release from cultured astrocytes. Am J Physiol Cell Physiol 2002;283(2):C569–78.
[105] Abdullaev IF, Rudkouskaya A, Schools GP, Kimelberg HK, Mongin AA. Pharmacological comparison of swelling-activated excitatory amino acid release and Cl^- currents in cultured rat astrocytes. J Physiol 2006;572(Pt 3):677–89.
[106] Takano T, Kang J, Jaiswal JK, Simon SM, Lin JH, Yu Y, et al. Receptor-mediated glutamate release from volume sensitive channels in astrocytes. Proc Natl Acad Sci USA 2005;102(45):16466–71.
[107] Basarsky TA, Feighan D, MacVicar BA. Glutamate release through volume-activated channels during spreading depression. J Neurosci 1999;19(15):6439–45.
[108] Haskew-Layton RE, Rudkouskaya A, Jin Y, Feustel PJ, Kimelberg HK, Mongin AA. Two distinct modes of hypoosmotic medium-induced release of excitatory amino acids and taurine in the rat brain in vivo. PloS one 2008;3(10):e3543.
[109] Eskandari S, Zampighi GA, Leung DW, Wright EM, Loo DD. Inhibition of gap junction hemichannels by chloride channel blockers. J Membr Biol 2002;185(2):93–102.
[110] Stout CE, Costantin JL, Naus CC, Charles AC. Intercellular calcium signaling in astrocytes via ATP release through connexin hemichannels. J Biol Chem 2002;277(12):10482–8.
[111] Spray DC, Ye ZC, Ransom BR. Functional connexin "hemichannels": a critical appraisal. Glia 2006;54(7):758–73.
[112] Cotrina ML, Lin JH, Alves-Rodrigues A, Liu S, Li J, Azmi-Ghadimi H, et al. Connexins regulate calcium signaling by controlling ATP release. Proc Natl Acad Sci USA 1998;95(26):15735–40.
[113] Cotrina ML, Lin JH, López-Garcia JC, Naus CC, Nedergaard M. ATP-mediated glia signaling. J Neurosci 2000;20(8):2835–44.
[114] Pearson RA, Dale N, Llaudet E, Mobbs P. ATP released via gap junction hemichannels from the pigment epithelium regulates neural retinal progenitor proliferation. Neuron 2005;46(5):731–44.
[115] Ye ZC, Wyeth MS, Baltan-Tekkok S, Ransom BR. Functional hemichannels in astrocytes: a novel mechanism of glutamate release. J Neurosci 2003;23(9):3588–96.
[116] Zhao HB, Yu N, Fleming CR. Gap junctional hemichannel-mediated ATP release and hearing controls in the inner ear. Proc Natl Acad Sci USA 2005;102(51):18724–9.

[117] Cherian PP, Siller-Jackson AJ, Gu S, Wang X, Bonewald LF, Sprague E, et al. Mechanical strain opens connexin 43 hemichannels in osteocytes: a novel mechanism for the release of prostaglandin. Mol Biol Cell 2005;16(7):3100–6.

[118] Gomes P, Srinivas SP, Van Driessche W, Vereecke J, Himpens B. ATP release through connexin hemichannels in corneal endothelial cells. Invest Ophthalmol Vis Sci 2005;46(4):1208–18.

[119] Montero TD, Orellana JA. Hemichannels: new pathways for gliotransmitter release. Neuroscience 2015;286:45–59.

[120] Wong CW, Christen T, Roth I, Chadjichristos CE, Derouette JP, Foglia BF, et al. Connexin37 protects against atherosclerosis by regulating monocyte adhesion. Nat Med 2006;12(8):950–4.

[121] Navarrete M, Araque A. Endocannabinoids potentiate synaptic transmission through stimulation of astrocytes. Neuron 2010;68(1):113–26.

[122] Panatier A, Theodosis DT, Mothet JP, Touquet B, Pollegioni L, Poulain DA, et al. Glia-derived D-serine controls NMDA receptor activity and synaptic memory. Cell 2006;125(4):775–84.

[123] Stellwagen D, Malenka RC. Synaptic scaling mediated by glial TNF-α. Nature 2006;440(7087):1054–9.

[124] Angulo MC, Kozlov AS, Charpak S, Audinat E. Glutamate released from glial cells synchronizes neuronal activity in the hippocampus. J Neurosci 2004;24(31):6920–7.

[125] Kozlov AS, Angulo MC, Audinat E, Charpak S. Target cell-specific modulation of neuronal activity by astrocytes. Proc Natl Acad Sci USA 2006;103(26):10058–63.

[126] Martineau M, Galli T, Baux G, Mothet JP. Confocal imaging and tracking of the exocytotic routes for D-serine-mediated gliotransmission. Glia 2008;56(12):1271–84.

[127] Martineau M, Shi T, Puyal J, Knolhoff AM, Dulong J, Gasnier B, et al. Storage and uptake of D-serine into astrocytic synaptic-like vesicles specify gliotransmission. J Neurosci 2013;33(8):3413–23.

[128] Mothet JP, Pollegioni L, Ouanounou G, Martineau M, Fossier P, Baux G. Glutamate receptor activation triggers a calcium-dependent and SNARE protein-dependent release of the gliotransmitter D-serine. Proc Natl Acad Sci USA 2005;102(15):5606–11.

[129] Fossat P, Turpin FR, Sacchi S, Dulong J, Shi T, Rivet JM, et al. Glial D-serine gates NMDA receptors at excitatory synapses in prefrontal cortex. Cereb Cortex 2012;22(3):595–606.

[130] Mothet JP, Parent AT, Wolosker H, Brady Jr RO, Linden DJ, Ferris CD, et al. D-serine is an endogenous ligand for the glycine site of the N-methyl-D-aspartate receptor. Proc Natl Acad Sci USA 2000;97(9):4926–31.

[131] Schell MJ, Molliver ME, Snyder SH. D-serine, an endogenous synaptic modulator: localization to astrocytes and glutamate-stimulated release. Proc Natl Acad Sci USA 1995;92(9):3948–52.

[132] Henneberger C, Papouin T, Oliet SH, Rusakov DA. Long-term potentiation depends on release of D-serine from astrocytes. Nature 2010;463(7278):232–6.

[133] Yang Y, Ge W, Chen Y, Zhang Z, Shen W, Wu C, et al. Contribution of astrocytes to hippocampal long-term potentiation through release of D-serine. Proc Natl Acad Sci USA 2003;100(25):15194–9.

[134] Serrano A, Haddjeri N, Lacaille JC, Robitaille R. GABAergic network activation of glial cells underlies hippocampal heterosynaptic depression. J Neurosci 2006;26(20):5370–82.

[135] Zhang Z, Chen G, Zhou W, Song A, Xu T, Luo Q, et al. Regulated ATP release from astrocytes through lysosome exocytosis. Nat Cell Biol 2007;9(8):945–53.

[136] Coco S, Calegari F, Pravettoni E, Pozzi D, Taverna E, Rosa P, et al. Storage and release of ATP from astrocytes in culture. J Biol Chem 2003;278(2):1354–62.

[137] Guthrie PB, Knappenberger J, Segal M, Bennett MV, Charles AC, Kater SB. ATP released from astrocytes mediates glial calcium waves. J Neurosci 1999;19(2):520–8.

[138] Newman EA. Glial cell inhibition of neurons by release of ATP. J Neurosci 2003;23(5):1659–66.

[139] Panatier A, Vallee J, Haber M, Murai KK, Lacaille JC, Robitaille R. Astrocytes are endogenous regulators of basal transmission at central synapses. Cell 2011;146(5):785–98.

[140] Gordon GR, Baimoukhametova DV, Hewitt SA, Rajapaksha WR, Fisher TE, Bains JS. Norepinephrine triggers release of glial ATP to increase postsynaptic efficacy. Nat Neurosci 2005;8(8):1078–86.

[141] Gordon GR, Iremonger KJ, Kantevari S, Ellis-Davies GC, MacVicar BA, Bains JS. Astrocyte-mediated distributed plasticity at hypothalamic glutamate synapses. Neuron 2009;64(3):391–403.

[142] Hamilton NB, Attwell D. Do astrocytes really exocytose neurotransmitters? Nat Rev Neurosci 2010;11(4):227–38.

[143] Nedergaard M, Verkhratsky A. Artifact versus reality—how astrocytes contribute to synaptic events. Glia 2012;60(7):1013–23.

[144] Sloan SA, Barres BA. Looks can be deceiving: reconsidering the evidence for gliotransmission. Neuron 2014;84(6):1112–5.

[145] Stout C, Charles A. Modulation of intercellular calcium signaling in astrocytes by extracellular calcium and magnesium. Glia 2003;43(3):265–73.

[146] Pascual O, Casper KB, Kubera C, Zhang J, Revilla-Sanchez R, Sul JY, et al. Astrocytic purinergic signaling coordinates synaptic networks. Science 2005;310(5745):113–6.

[147] Deng Q, Terunuma M, Fellin T, Moss SJ, Haydon PG. Astrocytic activation of A1 receptors regulates the surface expression of NMDA receptors through a Src kinase dependent pathway. Glia 2011;59(7):1084–93.

[148] Fellin T, Halassa MM, Terunuma M, Succol F, Takano H, Frank M, et al. Endogenous nonneuronal modulators of synaptic transmission control cortical slow oscillations in vivo. Proc Natl Acad Sci USA 2009;106(35):15037–42.

[149] Fujita T, Chen MJ, Li B, Smith NA, Peng W, Sun W, et al. Neuronal transgene expression in dominant-negative SNARE mice. J Neurosci 2014;34(50):16594–604.

[150] Nedergaard M, Takano T, Hansen AJ. Beyond the role of glutamate as a neurotransmitter. Nat Rev Neurosci 2002;3(9):748–55.

CHAPTER 4

Types of Epilepsy

OUTLINE

OVERVIEW

It is estimated that one in 26 people will develop epilepsy in their lifetime, amounting to almost 12 million people in the United States alone [1]. Epilepsy is a group of conditions characterized by sporadic occurrence of seizures and unconsciousness. This severely limits the ability to perform everyday tasks and leads to increased difficulty with learning and memory, maintenance of steady employment, driving, and overall socioeconomic integration. Epilepsy is *not* a benign condition as it is associated with both increased mortality and decreased quality of life. A greater understanding of the cellular and molecular mechanisms underlying seizures and epilepsy is necessary, as it may lead to novel antiepileptic treatments.

In this chapter, we will provide a brief overview of seizure types, epilepsy classification, and common animal models of epilepsy.

SEIZURES AND EPILEPSY

A *seizure* is the clinical manifestation of abnormal excessive or synchronous neuronal activity in the brain. *Epilepsy* comprises a variety of syndromes characterized by recurrent seizures unprovoked by systemic or acute neurologic insults.

Astrocytes and Epilepsy
DOI: http://dx.doi.org/10.1016/B978-0-12-802401-0.00004-1

A summary of types of seizures is given in Table 4.1 [2]. *Partial* or *focal* seizures involve seizure onset from a focal area in one hemisphere. *Simple partial seizures* do not involve loss of consciousness whereas *complex partial seizures* involve some loss of consciousness or awareness.

Simple partial seizures often have focal or localized signs related to the area of cortex involved. In addition they may be associated with auras. Auras ("breezes" in Greek) have been defined as "that portion of the seizure which occurs before consciousness is lost and for which memory is retained afterwards" [3]. Distinct epileptic auras have been

TABLE 4.1 Overview of Seizure Types

Seizure Type	Brief Description	EEG Characteristic
Partial (or focal)	Seizure onset from one area of the brain and limited to one hemisphere. Divided into *simple partial seizures* (no loss of consciousness) and *complex partial seizures* (with loss of consciousness)	Start focally then may secondarily generalize
Neocortical	Seizure generation from the neocortex; manifestation depends on exact location of origin and pattern of spread	Focal onset from neocortex
Temporal lobe	Seizure generation from temporal lobe structures, such as the hippocampus; often consists of epigastric aura followed by automatisms, dystonia of contralateral hand, and postictal confusion	Focal onset from temporal lobe
Generalized	Seizure onset simultaneously from both hemispheres	Generalized
Absence	Brief loss of consciousness, eye blinking and staring, and/or facial movements with no postictal confusion	3-Hz generalized spike-and-slow-wave complexes
Myoclonic	Quick, repetitive, arrhythmic muscle twitching involving one or both sides of the body; consciousness remains intact	Generalized spike-and-wave discharge
Clonic	Seizures consist of rhythmic muscle jerks during impaired consciousness	Fast activity (≥10 Hz) and slow waves with occasional spike-wave patterns
Tonic	An increase in muscle tone causes flexion of head, trunk, and/or extremities for several seconds	Bilateral synchronous medium to high-voltage fast activity (10–25 Hz)
Tonic-clonic	Tonic extension of muscles followed by clonic rhythmic movements and postictal confusion	Tonic phase: Generalized rhythmic discharges decreasing in frequency and increasing in amplitude Clonic phase: Slow waves
Atonic	Brief loss of postural tone, which can result in falls and injuries	Slow rhythmic (1–2 Hz) spike-and-wave complexes or more rapid, irregular multifocal spike-and-wave activity

Modified with permission from Hubbard et al. Glial cell changes in epilepsy: overview of the clinical problem and therapeutic opportunities. Neurochem Int 2013; 63:638-651 (Table 1).

categorized as: (1) epigastric; (2) fear and anxiety; (3) experiential (including déjà vu/jamais vu); (4) olfactory-gustatory ("uncinate fits"); (5) autonomous/vegetative; and (6) nonspecific [4]. The most common aura in temporal lobe epilepsy (TLE, the most common type of partial epilepsy) is rising epigastric aura [5]; interestingly, a single localized epileptogenic lesion can lead to multiple aura types [6].

Complex partial seizures by definition lead to some impairment of consciousness or awareness during the seizure. Clinical manifestations vary with site of origin and degree of spread. Complex partial seizures are often associated with auras, and also automatisms (eg, lip smacking, fumbling). Postictal confusion is very common. The duration of complex partial seizures is usually about 1–2 minutes.

Generalized seizures, by contrast, do not have focal onset. Generalized seizures are divided into absence, myoclonic, clonic, tonic, tonic-clonic, and atonic seizures. *Absence seizures* (petit mal) involve brief staring spells with impairment of awareness, usually lasting 3–20 seconds, with sudden onset and sudden resolution. Often provoked by hyperventilation, they have onset typically between 4 and 14 years of age, and often resolve by 18 years of age. They are usually associated with normal development and intelligence but episodes in school of "tuning out" have occasionally led to misdiagnosis as attention deficit-hyperactivity disorder (ADHD). *Myoclonic seizures* are characterized by brief, shock-like jerks of a muscle or group of muscles, typically bilateral, and so brief (seizures <1 second) that impairment of consciousness is difficult to assess. These must be differentiated from benign, non-epileptic myoclonus (eg, while falling asleep). *Clonic seizures* are similar to myoclonic but repetition rate is slower; *tonic seizures* involve rigid posturing of the limbs or torso often with deviation of the head or eyes toward one side; and *tonic-clonic* (grand mal) seizures involve both tonic and clonic phases. During the tonic phase, the patient may cry out, become rigid, and exhibit signs such as respiratory impairment, cyanosis, pupillary dilation, and increase in heart rate and blood pressure. During the clonic phase, the seizure evolves into clonic movement characterized by jerking of the extremities. Tonic-clonic seizures may extend for several minutes prior to termination or may evolve into *status epilepticus* (SE; continuous seizures). *Atonic seizures* involve sudden loss of postural tone which when severe results in falls (drop attacks), when milder may produce head nods or jaw drops. Consciousness is impaired but the duration is usually few seconds (<1 minute).

Seizure phases include *preictal* (period before seizure), *ictal* (period during the seizure), *interictal* (period between seizures), and *postictal* (period of transition from the ictal phase to the patient's normal level of awareness and function—may be minutes to hours). Ictal and postictal assessment includes whether the patient can speak and follow commands, head or eye deviation, posturing of extremities, whether the patient sustained injuries (eg, falls, tongue biting), behavioral changes during and after seizure, incontinence, weakness (eg, Todd's paralysis [7]), aphasia, and postictal confusion. Typical diagnostic evaluation of seizure will include a good history (family history, history of head trauma, febrile seizures, meningitis/encephalitis, medication history, illicit drug use) and other tests such as blood tests (complete blood count, electrolytes, glucose, calcium, magnesium, phosphate, hepatic, and renal function tests), lumbar puncture (limited to cases in which infectious etiology such as meningitis or encephalitis suspected), blood or urine screen for drugs, electroencephalogram (EEG), and brain MRI. The key goal is to diagnose the type of epilepsy if present, determine etiology and comorbid problems, establish prognosis, and decide treatment options. Definitive diagnosis is obtained often with the gold-standard test of video-EEG monitoring in an inpatient epilepsy monitoring unit. This is the cornerstone of diagnosing seizures (vs pseudoseizures or other syndromes) and epilepsies, to determine epileptogenic zone anatomy and to assess for candidacy for epilepsy surgery. Additional important tests employed may include: Wada test (intracarotid sodium amobarbital procedure, which anesthetizes one half of the brain temporarily to assess language and memory function localization), interictal PET (positron emission tomography), ictal or interictal SPECT (single photon emission computed tomography), MEG (magnetoencephalography), neuropsychological testing, and intracranial video-EEG monitoring.

INTERNATIONAL LEAGUE AGAINST EPILEPSY CLASSIFICATION

> All attempts at classification of seizures are hampered by our limited knowledge of the underlying pathological processes within the brain and that any classification must of necessity be a tentative one and will be subject to change with every advance in scientific understanding of epilepsy. ***Source: Ref. [8].***

The International League Against Epilepsy (ILAE) (www.ilae.org) has made many attempts over the years to standardize and refine a classification of epilepsies and epileptic syndromes. In 1981, they issued a classification for "epileptic seizures" and in 1989 for "epilepsies and epileptic syndromes." Since 1997, the ILAE Task Force on Classification has made further significant efforts including the reports of 2001 and 2006 [9,10]. In 2010 [11] and 2014 [12], the ILAE issued other reports adding new terminology and concepts. For example, the previous etiologic categorization of epilepsies as "idiopathic," "symptomatic," or "cryptogenic" was replaced with "genetic," "structural/metabolic," or "unknown cause" [11]. However, these reports have generated significant controversy [13–16]. The complexity of classification of epilepsies, whether genetic, structural/metabolic, or of unknown cause is clear to see in Table 4.2. In a new, as yet unadopted classification, the ILAE commission (Scheffer et al. [12], www.ilae.org) proposes expanding the etiologic subtypes into six categories: genetic, structural, metabolic, immune, infectious, and unknown. Ultimately, as research progresses and understanding deepens, epilepsies will be increasingly categorized by etiology rather than (as traditionally done) by clinical semiology or electrographic (EEG) characteristics. Recently, the ILAE has published a new definition and classification scheme for SE as well, a topic omitted in many previous ILAE reports [17]. For updated detailed classification data and reports on the many types of epilepsy, the reader is encouraged to consult www.EpilepsyDiagnosis.org, an online diagnostic manual of the epilepsies published by the ILAE Commission on Classification and Terminology.

ANIMAL MODELS OF EPILEPSY

Critical to the neuroscientific researcher attempting to identify new concepts and targets for treatment of various forms of epilepsy are animal models. Comparison of various animal models of epilepsy for the development of antiepileptogenic and disease-modifying drugs has been previously extensively reviewed [18–20] and is itself the topic of many books and

TABLE 4.2 Electroclinical Syndromes and Other Epilepsies

Electroclinical Syndromes Arranged by Age at Onset[a]
Neonatal period
Benign familial neonatal epilepsy (BFNE)
Early myoclonic encephalopathy (EME)
Ohtahara syndrome
Infancy
Epilepsy of infancy with migrating focal seizures
West syndrome
Myoclonic epilepsy in infancy (MEI)
Benign infantile epilepsy
Benign familial infantile epilepsy
Dravet syndrome
Myoclonic encephalopathy in nonprogressive disorders
Childhood
Febrile seizures plus (FS+) (can start in infancy)
Panayiotopoulos syndrome
Epilepsy with myoclonic atonic (previously astatic) seizures
Benign epilepsy with centrotemporal spikes (BECTS)
Autosomal dominant nocturnal frontal lobe epilepsy (ADNFLE)
Late-onset childhood occipital epilepsy (Gastaut type)
Epilepsy with myoclonic absences
Lennox-Gastaut syndrome
Epileptic encephalopathy with continuous spike-and-wave during sleep (CSWS)[b]
Landau-Kleffner syndrome (LKS)
Childhood absence epilepsy (CAE)
Adolescence—Adult
Juvenile absence epilepsy (JAE)
Juvenile myoclonic epilepsy (JME)
Epilepsy with generalized tonic-clonic seizures alone
Progressive myoclonus epilepsies (PME)
Autosomal dominant epilepsy with auditory features (ADEAF)
Other familial temporal lobe epilepsies

(*Continued*)

TABLE 4.2 (Continued)

Electroclinical Syndromes Arranged by Age at Onset[a]
Less specific age relationship
Familial focal epilepsy with variable foci (childhood to adult)
Reflex epilepsies
Distinctive constellations
Mesial temporal lobe epilepsy with hippocampal sclerosis (MTLE with HS)
Rasmussen syndrome
Gelastic seizures with hypothalamic hamartoma
Hemiconvulsion–hemiplegia–epilepsy
Epilepsies that *do not* fit into any of these diagnostic categories can be distinguished first on the basis of the presence or absence of a known structural or metabolic condition (presumed cause) and then on the basis of the primary mode of seizure onset (generalized vs focal)
Epilepsies attributed to and organized by structural–metabolic causes
Malformations of cortical development (hemimegalencephaly, heterotopias, etc.)
Neurocutaneous syndromes (tuberous sclerosis complex, Sturge-Weber, etc.)
Tumor
Infection
Trauma
Angioma
Perinatal insults
Stroke
Etc.
Epilepsies of unknown cause
Conditions with epileptic seizures that are traditionally not diagnosed as a form of epilepsy per se
Benign neonatal seizures (BNS)
Febrile seizures (FS)

[a]*The arrangement of electroclinical syndromes does not reflect etiology.*
[b]*Sometimes referred to as Electrical Status Epilepticus during Slow Sleep (ESES).*
Reproduced with permission from Berg et al. Revised terminology and concepts for organization of seizures and epilepsies: report of the ILAE Commission on Classification and Terminology, 2005–2009. Epilepsia 2010; 51:676–685 (Table 3).

book chapters [20,21]. Some common (not an exhaustive list) animal models of epilepsy together with the types of human epilepsy they purport to model are listed in Table 4.3 [22]. The "perfect animal model" of a specific epileptic syndrome would recapitulate: (1) age of onset; (2) etiology; (3) seizure phenotype and EEG characteristics; and (4) comorbidities or other long-term consequences [18].

TABLE 4.3 Summary of Animal Models

Model	Induction	Manifestations	Human Relevance	Use	Limitations
CHEMOCONVULSANTS					
KA-SE	Systemic or intrahippocampal injection	Limbic SE and chronic seizures	TLE with hippocampal sclerosis	AED screening, mechanisms of epileptogenesis	High mortality; variable frequency and severity of spontaneous seizures; not all neural damage comes from seizures
Pilo-SE	Systemic or intrahippocampal injection	Limbic SE and chronic seizures	TLE with hippocampal sclerosis	AED screening, mechanisms of epileptogenesis, and cognitive/psychiatric comorbidities	High mortality; variable frequency and severity of spontaneous seizures; neocortical lesions
Acute chemical models	Systemic or intrahippocampal injection	Nonconvulsive absence or generalized tonic-clonic seizures, depending on the drug and dose	Acute and repetitive seizures	Rapid screening of AED, effect of repetitive seizures	Lack of spontaneous recurrent seizures (SRS) and of neuronal loss or other neuropathological hallmarks
SE in immature rodents	Systemic injection of KA or Pilo	Tonic-clonic seizures	Prolonged seizures during development	Epileptogenesis and long-term consequences	KA and Pilo are not clinical causes of seizures; more extensive damage compared with other models
Repetitive seizures in immature rodents	Systemic injection of PTZ or flurothyl inhalation	Myoclonic and generalized tonic-clonic seizures	Repetitive brief seizures during development	AED screening and cognitive deficits	No spontaneous seizures in adulthood
ELECTRICAL STIMULATION					
Electroshock-induced seizures	Corneal or auricular stimulation	Generalized tonic-clonic seizures	Tonic-clonic seizures	AED screening and molecular and physiological alternations related to epileptiform activities	Low-predictive validity for some AEDs
After discharges	Focal electrical stimulation	Complex partial seizures and myoclonic seizures	Focally generated seizure-like patterns that often spread to other regions	Electrophysiological and behavioral changes caused by focally generated seizure patterns	No specificity for groups of neurons; studying neuronal activity during stimulation is difficult

(*Continued*)

TABLE 4.3 (Continued)

Model	Induction	Manifestations	Human Relevance	Use	Limitations
Kindling	Repeated afterdischarge induction	Partial seizures evolving into secondary generalization and, eventually, spontaneous seizures	Consequences of poorly controlled seizures and dynamics of epileptogenic processes	Prevention of epileptogenesis processes and treatment of pharmacoresistant epilepsies	Costly and time-consuming procedure
BRAIN PATHOLOGY					
Hyperthermic seizures	Increase of body temperature in immature rodents through stream of heated air	Immobility, facial automatism, myoclonic jerks	Febrile seizure	Epileptogenesis mechanisms and cognitive consequences	Subtle behavioral seizures, necessity of EEG recording, possible morbidity from heat exposure
Hypoxia model	Exposure to air with low O_2 concentration in immature rodents	Brief and repetitive tonic-clonic seizures	Neonatal hypoxic encephalopathy	AED screening, long-term consequences, and epileptogenesis mechanisms	Susceptibility for seizures varies with the strain and age of rodents
Posttraumatic epilepsy	Rostral parasagittal fluid percussion injury	Generalized tonic-clonic seizures in the long term, with low frequency	Posttraumatic epilepsy	AED screening, mechanisms of epileptogenesis, and hippocampal sclerosis with dual pathology	Laborious induction, long latency periods, mild seizures during the first 4 months of posttrauma
GENETIC MODELS					
Audiogenic models	Acoustic stimulation in genetically prone rats	Wild running and tonic-clonic seizures	Reflex epilepsy and TLE studies	Epileptogenesis mechanisms and comorbidities associated with epilepsies	Necessity of a trigger to evoke seizures; lack of SRS
GAERS, WAG/Rij, and mouse models of absence seizures	Spontaneous seizures	SWD generalization, behavioral arrest	Generalized idiopathic epilepsies	Electrographic and behavioral similarity to human absence seizures, response to AEDs	Diverse (and not fully known) genetic alterations

KA, kainic acid; *SE*, status epilepticus; *TLE*, temporal lobe epilepsy; *AED*, antiepileptic drugs; *Pilo*, pilocarpine; *PTZ*, pentylenetetrazol; *EEG*, electroencephalography; *GAERS*, Genetic absence epilepsy in rats from Strasbourg; *WAG/Rij*, Wistar Albino Glaxo/Rijswijk rats; *SWD*, spike-and-wave discharges.
Reproduced with permission from Kandratavicius et al. Animal models of epilepsy: use and limitations. Neuropsychiatr Dis Treat 2014; 10:1693–1705.

Animal models of epilepsy have been used for distinct goals:

1. Discovery of new antiepileptic drugs (AEDs).
2. Characterization of anticonvulsant activity of new AEDs.
3. Comparison of adverse effects of AEDs in epileptic versus nonepileptic animals.
4. Estimation of effective plasma concentrations of new AEDs for clinical trial planning.
5. Identification of *antiepileptogenic* drugs that abrogate the underlying process of epileptogenesis.
6. Identification of "disease-modifying" drugs.

Both acute and chronic models of epilepsy have been developed (Fig. 4.1). Acute models to generate seizures include maximal electroshock seizure test, pentylenetetrazol (PTZ) administration, and flurothyl administration. These models have been used effectively to study processes around the time of seizure onset, and to screen for drugs that suppress seizures acutely. Chronic models of epilepsy include the kindling model of epilepsy (can be done electrically or chemically [23]), post-SE models in which epilepsy develops after a sustained SE (such as the intrahippocampal kainic acid model [24]), and genetic models of different types

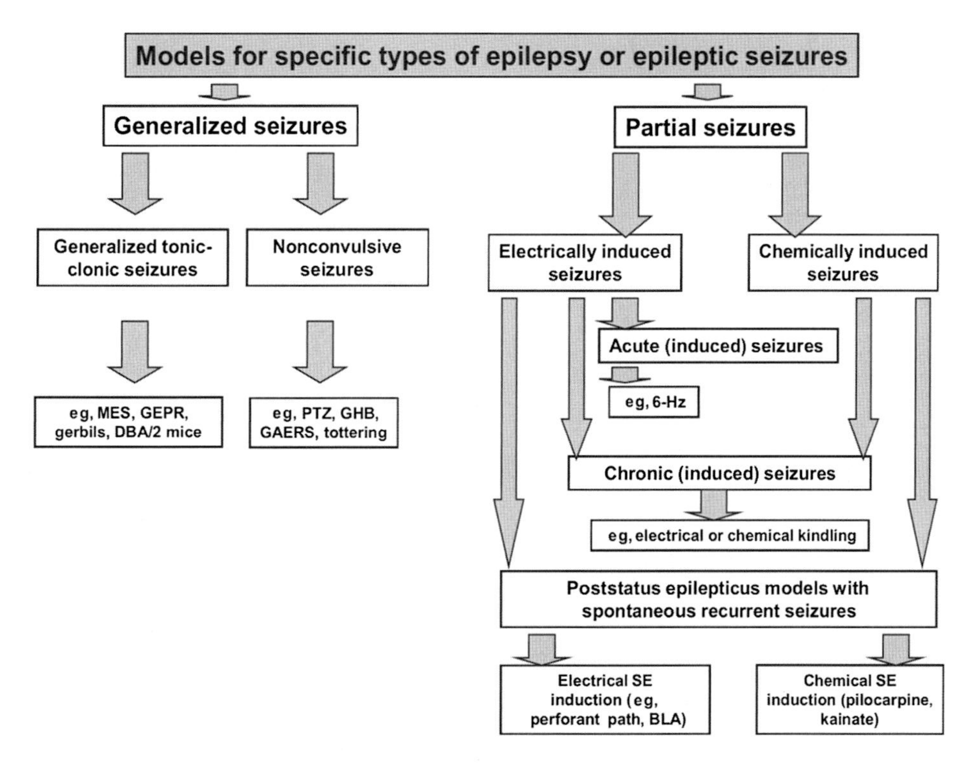

FIGURE 4.1 **Overview of models for specific types of epilepsy or epileptic seizures.** There are numerous models not shown in this figure, including chronic epilepsy models in which spontaneous partial seizures develop after traumatic brain injury, ischemic brain damage, or febrile seizures. *Source: Reproduced with permission from Löscher W. Critical review of current animal models of seizures and epilepsy used in the discovery and development of new antiepileptic drugs. Seizure 2011; 20:359–68 (Figure 3).*

of epilepsy. Kindling models and post-SE models (such as kainic acid and pilocarpine in both mice and rats) are commonly used to study epileptogenic processes and potential drug targets. It is notable that much AED drug development has occurred in acute seizure models in previously healthy (nonepileptic) animals (Fig. 4.2), whereas the pharmacology of chronic seizure models and the anticonvulsant efficacy of a specific agent in an epileptic animal versus a nonepileptic animal may be quite different [20]. Thus, the point has been made that, while more time consuming, chronic models should be used earlier in drug development to minimize false positives [20]. Marked differences among models provide a challenge as well, and therefore promising agents should probably be tested in a variety of models if feasible. This was done recently with retigabine, which effectively suppressed seizures in many distinct preclinical models (electroshock, electrical kindling of the amygdala, pentylenetetrazol, kainate, *N*-methyl-D-aspartate (NMDA), and picrotoxin) prior to entering clinical trials.

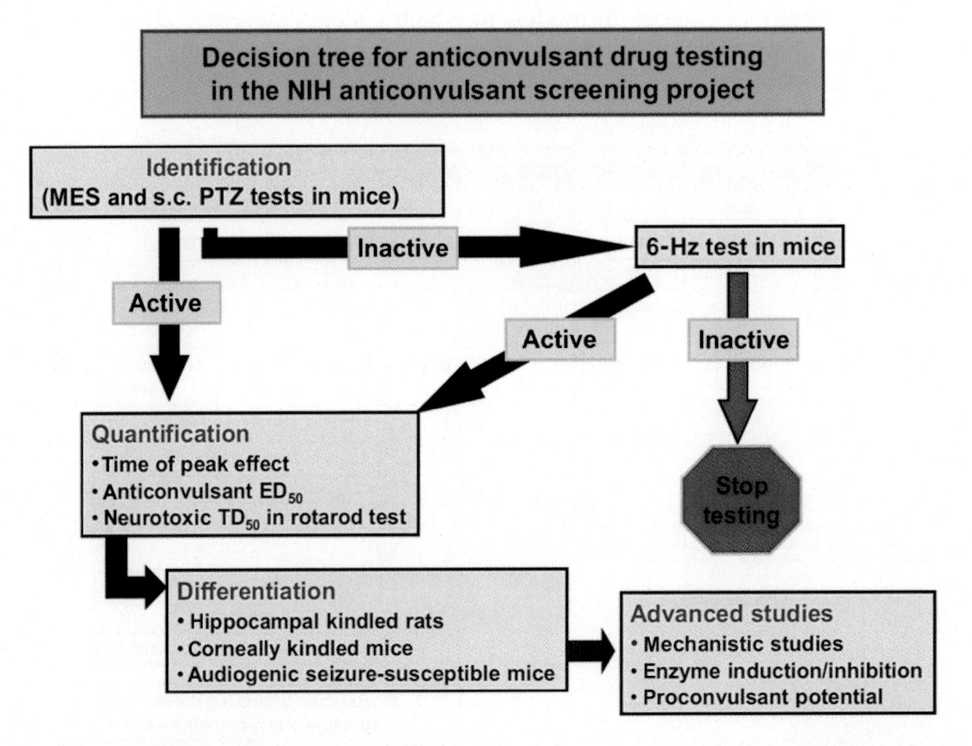

FIGURE 4.2 **Schematic diagram illustrating the initial screen of the NINDS-sponsored University of Utah Anticonvulsant Drug Development (ADD) Program.** An investigational compound is initially screened for efficacy in the MES and s.c. PTZ tests. The activity of those compounds with demonstrated efficacy and minimal behavioral toxicity is subsequently quantitated (ED50 and TD50) at the time of peak anticonvulsant effect. Compounds found *inactive* in the MES and s.c. PTZ tests are evaluated in the LEF-sensitive 6-Hz seizure test in mice. For those compounds that are found to be active in the 6-Hz test, their activity is quantitated at their respective time of peak effect. All compounds found to be active in one or more of these three identification screens are then differentiated on the basis of their activity in additional seizure models, including the hippocampal kindled rate model of TLE. *Source: Reproduced with permission from Löscher W. Critical review of current animal models of seizures and epilepsy used in the discovery and development of new antiepileptic drugs. Seizure 2011; 20:359–68 (Figure 5).*

Perhaps the most exciting area of current research is the discovery or development of antiepileptogenic or disease-modifying treatments. The traditional approach has been to screen for compounds that possess acute seizure suppressive effects; by contrast, the ideal antiepileptogenic agent would prevent the development of epilepsy in those at risk. Currently, no AED is capable of preventing or modifying epilepsy after brain insults, such as traumatic brain injury (TBI) [25,26]. Prevention of epilepsy (antiepileptogenesis), however, can be fruitfully explored in certain chronic animal models of epilepsy in which the animal starts out nonepileptic, proceeds through the epileptogenic process, and then becomes epileptic (exhibits spontaneous recurrent seizures or SRS). Models used for this purpose include post-SE models of TLE, models of TBI, and kindling. In this scheme, potential antiepileptogenic agents are tested after the initial insult (such as SE or TBI) and given chronically while the animals are monitored for the development of epilepsy/SRS (usually with chronic in vivo video-EEG monitoring) (Fig. 4.3) [19,27].

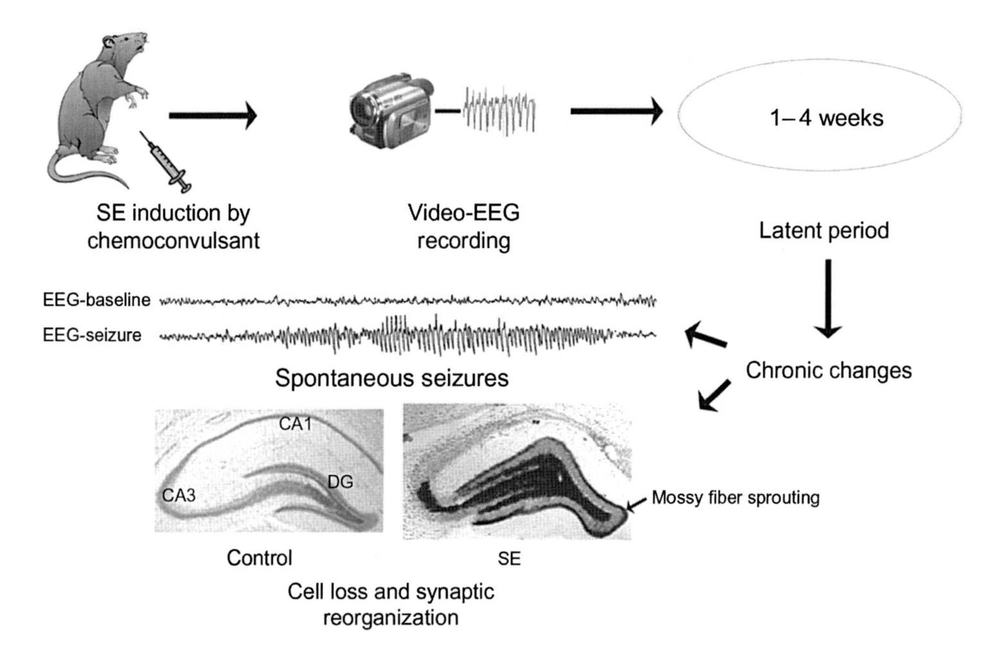

FIGURE 4.3 **Schematic diagram depicting the typical protocol to study TLE using chemoconvulsant-induced status epilepticus (SE) model.** SE is induced in a rat or mouse by a chemoconvulsant such as kainic acid or pilocarpine. The animal is implanted with intracranial electrodes before or after SE induction for video-EEG monitoring to determine occurrence of SRS (epilepsy) following SE. Approximately 1–4 weeks after kainic acid or pilocarpine-induced SE, animals demonstrate spontaneous seizures (percentage of animals developing spontaneous seizures depends on chemoconvulsant used, severity of SE, and age at SE induction). The "latent period" between SE and onset of spontaneous seizures has been actively studied by epilepsy researchers in order to identify changes that occur after SE, which may play a crucial role in epilepsy development. Cell loss in multiple brain regions including the hippocampus (as seen in CA1 and CA3 areas of hippocampal sections obtained from animal that had experienced SE) and synaptic reorganization of mossy fibers in the dentate gyrus (DG) of the hippocampus, similar to what has been observed in patients with TLE, occurs in many animals that have experienced SE. *Source: Reproduced with permission from Raol YH, Brooks-Kayal AR. Experimental models of seizures and epilepsies. Prog Mol Biol Transl Sci 2012; 105:57–82.*

MEDICAL MANAGEMENT OF EPILEPSY

Management of epilepsy involves a multidisciplinary collaboration, best done at epilepsy centers with fellowship-trained neurologists, neurosurgeons trained in epilepsy surgery, neuropsychologists, nurses, and EEG technicians and radiology and other staff dedicated to the highest standards of epilepsy care. Medical management may include medication treatment with AEDs as well as dietary or other lifestyle changes (eg, ketogenic diet).

Workup for epilepsy surgery in properly selected cases may involve: (1) establishing the presence of AED resistance; (2) delineating the epileptogenic zone within the brain; and (3) estimating the risk which might occur for postoperative neurologic or cognitive deficits. Types of epilepsy surgery include: (1) intracranial video-EEG monitoring ("Phase 2" monitoring, with implantation of grid, strip, and/or depth electrodes); (2) resective surgery (eg, temporal lobectomy, selective amygdalohippocampectomy); (3) disconnective surgery (eg, corpus callosotomy, multiple subpial transection); and (4) neuromodulation (eg, vagus nerve stimulator implantation, NeuroPace device implantation, deep brain stimulator implantation) (Fig. 4.4). Best candidates for resective surgery to cure epilepsy have (1) EEG seizure onset from a focal area; (2) MRI abnormality in the same region; and (3) likelihood of being able to remove that region without significant neurologic or cognitive deficits. These criteria are most commonly satisfied by patients with medically intractable TLE. Randomized trials have clearly shown class I evidence of benefit of surgery for TLE [28]. Certain procedures (in particular TLE surgery and hemispherotomy) have excellent seizure outcome (at least 70% seizure-free postoperatively) [29,30].

The mainstay of treatment of most patients with epilepsy remains AEDs (Table 4.4). Choice of AEDs involves consideration of seizure type, epilepsy syndrome, pharmacokinetic profile, interactions/other medical conditions, efficacy, expected adverse effects, and cost. Most current AEDs target neuronal voltage-gated sodium channels and calcium channels, glutamate receptors, or γ-aminobutyric acid (GABA) systems [31,32] (Table 4.5). For example, Na^+ channel blockers such as phenytoin and carbamazepine reduce the frequency of neuronal action potentials, and GABA transaminase (GABAT) inhibitors, such as vigabatrin, increase GABA-mediated inhibition [31]. The mode of action of several commonly prescribed AEDs, such as valproate, is not entirely understood [31,33,34]. One of the newest AEDs, retigabine (ezogabine), has a distinct method of action as a K^+ channel opener.

However, there are several major drawbacks to the current AEDs. First, currently used AEDs often act as general CNS depressants and cause cognitive impairment, including impaired learning, memory loss, and mental slowing [35]. Cognitive impairment becomes particularly important in patients being treated with chronic AEDs. Moreover, polypharmacy has a more severe impact on cognitive function when compared to monotherapy, regardless of which type of AEDs are being used [35]. Second, about 30% of patients being treated with AEDs, even with optimal current therapy, have poor seizure control and become medically refractory. In addition, adverse effects are frequently observed at drug doses within the recommended range [34]. Third, several studies have shown that there is an increased risk of teratogenicity in women with epilepsy who are receiving pharmacological treatment [36,37]. For women taking enzyme-inducing AEDs, such as phenytoin or carbamazepine, hormonal forms of contraception are affected and the efficacy of oral

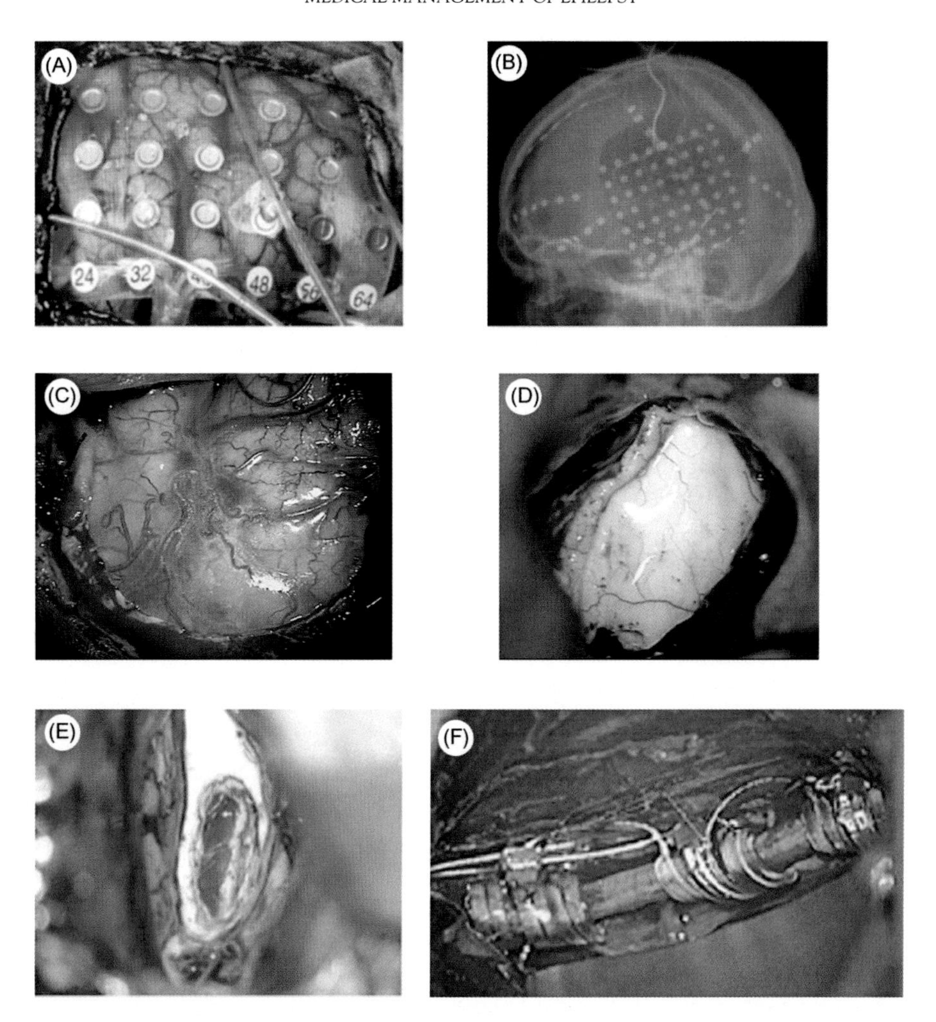

FIGURE 4.4 **Types of epilepsy surgery.** (A) Intracranial implantation of subdural grid and strip electrodes. (B) Postoperative lateral skull radiograph indicating extensive coverage of the brain by grid and strip electrodes. Subsequent video-EEG monitoring will be performed to delineate epileptogenic zone(s) in preparation for tailored surgical resection. (C) Surgical exposure for a transsylvian selective amygdalohippocampectomy procedure. Visible within the field is the Sylvian fissure with arteries and veins interposed between the cortical surfaces of the left frontal and temporal lobes. (D) Exposure of the hippocampus during the selective amygdalohippocampectomy procedure. (E) Exposure of the corpus callosum during a corpus callosotomy. Seen in a trajectory down the interhemispheric fissure, the corpus callosum appears snow-white. Midline disconnection of the corpus callosum is shown down to the level of the midline septum pellucidum at the level of the lateral ventricles. Corpus callosotomy is very effective against atonic seizures (drop attacks). (F) Vagus nerve stimulator electrodes have been placed around the vagus nerve (tenth cranial nerve) in the neck. These are connected to an implantable pulse generator (battery) for palliation of seizures (VNS therapy). *All images courtesy of Devin Binder.*

TABLE 4.4 Antiepileptic Drugs: Timeline of Introduction

- 1857—Bromides
- 1912—Phenobarbital (PB)
- 1937—Phenytoin (PHT)
- 1944—Trimethadione
- 1954—Primidone
- 1958—ACTH
- 1960—Ethosuximide (ESM)
- 1963—Diazepam
- 1974—Carbamazepine (CBZ)
- 1975—Clonazepam (CZP)
- 1978—Valproate (VPA)
- 1993—Felbamate (FBM), gabapentin (GBP)
- 1995—Lamotrigine (LTG)
- 1997—Topiramate (TPM), tiagabine (TGB)
- 1999—Levetiracetam (LEV)
- 2000—Oxcarbazepine (OXC), zonisamide (ZNS)
- 2005—Pregabalin (PGB)
- 2008—Lacosamide (LCM), rufinamide (RUF)
- 2009—Vigabatrin (VGB)
- 2011—Retigabine
- 2011—Clobazam
- 2012—Perampanel
- 2013—Eslicarbazepine

contraceptives cannot be guaranteed [36], thus complicating family planning. Finally, AEDs are associated with a number of adverse effects including mood alteration, suicidality, severe mucocutaneous reactions, hepatotoxic effects, decreased bone mineral density (osteoporosis), weight management difficulties, skin rash, pseudolymphoma, and many others, which often leads to treatment failure [38]. "Older" versus "newer" AEDs may have a difference in side effect profiles, but not necessarily a difference in efficacy.

Thus, it is clear that there is a clinical unmet need for new AEDs, in particular those that may have antiepileptogenic effects or increased tolerability. Given the fact that many of the adverse neuropsychological effects of current AEDs may relate to negative effects on normal synaptic function, it is clear that nonneuronal targets may be of interest in future development of AEDs.

WHY GLIAL CELLS?

Over the last two decades, several lines of evidence have suggested that glial cells are potential therapeutic targets for the treatment of epilepsy and other central nervous system (CNS) diseases [39,40]. Glia are involved in many important physiological functions. For example, astrocytes play an established role in removal of glutamate at synapses, neuronal pathfinding, and the sequestration and redistribution of K^+ during neural activity [41]. Microglia are the resident CNS immune cells and are important for initiating the

TABLE 4.5 Antiepileptic Drugs (AEDs): Mechanisms of Action

Antiepileptic drug (AED)	Voltage-dependent channels		Neuromodulation			
	Na^+	Ca^{2+}	GABAergic	Glutamatergic	Carbonic Anhydrase	Novel Target
Established AEDs						
Carbamazepine	↓					
Ethosuximide		↓				
Phenobarbital			↑			
Phenytoin	↓					
Valproic acid	↓	↓	↑			
Second-Generation AEDs						
Felbamate	↓	↓	↑	↓ (NMDA)		
Gabapentin		↓				α2δ subunit
Lacosamide	↓ (slow inactivation)					
Lamotrigine	↓					h-current
Levetiracetam						SV2A
Oxcarbazepine	↓					
Pregabalin		↓				
Retigabine						K_v7 potassium channels
Rufinamide	↓					
Tiagabine			↑ (decrease reuptake)			
Topiramate	↓	↓	↑	↓ (AMPA/ kainate)	↓	
Vigabatrin			↑ (decrease metabolism)			
Zonisamide	↓	↓	↑ (?)		↓	

inflammatory response to brain injury and infection [42]. Moreover, it is becoming increasingly clear that glial cells play a role in seizure susceptibility and the development of epilepsy [39,40,43–48]. Direct stimulation of astrocytes leads to prolonged neuronal depolarization and epileptiform discharges [46]. Glial cells can release neuroactive molecules and also modulate synaptic transmission through modifications in channels, gap junctions, receptors, and transporters [39,43,45,46,49–54]. Furthermore, striking changes in glial cell shape and function occur in various forms of epilepsy which may contribute to increased neuronal excitability and the development of epilepsy. Some of these changes include astroglial proliferation, dysregulation of water and ion channel expression, alterations in secretion of neuroactive molecules, and increased activation of inflammatory pathways [43,44,48,55–59].

Evidence from studies in human tissue further suggests an important role for astrocytes in epilepsy. Astrocytes undergo activation to become reactive astrocytes in the epileptic brain [44,57]. Changes in the expression of various astrocytic enzymes, such as adenosine kinase [60] and glutamine synthetase [61], contribute to the increased neuronal excitability found in epileptic tissue. In addition, microglia and inflammatory pathways contribute to the pathogenesis of seizures in various forms of epilepsy [62].

In this book, we review the current literature surrounding the involvement of astrocytes in epilepsy. Individual chapters will consider functional changes in astrocyte calcium signaling, potassium channels, water channels, glutamate metabolism, adenosine metabolism, gap junctions, blood–brain barrier disruption, and inflammation. Based on the current evidence, there is strong rationale for the development of novel astrocyte-centered therapeutic opportunities for the treatment of epilepsy.

References

[1] Hesdorffer DC, Logroscino G, Benn EK, Katri N, Cascino G, Hauser WA. Estimating risk for developing epilepsy: a population-based study in Rochester, Minnesota. Neurology 2011;76(1):23–7.

[2] Hubbard JA, Hsu MS, Fiacco TA, Binder DK. Glial cell changes in epilepsy: overview of the clinical problem and therapeutic opportunities. Neurochem Int 2013;63(7):638–51.

[3] International CoCaTot Epilepsy LA. Proposal for revised clinical and electroencephalographic classification of epileptic seizures. Epilepsia 1981;22:489–501.

[4] Wieser HG. ILAE commission report: mesial temporal lobe epilepsy with hippocampal sclerosis. Epilepsia 2004;45:695–714.

[5] Binder DK, Garcia PA, Elangovan GK, Barbaro NM. Characteristics of auras in patients undergoing temporal lobectomy. J Neurosurg 2009;111(6):1283–9.

[6] Lee DJ, Owen CM, Khanifar E, Kim RC, Binder DK. Isolated amygdala neurocysticercosis in a patient presenting with déjà vu and olfactory auras: case report. J Neurosurg Pediatr 2009;3(6):538–41.

[7] Binder DK. A history of Todd and his paralysis. Neurosurgery 2004;54(2):480–6. discussion 6–7.

[8] Gastaut H, Caveness WF, Landolt W, Lorentz de Haas AM, McNaughton FL, Magnus O, et al. A proposed international classification of epileptic seizures. Epilepsia 1964;5:297–306.

[9] Engel J. A proposed diagnostic scheme for people with epileptic seizures and with epilepsy: report of the ILAE Task Force on Classification and Terminology. Epilepsia 2001;42:796–803.

[10] Engel J. Report of the ILAE classification core group. Epilepsia 2006;47:1558–68.

[11] Berg AT, Berkovic SF, Brodie MJ, Buchhalter J, Cross JH, van Emde Boas W, et al. Revised terminology and concepts for organization of seizures and epilepsies: report of the ILAE Commission on Classification and Terminology, 2005–2009. Epilepsia 2010;51(4):676–85.

[12] Scheffer IE, Berkovic SF, Capovilla P, Connolly MB, Guilhoto L, Hirsch E, et al. The organization of the epilepsies: report of the ILAE Commission on Classification and Terminology. ILAE website, <www.ilae.org>; 2014.

[13] Panayiotopoulos CP. The new ILAE report on terminology and concepts for organization of epileptic seizures: a clinician's critical view and contribution. Epilepsia 2011;52(12):2155–60.

[14] Panayiotopoulos CP. The new ILAE report on terminology and concepts for the organization of epilepsies: critical review and contribution. Epilepsia 2012;53(3):399–404.

[15] Shorvon SD. The etiologic classification of epilepsy. Epilepsia 2011;52:1052–7.

[16] Wolf P. Networks and systems, conceptualizations, and research. Epilepsia 2011;52:1198–200.

[17] Trinka E, Cock H, Hesdorffer D, Rossetti AO, Scheffer IE, Shinnar S, et al. A definition and classification of status epilepticus—report of the ILAE Task Force on Classification of Status Epilepticus. Epilepsia 2015;56(10):1515–23.

[18] Raol YH, Brooks-Kayal AR. Experimental models of seizures and epilepsies. Prog Mol Biol Transl Sci 2012;105:57–82.

[19] Löscher W. Critical review of current animal models of seizures and epilepsy used in the discovery and development of new antiepileptic drugs. Seizure 2011;20(5):359–68.

[20] Löscher W. Animal models of epilepsy for the development of antiepileptogenic and disease-modifying drugs. A comparison of the pharmacology of kindling and post-status epilepticus models of temporal lobe epilepsy. Epilepsy Res 2002;50(1-2):105–23.

[21] Pitkänen A, Schwartzkroin PA, Moshé SL. Models of seizures and epilepsy. Amsterdam: Elsevier; 2006.

[22] Kandratavicius L, Balista PA, Lopes-Aguiar C, Ruggiero RN, Umeoka EH, Garcia-Cairasco N, et al. Animal models of epilepsy: use and limitations. Neuropsychiatr Dis Treat 2014;10:1693–705.

[23] Binder DK, McNamara JO. Kindling: a pathologic activity-driven structural and functional plasticity in mature brain Corcoran M, Moshé S, editors. Kindling 5. New York: Plenum Press; 1998. p. 245–54.

[24] Arabadzisz D, Antal K, Parpan F, Emri Z, Fritschy JM. Epileptogenesis and chronic seizures in a mouse model of temporal lobe epilepsy are associated with distinct EEG patterns and selective neurochemical alterations in the contralateral hippocampus. Exp Neurol 2005;194(1):76–90.

[25] Temkin NR, Dikmen SS, Wilensky AJ, Keihm J, Chabal S, Winn HR. A randomized, double-blind study of phenytoin for the prevention of post-traumatic seizures. N Engl J Med 1990;323(8):497–502.
[26] Chang BS, Lowenstein DH, Quality Standards Subcommittee of the American Academy of Neurology. Practice parameter: antiepileptic drug prophylaxis in severe traumatic brain injury. Report of the Quality Standards Subcommittee of the American Academy of Neurology. Neurology 2003;60(1):10–16.
[27] Löscher W, Brandt C. Prevention or modification of epileptogenesis after brain insults: experimental approaches and translational research. Pharmacol Rev 2010;62(4):668–700.
[28] Wiebe S, Blume WT, Girvin JP, Eliasziw M, Effectiveness, Efficiency of Surgery for Temporal Lobe Epilepsy Study Group. A randomized, controlled trial of surgery for temporal-lobe epilepsy. N Engl J Med 2001;345(5):311–8.
[29] Binder DK, Schramm J. Multilobar resections and hemispherectomy Engel J, Pedley TA, editors. Epilepsy: a comprehensive textbook (2nd ed.). Philadelphia, PA: Lippincott-Raven; 2007. p. 1879–90.
[30] Binder DK, Schramm J. Resective surgical techniques: mesial temporal lobe epilepsy Lüders HO, Najm I, Bingaman W, editors. Textbook of epilepsy surgery (3rd ed.). Taylor & Francis; 2008. p. 1083–92.
[31] Rogawski MA, Loscher W. The neurobiology of antiepileptic drugs. Nat Rev Neurosci 2004;5(7):553–64.
[32] White HS, Rho JM. Mechanisms of action of antiepileptic drugs. West Islip, NY: Professional Communications, Inc.; 2010.
[33] Kwan P, Schachter SC, Brodie JM. Drug-resistant epilepsy. N Engl J Med 2012;365:919–26.
[34] Perucca E. An introduction to antiepileptic drugs. Epilepsia 2005;46(Suppl. 4):31–7.
[35] Aldenkamp AP, De Krom M, Reijs R. Newer antiepileptic drugs and cognitive issues. Epilepsia 2003;44:21–9.
[36] Crawford P. Best practice guidelines for the management of women with epilepsy. Epilepsia 2005;46(Suppl. 9).117–24.
[37] Wlodarczyk BJ, Palacios AM, George TM, Finnell RH. Antiepileptic drugs and pregnancy outcomes. Am J Med Genet 2012;158A(8):2071–90.
[38] Perucca P, Gilliam FG. Adverse effects of antiepileptic drugs. Lancet Neurol 2012;11(9):792–802.
[39] Binder DK, Steinhäuser C. Functional changes in astroglial cells in epilepsy. Glia 2006;54:358–68.
[40] Friedman A, Kaufer D, Heinemann U. Blood-brain barrier breakdown-inducing astrocytic transformation: novel targets for the prevention of epilepsy. Epilepsy Res 2009;85:142–9.
[41] Ransom B, Behar T, Nedergaard M. New roles for astrocytes (stars at last). Trends Neurosci 2003;26:520–2.
[42] Carson MJ, Thrash JC, Walter B. The cellular response in neuroinflammation: the role of leukocytes, microglia and astrocytes in neuronal death and survival. Clin Neurosci Res 2006;6(5):237–45.
[43] Binder DK, Nagelhus EA, Ottersen OP. Aquaporin-4 and epilepsy. Glia 2012;60:1203–14.
[44] Clasadonte J, Haydon PG. Astrocytes and epilepsy Noebles JL, Avoli M, Rogawski MA, Olsen RW, Delgado-Escueta AV, editors. Jasper's basic mechanisms of the epilepsies (4th ed.); 2012 p. 19.
[45] Hsu MS, Lee DJ, Binder DK. Potential role of the glial water channel aquaporin-4 in epilepsy. Neuron Glia Biol 2007;3(4):287–97.
[46] Tian G, Azmi H, Takano T, Xu Q, Peng W, Lin J, et al. An astrocytic basis of epilepsy. Nat Med 2005;11(9):973–81.
[47] Seifert G, Carmignoto G, Steinhäuser C. Astrocyte dysfunction in epilepsy. Brain Res Rev 2010;63:212–21.
[48] Seifert G, Schilling K, Steinhäuser C. Astrocyte dysfunction in neurological disorders: a molecular perspective. Nat Rev Neurosci 2006;7:194–206.
[49] Beenhakker MP, Huguenard JR. Astrocytes as gatekeepers of $GABA_B$ receptor function. J Neurosci 2010;30(45):15262–76.
[50] Wang F, Smith NA, Xu Q, Fujita T, Baba A, Matsuda T, et al. Astrocytes modulate neural network activity by Ca^{2+}-dependent uptake of extracellular K^+. Sci Signal 2012;5(218):ra26.
[51] Santello M, Bezzi P, Volterra A. TNFalpha controls glutamatergic gliotransmission in the hippocampal dentate gyrus. Neuron 2011;69(5):988–1001.
[52] Rouach N, Koulakoff A, Abudara V, Willecke K, Giaume C. Astroglial metabolic networks sustain hippocampal synaptic transmission. Science 2008;322(5907):1551–5.
[53] Volterra A, Steinhäuser C. Glial modulation of synaptic transmission in the hippocampus. Glia 2004;47:249–57.
[54] Halassa MM, Fellin T, Haydon PG. The tripartite synapse: roles for gliotransmission in health and disease. Trends Mol Med 2007;13(2):54–63.
[55] Steinhäuser C, Seifert G. Glial membrane channels and receptors in epilepsy: impact for generation and spread of seizure activity. Eur J Pharmacol 2002;447:227–37.
[56] de Lanerolle NC, Lee T. New facets of the neuropathology and molecular profile of human temporal lobe epilepsy. Epilepsy Behav 2005;7:190–203.

[57] Heinemann U, Jauch GR, Schulze JK, Kivi A, Eilers A, Kovacs R, et al. Alterations of glial cell functions in temporal lobe epilepsy. Epilepsia 2000;41(Suppl. 6):S185–9.
[58] Hinterkeuser S, Schröder W, Hager G, Seifert G, Blümcke I, Elger CE, et al. Astrocytes in the hippocampus of patients with temporal lobe epilepsy display changes in potassium conductances. Eur J Neurosci 2000;12:2087–96.
[59] Kivi A, Lehmann TN, Kovács R, Eilers A, Jauch R, Meencke HJ, et al. Effects of barium on stimulus-induced rises of $[K^+]_o$ in human epileptic non-sclerotic and sclerotic hippocampal area CA1. Eur J Neurosci 2000;12:2039–48.
[60] Aronica E, Zurolo E, Iyer A, de Groot M, Anink J, Carbonell C, et al. Upregulation of adenosine kinase in astrocytes in experimental and human temporal lobe epilepsy. Epilepsia 2011;52(9):1645–55.
[61] Coulter DA, Eid T. Astrocytic regulation of glutamate homeostasis in epilepsy. Glia 2012;60:1215–26.
[62] Ravizza T, Gagliardi B, Noé F, Boer K, Aronica E, Vezzani A. Innate and adaptive immunity during epileptogenesis and spontaneous seizures: evidence from experimental models and human temporal lobe epilepsy. Neurobiol Dis 2008;29(1):142–60.

CHAPTER

5

Neuropathology of Human Epilepsy

OUTLINE

OVERVIEW

In the 2010 International League Against Epilepsy (ILAE) classification of electroclinical syndromes and other epilepsies, the ILAE report listed epilepsies attributed to structural–metabolic causes [1]. These include:

- Malformations of cortical development (hemimegalencephaly, heterotopias, etc.)
- Neurocutaneous syndromes (tuberous sclerosis (TS) complex, Sturge-Weber, etc.)
- Tumor
- Infection
- Trauma
- Perinatal insults
- Stroke
- Vascular malformation.

Astrocytes and Epilepsy
DOI: http://dx.doi.org/10.1016/B978-0-12-802401-0.00005-3

These comprise known epilepsy-related structural causes that are amenable to pathologic evaluation, diagnosis, and classification. In this chapter, we will review some of the features of human epilepsy neuropathology. In particular, we will focus on hippocampal sclerosis (HS), long-term epilepsy-associated tumors (LEAT) and focal cortical dysplasias (FCD). In a large German series of 4512 epilepsy tissue samples, these three pathologies were the most common: HS (40%), LEAT (27%), and FCD (13%) [2]. Therefore, for the purposes of this chapter, we will first review the main features of the neuropathology of HS, LEAT, and FCD; then we will describe possible astrocytic contributions to TS, tumor-associated epilepsy, posttraumatic epilepsy (PTE), and poststroke epilepsy (PSE).

HIPPOCAMPAL SCLEROSIS

Temporal lobe Epilepsy and the Development of Epilepsy Surgery

Affecting over 40 million people worldwide [3], temporal lobe epilepsy (TLE) is characterized by recurrent seizure activity originating within the temporal lobe. TLE is the most common form of epilepsy found in adults and seizures are medically intractable in about 40% of patients suffering from this disease [4,5]. Depth electrode studies have shown that most seizures originate from the hippocampus [6]. The most common pathological feature found in these patients is HS, characterized by a pronounced loss of pyramidal neurons, granule cell dispersion, mossy fiber sprouting, astrogliosis, and microvascular proliferation [3,7–10].

The discovery of the importance of HS has paralleled developments in epilepsy surgery and in pathological evaluation.

> In case after case we found the cortex to be tough in the anterior and deep portion of the first temporal convolution. This abnormality extended into, and grew more marked in, the uncus and hippocampal gyrus. The tissue was tough, rubbery, and slightly yellow…We have gradually realized the importance of this discovery, as the epileptogenic focus was often shown by electrocorticography to be situated here, and furthermore patients returned with continuing seizure when we had made anterior temporal removals without excision of this area [11].

The earliest resections for epilepsy, from Macewen [12] and Horsley [13] to Krause [14] and Foerster [15], were largely designed to remove cortical areas reflected in patients' ictal semiology. Following Hans Berger's discovery of the human electroencephalogram (EEG) in 1929 [16], EEG was adapted to identify and better localize epileptiform abnormalities, in particular in the temporal lobe. In the late 1930s, Wilder Penfield and Herbert Jasper at the Montreal Neurological Institute developed the use of EEG in combination with electrocorticography and functional mapping of "eloquent" brain areas to tailor epilepsy resections [11,17]. The developing concept of "psychomotor" or TLE and its surgical treatment [18–20] led to the advent of anatomically standardized en bloc temporal lobe resection [21–23]. Soon after, it became clear that pure lateral temporal cortical removal was associated with unsatisfactory seizure outcome, and removal of the deep structures including hippocampus and amygdala was required [11,24–28] (see quote above). Bilateral resections were abandoned early since they resulted in dense anterograde amnesia, such as in the famous case of H.M. [29].

Over the last half-century, many modifications to either "tailored" or "anatomic" temporal lobe resections have been adopted. Anatomically standardized resections rely on the concepts that resection of pathology seen on imaging studies will include the epileptogenic zone, and that resection of eloquent areas will be avoided by conforming to certain anatomic boundaries. In contrast, tailored resections emphasize altering degree of resection based on individual pathophysiology, functional mapping of eloquent cortex, and intraoperative electrocorticography [30–33]. Of course, the goal of all epilepsy surgery remains extirpation of the epileptogenic zone without producing neurological or cognitive deficits.

There are several differences among current techniques for TLE surgery. The first is the relative extent of mesial temporal versus neocortical resection, from neocortical resection without mesial resection [34,35] to "anteromedial temporal lobectomy" (limited cortical removal followed by extended hippocampal resection) [36–38] to mesial resection without neocortical resection (selective amygdalohippocampectomy) [39–41]. The second is the exact extent of mesial resection, with variations in amygdalar resection [42,43] and hippocampal and parahippocampal gyrus resection [38,42,44–47]. The third difference is surgical approach (eg, transsylvian [40,41,48] versus transcortical [39,44,49–52] versus subtemporal [53,54] selective amygdalohippocampectomy). A final difference is whether the resection is anatomic or "tailored," as described above [30–33]. There has been a trend toward reducing the extent of neocortical resection and increasing the extent of resection of mesial structures. Nevertheless, presence of lateral pathology or dual pathology clearly requires either standard temporal lobectomy or tailored approaches.

Patient selection, preoperative work-up [55], anesthesia [56], outcome and complications of TLE surgery will not be considered in detail here. Suffice to say here that the efficacy of TLE surgery is now very well established. Randomized trials have demonstrated clear superiority of surgery versus best medical treatment for TLE [57–59]. Approximately two-third of patients will be seizure-free postoperatively [60], a figure which is remarkably consistent across large-volume epilepsy centers. The interested reader is also referred to other references that present detailed anatomic discussions of the intricate surgical and vascular anatomy of the temporal lobe [37,48,50,61–72].

Development of the Concept of HS

The history of neuropathology of HS provides perspective on the modern concept of HS. In 1825, Bouchet and Cazauvieilh [73] described the consistency of the hippocampus in 18 postmortem evaluations of patients with epilepsy. Induration was noted in four cases, softening in one (also see [74]). However, it was Sommer in 1880 who provided a detailed description of HS in 90 postmortem cases [75]. Sommer found that "all pyramidal cells... which are doubtlessly of greatest importance for the function of Ammon's horn are missing" [75]. In particular, they were missing in an area which came to be called "Sommer's sector," corresponding to the CA1 subfield, which usually exhibits the greatest degree of neuronal loss in HS (Fig. 5.1). Later, Bratz published a series of findings in 70 patients with epilepsy [76] (Fig. 5.2). In addition to confirming and extending the observations of Sommer, Bratz recognized an area that was resistant to degeneration (modern-day CA2). Spielmeyer later reported that in severe cases, all sectors could exhibit degeneration and loss of pyramidal cells (total HS) [77]. In 1966, Margerison and Corsellis reported a postmortem study

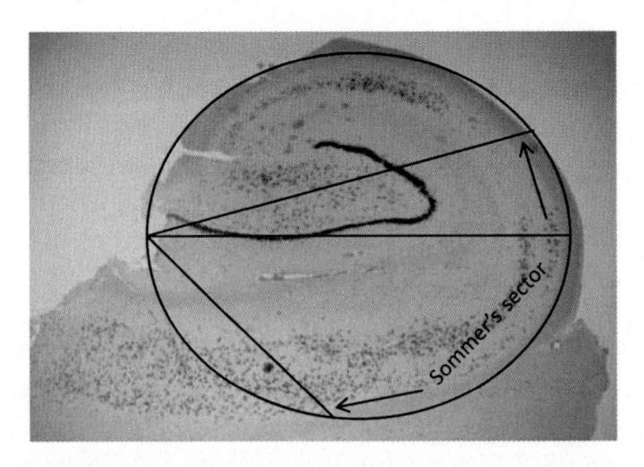

FIGURE 5.1 **Sommer's sector.** Sommer described a vulnerable region of neuronal loss in Ammon's horn sclerosis that became known as Sommer's sector (modern-day CA1). In his original paper he describes this as "*The whole part of the gray matter that belongs to the medial wall of the lower horn, approximately from the fold-over locus of the ventricular endothelium over the middle of the curvature...or if one considers the Ammon's horn as an ellipsoid, whose biggest axis goes parallel to the horizontal plane, in a sector in which the lower exterior quadrant (calculated from the medial plane of the brain) covers approximately 60 degrees of the horizontal axis towards the smaller one, and 20 degrees in the upper exterior.*" In the figure, these parameters have been superimposed on a surgical resection specimen immunostained with neuronal nuclear antigen to highlight Sommer's description. *Source: Reproduced with permission from Thom M. Hippocampal sclerosis: progress since Sommer. Brain Pathol 2009;19:565–72 (Figure 1).*

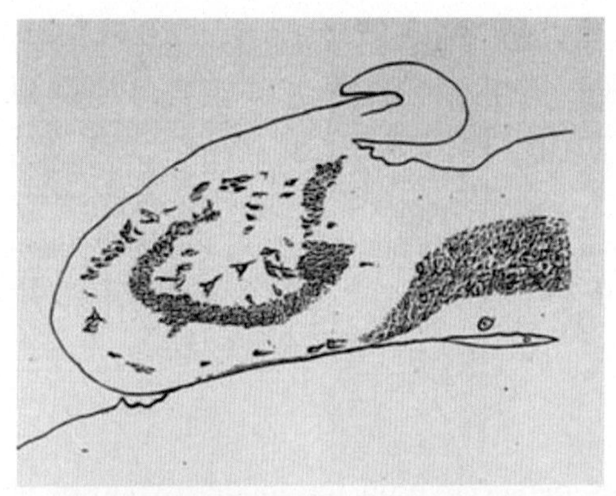

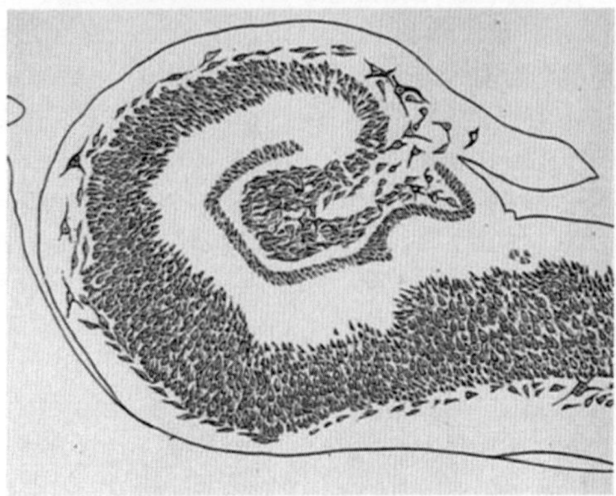

FIGURE 5.2 **Early illustration of hippocampal sclerosis.** From an original paper published in 1899 by Bratz depicting the regions of neuronal loss and reduction in volume in the hippocampus compared with a control. Note that the granule cell layer also appears somewhat broader than in the control although the phenomena of granule cell dispersion in the dentate gyrus remained to be described until 1990. *Source: Reproduced with permission from Thom M. Hippocampal sclerosis: progress since Sommer. Brain Pathol 2009;19:565–72 (Figure 2).*

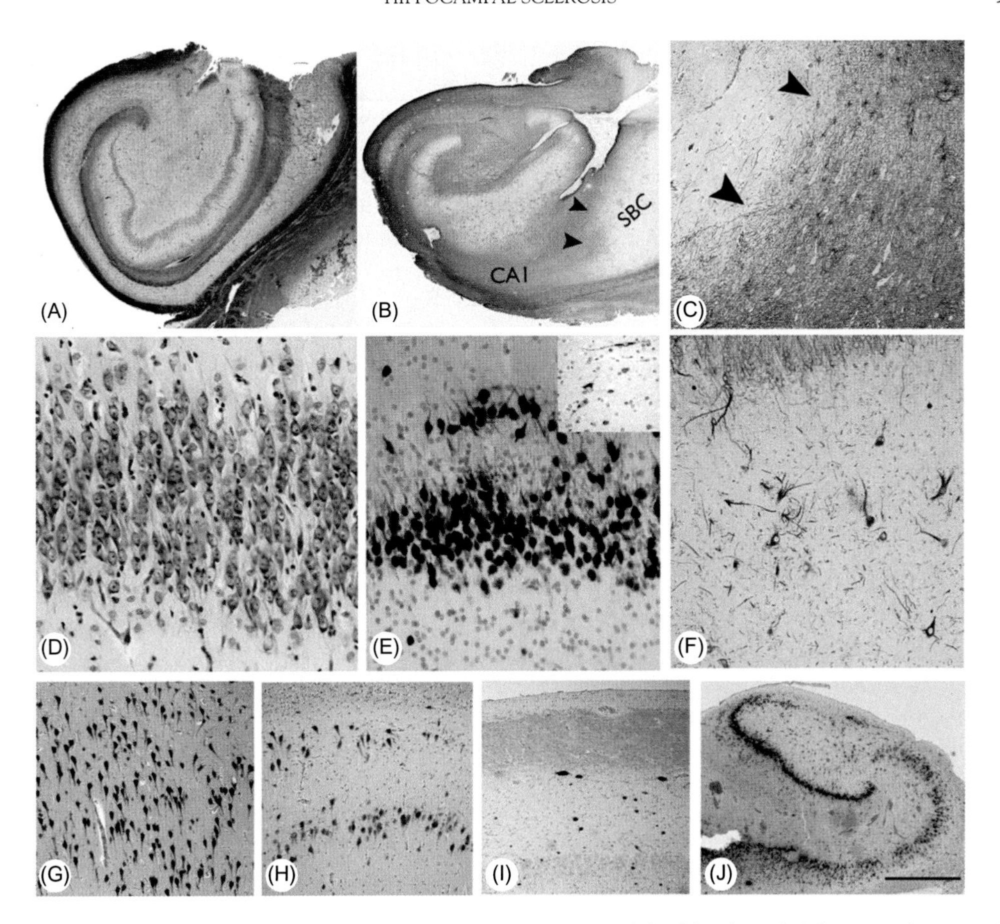

FIGURE 5.3 **Typical pathology of hippocampal sclerosis.** (A) Luxol-fast blue/cresyl violet-stained preparation of a surgical sample with subfield neuronal loss involving CA1 and CA4. (B) GFAP preparation confirming dense fibrillary gliosis in CA1 with a sharp cut-off point (arrows) with the adjacent subiculum (SBC) and (C) higher magnification detail of the abrupt transition of gliosis at the CA1/subiculum (arrows). (D) Granule cell dispersion as seen with cresyl violet and (E) with NeuN showing clusters of dispersed granule cells (inset shows reelin-positive interneurons and bipolar cells in the molecular layer). (F) Neurofilament-positive neurons with enlarged cell bodies in CA4 of the sclerotic hippocampus. (G) CA1 pyramidal cell layer in a patient with MTLE but no evidence of neuronal loss. (H) CA1 with evidence of partial neuronal loss from mid zone and (I) CA1 with severe neuronal depletion and collapse of the layer. (J) Section of the pes hippocampus from a patient with confirmed classical ILAE type 1 sclerosis in the body; in the pes neuronal loss is limited to the endplate in this case as an illustration of variability in the pattern of atrophy that may occur along the longitudinal hippocampal axis. Bar is equivalent to approximately 1 mm (A, B, J), 250 μm (C, F, G, H, I), and 100 μm (D, E). *Source: Reproduced with permission from Thom M. Hippocampal sclerosis in epilepsy: a neuropathology review. Neuropathol Appl Neurobiol 2014;40:520–43 (Figure 1).*

of 55 epilepsy patients and described another pattern with cell loss and gliosis in the hilus with sparing of Sommer's sector, termed "endfolium sclerosis" [9]. More recently yet another pattern with more selective CA1 loss has been described [6,10,78–80]. To this day, HS remains one of the most common pathologies seen in patients with medically-intractable focal epilepsy, based on both epilepsy surgery and autopsy data (Fig. 5.3). Obviously, the availability of resected hippocampal tissue from epilepsy surgery has greatly added to the previous postmortem studies. Consistently, HS was present in 30–66% of temporal lobe cases in both early and more recent studies [6,9,75,76,78,81,82]. When present, the sclerotic hippocampus is the predominant focus of seizure origin [83].

HS is related to the broader concept of mesial temporal sclerosis (MTS). The hippocampus often shows the most striking histopathologic changes in TLE, but other associated temporal and limbic areas can be affected as well (hence the term MTS). Sommer and Bratz both described cases of histological damage outside of the hippocampus, such as parahippocampal gyrus and temporal neocortex. Meyer and Beck described amygdala sclerosis [84], which was also recognized later by Margerison and Corsellis [9] and others [85]. Damage to entorhinal cortex and other cortical areas has also been well described [85–88]. Figs. 5.3 and 5.4 demonstrate typical pathological features of HS, including pyramidal cell loss, mossy fiber sprouting (originally described by Sutula [89]), and granule cell dispersion (originally described by Houser [90]).

Let us now consider astrogliosis and microvascular proliferation, the two remaining hallmarks of HS (Fig. 5.5).

Astrogliosis

One striking hallmark of the sclerotic hippocampus is that while there is a specific pattern of neuronal loss, there is also "reactive gliosis" with hypertrophic glial cells exhibiting prominent GFAP staining and long, thick processes. In contrast with all the studies of hippocampal neuron loss, only a few studies have attempted to quantitate changes in astrocyte numbers and densities in epileptic tissue [91–94]. Previously, it was believed that neuronal loss led to astroglial phagocytosis and, consequently, increased gliosis [5]. This view has changed, however, as reactive changes in astrocytes, or *reactive gliosis (astrogliosis)*, is commonly found in both sclerotic and nonsclerotic hippocampal tissue specimens. Gliosis commonly involves fibrillary gliosis in CA1 and radial gliosis in the dentate gyrus [10]. These glial cell changes have become a hallmark of the sclerotic hippocampus and involve astrocyte hypertrophy [4,91,94–98], increased expression of glial fibrillary acidic protein (GFAP) [4] and vimentin [7,99,100], and changes in glial specific proteins [7,95,100]. Thus, it is becoming increasingly clear that multifaceted changes in astrocyte phenotype may play more of a causative role in the overall pathology and seizure susceptibility [5,95,101–103] (topic of the remainder of this book). A recent report indicates a positive correlation between chronic seizure burden and degree of gliosis, and interestingly an increase in reactive astrocyte number in CA3 was the strongest predictor of poor postoperative seizure outcome [104].

Microvascular Proliferation

Interestingly, both Sommer and Bratz commented on abnormalities of the microvasculature in epileptic specimens. In sclerotic areas the vessels were "thick walled and glossy" [76]. Abnormalities of microvasculature, however, were subsequently mostly overlooked until recently. It is now known that there is proliferation of microvessels, increased vascular endothelial growth factor (VEGF) expression, and loss of blood–brain barrier (BBB) integrity in HS [105,106]. Vascular leakage of proteins such as albumin or IgG may also promote

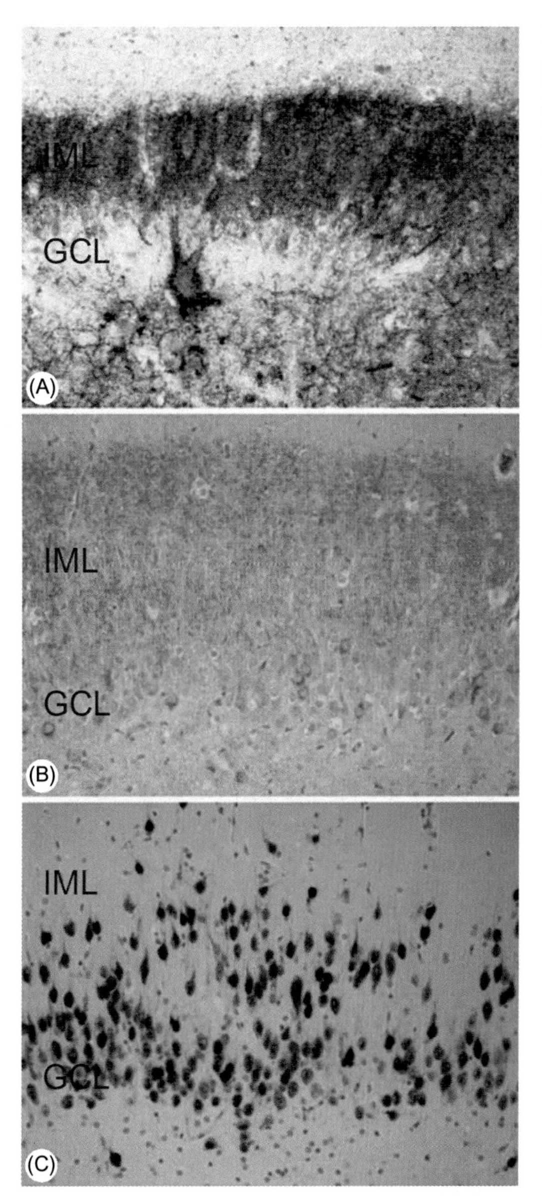

FIGURE 5.4 **Mossy fiber sprouting and dentate granule cell dispersion in hippocampal sclerosis (HS).** Mossy fiber sprouting, as described by Sutula in 1989, can be seen in the molecular layer of the dentate gyrus in HS using Timm's method (A) or dynorphin immunohistochemistry (B). Dispersion of the granule cells of the dentate gyrus in HS is also a common finding and was first described by Houser in 1990 and is illustrated here with neuronal nuclear antigen staining (C). *GCL*, granule cell layer; *IML*, inner molecular layer. *Source: Reproduced with permission from: Thom M. 2009. Hippocampal sclerosis: progress since Sommer. Brain Pathol 19:565–72 (Figure 4).*

neuronal dysfunction or contribute to epileptogenesis [107,108]. Indeed, it has been shown that focal opening of the BBB can promote epileptogenesis, not only in HS but in other focal epilepsies [105,106,109]. In addition, in HS, overexpression of drug transporter proteins such as P-glycoprotein at the BBB facilitate drug efflux, contributing to treatment failure [110,111].

Recently, the Neuropathology Task Force of the ILAE has issued a reclassification of HS. The aim was to synthesize previous attempts to classify the various patterns

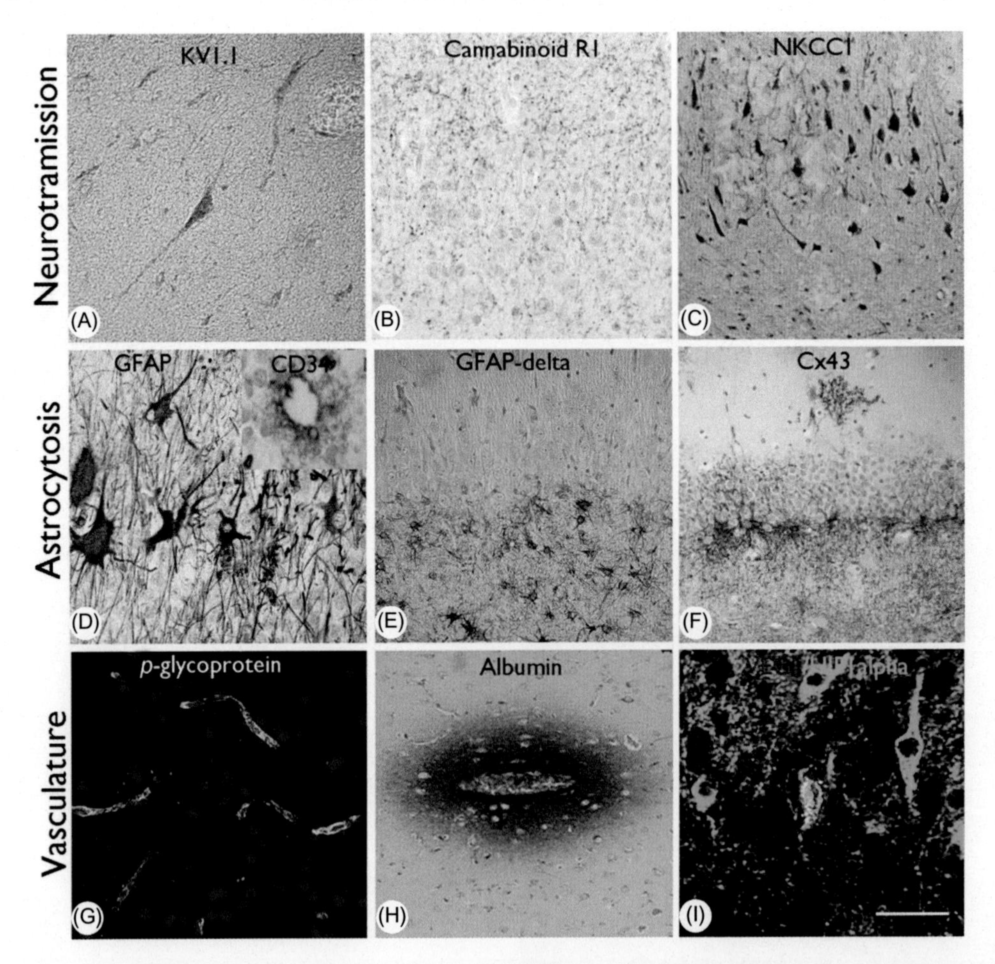

FIGURE 5.5 **Pathological alterations potentially contributing to epileptogenesis in hippocampal sclerosis (HS).** (A) Prominent labeling of voltage-gated potassium channel (Kv1.1) in residual neurons in CA1 in HS; differences were noted in the patterns of expression between hippocampal subfields in HS compared with controls (M. Thom, unpub. obs.). (B) Prominent labeling of cannabinoid receptor is noted in the plexus in the molecular layer of the dentate gyrus in HS. (C) Intense labeling of CA1 neurons with cation-chloride cotransporter (NKCC) in HS. (D) In some cases of HS in epilepsy, particularly in adult onset cases or in the context of recent encephalitis, striking reactive, and "balloon cell" gliosis can be seen in the granule cell layer with a proportion of these cells showing membranous CD34 staining, mimicking a focal cortical dysplasia (inset). (E) Immunolabeling with isoform GFAP-delta highlights prominent numbers of small astroglia, particularly in the subgranular zone, in HS. (F) Immunostaining for gap junction protein connexin 43 (Cx43) demonstrated prominent labeling of astrocytic cells in the subgranular zone and occasionally in the molecular layer in a *postmortem* HS case. (G) Expression of drug transporter protein, *p*-glycoprotein, is shown on capillaries within hippocampus. (H) Albumin staining demonstrating focal leakage from small capillary vessels in a case of HS/TLE. (I) Coexpression of vascular endothelial growth factor (VEGF) and hypoxia inducible factor 1 alpha (HIF1-alpha) is shown in pyramidal neurons in HS. Bar is equivalent to approximately 35 μm (A, B, D, E, I); 60 μm (C, E, F, H). *Source: Reproduced with permission from Thom M. Hippocampal sclerosis in epilepsy: a neuropathology review. Neuropathol Appl Neurobiol 2014;40:520–43 (Figure 4).*

of hippocampal neuronal loss and to correlate subtypes with postsurgical outcomes. Ultimately, they reached consensus on the following classification [112]:

ILAE type 1: Severe neuronal loss and gliosis predominantly in CA1 and CA4 regions;
ILAE type 2: Predominant neuronal cell loss and gliosis in CA1;
ILAE type 3: Predominant neuronal cell loss and gliosis in CA4;
No HS: Normal content of neurons with reactive gliosis only.

HS ILAE type 1 is more often associated with injury before age 5, early seizure onset, and favorable postoperative seizure control. ILAE types 2 and 3 are possibly associated with poorer postoperative seizure control [112]. Given the heterogeneity of HS pathology (Fig. 5.6) [10], further clinical studies will be needed based on the new ILAE classification to stratify epidemiologic factors, imaging data, and surgical outcome data.

TUMOR-ASSOCIATED EPILEPSY AND LEATs

Tumor-Associated Epilepsy

Tumor-associated epilepsy occurs in 20–45% of patients with primary brain tumors (PBTs) [113–115]. These seizures typically manifest as focal seizures with or without secondary generalization and are commonly refractory to antiepileptic drug (AED) treatment. Surgical removal of tumors usually results in seizure control, but many tumors cannot safely be resected. The underlying pathophysiology of seizures secondary to brain tumors is poorly understood. However, a variety of hypotheses have been proposed including altered neuronal regulation and connections, deranged vascular permeability, abnormal BBB, and impaired glial cell function. The tumor itself may be the seizure focus or the tumor may cause secondary perilesional tissue alterations such as growth, inflammation, edema, or necrosis triggering seizure activity.

Tumor histology influences the propensity to generate seizures. Low-grade PBT grow slowly and invade normal surrounding tissue and have a high frequency of epilepsy. For example, dysembryoblastic neuroepithelial tumors (DNTs) and Grade II astrocytomas exhibit this kind of activity [116,117]. Glioneuronal malformations such as gangliogliomas and FCDs lack normal neuronal organization as well as connections and can form epileptic foci [118]. High-grade gliomas such as glioblastoma multiforme (GBM) exhibit a lower frequency of seizures [116,117]. The seizures in these lesions may originate in areas of necrosis and hemosiderin deposition [119].

Tumor location also plays a critical role. Both intra-axial as well as extra-axial PBTs may be associated with seizures. Hence, a variety of mechanisms may contribute to seizure generation, as extra-axial tumors compress normal brain tissue, whereas intra-axial tumors infiltrate normal brain tissue. Intra-axial location may also influence epileptogenicity, with temporal or frontal lobe location being more associated with seizures than other lobes [113,120–122]. Infratentorial as well as sellar tumors seldom cause seizures [113].

Alterations in the peritumoral brain microenvironment may predispose to seizure generation. Some of the common changes seen in PBT are altered BBB function, enzymatic changes, and impaired intercellular connections. BBB alteration or disruption may lead to abnormal

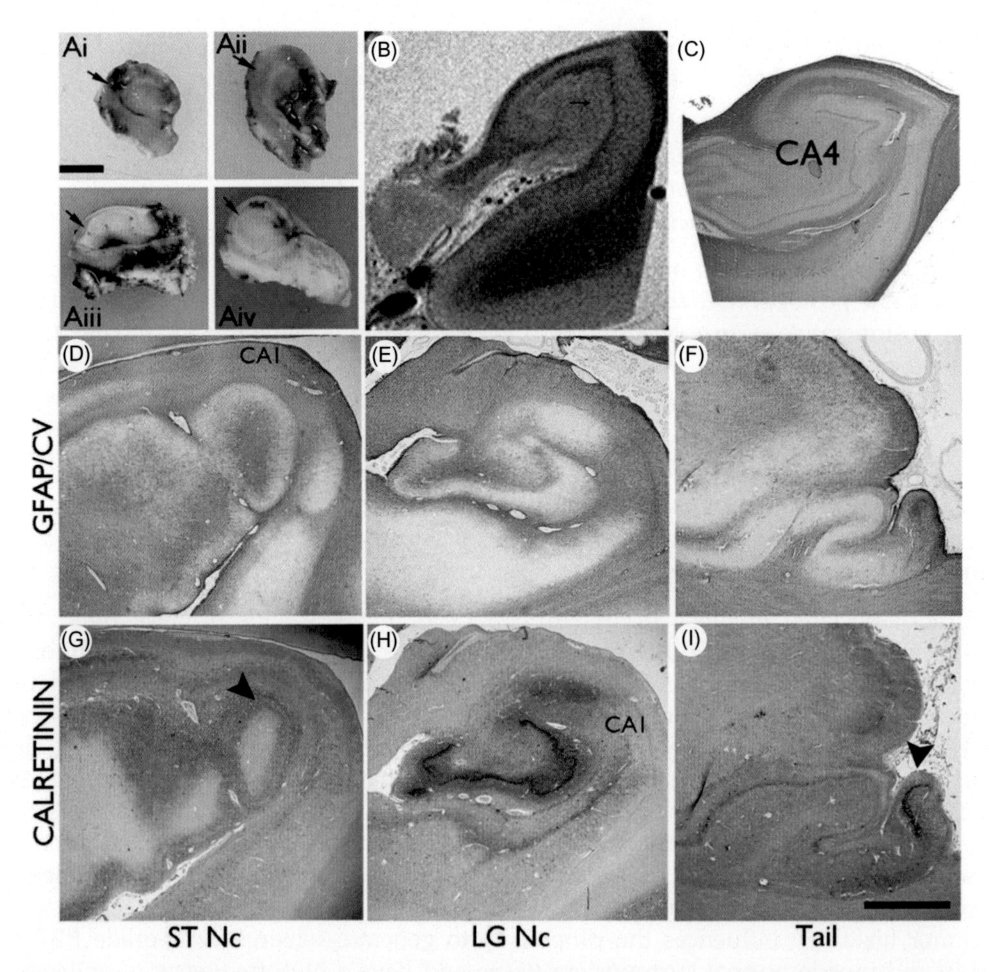

FIGURE 5.6 **Variability of hippocampal sclerosis (HS) in *postmortem* and surgical samples.** (A) Surgical hippocampectomy specimens, which on histological examination correlated to (Ai) endfolium gliosis with no evidence of sclerosis; (Aii) ILAE Type 3 HS (endfolium sclerosis); (Aiii) ILAE type 2 HS (CA1 predominant sclerosis); and (Aiv) ILAE type 1 HS (classical hippocampal sclerosis). In all of the images, the arrow indicates the pyramidal cell layer of CA1. (B) A 9.4 T MRI image of a *postmortem* hippocampus from a patient with longstanding epilepsy and ILAE type 2 (CA1 predominant pattern) of HS at this level (shown in (C) in a Luxol-fast blue/cresyl violet preparation), although in other levels the pattern was type 1 (classical). The MRI has the ability to identify subfields and white matter tracts and, indistinctly (arrowed), the dentate gyrus. Improved high-fields sequences in the future may be able to identify and define patterns of HS preoperatively. The MRI image was provided with courtesy of Dr Sofia Eriksson at the Department of Clinical and Experimental Epilepsy, UCL, Institute of Neurology. (D–I) Paired sections of hippocampus from one hemisphere labeled with (D–F) GFAP/counterstained with cresyl violet and (G–I) calretinin: at the level of the subthalamic nucleus (STNc; D, G), lateral geniculate nucleus (LGNc; E, H), and hippocampal tail (F, I). Classical pattern (ILAE type 1) HS is seen in the anterior levels with sprouting of calretinin-positive fibers visible in the dentate gyrus (arrow) at this low magnification; in the tail gliosis and neuronal loss is visible in the CA4 region of the hippocampal tail. Bar is equivalent to approximately 3 mm (D–I). *Source: Reproduced with permission from Thom M. Hippocampal sclerosis in epilepsy: a neuropathology review. Neuropathol Appl Neurobiol 2014;40:520–43 (Figure 5).*

leakiness and extravasation of plasma proteins or other substances which may lead to increased excitability and seizures. For example, it is known that focal disruption of the BBB can lead to the development of a seizure focus [123]. The peritumoral microenvironment is also associated with altered levels of enzymes such as lactate dehydrogenase, cAMP phosphodiesterase, enolase, and thymidine kinase which may cause metabolic imbalance [124]. Intercellular connections between adjacent glial cells occur via connexin transmembrane gap junction proteins [125]. Altered expression of connexins has been found in epilepsy-associated brain tumors [125] which may also predispose the perilesional epileptic cortex to hyperexcitability.

Other alterations in the peritumoral microenvironment may contribute to seizure generation. PBT have a relatively higher metabolic rate than normal brain tissue. This creates a relative hypoxia and interstitial acidosis. Marked vasogenic brain edema is commonly seen around brain tumors, and increased tissue water content may also modulate neural excitability [126]. In addition to increased water content, an increase in sodium, calcium, and serum proteins have been seen in quantitative studies of peritumoral tissue [127].

Glutamate, the major excitatory neurotransmitter in the brain, may contribute to perilesional seizure generation. A "glutamate hypothesis" of tumor-associated epilepsy has been advanced which suggests that tumors excite surrounding tissue by glutamate overstimulation. Several lines of evidence are relevant to this hypothesis. (1) Microdialysis studies of gliomas have revealed reduced glutamate in the tumor compared to peritumoral tissue, consistent with glutamate release from the tumor [128]. (2) Changes in ionotropic and metabotropic glutamate receptors (mGluRs) may contribute. Gliomas exhibit a high concentration of ionotropic glutamate receptors (AMPA, NMDA and kainate receptors). The glutamate receptor subunit GluR2 has been found to be underedited at the Q/R site in gliomas, which would increase AMPA receptor Ca^{2+} permeability and potentially result in increased glutamate release by glioma cells [129]. Alternatively, sustained NMDA receptor activation may represent a pathological mechanism for seizure generation [130]. Astrocytes in perilesional areas demonstrate increased expression of kainate receptors [131]. Activation of kainate receptors has been demonstrated to downregulate inhibitory stimuli which may predispose to formation of an epileptic focus in the perilesional area [132]. Metabotropic glutamate receptors (mGluR) lead to intracellular signaling through guanosine triphosphate–related (GTP) proteins and protein kinase cascades resulting in long-term neuromodulation [131]. Multiple subtypes of mGluRs have been found to be overexpressed in reactive astrocytes in the perilesional zone around tumors compared to normal cortex [131]. (3) Sontheimer's group found that glioma cells release larger than normal amounts of glutamate in vitro [133]. The release of glutamate from glioma cells was accompanied by a marked deficit in Na^+-dependent glutamate uptake, reduced expression of astrocytic glutamate transporters, and upregulation of cystine–glutamate exchange [134]. Hence, glioma cell release glutamate at the margins of the tumor may initiate seizures in peritumoral neurons.

Glutamate may also be involved in tumor invasion. One recent hypothesis of tumor invasion proposes that glioma cells use release of glutamate to induce excitotoxic destruction of neurons to create new micro-passages for migration of tumor cells [134–137]. Neoplastic cells have inherently higher glutamate concentrations that serve as an autocrine trigger to increase rhythmic oscillations of calcium usually necessary for cellular migration [138]. These intrinsically higher glutamate concentrations may also lead to local hyperexcitability along the paths of glioma cell migration.

Alterations in GABA, the main inhibitory neurotransmitter in the brain, may also contribute to tumor-associated seizures. It would be simplistic to assume that increased GABA activity would always suppress epileptic activity. GABA does not always hyperpolarize neurons; based on the GABA reversal potential (itself determined by expression of chloride transporters), GABA can depolarize neurons [139–141]. Studies of PBTs demonstrate increased GABA immunoreactivity [142,143], but further studies will be necessary to determine the exact role of the GABA system in contributing to the hyperexcitability of perilesional areas. Recent data from the Sontheimer laboratory in a validated mouse glioma model demonstrated reduction in peritumoral parvalbumin-positive GABAergic inhibitory interneurons (which would *disinhibit* the peritumoral neuronal networks) [144].

A final distinct potential mechanism underlying tumor-associated epilepsy is altered K^+ homeostasis. In support of this hypothesis, both reduced K_{ir} currents [145] and mislocalization of $K_{ir}4.1$ channels [146] have been found in malignant astrocytes.

Long-Term Epilepsy-Associated Tumors

Any brain tumor can generate seizures, but the term "long-term epilepsy-associated tumors" (LEAT) has been applied to various lesions found in patients with long histories (>2 years) of medically-intractable epilepsy [147–149]. LEATs differ from most other PBTs in: young age of onset, slow growth, presence of some neuronal (in addition to glial) differentiation within the tumor, and neocortical location (often temporal lobe) [148]. The most common LEATs are gangliogliomas and dysembryoplastic neuroepithelial tumors (DNTs). Gangliogliomas represent the most frequent tumor in patients with focal epilepsy [150–153]. These lesions contain both a highly-differentiated glial component and also a dysplastic neuronal component (Fig. 5.7). DNTs, by contrast, are composed of floating neurons and oligodendroglia-like elements [154,155]. These lesions are on a continuum with types of FCDs associated with epilepsy [156], the main distinction being that FCDs have no proliferative component [153]. In the 2011 ILAE FCD classification [157], tumor-associated cytopathological changes are categorized as type IIIb (see below and Table 5.1).

The exact cellular and molecular reasons why glioneuronal tumors are associated with epilepsy are not completely clear. In addition to the pathophysiological mechanisms listed above, dysplasia-like neuronal/glial disorganization is a key factor. Biochemical alterations in phosphatidylinositol 3-kinase pathway components in epilepsy-associated glioneuronal lesions have also been found [118,149,151]. In addition, glioneuronal tumors overexpress neurotransmitter producing enzymes, neurotransmitter receptors, and neuropeptides, all of which may contribute to the formation of a hyperexcitable focus [158,159]. An extensive perilesional inflammatory reaction with accumulation of activated microglial cells may also contribute to focal seizures arising from these lesions [160–162]. Studies have shown upregulation of interleukin-1β, activation of the complement cascade and the Toll-like receptor pathway [163,164]. Activation of inflammatory cascades could cause or contribute to the other peritumoral changes responsible for tumor epileptogenicity [165] (see also *Chapter 12: Blood—Brain Barrier Disruption*).

One of the largest series of tumor-associated epilepsy cases is the Bonn series. Between 1988 and 1999, 229 patients were treated for PBT associated with epilepsy [147]. These patients had both pharmacoresistant epilepsy and a PBT. Inclusion criteria included

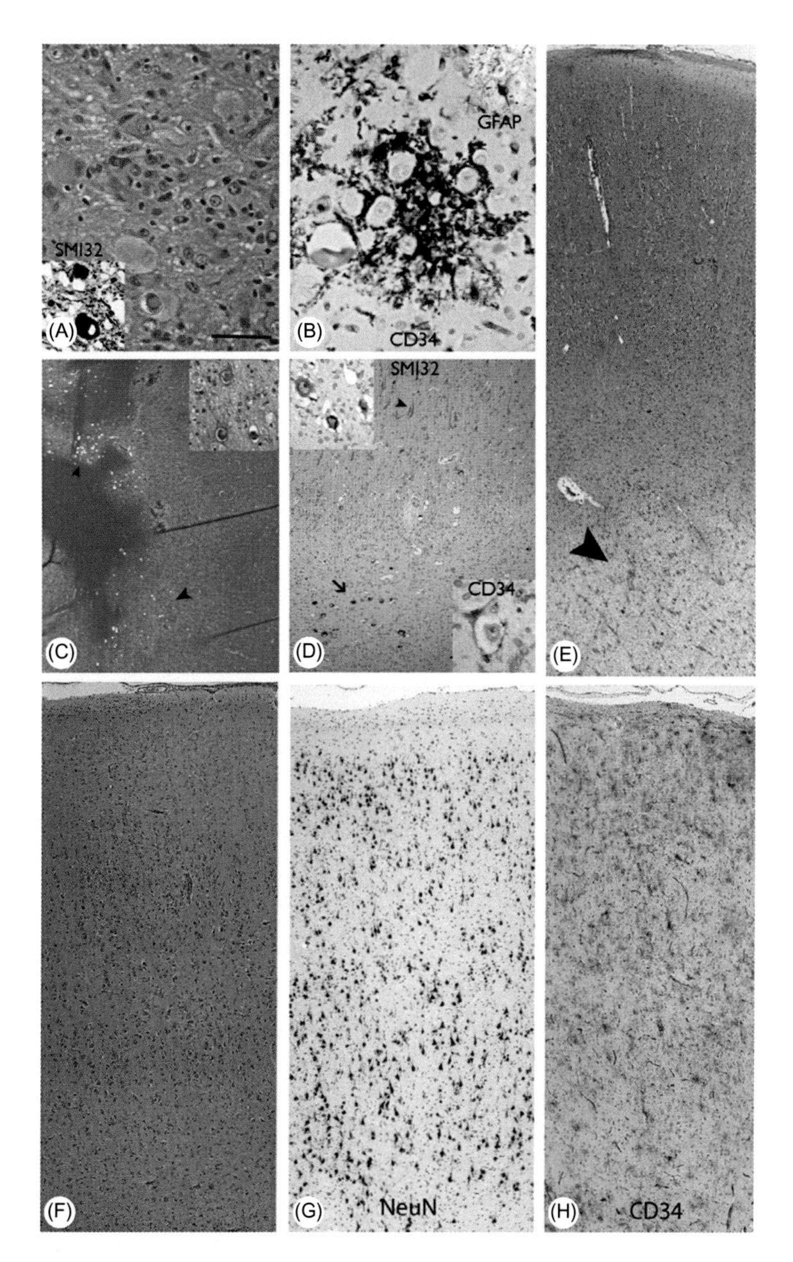

FIGURE 5.7 **Pathology of ganglioglioma.** (A) Aggregates of atypical ganglion cells of varying size typically with neurofilament positivity demonstrated (inset) are diagnostic criteria. (B) Clusters of CD34 positivity cells around dysplastic neurons are often identified with scattered intermingled astrocytic cells (inset). (C) and (D) LEAT with features of diffuse ganglion cell tumor in the temporal lobe with cortex on right side comprising diffuse and subcortical nodular aggregates (arrowhead in C) of neuronal islands with atypical neurons (inset in C) within islands but no glial component (Luxol-fast blue/cresyl violet preparation). In the same case (D), neurofilament highlights normal orientation of cortical neurons (arrowhead) as well as the single dysmorphic white matter neurons (arrow in main picture and top inset); CD34 was focally expressed around abnormal neurons (bottom inset). (E) A mixed glioneuronal tumor with rarefaction and pallor of the white matter beneath a diffusely infiltrated cortex (arrowhead) but without cavitation or cystic change. (F–H) Cortex adjacent to a ganglioglioma. (F). H&E shows a disrupted cortical architecture which may mimic a cortical dysplasia. (G) NeuN reveals the vestiges of an overrun cortex with residual lamination, including layers II and IV, recognizable. (H) CD34 staining reveals scattered multipolar cells, including in layer I. Scale bar (A), (B) = 40 mm; (C) = 100 mm; (D–H) = 115 mm. *H&E* = hematoxylin and eosin; *LEAT* = long-term epilepsy-associated tumor; *NeuN* = neuronal nuclear antigen. *Source: Reproduced with permission from Thom M. et al. Long-term epilepsy-associated tumors. Brain Pathol 2012;22:350–79 (Figure 1).*

TABLE 5.1 ILAE Classification of Focal Cortical Dysplasias (FCD)

FCD Type I (isolated)	Focal Cortical Dysplasia with abnormal radial cortical lamination (FCD Ia)	Focal Cortical Dysplasia with abnormal tangential cortical lamination (FCD Ib)		Focal Cortical Dysplasia with abnormal radial and tangential cortical lamination (FCD Ic)
FCD Type II (isolated)	Focal Cortical Dysplasia with dysmorphic neurons (FCD IIa)		Focal Cortical Dysplasia with dysmorphic neurons and balloon cells (FCD IIb)	
FCD Type III (associated with principal lesion)	Cortical lamination abnormalities in the temporal lobe associated with hippocampal sclerosis (FCD IIIa)	Cortical lamination abnormalities adjacent to a glial or glioneuronal tumor (FCD IIIb)	Cortical lamination abnormalities adjacent to vascular malformation (FCD IIIc)	Cortical lamination abnormalities adjacent to any other lesion acquired during early life, for example, trauma, ischemic injury, encephalitis (FCD IIId)

FCD Type III (not otherwise specified, NOS): if clinically/radiologically suspected principal lesion is not available for microscopic inspection.

Reproduced with permission from Blümcke I. et al. The clinicopathologic spectrum of focal cortical dysplasias: a consensus classification proposed by an ad hoc Task Force of the ILAE Diagnostic Methods Commission. Epilepsia 2011;52:158–174 (Table 1).

intractable epilepsy for more than 2 years, supratentorial and hemispheric in location, presurgical evaluation, resection of more than 50% of the tumor, and the availability of postoperative MRI. Of 229, 207 were available for follow-up. Median follow-up was 8 years (range 2–14 years). Histopathology revealed 144 (70%) World Health Organization (WHO) grade I tumors (82 gangliogliomas, 33 pilocytic astrocytomas, 29 dysembryoplastic neuroepithelial tumors), 59 (29%) WHO grade II lesions (35 astrocytomas, 15 oligodendrogliomas, 5 pleomorphic xanthoastrocytomas, 4 gangliogliomas), and 4 (1%) WHO grade III tumors (3 astrocytomas, 1 ganglioglioma). Most tumors were in the temporal lobe (83%).

Patients in this study had pharmacoresistant seizures for more than 2 years with median onset of seizures at the age of 13 years and were operated at the median age of 28 years. Preoperative EEG revealed an additional seizure focus other than the tumor in 13% of the patients. 203 (97.5%) of the patients had complex partial seizures, 62% of which progressed to secondary generalization. One year after surgery, 169 (82%) were seizure free (Engel class I), 7 (3%) had rare seizures (Engel class II), 20 (10%) had worthwhile improvement (Engel class III) and 11 (5%) had no benefit (Engel class IV). In 67 (40%) of the seizure-free patients, AEDs could be discontinued. Epilepsy recurred in only 18 (11%) of the 169 seizure-free patients. 11 (29%) of the 38 patients originally classified into Engel class II–IV eventually improved to Engel class I at median follow-up of 3 years from surgery.

Overall, this large study demonstrates that the majority of tumor-associated pharmacoresistant epilepsy occurs with lower-grade tumors, and that there is a temporal lobe predominance. These authors also found that improvement of seizure outcome following tumor resection was dependent on histology, complete resection of the tumor, and early removal of the tumor. Age greater than 40 years, frontal tumors, gemistocytic differentiation, and incomplete resection were associated with poorer seizure outcome [147].

In a related study, the Bonn group presented evidence for a subtype of grade II astrocytomas in long-term epilepsy with a different prognosis and unique histological characteristics [166]. Two groups were compared: a first group of LEAT astrocytomas ($n = 19$) with a mean seizure duration of 12.5 years and a second group of ordinary astrocytomas ($n = 87$) with a mean seizure duration of 1.5 years. All patients were operated between 1988 and 1999 and followed for 2–13 years (median 7 years). All tumor cases were reviewed and partly reclassified as a result of the use of modern immunohistochemical techniques. Histological subtyping revealed a possible new isomorphic astrocytoma in seven patients. Cytological hallmarks consisted of low cellularity, lack of mitotic activity, and highly-differentiated astroglial elements infiltrating into adjacent brain parenchyma [167]. Compared to the classical variant of diffuse astrocytoma, immunohistochemical reactions revealed a cellular proliferation below 1%, absence of nuclear p53 accumulation, and a lack of glial MAP2 and CD34 expression. The isomorphic subtype had 50% fewer recurrences at 7.5 years and an estimated long-term survival of 80%. LEAT astrocytomas differed from ordinary non-LEAT astrocytomas in overall length of history, younger age at first seizure, and high percentage of 10-year survivors (80%). The authors postulated a new subtype of epilepsy-associated astrocytoma, provisionally called isomorphic LEA astrocytoma, with significantly better survival and lower recurrence rate.

In another large study of patients undergoing TLE surgery, some of whom with tumor pathologies, prognostic factors, and outcome were evaluated [60]. A series of 321 patients underwent surgery for TLE between 1989 and 1997. Mean follow-up duration was 38 months. Of the 321 patients, only 312 patients had a clear histopathological diagnosis. 116 patients (36.1%) had neoplastic pathology. Among the neoplasias, there were 56 gangliogliomas, 16 DNTs, and 44 other gliomas. Seizure outcome was better with gangliogliomas or DNT (94.4% good seizure control) compared to other gliomas (79.6% good seizure control) and least favorable in cases of cortical dysplasia (68.3% good seizure control).

Another study of epileptogenic parietal and occipital lobe lesions correlated MRI and histopathology to seizure control [168]. In this retrospective study, 42 patients who underwent resective epilepsy surgery for parietal and occipital lobe lesions between 1998 and 2003 were analyzed. Histopathological diagnoses included 5 gangliogliomas, 5 DNT, 16 FCD, 4 vascular malformations, 6 gyral scars, and 1 infection. Postoperative seizure-free outcome differed by histology: 62% seizure-free for glioneuronal tumors, 69% for FCD, 71% for vascular malformations, and 40% for gyral scars. Overall, this and other studies of parietal and occipital epilepsy indicate a seizure-free outcome greater than 60% following resection of epilepsy-associated glioneuronal tumors [169,170].

In summary, tumor-associated epilepsy is an important clinical problem. Clinical studies suggest that intractable seizures are generally associated with early-onset and lower-grade lesions that may have a more benign tumor histology but nevertheless cause chronic seizures (LEATs). This is true of low-grade tumors such as oligodendrogliomas and low-grade astrocytomas as well as glioneuronal tumors such as gangliogliomas and DNTs. Following tumor resection, seizure freedom is usually excellent (~2/3 of patients seizure free). Across all studies, the most significant factors associated with seizure freedom are completeness of tumor resection and duration of tumor-associated epilepsy. However, the pathogenesis of tumor-associated seizures is still unclear. Further understanding of the dynamic processes at the tumor-brain interface may lead to novel concepts and treatment targets for the control of tumor-associated epilepsy.

FOCAL CORTICAL DYSPLASIAS

FCDs are localized regions of malformed cortex, and a wide variety of histopathology from ectopic neurons to wholesale cortical dyslamination has been included under the category of FCD. More broadly, malformations of cortical development (MCD), which include FCDs as well as other more diffuse pathologies, are an important form of pathology in patients with intractable focal epilepsies [171,172]. While large or diffuse malformations (such as Sturge-Weber syndrome, TS, hemimegalencephaly, polymicrogyria, and nodular or band heterotopias) may be diagnosed by MRI, smaller FCDs may evade diagnosis by imaging and require pathologic verification. In a large German epilepsy database, FCD was noted in 13% of pathologic specimens [2].

The Neuropathology Task Force of the ILAE published a revised classification of FCDs in 2011 [157]. They proposed a three-tiered classification system (Table 5.1), which distinguishes isolated forms (FCD Type I and II) from those associated with another principal lesion, for example, HS (FCD Type IIIa), tumors (FCD Type IIIb), vascular malformations (FCD Type IIIc), or lesions acquired during early life (eg, traumatic injury, ischemic injury or encephalitis, FCD Type IIId) [157]. An example of FCD type IIb is shown in Fig. 5.8. This revised classification system will serve as a basis to stratify further studies of clinical

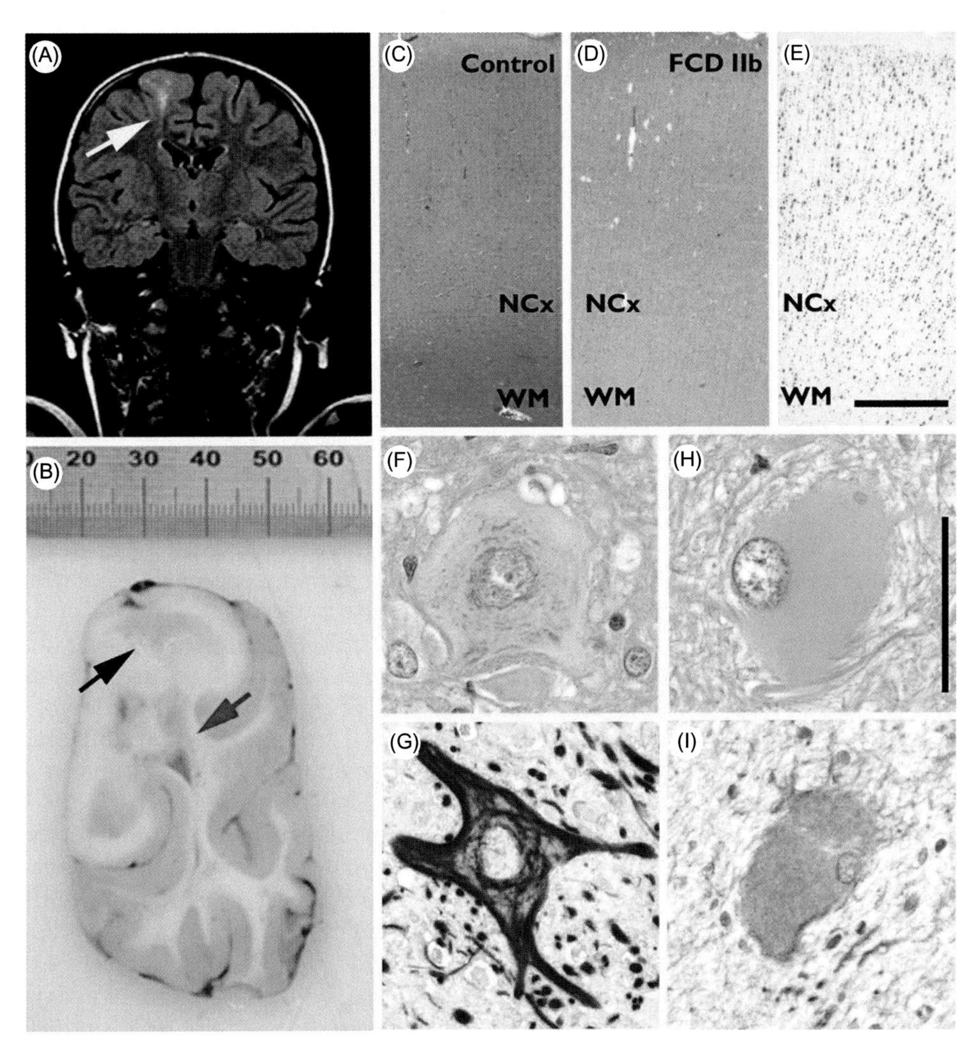

FIGURE 5.8 **Imaging and histopathological findings in focal cortical dysplasia (FCD) Type IIb.** (A) The "transmantle-sign" in T2 FLAIR imaging is characterized by a funnel-like hyperintensity (arrow) tapering from the gyrus to the ventricle. (B) Inspection of the surgical specimen reveals a distinct correlation between T2 FLAIR hyperintensity and lack of normal myelin content (black arrow points to grayish subcortical areas), which can be identified from the subcortical white matter to the ventricle (red arrow). (C) H&E staining combined with Luxol-fast blue (H&E-LFB) allows visualization of a sharp boundary between neocortex (NCX) and white matter (WM) in a control subject. (D) H&E-LFB. In this FCD Type IIb specimen, the myelin content is significantly reduced (see also macroscopic image in B). (E) NeuN immunohistochemistry, 4 μm paraffin embedded serial section to D. Severe cortical dyslamination is visible (with the exception of Layer 1). In addition, cortical thickness is increased and not distinguishable from WM border (same magnification as (C) and (D)). Scale bar= 1 mm. (F) In FCD Type IIb, enlarged dysmorphic neurons present with a huge nucleus and abnormal intracytoplasmic Nissl aggregates. (G) Antibodies to nonphosphorylated neurofilament proteins (SMI32) reveal aberrant NFP accumulation in a dysmorphic neuron. (H) Balloon cells are another hallmark of this FCD variant. Scale bar= 50 μm, applies also to (F), (G), and (I). (I) Balloon cells express the intermediate filament vimentin. (E), (G), and (I) 4 μm paraffin embedded sections, counterstained with haematoxylin. *Source: Reproduced with permission from Blümcke I. et al. The clinicopathologic spectrum of focal cortical dysplasias: a consensus classification proposed by an ad hoc Task Force of the ILAE Diagnostic Methods Commission. Epilepsia 2011;52:158–74 (Figure 4).*

features, imaging, and postoperative seizure control to specific types of FCDs [172]. A subsequent publication found good interobserver and intraobserver agreement in applying the new classification system [173]. Representative clinical outcomes following surgery for FCDs are given above (in the LEAT discussion, as they are often discussed together); in general, seizure-free outcomes following surgery for FCDs are inferior to those following surgery for HS or LEATs [60,147].

TUBEROUS SCLEROSIS

TS is a prototypical example of a "neurocutaneous" disorder or "phakomatosis," a group of pathologies that also includes Sturge-Weber disease (a cause of catastrophic unihemispheric pediatric epilepsy often requiring hemispherotomy), von Hippel-Lindau disease, ataxia-telangiectasia, and neurofibromatosis (type I, type II, and schwannomatosis). TS is a multisystem genetic disorder resulting from autosomal dominant mutations of either the TSC1 or TSC2 genes. The TSC1 gene encodes the protein hamartin and TSC2 encodes tuberin, which are thought to be regulators of cell signaling and growth [174]. Epilepsy occurs in 80–90% of cases of TS, frequently involves multiple seizure types and is often medically refractory [175]. Cortical tubers represent the pathologic substrate of TS, and microscopically consist of a specific type of dysplastic lesion with astrocytosis and abnormal giant cells [176]. While this suggests that astrocytes are involved in the pathologic lesion, in itself this is not evidence for a causative role of astrocytes in TS epileptogenesis. However, recent evidence using astrocyte-specific TSC1 conditional knockout mice has provided insight into a potential role of astrocytes in the etiology of TS. These mice, which have conditional inactivation of the TSC1 gene in GFAP-expressing cells ($Tsc1^{GFAP}$CKO mice), develop severe spontaneous seizures by 2 months of age and die prematurely [177]. Intriguingly, the time point of onset of spontaneous seizures in these mice is concordant with increased astroglial proliferation. Furthermore, two functions of astrocytes—glutamate and K^+ reuptake—are impaired in these mice. These mice display reduced expression of the astrocyte glutamate transporters GLT1 and GLAST [178]. In addition, recent evidence indicates that astrocytes from $Tsc1^{GFAP}$CKO mice exhibit reduced K_{ir} channel activity, and hippocampal slices from these mice demonstrated increased sensitivity to K^+-induced epileptiform activity [179]. Astrocyte proliferation also contributes to the development of subependymal giant cell astrocytomas (SEGAs), another pathologic hallmark of TS. Together, these studies demonstrate that changes in astrocyte properties may be a direct cause of TS epileptogenesis [180].

POSTTRAUMATIC EPILEPSY

PTE refers to a recurrent seizure disorder caused by traumatic brain injury. It is a common and important form of epilepsy [181,182], and develops in a variable proportion of traumatic brain injury survivors depending on the severity of the injury and the time after injury [183,184]. Anticonvulsant prophylaxis is ineffective at preventing the occurrence of late seizures [185–187].

Hippocrates (460–357 BC), in *On Injuries to the Head*, noted that a wound of the left temporal region would cause convulsions on the right side of the body and vice versa. John Hughlings Jackson (end 19th century), hypothesized that a focal injury creates a "discharging lesion." Pío Del Río-Hortega and Wilder Penfield held that a brain scar, "meningocerebral cicatrix," causes the discharging lesion [188] (see *Chapter 1: History of Astrocytes*):

> The brain may be injured by contusion, laceration, compression, and it is well known that these insults may result in epilepsy after a silent period of strange ripening...when there is widespread injury of a man's brain, epileptic discharge may develop in one area of that brain and not in another. Our attention should therefore be directed toward the discovery of this mysterious difference [189].

Military and civilian data have extensively documented the risk of PTE after TBI. Credner [190] studied 1990 German cases of World War I injuries with head trauma (examined 1914–28). PTE was reported in 38%. In those with intact dura, PTE incidence was 19%; with dural penetration PTE incidence was 49%. Ascroft [191] studied 317 cases of gunshot wounds to the head during World War I. With 7–20 year follow-up, PTE was reported in 34%. In those with intact dura, PTE incidence was 23% and with dural penetration 45% (remarkably similar to Credner's data). Russell and Whitty reported a 43% incidence of PTE in 820 World War II veterans [192]. Caveness et al. compared PTE incidence following World War I (32%) versus World War II (34%) versus the Korean campaign (31%) [193]. Subsequently Caveness reported on 356 Korean War Naval and Marine casualties, aged 17–24, evaluated between 1951 and 1954 (8–11 year follow-up) [194]. 56% had missile injuries, 44% had nonmissile head injuries. 109/356 (31%) developed PTE; in the subgroup with dural penetration, PTE incidence was 55% (61/110). A later analysis of Vietnam War cases included a 1967–70 roster of 1250 head-injured Vietnam veterans, with registry of type and extent of head injury [184]. 1030 cases from July 1976 to July 1978 were analyzed in detail, and PTE incidence was 344/1030 (33%). Latency to seizure onset was 40–50% by 6 months, 70% by 1 year, and 80% by 2 years. A later publication based on the Vietnam Head Injury Study (VHIS) with longer follow-up of 421 Vietnam War veterans who had penetrating brain wounds found a slightly higher incidence of PTE (53%) [195]. The remarkable length of time that can pass prior to the development of PTE after TBI was emphasized in a recent analysis of the VHIS, in which 11/87 patients (12.6%) with PTE reported "very late" onset of PTE (more than 14 years after injury) [196]. In summary, in spite of changes in medical care from 1914–70, a similar incidence of PTE was observed in all 20th century wars.

Civilian data are also available to delineate the risk of PTE. A population-based study from Olmsted County, Minnesota of 4541 children and adults with TBI from 1935–84 divided TBI into *mild* (loss of consciousness or amnesia <30 minutes), *moderate* (loss of consciousness 30 minutes to 24 hours or skull fracture) and *severe* (loss of consciousness or amnesia >24 hours, subdural hematoma, or contusion) [183]. In a multivariate analysis, significant risk factors for later seizures were brain contusion with subdural hematoma, skull fracture, loss of consciousness or amnesia for more than 1 day, and age of 65 years or older.

The above (military and civilian) studies led to identification of the main risk factors for PTE: severity of trauma, penetrating head injuries, intracranial hematoma, depressed skull fracture, prolonged unconsciousness (>24 hours) and presence of early posttraumatic seizures (PTS).

Efforts at pharmacological prophylaxis of PTE have so far been unsuccessful. In 1990, Temkin et al. performed a double-blinded randomized trial of phenytoin for the prevention of PTE [185]. Incidence of early PTS was reduced from 14.2% in placebo-treated patients to 3.6% in phenytoin-treated patients. Later studies with valproate [186] and magnesium sulfate [197] were also ineffective. More recently, studies are underway with the newer AED levetiracetam [198–203], but no clear evidence of antiepileptogenic efficacy has yet been established.

Interestingly, surgical treatment for PTE is probably underutilized, as in selected patients it can be highly effective (Fig. 5.9). The Johns Hopkins neurosurgeon A. Earl Walker published an early monograph entitled *Posttraumatic epilepsy* in 1949 [204] in which he detailed gross pathological and surgical findings in cases of PTE. In a publication in 1948 he summarized part of his findings as follows:

> In the group of medical failures some 40 cases were subjected to surgical removal of an epileptogenic focus. The remaining 26 cases serve as a control group for the surgical treatment. Approximately one year after operation one-third of the cases treated by cortical excision have had no further attacks and another fifth have had only one attack or the aura of their attacks. In the control group only 3 patients of 20 have had one or no attacks, and there is doubt of the reliability of the report on one of the cases [205].

Indeed, other more recent studies have shown efficacy at surgical treatment of PTE in select cases. In a series of 25 patients with intractable complex partial seizures following head trauma, MRI and EEG monitoring led to seizure localization in 9 patients; all 9 underwent surgery and were seizure free at 1 year [206]. Another study of 17 patients with intractable epilepsy underwent resection of frontal lobe encephalomalacias; 12 became seizure free at 3 years [207]. Finally, a single-institution retrospective study of all PTE patients treated over a 17-year time span identified 21 patients who met inclusion and exclusion criteria [208]. In long-term follow-up, 6 patients (28%) were seizure free and another 6 (28%) had reduction in seizures. In 8 patients with the combination of encephalomalacia and invasive intracranial EEG evaluation for tailoring surgery, 5 (62.5%) were rendered seizure free. Thus, a significant number of patients with PTE can achieve good to excellent seizure control with epilepsy surgery, especially if guided by intracranial EEG analysis to define extent of resection [208].

The mysterious events that occur during Penfield's "silent period of strange ripening" are the subject of modern-day PTE translational research. In general, early PTS are thought to be a relatively nonspecific response to the physical insult; whereas late PTS (ie, PTE) may be related to structural and functional changes in neuronal–glial networks. Weight-drop and fluid-percussion injury (FPI) animal models of PTE have demonstrated characteristic structural and functional changes in the hippocampus, such as death of dentate hilar neurons and mossy fiber sprouting [209–211]. Recently, studies have also implicated altered astrocyte function in PTE models. Recordings from glial cells in hippocampal slices 2 days after FPI demonstrated reduction in transient outward and inward K^+ currents, and antidromic stimulation of CA3 led to abnormal extracellular K^+ accumulation in posttraumatic slices compared to controls [212]. This was accompanied by the appearance of electrical afterdischarges in CA3. Thus, this study suggests impaired K^+ homeostasis in posttraumatic hippocampal glia. Another study demonstrated reduction in expression of the astrocyte glutamate transporter GLT1 in a posttraumatic epilepsy model induced by intracortical ferrous chloride injection, suggesting impaired glutamate transport [213]. Restoration of GLT1 expression after lateral FPI by the β-lactam antibiotic ceftriaxone

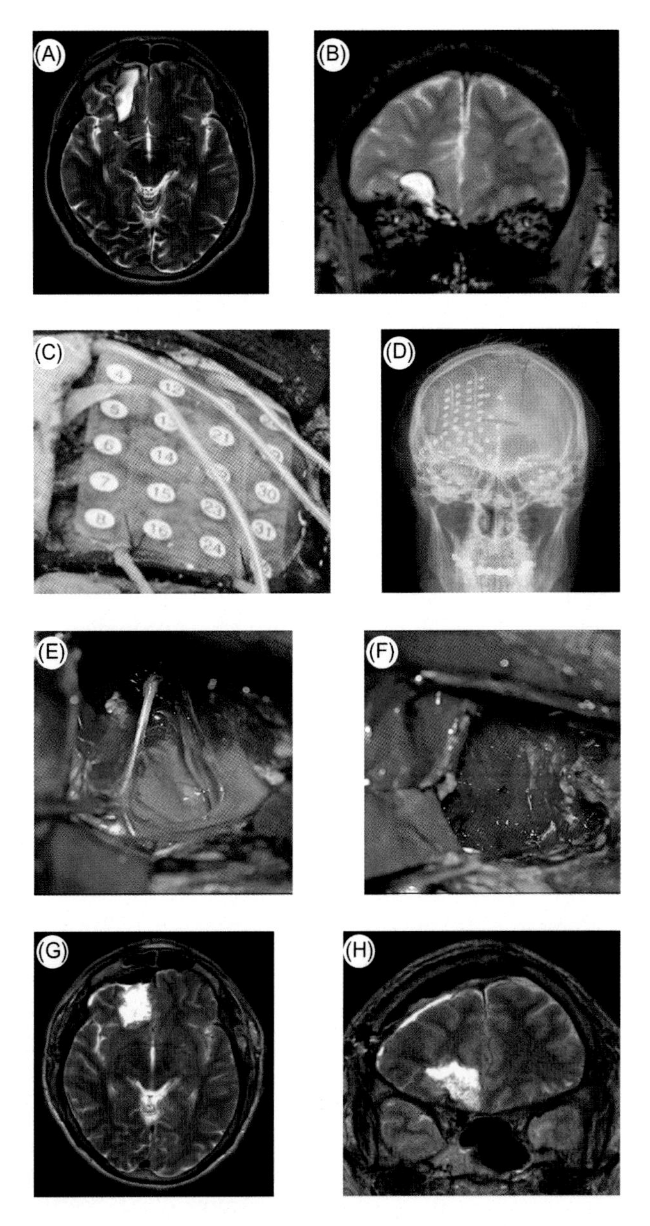

FIGURE 5.9 **Case of posttraumatic epilepsy successfully treated with surgery.** A 44-year-old man with medically-intractable focal epilepsy was found to have had a severe closed head injury at age 16 from a bicycle accident. At the time, he was in a coma for 3 days. He started to have recurrent stereotyped seizures at age 19. He had an aura consisting of an unpleasant "chemical" taste and smell, fear, and binaural auditory distortion followed by impairment of consciousness associated with behavioral arrest, unresponsiveness, and postictal confusion. His frequency of these complex partial seizures (CPSs) on presentation was 4–6 per week, with 1–2 generalized tonic-clonic (GTC) seizures per month. (A) Axial T2 MRI shows focal T2 abnormality in the right frontal lobe. (B) Coronal gradient-echo (GRE) MRI demonstrates hypointense hemosiderin rim around the posttraumatic cyst. (C) Placement of intracranial EEG electrodes. (D) Anteroposterior radiograph demonstrating electrode coverage. Intracranial video-EEG (Phase 2) monitoring was performed, and his typical seizures were found to arise around the epileptogenic cyst. (E) Intraoperative exposure of the epileptogenic cyst following electrode removal. Yellow-stained hemosiderin rim is evident. (F) View of right frontal cyst cavity following cyst and hemosiderin rim removal. (G) Postresection axial T2 MRI. (H) Postresection coronal T2 MRI. The patient was discharged home postresection day #2 with no neuropsychological deficit, and since the procedure (>2 years ago) has had two auras and no CPSs or GTC seizures. *All images courtesy of Devin Binder.*

decreased gliosis and reduced cumulative posttraumatic seizure duration in rats, providing proof-of-principle for the idea of restoration of glutamate homeostasis as an antiepileptogenic strategy against PTE [214]. Further studies of the role of glial cells in posttraumatic epilepsy appear warranted now that more reliable posttraumatic epilepsy animal models have been developed [215]; however, these models can be low-yield and require chronic in vivo video-EEG monitoring in large cohorts of animals [216–218]. In addition, newer models of blast TBI may be more relevant for particular clinical (such as military) environments, but reproducibility in blast TBI models are not yet at the level of CCI (controlled cortical impact) and FPI models [219]. Despite the multiple challenges of PTE research, investigators are gradually identifying disease-modifying effects of different treatments after TBI in experimental models [220].

The most important recent clinical study of PTE was a genetics and biomarker cohort study of 256 patients with moderate to severe TBI [221]. The authors' hypothesis was that TBI-induced inflammation contributes to PTE. They assessed whether genetic variation in the IL-1β gene, IL-1β levels in cerebrospinal fluid (CSF) and serum would predict PTE development after TBI. They found that higher CSF/serum IL-1β ratios were associated with increased risk for PTE (as assessed 1 week to 3 years postinjury). This is the first clinical report clearly linking IL-1β to PTE risk, and provides rationale for testing targeted IL-1β therapies as prophylaxis against PTE [221].

In summary, (1) PTE will develop in a significant proportion of patients who survive severe TBI; (2) prophylaxis of early PTS is effective in severe TBI patients; (3) so far, however, there is no antiepileptogenic therapy that can prevent late PTS/PTE; (4) surgery may be an underutilized therapy in certain PTE patients with focal seizure onset and with concordant MRI and EEG; and (5) new therapeutic approaches will arise from better understanding of relevant biomarkers and mechanisms of PTE in animal models.

POSTSTROKE EPILEPSY

Ischemic stroke is a known risk factor for development of epilepsy [222]. PSE develops in 3–30% of patients after stroke [223–226]. Stroke severity [225,226], cortical involvement [224,227], and hemorrhagic versus ischemic stroke [223] are independent predictors for PSE. In addition to the ischemic injury to the brain itself, histopathological changes in blood vessels, white matter, and astrocytes in the perilesional area may contribute to poststroke epileptogenesis. It has been shown in human pathologic specimens that astrogliosis and glial scar formation occur in the peri-infarct region [228]. Similarly, photothrombotic stroke induces astrogliosis in mouse cortex [229]. The dual role of astrocytes in repair after brain injury (such as stroke) versus potential deleterious consequences of reactive gliosis has been discussed [230,231]. Also, it is known that the transcriptome of reactive astrocytes varies by type of insult (eg, LPS vs ischemia) [232]. Based on dysfunction of the BBB initiated by stroke, other processes such as serum albumin extravasation can result; and Daniela Kaufer and colleagues have worked out a mechanism by which albumin induces excitatory synaptogenesis through astrocyte TGF-β/ALK5 signaling [233]. This mechanism provides an astrocytic basis for PSE, and suggests possible therapeutic approaches (TGF-β inhibition) to prevent PSE. It is interesting to note, however, that other investigators studying mice with induced deficiency of TGF-β signaling in astrocytes exhibited greater neuroinflammation following photothrombotic stroke than wild-type controls, worse motor outcomes, and late infarct expansion [234]. Thus, immunoregulatory functions of astrocytes must be balanced against epileptogenicity to optimize therapeutic targets. In addition, further development of animal models is required, as PSE models are in their infancy [222] and different types of stroke (eg, ischemic versus hemorrhagic versus subarachnoid hemorrhage) will doubtless lead to distinct patterns of astrocyte phenotypic changes [235].

SUMMARY: VALUE OF HUMAN TISSUE DATA AND RESEARCH

It is clear that the wealth of neuropathology research over the past few decades has led to a much better understanding of the histopathological findings in a variety of human epilepsies. However, while a great deal of information has emerged from large clinical studies, correlation of exact histopathological findings to imaging findings and electroclinical syndromes is still incomplete. For the future, human tissue research will remain critically important as it has a number of major advantages, including:

1. Allows exploration of the true complexity and variety of human histopathological changes. A good example is the variability in pathologic findings among different cases of HS (Fig. 5.6).
2. Allows comparison to animal models. Specific animal models may recapitulate some but not all of the human histopathology *or* may be good models for *subtypes* of human epilepsy. For example, the clear pattern of CA1 pyramidal cell loss seen in the mouse intrahippocampal kainic acid model [236,237] may be a good model for ILAE type 2 HS but clearly not for isolated endfolium sclerosis (ILAE type 3 HS) or non-HS pathology.
3. Allows comparison of epileptogenic versus normal regions in focal epilepsies.
4. Precise spatial localization. Human histopathology data can localize pathology to region and cortical layer.
5. Definition of cell types involved. Status of neurons, glia, inflammatory cells, BBB, extracellular matrix (ECM) components can all be assessed independently to form hypotheses about mechanisms of pathology.
6. As human tissue specimens reflect the chronically epileptic state, they allow study of the long-term effects of seizures in humans.
7. Highest validity. Compared with even the best modern neuroimaging, the "gold standard" remains pathologic diagnosis.
8. Comparison of epilepsy surgery specimens to postmortem specimens is likely to yield separate and useful information.

References

[1] Berg AT, Berkovic SF, Brodie MJ, Buchhalter J, Cross JH, van Emde Boas W, et al. Revised terminology and concepts for organization of seizures and epilepsies: report of the ILAE Commission on Classification and Terminology, 2005–2009. Epilepsia 2010;51(4):676–85.
[2] Blümcke I. Neuropathology of focal epilepsies: a critical review. Epilepsy Behav 2009;15(1):34–9.

[3] de Lanerolle NC, Lee T, Spencer DD. Histopathology of human epilepsy Noebels JL, Avoli M, Rogawski MA, Olsen RW, Delgado-Escueta AV, editors. Jasper's basic mechanisms of the epilepsies (4th ed.). Bethesda (MD): National Center for Biotechnology Information (US); 2012.
[4] Das A, Wallace GC, Holmes C, McDowell ML, Smith JA, Marshall JD, et al. Hippocampal tissue of patients with refractory temporal lobe epilepsy is associated with astrocyte activation, inflammation, and altered expression of channels and receptors. Neuroscience 2012;220:237–46.
[5] de Lanerolle NC, Lee T. New facets of the neuropathology and molecular profile of human temporal lobe epilepsy. Epilepsy Behav. 2005;7:190–203.
[6] de Lanerolle NC, Kim JH, Williamson A, Spencer SS, Zaveri HP, Eid T, et al. A retrospective analysis of hippocampal pathology in human temporal lobe epilepsy: evidence for distinctive patient subcategories. Epilepsia 2003;44(5):677–87.
[7] Clasadonte J, Haydon PG. Astrocytes and epilepsy. In: Noebles JL, Avoli, M, Rogawski, MA, Olsen, RW, Delgado-Escueta, AV, editors. Jasper's basic mechanisms of the epilepsies. 4th ed. Bethesda (MD): National Center for Biotechnology Information (US); 2012. p. 19.
[8] Blümcke I, Beck H, Lie AA, Wiestler OD. Molecular neuropathology of human mesial temporal lobe epilepsy. Epilepsy Res 1999;36:205–23.
[9] Margerison JH, Corsellis JA. Epilepsy and the temporal lobes: a clinical, electroencephalographic and neuropathological study of the brain in epilepsy, with particular reference to the temporal lobes. Brain 1966;89:499–530.
[10] Thom M. Hippocampal sclerosis in epilepsy: a neuropathology review. Neuropathol Appl Neurobiol 2014;40(5):520–43.
[11] Penfield W, Jasper H. Epilepsy and the functional anatomy of the human brain; London: J. and A. Churchill Ltd. 1954.
[12] Macewen W. The surgery of the brain and spinal cord. Br Med J 1888;2:302–9.
[13] Horsley V. Brain surgery. Br Med J 1886;2:670–5.
[14] Krause F. Die operative Behandlung der Epilepsie. Med Klin Berlin 1909;5:1418–22.
[15] Foerster O. Zur Pathogenese und chirurgischen Behandlung der Epilepsie. Zentralbl Chir 1925;52:531–49.
[16] Berger H. Über das Elektroenkephalogramm des Menschen. Archiv für Psychiatrie und Nervenkrankheiten 1929;87:527–70.
[17] Jasper HH. History of the early development of electroencephalography and clinical neurophysiology at the Montreal Neurological Institute: the first 25 years 1939-1964. Can J Neurol Sci 1991;18(4 Suppl):533–48.
[18] Jackson JH, Colman WS. Case of epilepsy with tasting movements and "dreamy state" with very small patch of softening in the left uncinate gyrus. Brain 1898;21:580–90.
[19] Penfield W, Flanigin HF. Surgical therapy of temporal lobe seizures. Arch Neurol Psychiatry 1950;64:491–500.
[20] Bailey P, Gibbs FA. The surgical treatment of psychomotor epilepsy. JAMA 1951;145:365–70.
[21] Falconer MA. Discussion on the surgery of temporal lobe epilepsy. Proc R Soc Med 1953;46:971–5.
[22] Meyer A, Falconer MA, Beck C. Pathological findings in temporal lobe epilepsy. J Neurol Neurosurg Psychiatry 1954;17:276–85.
[23] Falconer MA, Hill D, Myer A, et al. Treatment of temporal lobe epilepsy by temporal lobectomy—a survey of findings and results. Lancet 1955;1:827–35.
[24] Penfield W, Paine K. Results of surgical therapy for focal epileptic seizures. Can Med Assoc J 1955;73:515–30.
[25] Morris AA. Temporal lobectomy with removal of uncus, hippocampus and amygdala. Arch Neurol Psychiatry 1956;76:479–96.
[26] Bailey P. Surgical treatment of psychomotor epilepsy: five-year follow-up. Southern Med J 1961;54:299–301.
[27] Feindel W, Penfield W. Localization of discharge in temporal lobe automatism. Arch Neurol Psychiatry 1954;72:605–30.
[28] Polkey CE. Temporal lobe resections Oxbury JM, Polkey CE, Duchowny M, editors. Intractable focal epilepsy. London: W.B. Saunders; 2000. p. 667–95.
[29] Scoville WB, Milner B. Loss of recent memory after bilateral hippocampal lesions. J Neurol Neurosurg Psychiatry 1957;20:11–21.
[30] Silbergeld DL, Ojemann GA. The tailored temporal lobectomy. Neurosurg Clin N Am 1993;4(2):273–81.
[31] Ojemann GA. Intraoperative tailoring of temporal lobe resections Engel J, editor. Surgical treatment of the epilepsies (2nd ed.). Philadelphia: Lippincott-Raven; 1996. p. 481–8.
[32] Miles AN, Ojemann GA. Tailored resections for epilepsy Winn HR, editor. Youmans Neurological Surgery (5th ed.). Philadelphia: W.B. Saunders; 2004. p. 2615–28.

[33] McKhann II GM, Schoenfeld-McNeill J, Born DE, Haglund MM, Ojemann GA. Intraoperative hippocampal electrocorticography to predict the extent of hippocampal resection in temporal lobe epilepsy surgery. J Neurosurg 2000;93(1):44–52.
[34] Hardiman O, Burke T, Phillips J, Murphy S, O'Moore B, Staunton H, et al. Microdysgenesis in resected temporal neocortex: incidence and clinical significance in focal epilepsy. Neurology 1988;38(7):1041–7.
[35] Schramm J, Kral T, Grünwald T, Blümcke I. Surgical treatment for neocortical temporal lobe epilepsy: clinical and surgical aspects and seizure outcome. J Neurosurg 2001;94(1):33–42.
[36] Spencer DD, Spencer SS, Mattson RH, Williamson PD, Novelly RA. Access to the posterior medial temporal lobe structures in the surgical treatment of temporal lobe epilepsy. Neurosurgery 1984;15(5):667–71.
[37] Fried I. Anatomic temporal lobe resections for temporal lobe epilepsy. Neurosurg Clin N Am 1993;4(2): 233–42.
[38] Spencer DD, Doyle WK. Temporal lobe operations for epilepsy: radical hippocampectomy Schmidek HH, Sweet WH, editors. Operative neurosurgical techniques: indications, methods, and results (3rd ed.). Philadelphia: W.B. Saunders; 1995. p. 1305–16.
[39] Niemeyer P. The transventricular amygdalohippocampectomy in temporal lobe epilepsy Baldwin M, Bailey P, editors. Temporal lobe epilepsy. Springfield, IL: Charles C. Thomas; 1958. p. 461–82.
[40] Wieser HG, Yasargil MG. Die "selektive Amygdala-Hippokampektomie" als chirurgische Behandlungsmethode der mediobasal-limbischen Epilepsie. Neurochirurgia (Stuttg) 1982;25:39–50.
[41] Wieser HG, Yasargil MG. Selective amygdalohippocampectomy as a surgical treatment of mediobasal limbic epilepsy. Surg Neurol 1984;17:445–57.
[42] Feindel W, Rasmussen T. Temporal lobectomy with amygdalectomy and minimal hippocampal resection: review of 100 cases. Can J Neurol Sci 1991;18(4 Suppl):603–5.
[43] Goldring S, Edwards I, Harding GW, Bernardo KL. Results of anterior temporal lobectomy that spares the amygdala in patients with complex partial seizures. J Neurosurg 1992;77(2):185–93.
[44] Olivier A. Commentary: cortical resections Engel J, editor. Surgical treatment of the epilepsies. New York: Raven Press; 1987. p. 405–16.
[45] Davidson S, Falconer MA. Outcome of surgery in 40 children with temporal-lobe epilepsy. Lancet 1975;1(7919):1260–3.
[46] Crandall PH. Standard en bloc anterior temporal lobectomy Spencer SS, Spencer DD, editors. Surgery for epilepsy. Boston, MA: Blackwell Scientific Publications; 1991. p. 118–29.
[47] Nayel MH, Awad IA, Luders H. Extent of mesiobasal resection determines outcome after temporal lobectomy for intractable complex partial seizures. Neurosurgery 1991;29(1):55–60. discussion -1.
[48] Yasargil MG, Teddy PJ, Roth P. Selective amygdalo-hippocampectomy: operative anatomy and surgical technique. Adv Tech Stand Neurosurg 1985;12:93–123.
[49] Niemeyer P, Bello H. Amygdalo-hippocampectomy in temporal lobe epilepsy: microsurgical technique. Excerpta Med 1973;293:20. (abstract 48).
[50] Olivier A. Surgery of epilepsy: overall procedure Apuzzo MLJ, editor. Neurosurgical aspects of epilepsy (Neurosurgical Topics). Park Ridge, IL: American Association of Neurological Surgeons; 1990. p. 117–48.
[51] Olivier A. Surgical techniques in temporal lobe epilepsy. Clin Neurosurg 1997;44:211–41.
[52] Olivier A. Transcortical selective amygdalohippocampectomy in temporal lobe epilepsy. Can J Neurol Sci 2000;27(Suppl 1):S68–76.
[53] Hori T, Tabuchi S, Kurosaki M, Kondo S, Takenobu A, Watanabe T. Subtemporal amygdalohippocampectomy for treating medically intractable temporal lobe epilepsy. Neurosurgery 1993;33(1):50–6. discussion 6–7.
[54] Hori T, Yamane F, Ochiai T, Kondo S, Shimizu S, Ishii K, et al. Selective subtemporal amygdalohippocampectomy for refractory temporal lobe epilepsy: operative and neuropsychological outcomes. J Neurosurg 2007;106(1):134–41.
[55] Kral T, Clusmann H, Urbach J, Schramm J, Elger CE, Kurthen M, et al. Preoperative evaluation for epilepsy surgery (Bonn Algorithm). Zentralbl Neurochir 2002;63(3):106–10.
[56] Herrick IA, Gelb AW. Anesthesia for temporal lobe epilepsy surgery. Can J Neurol Sci 2000;27(Suppl 1):S64–7. discussion S92–6.
[57] Josephson CB, Dykeman J, Fiest KM, Liu X, Sadler RM, Jette N, et al. Systematic review and meta-analysis of standard vs selective temporal lobe epilepsy surgery. Neurology 2013;80(18):1669–76.
[58] Wiebe S. Epilepsy. Outcome patterns in epilepsy surgery—the long-term view. Nat Rev Neurol 2011; 8(3):123–4.

[59] Wiebe S, Blume WT, Girvin JP, Eliasziw M, Effectiveness and Efficiency of Surgery for Temporal Lobe Epilepsy Study Group. A randomized, controlled trial of surgery for temporal-lobe epilepsy. N Engl J Med 2001;345(5):311–8.

[60] Clusmann H, Schramm J, Kral T, Helmstaedter C, Ostertun B, Fimmers R, et al. Prognostic factors and outcome after different types of resection for temporal lobe epilepsy. J Neurosurg 2002;97(5):1131–41.

[61] Erdem A, Yasargil G, Roth P. Microsurgical anatomy of the hippocampal arteries. J Neurosurg 1993;79(2):256–65.

[62] Roper SN, Rhoton Jr. AL. Surgical anatomy of the temporal lobe. Neurosurg Clin N Am 1993;4(2):223–31.

[63] Duvernoy HM. The human hippocampus, 3rd ed. Berlin: Springer; 2005.

[64] Yasargil MG, Wieser HG, Valavanis A, von Ammon K, Roth P. Surgery and results of selective amygdala-hippocampectomy in one hundred patients with nonlesional limbic epilepsy. Neurosurg Clin N Am 1993;4(2):243–61.

[65] Wen HT, Rhoton Jr. AL, de Oliveira E, Cardoso AC, Tedeschi H, Baccanelli M, et al. Microsurgical anatomy of the temporal lobe: part 1: mesial temporal lobe anatomy and its vascular relationships as applied to amygdalohippocampectomy. Neurosurgery 1999;45(3):549–91.

[66] Krolak-Salmon P, Guenot M, Tiliket C, Isnard J, Sindou M, Mauguiere F, et al. Anatomy of optic nerve radiations as assessed by static perimetry and MRI after tailored temporal lobectomy. Br J Ophthalmol 2000;84(8):884–9.

[67] Sindou M, Guenot M. Surgical anatomy of the temporal lobe for epilepsy surgery. Adv Tech Stand Neurosurg 2003;28:315–43.

[68] Sincoff EH, Tan Y, Abdulrauf SI. White matter fiber dissection of the optic radiations of the temporal lobe and implications for surgical approaches to the temporal horn. J Neurosurg 2004;101(5):739–46.

[69] Peuskens D, van Loon J, Van Calenbergh F, van den Bergh R, Goffin J, Plets C. Anatomy of the anterior temporal lobe and the frontotemporal region demonstrated by fiber dissection. Neurosurgery 2004;55(5):1174–84.

[70] Yasargil MG, Ture U, Yasargil DCH. Impact of temporal lobe surgery. J Neurosurg 2004;101(5):725–38.

[71] Kier EL, Staib LH, Davis LM, Bronen RA. MR imaging of the temporal stem: anatomic dissection tractography of the uncinate fasciculus, inferior occipitofrontal fasciculus, and Meyer's loop of the optic radiation. AJNR Am J Neuroradiol 2004;25(5):677–91.

[72] Rubino PA, Rhoton Jr. AL, Tong X, Oliveira E. Three-dimensional relationships of the optic radiation. Neurosurgery 2005;57(4 Suppl):219–27. discussion -27.

[73] Bouchet C, Cazauvieilh CA. De l'épilepsie considéréé dans ses rapports avec l'aliénation mentale: recherches sur la nature et le siège de ces deux maladies. Arch Gen Med 1825;9:510–42.

[74] Temkin O. The falling sickness: a history of epilepsy from the Greeks to the beginnings of modern neurology, 2nd ed. Baltimore, MD: The Johns Hopkins University Press; 1971.

[75] Sommer W. Erkrankung des Ammonshorn als aetiologisches Moment der Epilepsie. Arch Psychiatr Nervenkr 1880;10:631–75.

[76] Bratz E. Ammonshornbefunde bei epileptikern. Arch Psychiatr Nervenkr 1899;32:820–35.

[77] Spielmeyer W. Die Pathogenese des Epileptisches Krampfes: Histopathologischer Teil. Ztschr Neurol Pscyhiat 1927;109:501–19.

[78] Blümcke I, Coras R, Miyata H, Ozkara C. Defining clinico-neuropathological subtypes of mesial temporal lobe epilepsy with hippocampal sclerosis. Brain Pathol 2012;22(3):402–11.

[79] Blümcke I, Pauli E, Clusmann H, Schramm J, Becker A, Elger C, et al. A new clinico-pathological classification system for mesial temporal sclerosis. Acta Neuropathol 2007;113(3):235–44.

[80] Sagar HJ, Oxbury JM. Hippocampal neuron loss in temporal lobe epilepsy: correlation with early childhood convulsions. Ann Neurol 1987;22(3):334–40.

[81] Bruton CJ. The neuropathology of temporal lobe epilepsy. New York: Oxford University Press; 1988.

[82] de Tisi J, Bell GS, Peacock JL, McEvoy AW, Harkness WF, Sander JW, et al. The long-term outcome of adult epilepsy surgery, patterns of seizure remission, and relapse: a cohort study. Lancet 2011;378(9800):1388–95.

[83] Babb TL, Lieb JP, Brown WJ, Pretorius J, Crandall PH. Distribution of pyramidal cell density and hyperexcitability in the epileptic human hippocampal formation. Epilepsia 1984;25(6):721–8.

[84] Meyer A, Beck E. The hippocampal formation in temporal lobe epilepsy. Proc R Soc Med 1955;48:457–62.

[85] Yilmazer-Hanke DM, Wolf HK, Schramm J, Elger CE, Wiestler OD, Blümcke I. Subregional pathology of the amygdala complex and entorhinal region in surgical specimens from patients with pharmacoresistant temporal lobe epilepsy. J Neuropathol Exp Neurol 2000;59(10):907–20.

[86] Babb TL, Brown WJ, Pretorius J, Davenport C, Lieb JP, Crandall PH. Temporal lobe volumetric cell densities in temporal lobe epilepsy. Epilepsia 1984;25(6):729–40.

[87] Dawodu S, Thom M. Quantitative neuropathology of the entorhinal cortex region in patients with hippocampal sclerosis and temporal lobe epilepsy. Epilepsia 2005;46(1):23–30.
[88] Du F, Whetsell Jr. WO, Abou-Khalil B, Blumenkopf B, Lothman EW, Schwarcz R. Preferential neuronal loss in layer III of the entorhinal cortex in patients with temporal lobe epilepsy. Epilepsy Res 1993;16(3):223–33.
[89] Sutula T, Cascino G, Cavazos J, Parada I, Ramirez L. Mossy fiber synaptic reorganization in the epileptic human temporal lobe. Ann Neurol 1989;26(3):321–30.
[90] Houser CR. Granule cell dispersion in the dentate gyrus of humans with temporal lobe epilepsy. Brain Res 1990;535(2):195–204.
[91] Briellmann RS, Kalnins RM, Berkovic SF, Jackson GD. Hippocampal pathology in refractory temporal lobe epilepsy: T2-weighted signal change reflects dentate gliosis. Neurology 2002;58(2):265–71.
[92] Krishnan B, Armstrong DL, Grossman RG, Zhu ZQ, Rutecki PA, Mizrahi EM. Glial cell nuclear hypertrophy in complex partial seizures. J Neuropathol Exp Neurol 1994;53(5):502–7.
[93] Mitchell LA, Jackson GD, Kalnins RM, Saling MM, Fitt GJ, Ashpole RD, et al. Anterior temporal abnormality in temporal lobe epilepsy: a quantitative MRI and histopathologic study. Neurology 1999;52(2):327–36.
[94] Van Paesschen W, Revesz T, Duncan JS, King MD, Connelly A. Quantitative neuropathology and quantitative magnetic resonance imaging of the hippocampus in temporal lobe epilepsy. Ann Neurol 1997;42(5):756–66.
[95] Binder DK, Steinhäuser C. Functional changes in astroglial cells in epilepsy. Glia 2006;54:358–68.
[96] Borges K, Gearing M, McDermott DL, Smith AB, Almonte AG, Wainter BH, et al. Neuronal and glial pathological changes during epileptogenesis in the mouse pilocarpine model. Exp Neurol 2003;182:21–34.
[97] Shaprio L, Wang L, Ribak CE. Rapid astrocyte and microglial activation following pilocarpine-induced seizures in rats. Epilepsia 2008;49(Suppl. 2):33–41.
[98] Novozhilova AP, Gaikova ON. Cellular gliosis of the human brain white matter and its significance in pathogenesis of focal epilepsy. Morfologiia 2011;119(2):20–4.
[99] Pekny M, Nilsson M. Astrocyte activation and reactive gliosis. Glia 2005;50:427–34.
[100] Wilhelmsson U, Li L, Pekna M, Berthold C, Blom S, Eliasson C, et al. Absence of glial fibrillary acidic protein and vimentin prevents hypertrophy of astrocytic processes and improves post-traumatic regeneration. J Neurosci 2004;24(21):5016–21.
[101] Bedner P, Dupper A, Huttmann K, Muller J, Herde MK, Dublin P, et al. Astrocyte uncoupling as a cause of human temporal lobe epilepsy. Brain 2015;138(Pt 5):1208–22.
[102] Lee TS, Mane S, Eid T, Zhao H, Lin A, Guan Z, et al. Gene expression in temporal lobe epilepsy is consistent with increased release of glutamate by astrocytes. Mol Med 2007;13(1–2):1–13.
[103] Schwarcz R. Early glial dysfunction in epilepsy. Epilepsia 2008;49(Suppl 2):1–2.
[104] Johnson AM, Sugo E, Barreto D, Hiew CC, Lawson JA, Connolly AM, et al. The severity of gliosis in hippocampal sclerosis correlates with pre-operative seizure burden and outcome after temporal lobectomy. Mol Neurobiol 2015, Oct 9 [Epub ahead of print].
[105] Rigau V, Morin M, Rousset MC, de Bock F, Lebrun A, Coubes P, et al. Angiogenesis is associated with blood-brain barrier permeability in temporal lobe epilepsy. Brain 2007;130(Pt 7):1942–56.
[106] van Vliet EA, da Costa Araujo S, Redeker S, van Schaik R, Aronica E, Gorter JA. Blood-brain barrier leakage may lead to progression of temporal lobe epilepsy. Brain 2007;130(Pt 2):521–34.
[107] Michalak Z, Lebrun A, Di Miceli M, Rousset MC, Crespel A, Coubes P, et al. IgG leakage may contribute to neuronal dysfunction in drug-refractory epilepsies with blood-brain barrier disruption. J Neuropathol Exp Neurol 2012;71(9):826–38.
[108] Kovacs R, Heinemann U, Steinhäuser C. Mechanisms underlying blood-brain barrier dysfunction in brain pathology and epileptogenesis: role of astroglia. Epilepsia 2012;53(Suppl 6):53–9.
[109] Seiffert E, Dreier JP, Ivens S, Bechmann I, Tomkins O, Heinemann U, et al. Lasting blood-brain barrier disruption induces epileptic focus in the rat somatosensory cortex. J Neurosci 2004;24(36):7829–36.
[110] Aronica E, Sisodiya SM, Gorter JA. Cerebral expression of drug transporters in epilepsy. Adv Drug Deliv Rev 2012;64(10):919–29.
[111] Feldmann M, Asselin MC, Liu J, Wang S, McMahon A, Anton-Rodriguez J, et al. P-glycoprotein expression and function in patients with temporal lobe epilepsy: a case-control study. Lancet Neurol 2013;12(8):777–85.
[112] Blümcke I, Thom M, Aronica E, Armstrong DD, Bartolomei F, Bernasconi A, et al. International consensus classification of hippocampal sclerosis in temporal lobe epilepsy: a task force report from the ILAE Commission on Diagnostic Methods. Epilepsia 2013;54(7):1315–29.

[113] Liigant A, Haldre S, Õun A, Linnamägi U, Saar A, Asser T, et al. Seizure disorders in patients with brain tumors. Eur Neurol 2001;45(1):46–51.
[114] Ettinger AB. Structural causes of epilepsy. Neurol Clin 1994;12:41–56.
[115] Rasmussen T. Surgery of epilepsy associated with brain tumors. Adv Neurol 1975;8:227–39.
[116] Herman ST. Epilepsy after brain insult: targeting epileptogenesis. Neurology 2002;59(9 Suppl 5):S21–6.
[117] Cascino GD. Epilepsy and brain tumors: implications for treatment. Epilepsia 1990;31(Suppl 3):S37–44.
[118] Schick V, Majores M, Koch A, Elger CE, Schramm J, Urbach H, et al. Alterations of phosphatidylinositol 3-kinase pathway components in epilepsy-associated glioneuronal lesions. Epilepsia 2007;48(Suppl 5):65–73.
[119] Riva M. Brain tumoral epilepsy: a review. Neurol Sci 2005;26(Suppl 1):S40–2.
[120] Lynam LM, Lyons MK, Drazkowski JF, Sirven JI, Noe KH, Zimmerman RS, et al. Frequency of seizures in patients with newly diagnosed brain tumors: a retrospective review. Clin Neurol Neurosurg 2007;109(7):634–8.
[121] van Breemen MS, Vecht CJ. Optimal seizure management in brain tumor patients. Curr Neurol Neurosci Rep 2005;5(3):207–13.
[122] van Breemen MS, Wilms EB, Vecht CJ. Epilepsy in patients with brain tumours: epidemiology, mechanisms, and management. Lancet Neurol 2007;6(5):421–30.
[123] Ivens S, Kaufer D, Flores LP, Bechmann I, Zumsteg D, Tomkins O, et al. TGF-beta receptor-mediated albumin uptake into astrocytes is involved in neocortical epileptogenesis. Brain 2007;130(Pt 2):535–47.
[124] Kim E, Lowenson JD, MacLaren DC, Clarke S, Young SG. Deficiency of a protein-repair enzyme results in the accumulation of altered proteins, retardation of growth, and fatal seizures in mice. Proc Natl Acad Sci USA 1997;94(12):6132–7.
[125] Aronica E, Gorter JA, Jansen GH, Leenstra S, Yankaya B, Troost D. Expression of connexin 43 and connexin 32 gap junction proteins in epilepsy-associated brain tumors and in the perilesional epileptic cortex. Acta Neuropathol 2001;101(5):449–59.
[126] Schwartzkroin PA, Baraban SC, Hochman DW. Osmolarity, ionic flux, and changes in brain excitability. Epilepsy Res 1998;32(1–2):275–85.
[127] Hossmann KA, Seo K, Szymas J, Wechsler W. Quantitative analysis of experimental peritumoral edema in cats. Adv Neurol 1990;52:449–58.
[128] Bianchi L, De Micheli E, Bricolo A, Ballini C, Fattori M, Venturi C, et al. Extracellular levels of amino acids and choline in human high grade gliomas: an intraoperative microdialysis study. Neurochem Res 2004;29(1):325–34.
[129] Maas S, Patt S, Schrey M, Rich A. Underediting of glutamate receptor GluR-B mRNA in malignant gliomas. Proc Natl Acad Sci USA 2001;98(25):14687–92.
[130] Mody I, Heinemann U. NMDA receptors of dentate gyrus granule cells participate in synaptic transmission following kindling. Nature 1987;326(6114):701–4.
[131] Aronica E, Yankaya B, Jansen GH, Leenstra S, van Veelen CW, Gorter JA, et al. Ionotropic and metabotropic glutamate receptor protein expression in glioneuronal tumours from patients with intractable epilepsy. Neuropathol Appl Neurobiol 2001;27(3):223–37.
[132] Rodríguez-Moreno A, Herreras O, Lerma J. Kainate receptors presynaptically downregulate GABAergic inhibition in the rat hippocampus. Neuron 1997;19(4):893–901.
[133] Ye ZC, Sontheimer H. Glioma cells release excitotoxic concentrations of glutamate. Cancer Res 1999;59(17):4383–91.
[134] Ye ZC, Rothstein JD, Sontheimer H. Compromised glutamate transport in human glioma cells: reduction-mislocalization of sodium-dependent glutamate transporters and enhanced activity of cystine-glutamate exchange. J Neurosci 1999;19(24):10767–77.
[135] Sontheimer H. An unexpected role for ion channels in brain tumor metastasis. Exp Biol Med (Maywood) 2008;233(7):779–91.
[136] Sontheimer H. A role for glutamate in growth and invasion of primary brain tumors. J Neurochem 2008;105(2):287–95.
[137] de Groot J, Sontheimer H. Glutamate and the biology of gliomas. Glia 2011;59(8):1181–9.
[138] Lyons SA, Chung WJ, Weaver AK, Ogunrinu T, Sontheimer H. Autocrine glutamate signaling promotes glioma cell invasion. Cancer Res 2007;67(19):9463–71.
[139] Kohling R, Vreugdenhil M, Bracci E, Jefferys JG. Ictal epileptiform activity is facilitated by hippocampal GABAA receptor-mediated oscillations. J Neurosci 2000;20(18):6820–9.
[140] Kohling R. Neuroscience. GABA becomes exciting. Science 2002;298(5597):1350–1.

[141] Palma E, Amici M, Sobrero F, Spinelli G, Di Angelantonio S, Ragozzino D, et al. Anomalous levels of Cl^- transporters in the hippocampal subiculum from temporal lobe epilepsy patients make GABA excitatory. Proc Natl Acad Sci USA 2006;103(22):8465–8.
[142] Bateman DE, Hardy JA, McDermott JR, Parker DS, Edwardson JA. Amino acid neurotransmitter levels in gliomas and their relationship to the incidence of epilepsy. Neurol Res 1988;10(2):112–4.
[143] Haglund MM, Berger MS, Kunkel DD, Franck JE, Ghatan S, Ojemann GA. Changes in γ-aminobutyric acid and somatostatin in epileptic cortex associated with low-grade gliomas. J Neurosurg 1992;77(2):209–16.
[144] Campbell SL, Robel S, Cuddapah VA, Robert S, Buckingham SC, Kahle KT, et al. GABAergic disinhibition and impaired KCC2 cotransporter activity underlie tumor-associated epilepsy. Glia 2015;63(1):23–36.
[145] Bordey A, Sontheimer H. Electrophysiological properties of human astrocytic tumor cells in situ: enigma of spiking glial cells. J Neurophysiol 1998;79(5):2782–93.
[146] Olsen ML, Sontheimer H. Mislocalization of K_{ir} channels in malignant glia. Glia 2004;;46(1):63–73.
[147] Luyken C, Blümcke I, Fimmers R, Urbach H, Elger CE, Wiestler OD, et al. The spectrum of long-term epilepsy-associated tumors: long-term seizure and tumor outcome and neurosurgical aspects. Epilepsia 2003;44(6):822–30.
[148] Thom M, Blümcke I, Aronica E. Long-term epilepsy-associated tumors. Brain Pathol 2012;22(3):350–79.
[149] Becker AJ, Blümcke I, Urbach H, Hans V, Majores M. Molecular neuropathology of epilepsy-associated glioneuronal malformations. J Neuropathol Exp Neurol 2006;65(2):99–108.
[150] Blümcke I, Giencke K, Wardelmann E, Beyenburg S, Kral T, Sarioglu N, et al. The CD34 epitope is expressed in neoplastic and malformative lesions associated with chronic, focal epilepsies. Acta Neuropathol 1999;97(5):481–90.
[151] Becker AJ, Lobach M, Klein H, Normann S, Nothen MM, von Deimling A, et al. Mutational analysis of TSC1 and TSC2 genes in gangliogliomas. Neuropathol Appl Neurobiol 2001;27(2):105–14.
[152] Luyken C, Blümcke I, Fimmers R, Urbach H, Wiestler OD, Schramm J. Supratentorial gangliogliomas: histopathologic grading and tumor recurrence in 184 patients with a median follow-up of 8 years. Cancer 2004;101(1):146–55.
[153] Palmini A, Najm I, Avanzini G, Babb T, Guerrini R, Foldvary-Schaefer N, et al. Terminology and classification of the cortical dysplasias. Neurology 2004;62(6 Suppl 3):S2–S8.
[154] Daumas-Duport C, Scheithauer BW, Chodkiewicz JP, Laws Jr. ER, Vedrenne C. Dysembryoplastic neuroepithelial tumor: a surgically curable tumor of young patients with intractable partial seizures. Report of thirty-nine cases. Neurosurgery 1988;23(5):545–56.
[155] Wolf HK, Buslei R, Blumcke I, Wiestler OD, Pietsch T. Neural antigens in oligodendrogliomas and dysembryoplastic neuroepithelial tumors. Acta Neuropathol 1997;94(5):436–43.
[156] Taylor DC, Falconer MA, Bruton CJ, Corsellis JA. Focal dysplasia of the cerebral cortex in epilepsy. J Neurol Neurosurg Psychiatry 1971;34(4):369–87.
[157] Blümcke I, Thom M, Aronica E, Armstrong DD, Vinters HV, Palmini A, et al. The clinicopathologic spectrum of focal cortical dysplasias: a consensus classification proposed by an ad hoc Task Force of the ILAE Diagnostic Methods Commission. Epilepsia 2011;52(1):158–74.
[158] Wolf HK, Birkholz T, Wellmer J, Blümcke I, Pietsch T, Wiestler OD. Neurochemical profile of glioneuronal lesions from patients with pharmacoresistant focal epilepsies. J Neuropathol Exp Neurol 1995;54(5): 689–97.
[159] Wolf HK, Roos D, Blümcke I, Pietsch T, Wiestler OD. Perilesional neurochemical changes in focal epilepsies. Acta Neuropathol 1996;91(4):376–84.
[160] Aronica E, Gorter JA, Redeker S, Ramkema M, Spliet WG, van Rijen PC, et al. Distribution, characterization and clinical significance of microglia in glioneuronal tumours from patients with chronic intractable epilepsy. Neuropathol Appl Neurobiol 2005;31(3):280–91.
[161] Aronica E, Crino PB. Inflammation in epilepsy: clinical observations. Epilepsia 2011;52(Suppl 3):26–32.
[162] Vezzani A, Aronica E, Mazarati A, Pittman QJ. Epilepsy and brain inflammation. Exp Neurol 2013;244:11–21.
[163] Aronica E, Boer K, Becker A, Redeker S, Spliet WG, van Rijen PC, et al. Gene expression profile analysis of epilepsy-associated gangliogliomas. Neuroscience 2008;151(1):272–92.
[164] Ravizza T, Boer K, Redeker S, Spliet WG, van Rijen PC, Troost D, et al. The IL-1β system in epilepsy-associated malformations of cortical development. Neurobiol Dis 2006;24(1):128–43.
[165] Shamji MF, Fric-Shamji EC, Benoit BG. Brain tumors and epilepsy: pathophysiology of peritumoral changes. Neurosurg Rev 2009;32(3):275–84. discussion 84–6.

[166] Schramm J, Luyken C, Urbach H, Fimmers R, Blümcke I. Evidence for a clinically distinct new subtype of grade II astrocytomas in patients with long-term epilepsy. Neurosurgery 2004;55(2):340–7. discussion 7–8.
[167] Blümcke I, Luyken C, Urbach H, Schramm J, Wiestler OD. An isomorphic subtype of long-term epilepsy-associated astrocytomas associated with benign prognosis. Acta Neuropathol 2004;107(5):381–8.
[168] Urbach H, Binder D, von Lehe M, Podlogar M, Bien CG, Becker A, et al. Correlation of MRI and histopathology in epileptogenic parietal and occipital lobe lesions. Seizure 2007;16(7):608–14.
[169] Binder DK, Podlogar M, Clusmann H, Bien C, Urbach H, Schramm J, et al. Surgical treatment of parietal lobe epilepsy. J Neurosurg 2009;110(6):1170–8.
[170] Binder DK, Von Lehe M, Kral T, Bien CG, Urbach H, Schramm J, et al. Surgical treatment of occipital lobe epilepsy. J Neurosurg 2008;109(1):57–69.
[171] Barkovich AJ, Kuzniecky RI, Jackson GD, Guerrini R, Dobyns WB. A developmental and genetic classification for malformations of cortical development. Neurology 2005;65(12):1873–87.
[172] Guerrini R, Duchowny M, Jayakar P, Krsek P, Kahane P, Tassi L, et al. Diagnostic methods and treatment options for focal cortical dysplasia. Epilepsia 2015;56(11):1669–86.
[173] Coras R, de Boer OJ, Armstrong D, Becker A, Jacques TS, Miyata H, et al. Good interobserver and intraobserver agreement in the evaluation of the new ILAE classification of focal cortical dysplasias. Epilepsia 2012;53(8):1341–8.
[174] Au KS, Williams AT, Gambello MJ, Northrup H. Molecular genetic basis of tuberous sclerosis complex: from bench to bedside. J Child Neurol 2004;19(9):699–709.
[175] Thiele EA. Managing epilepsy in tuberous sclerosis complex. J Child Neurol 2004;19(9):680–6.
[176] Trombley IK, Mirra SS. Ultrastructure of tuberous sclerosis: cortical tuber and subependymal tumor. Ann Neurol 1981;9(2):174–81.
[177] Uhlmann EJ, Wong M, Baldwin RL, Bajenaru ML, Onda H, Kwiatkowski DJ, et al. Astrocyte-specific TSC1 conditional knockout mice exhibit abnormal neuronal organization and seizures. Ann Neurol 2002;52(3):285–96.
[178] Wong M, Ess KC, Uhlmann EJ, Jansen LA, Li W, Crino PB, et al. Impaired glial glutamate transport in a mouse tuberous sclerosis epilepsy model. Ann Neurol 2003;54(2):251–6.
[179] Jansen LA, Uhlmann EJ, Crino PB, Gutmann DH, Wong M. Epileptogenesis and reduced inward rectifier potassium current in tuberous sclerosis complex-1-deficient astrocytes. Epilepsia 2005;46(12):1871–80.
[180] Wong M, Crino PB. Tuberous sclerosis and epilepsy: role of astrocytes. Glia 2012;60(8):1244–50.
[181] Frey LC. Epidemiology of posttraumatic epilepsy: a critical review. Epilepsia 2003;44(Suppl 10):11–17.
[182] Garga N, Lowenstein DH. Posttraumatic epilepsy: a major problem in desperate need of major advances. Epilepsy Curr 2006;6(1):1–5.
[183] Annegers JF, Hauser WA, Coan SP, Rocca WA. A population-based study of seizures after traumatic brain injuries. N Engl J Med 1998;338(1):20–4.
[184] Caveness WF, Meirowsky AM, Rish BL, Mohr JP, Kistler JP, Dillon JD, et al. The nature of posttraumatic epilepsy. J Neurosurg 1979;50(5):545–53.
[185] Temkin NR, Dikmen SS, Wilensky AJ, Keihm J, Chabal S, Winn HR. A randomized, double-blind study of phenytoin for the prevention of post-traumatic seizures. N Engl J Med 1990;323(8):497–502.
[186] Temkin NR, Dikmen SS, Anderson GD, Wilensky AJ, Holmes MD, Cohen W, et al. Valproate therapy for prevention of posttraumatic seizures: a randomized trial. J Neurosurg 1999;91(4):593–600.
[187] D'Ambrosio R, Perucca E. Epilepsy after head injury. Curr Opin Neurol 2004;17(6):731–5.
[188] Del Río-Hortega P, Penfield WG. Cerebral cicatrix: the reaction of neuroglia and microglia to brain wounds. Bull Johns Hopkins Hosp 1927;41:278–303.
[189] Penfield W. Symposium on post-traumatic epilepsy. Epilepsia 1961;2:109–10.
[190] Credner L. Klinische und soziale auswirkungen von hirnschädigungen. Z Gesamte Neurologie et Psychiatrie 1930;126:721–57.
[191] Ascroft P. Traumatic epilepsy after gunshot wounds of the head. Br Med J 1941;1:739–44.
[192] Russell W, Whitty C. Studies in traumatic epilepsy. Part 1: factors influencing the incidence of epilepsy after brain wounds. J Neurol Neurosurg Psychiatry 1952;15:93–8.
[193] Caveness WF, Walker AE, Ascroft PB. Incidence of posttraumatic epilepsy in Korean veterans as compared with those from World War I and World War II. J Neurosurg 1962;19:122–9.
[194] Caveness WF. Onset and cessation of fits following craniocerebral trauma. J Neurosurg 1963;20:570–83.
[195] Salazar AM, Jabbari B, Vance SC, Grafman J, Amin D, Dillon JD. Epilepsy after penetrating head injury. I. Clinical correlates: a report of the Vietnam Head Injury Study. Neurology 1985;35(10):1406–14.

[196] Raymont V, Salazar AM, Lipsky R, Goldman D, Tasick G, Grafman J. Correlates of posttraumatic epilepsy 35 years following combat brain injury. Neurology 2010;75(3):224–9.
[197] Temkin NR, Anderson GD, Winn HR, Ellenbogen RG, Britz GW, Schuster J, et al. Magnesium sulfate for neuroprotection after traumatic brain injury: a randomised controlled trial. Lancet Neurol 2007;6(1):29–38.
[198] Chung MG, O'Brien NF. Prevalence of early posttraumatic seizures in children with moderate to severe traumatic brain injury despite levetiracetam prophylaxis. Pediatr Crit Care Med 2016;17(2):150–6.
[199] Cotton BA, Kao LS, Kozar R, Holcomb JB. Cost-utility analysis of levetiracetam and phenytoin for posttraumatic seizure prophylaxis. J Trauma 2011;71(2):375–9.
[200] Gabriel WM, Rowe AS. Long-term comparison of GOS-E scores in patients treated with phenytoin or levetiracetam for posttraumatic seizure prophylaxis after traumatic brain injury. Ann Pharmacother 2014;48(11):1440–4.
[201] Inaba K, Menaker J, Branco BC, Gooch J, Okoye OT, Herrold J, et al. A prospective multicenter comparison of levetiracetam versus phenytoin for early posttraumatic seizure prophylaxis. J Trauma Acute Care Surg 2013;74(3):766–71. discussion 71–3.
[202] Klein P, Herr D, Pearl PL, Natale J, Levine Z, Nogay C, et al. Results of phase 2 safety and feasibility study of treatment with levetiracetam for prevention of posttraumatic epilepsy. Arch Neurol 2012;69(10):1290–5.
[203] Pearl PL, McCarter R, McGavin CL, Yu Y, Sandoval F, Trzcinski S, et al. Results of phase II levetiracetam trial following acute head injury in children at risk for posttraumatic epilepsy. Epilepsia 2013;54(9):e135–7.
[204] Walker AE. Posttraumatic epilepsy. Springfield, IL: Charles C. Thomas; 1949.
[205] Walker AE, Quadfasel FA. Follow-up report on a series of posttraumatic epileptics. Am J Psychiatry 1948;104:781–2.
[206] Marks DA, Kim J, Spencer DD, Spencer SS. Seizure localization and pathology following head injury in patients with uncontrolled epilepsy. Neurology 1995;45(11):2051–7.
[207] Kazemi NJ, So EL, Mosewich RK, O'Brien TJ, Cascino GD, Trenerry MR, et al. Resection of frontal encephalomalacias for intractable epilepsy: outcome and prognostic factors. Epilepsia 1997;38(6):670–7.
[208] Hakimian S, Kershenovich A, Miller JW, Ojemann JG, Hebb AO, D'Ambrosio R, et al. Long-term outcome of extratemporal resection in posttraumatic epilepsy. Neurosurg Focus 2012;32(3):E10.
[209] Lowenstein DH, Thomas MJ, Smith DH, McIntosh TK. Selective vulnerability of dentate hilar neurons following traumatic brain injury: a potential mechanistic link between head trauma and disorders of the hippocampus. J Neurosci 1992;12(12):4846–53.
[210] Golarai G, Greenwood AC, Feeney DM, Connor JA. Physiological and structural evidence for hippocampal involvement in persistent seizure susceptibility after traumatic brain injury. J Neurosci 2001;21(21):8523–37.
[211] Santhakumar V, Ratzliff AD, Jeng J, Toth Z, Soltesz I. Long-term hyperexcitability in the hippocampus after experimental head trauma. Ann Neurol 2001;50(6):708–17.
[212] D'Ambrosio R, Maris DO, Grady MS, Winn HR, Janigro D. Impaired K^+ homeostasis and altered electrophysiological properties of post-traumatic hippocampal glia. J Neurosci 1999;19(18):8152–62.
[213] Samuelsson C, Kumlien E, Flink R, Lindholm D, Ronne-Engstrom E. Decreased cortical levels of astrocytic glutamate transport protein GLT1 in a rat model of posttraumatic epilepsy. Neurosci Lett 2000;289(3):185–8.
[214] Goodrich GS, Kabakov AY, Hameed MQ, Dhamne SC, Rosenberg PA, Rotenberg A. Ceftriaxone treatment after traumatic brain injury restores expression of the glutamate transporter, GLT1, reduces regional gliosis, and reduces post-traumatic seizures in the rat. J Neurotrauma 2013;30(16):1434–41.
[215] D'Ambrosio R, Fairbanks JP, Fender JS, Born DE, Doyle DL, Miller JW. Post-traumatic epilepsy following fluid percussion injury in the rat. Brain 2004;127(Pt 2):304–14.
[216] Kharatishvili I, Nissinen JP, McIntosh TK, Pitkänen A. A model of posttraumatic epilepsy induced by lateral fluid-percussion brain injury in rats. Neuroscience 2006;140(2):685–97.
[217] Pitkänen A, Immonen RJ, Grohn OH, Kharatishvili I. From traumatic brain injury to posttraumatic epilepsy: what animal models tell us about the process and treatment options. Epilepsia 2009;50(Suppl 2):21–9.
[218] Bolkvadze T, Pitkänen A. Development of post-traumatic epilepsy after controlled cortical impact and lateral fluid-percussion-induced brain injury in the mouse. J Neurotrauma 2012;29(5):789–812.
[219] Kovacs SK, Leonessa F, Ling GS. Blast TBI models, neuropathology, and implications for seizure risk. Front Neurol 2014;5:47.
[220] Pitkänen A, Immonen R. Epilepsy related to traumatic brain injury. Neurotherapeutics 2014;11(2):286–96.
[221] Diamond ML, Ritter AC, Failla MD, Boles JA, Conley YP, Kochanek PM, et al. IL-1β associations with posttraumatic epilepsy development: a genetics and biomarker cohort study. Epilepsia 2014;55(7):1109–19.
[222] Pitkänen A, Roivainen R, Lukasiuk K. Development of epilepsy after ischaemic stroke. Lancet Neurol 2015. Nov 13 [Epub ahead of print].

[223] Bladin CF, Alexandrov AV, Bellavance A, Bornstein N, Chambers B, Cote R, et al. Seizures after stroke: a prospective multicenter study. Arch Neurol 2000;57(11):1617–22.
[224] Menon B, Shorvon SD. Ischaemic stroke in adults and epilepsy. Epilepsy Res 2009;87(1):1–11.
[225] Jungehulsing GJ, Heuschmann PU, Holtkamp M, Schwab S, Kolominsky-Rabas PL. Incidence and predictors of post-stroke epilepsy. Acta Neurol Scand 2013;127(6):427–30.
[226] Graham NS, Crichton S, Koutroumanidis M, Wolfe CD, Rudd AG. Incidence and associations of poststroke epilepsy: the prospective South London Stroke Register. Stroke 2013;44(3):605–11.
[227] Zhang C, Wang X, Wang Y, Zhang JG, Hu W, Ge M, et al. Risk factors for post-stroke seizures: a systematic review and meta-analysis. Epilepsy Res 2014;108(10):1806–16.
[228] Huang L, Wu ZB, Zhuge Q, Zheng W, Shao B, Wang B, et al. Glial scar formation occurs in the human brain after ischemic stroke. Int J Med Sci 2014;11(4):344–8.
[229] Patience MJ, Zouikr I, Jones K, Clarkson AN, Isgaard J, Johnson SJ, et al. Photothrombotic stroke induces persistent ipsilateral and contralateral astrogliosis in key cognitive control nuclei. Neurochem Res 2015;40(2):362–71.
[230] Pekny M, Pekna M, Messing A, Steinhäuser C, Lee JM, Parpura V, et al. Astrocytes: a central element in neurological diseases. Acta Neuropathol 2016;131(3):323–45.
[231] Pekny M, Wilhelmsson U, Pekna M. The dual role of astrocyte activation and reactive gliosis. Neurosci Lett 2014;565:30–8.
[232] Zamanian JL, Xu L, Foo LC, Nouri N, Zhou L, Giffard RG, et al. Genomic analysis of reactive astrogliosis. J Neurosci 2012;32(18):6391–410.
[233] Weissberg I, Wood L, Kamintsky L, Vazquez O, Milikovsky DZ, Alexander A, et al. Albumin induces excitatory synaptogenesis through astrocytic TGF-β/ALK5 signaling in a model of acquired epilepsy following blood-brain barrier dysfunction. Neurobiol Dis 2015;78:115–25.
[234] Cekanaviciute E, Fathali N, Doyle KP, Williams AM, Han J, Buckwalter MS. Astrocytic transforming growth factor-β signaling reduces subacute neuroinflammation after stroke in mice. Glia 2014;62(8):1227–40.
[235] van Dijk BJ, Vergouwen MD, Kelfkens MM, Rinkel GJ, Hol EM. Glial cell response after aneurysmal subarachnoid hemorrhage: functional consequences and clinical implications. Biochim Biophys Acta. 2016;1862(3):492–505.
[236] Bouilleret V, Ridoux V, Depaulis A, Marescaux C, Nehlig A, Le Gal La Salle G. Recurrent seizures and hippocampal sclerosis following intrahippocampal kainate injection in adult mice: electroencephalography, histopathology and synaptic reorganization similar to mesial temporal lobe epilepsy. Neuroscience 1999;89(3):717–29.
[237] Lee DJ, Hsu MS, Seldin MM, Arellano JL, Binder DK. Decreased expression of the glial water channel aquaporin-4 in the intrahippocampal kainic acid model of epileptogenesis. Exp Neurol 2012;235(1):246–55.

CHAPTER

6

Astrocyte Calcium Signaling

OUTLINE

OVERVIEW

Astrocytes may release gliotransmitters in a calcium-dependent manner. Specifically, the mechanism is thought to mimic the vesicular release of neurotransmitters seen in neurons. Evidence for this includes: 1) Astrocytes respond to glutamate with propagating calcium waves; 2) Metabotropic glutamate receptors (mGluRs) are present on astrocytes and likely play a role in calcium signaling; 3) Stimulation of astrocytic calcium elevations leads to gliotransmitter release; and 4) Vesicular machinery necessary for exocytosis is expressed in astrocytes. Despite this growing body of evidence, calcium-induced exocytosis in astrocytes remains highly controversial. While the role of astrocyte calcium signaling is not fully understood, current evidence suggests that it is altered in the epileptic brain. More specifically, dysregulation of calcium channels and increased astrocytic calcium signaling have been observed in epileptic tissue. In this chapter, we summarize the evidence and controversy surrounding calcium-dependent gliotransmission and the role of astrocyte calcium signaling in epilepsy.

Astrocytes and Epilepsy
DOI: http://dx.doi.org/10.1016/B978-0-12-802401-0.00006-5

CALCIUM-INDUCED EXOCYTOSIS

Several lines of evidence have pointed toward calcium-dependent exocytosis as a common mechanism for calcium signaling in astrocytes. Initial studies found oscillatory calcium waves within astrocytes that led to gliotransmitter release. This mechanism involves metabotropic glutamate receptors (mGluRs) and inositol triphosphate (IP_3)-dependent release of internal calcium stores. Vesicular release proteins expressed by astrocytes suggested that gliotransmitters were released through exocytosis. Once released, gliotransmitters can then act on neighboring neurons and alter synaptic activity.

Elevated Intracellular Calcium in Astrocytes

Early evidence for astrocyte calcium signaling came from studies that demonstrated glutamate-triggered intracellular calcium wave oscillation and propagation in cultured astrocytes [1,2] and rat hippocampal slices [3–8]. Shortly after the initial discovery of glutamate-induced calcium waves in astrocytes [2], many other chemicals were found to produce the same effect including noradrenaline [9,10], histamine [11], acetylcholine [5,11,12], ATP [13–15], GABA [16], endocannabinoids [17], nitric oxide (NO) [18,19], and brain-derived neurotrophic factor (BDNF) [20]. Stimulation of mossy fibers originating in the dentate gyrus (DG) of rat organotypic hippocampal slice cultures led to delayed calcium waves in astrocytes within CA3 [21]. Long-lasting calcium oscillations were also seen after neuronal stimulation in both hippocampal and visual cortex transverse brain slices [22]. Propagating calcium elevations within astrocytes were induced by uncaging astrocytic IP_3 [3] or mechanical stimulation of a single cell [1,23] and these waves were spatially restricted to functionally independent microdomains of astrocytes [24]. Nearby neurons also responded with increases in cytosolic calcium levels [22,23] which was blocked with the gap junction inhibitor octanol [23]. This suggests that the astrocyte-neuron communication occurred through intracellular connections rather than synaptic connections.

Parri et al. [25] discovered that astrocytes in rat ventrobasal (VB) thalamus slices displayed intrinsic intracellular calcium oscillations independent of action potential-evoked transmitter release and neuronal firing. Di Castro et al. [26] studied endogenous calcium activity in adult mouse hippocampal slice preparations and observed both focal (confined to a specific subregion) and expanded (spread to 9 or more contiguous subregions) events in astrocyte processes. In contrast to results presented by Parri et al. [25] in the VB thalamus, Di Castro et al. [26] found that while the focal events were dependent on synaptic neurotransmitter release, expanded events depended on nearby, individual action potentials. Calcium activity was abolished with an IP_3 receptor blocker and overcome with subsequent bath application of thapsigargin (which empties intracellular calcium stores). Blocking astrocytic calcium activity decreased basal transmitter release at local synapses, suggesting that calcium elevations participate in local fine tuning of basal neurotransmission at excitatory synapses [26]. IP_3 receptor subtype 2 (IP_3R2) is the primary IP_3 receptor expressed by astrocytes. IP_3R2 knockout mice exhibited reduced calcium events [26], suggesting that activation of the IP_3-sensitive calcium channel may be a major contributor to intracellular calcium release in astrocytes.

Additional calcium sources besides IP_3 receptor-dependent release may contribute to cytosolic calcium levels within astrocytes. Diphenylboric acid 2-aminoethyl ester, a cell-permeable IP_3 receptor antagonist, reduced mechanically induced calcium accumulation and glutamate release only by ~32% in astrocyte cultures [27]. Caffeine exposure increased internal calcium levels and this effect was reduced by ryanodine, a blocker of calcium release from caffeine/ryanodine-sensitive stores. Incubation with cadmium, a broad-spectrum antagonist of calcium entry into the cell, reduced calcium accumulation and glutamate release by ~55% [27]. These data suggest that both internal and external calcium stores are involved in gliotransmission.

Astrocytes display heterogeneity [28], including different pattern of calcium excitation depending on brain region [29–31]. Intracellular calcium elevations are not stereotyped signals; instead, they exhibit varied spatiotemporal patterns, likely underlying different types of functions and generating distinct output signals [32]. For example, glutamate released from stimulation of Schaffer collateral axons of hippocampal neurons evoked intracellular calcium increases in astrocytes that then elicited *N*-methyl-D-aspartate (NMDA) receptor currents in pyramidal neurons; glutamate released by alveus stimulation, however, did not [5]. This illustrates that astrocyte responses may discriminate the activity of different synapses or axon pathways.

Increases in astrocytic internal calcium levels leads to NMDA receptor-mediated slow inward currents (SICs) in neighboring neurons, likely due to the astrocytic release of glutamate [25,33–41]. This phenomenon has been observed in various brain regions, including the nucleus accumbens [35], dorsal horn [34], and the medial nucleus of the trapezoid body [39]. Pharmacological evidence suggests that these astrocyte-induced SICs are mediated by NR2B-containing NMDA receptors [35–37]. Furthermore, activation of protease-activated receptor 1 (PAR-1), but not P2Y1 receptor (P2YR1), led to SICs in pyramidal neurons under conditions that isolated the NMDA receptor response [41].

While the majority of studies on intracellular calcium changes in astrocytes have been conducted in vitro or in situ, only a few studies have explored this topic in vivo. Spontaneous intracellular calcium fluctuations were observed in the rodent cortex [42–46], hippocampus [40], and locus coeruleus [47]. Astrocytic calcium kinetics differs from those seen in neurons. Astrocytic calcium signals were oscillatory, often not synchronized, and exhibited a slow onset and subsequent plateau in elevation. Neurons, on the other hand, exhibited fast, transient calcium signaling with a single-exponential decay [44].

Metabotropic Glutamate Receptors

Gq-linked protein coupled receptor (GPCR) activation of phospholipase C leads to the release of IP_3, activation of the IP_3 receptor, and subsequent release of calcium from internal stores. Calcium imaging techniques were used to demonstrate that astrocytes in vitro and in situ respond to neuronal activity with GPCR-mediated intracellular calcium increases [5,6,12,16,22,26,32,33,48,49]. Specifically, neuronal stimulation led to activation of astrocytic mGluRs [6,17,22,40,49–51] and involved mGluR subtype 5 (mGluR5) [4]. It is likely that the astrocytic mGluRs are activated by glutamate released from neurons [6]. Stimulation of the somatosensory cortex led to increased astrocytic calcium, a response that was mediated by synaptic glutamate release and activation of mGluRs [45,46]. In response, astrocytes are capable of increasing basal synaptic transmission [4]. Furthermore, NO may play a role in astrocytic calcium rises through the NO-G-kinase signaling pathway [52,53]. Conversely, the polyunsaturated fatty acids, arachidonic acid and docosahexaenoic acid, but not eicosapentaenoic acid, blocked intracellular calcium oscillations, inhibited store-operated calcium entry, and reduced the amplitudes of GPCR-calcium responses [54]. The metabolic gliotoxin fluorocitrate suppressed astrocytic intracellular calcium signaling and calcium-dependent glutamate release [50].

Astrocytic intracellular calcium increases can modulate inhibitory synaptic transmission. Photostimulation of astrocytes in the primary visual cortex increased the spontaneous firing of inhibitory neurons in an mGluR-dependent manner [55]. In hippocampal slices, uncaging astrocytic calcium increased the frequency of kainate receptor-depended spontaneous inhibitory postsynaptic potentials (sIPSCs) in nearby interneurons [56] but decreased the amplitude of evoked IPSCs and frequency of miniature IPSCs in an mGluR-dependent manner [57].

Calcium-Induced Gliotransmitter Release

Parpura et al. and others applied bradykinin to cultured astrocytes to stimulate internal calcium elevations, which resulted in astrocytic glutamate and aspartate release into the media [58–60]. Bradykinin also led to NMDA receptor-mediated neuronal calcium elevations in neuron-astrocyte cocultures, but not in neurons cultured alone [60]. Photolysis of caged astrocytic calcium increased astrocyte membrane capacitance [61] and potentiated transmitter release at synapses [49]. Innocenti et al. [62] used an enzyme-linked assay system involving glutamate dehydrogenase, an enzyme that metabolizes glutamate and consequently reduces NAD^+ to NADH, to measure extracellular glutamate levels in astrocyte cultures. Stimulation of intracellular calcium produced increased levels of NADH, corresponding to glutamate release. Similarly, mGluR agonist *trans*-aminocyclopentane-*trans*-1,3-dicarboxcylic acid (*trans*-ACPD)-induced release of glutamate evoked increases in extracellular NADH and, in most cells, an increase in whole cell capacitance [63]. Propagation of calcium waves occurred at a rate of 10–30 μm/s and glutamate release underlying NADH detection propagated at an average speed of 26 μm/s [64]. Depletion of internal calcium stores reduced the accumulation of NADH [62]. Other gliotransmitters may be released in a similar mGluR- and calcium-dependent manner including D-serine [65], ATP [66–68], and, in the rat olfactory bulb, GABA [69].

The release of gliotransmitters may play a role in plasticity. D-serine released from astrocytes acts as a coagonist for NMDA receptors; therefore, the degree of astrocytic coverage of neurons may govern the level of activation [70]. Clamping internal astrocytic calcium, depletion of D-serine, or disruption of D-serine release from astrocytes blocked long-term potentiation (LTP) in hippocampal slices from rats [71]. This could be reversed by endogenous application of D-serine or glycine and subsequent binding to the NMDA receptor coagonist sites [71]. An mGluR-mediated rise in intracellular calcium and subsequent ATP release multiplicatively scaled glutamate synapses [68]. This effect occurred quickly but was long lasting. Whisker stimulation induced delayed onset local field potentials (LFPs) in the barrel cortex whereas the combination of whisker stimulation with electrical stimulation of the nucleus basalis of Meynert (NBM) induced surges of intracellular calcium levels, elevated extracellular D-serine levels, and LTP in the cortex [72]. These effects were lost in the IP_3R2 knockout mice [72].

Exocytosis From Astrocytes

Astrocytes possess the machinery necessary for exocytosis of gliotransmitters, including the protein machinery necessary to form the soluble *N*-ethylmaleimide-sensitive factor attachment protein receptor (SNARE) complex [63,73–75]. ATPase, cellubrevin, ras-related protein rab3a, secretory carrier membrane protein, synapsin I, synaptic vesicle glycoprotein 2 A (SV2), synaptotagmin I, synaptobrevin II, synaptophysin, vesicular glutamate transporter 1 (VGLUT1), and VGLUT2 were all expressed in cultured astrocytes [58,63,76–82]. Astrocytic VGLUT3 expression was reported in slices from rats [83]. VGLUT and cellubrevin in astrocytic vesicles were found in close proximity to NMDA receptor-containing neuronal membranes [77]. Glutamate [76] and D-serine [65,84] colocalize with many of these exocytosis markers in cultured astrocytes. Pharmacological stimulation of intracellular calcium led to recruitment of synaptobrevin to the plasma membrane with concomitant disappearance of D-serine, suggesting calcium-dependent exocytosis [84].

The expression of neuronal syntaxin and SNAP-25 protein of the fusion complex have also been reported in cultured astrocytes [85] but their expression is controversial [58,80]. SNAP-23 is an analog of SNAP-25 and has been reported in both astrocyte cell cultures [79,86] and in the rat cerebellum [86]. Astrocytes analyzed immediately after isolation from 11-day-old animals were immunopositive for SNAP-23 and synaptobrevin II, almost entirely devoid of SNAP-25 and synaptophysin, and completely devoid of SV2 (Figs. 6.1 and 6.2) [82]. When cultured for various lengths of time, however, the expression of all proteins were abundant, suggesting that culturing astrocytes can induce the expression of certain proteins [82]. Jeftinija et al. [58] found that astrocyte cultures lacked SNAP-25 immunoreactivity, but pretreatment with BoTx-A (cleaves SNAP-25) and BoTx-C (cleaves syntaxins) decreased baseline and bradykinin-evoked glutamate release. After bradykinin-induced glutamate release in astrocyte cultures, α-latrotoxin stimulated calcium-independent glutamate release from astrocytes [59]. Since α-latrotoxin induces vesicle fusion and calcium-independent release of neurotransmitters from neurons, it suggests that astrocytes may release gliotransmitters in a similar manner. Furthermore, botulinum toxin B (cleaves SNAP-25) and tetanus toxin (cleaves synaptobrevin) decreased synaptobrevin II immunoreactivity and abolished glutamate release in cultured astrocytes [58,87]. The vacuolar type H^+-ATPase inhibitor bafilomycin also reduced glutamate release from astrocytes [27,78,87]. The inhibition of glutamate transporters, on the other hand, had no effect on calcium-dependent glutamate release in rat hippocampal cultured astrocytes [87].

Both ATP- and glutamate-storing vesicles may be present in astrocytes, but they have distinct properties. For example, ATP release was only partially sensitive to tetanus neurotoxin in cultured astrocytes whereas glutamate release was almost completely impaired [66,81]. In addition, lysosomes found in astrocytes contain ATP and execute calcium-dependent exocytosis of gliotransmitters [88–90]. ATP mediated by P2YR1 is thought to be the major determinant of astrocyte calcium wave propagation (Fig. 6.3) [91]. Blocking lysosome release of ATP prevented the spread of calcium waves [92].

Secretogranin II, a marker for dense-core granules, was detected in the Golgi complex [93] and in a population of dense-core vesicles [93,94] in cultured astrocytes. These vesicles also contained neuropeptide Y (NPY) and underwent botulinum toxin B-sensitive calcium-dependent exocytosis [94]. The activation of mGluRs may lead to calcium-dependent fusion

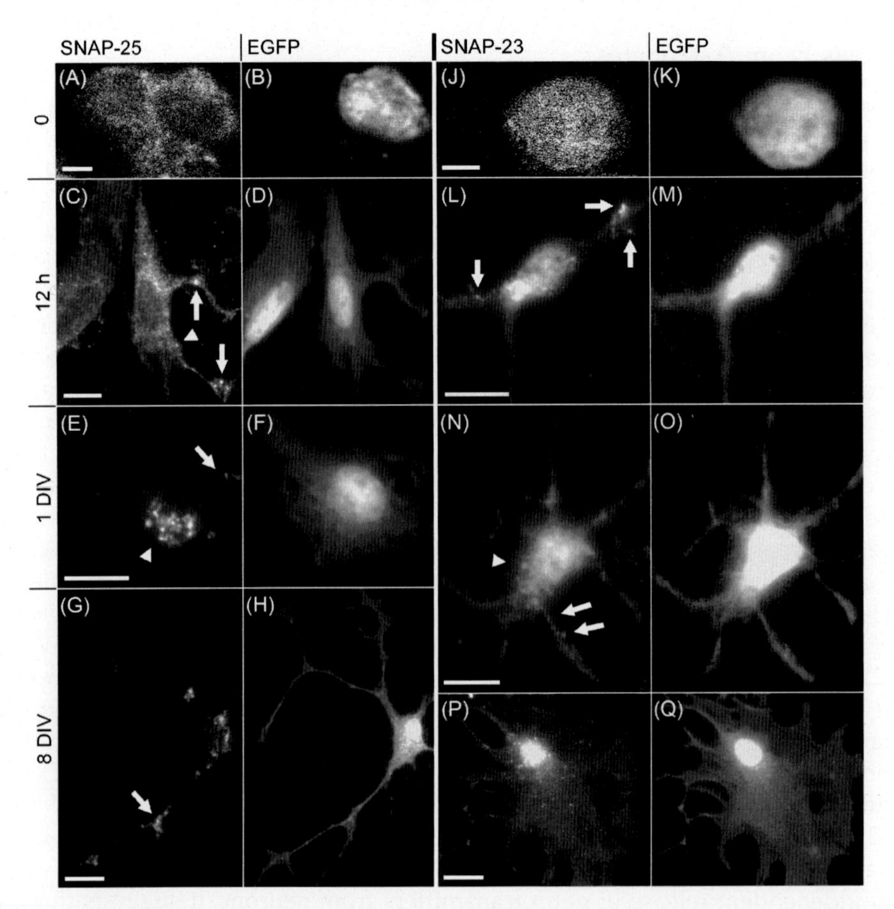

FIGURE 6.1 **Immunodetection of SNAP-25 and SNAP-23 in EGFP-expressing astrocytes isolated from 11-day-old animals.** Cells were fixed directly after isolation (A, B, J, K) or after culturing for 12h (C, D, L, M), 1 DIV (E, F, N, O), and 8 DIV (G, H, P, Q). A, C, E, G, distribution of SNAP-25; J, L, N, P, distribution of SNAP-23. B, D, F, H, K, M, O, Q represent the corresponding distribution of EGFP fluorescence, representing astrocytes. Arrows, punctuate immunofluorescence in cellular extensions; arrow heads, punctuate staining in cell bodies. Images were adjusted to improve brightness and contrast. Scale bars: 10 μm. *Source: Reproduced with permission from Wilhelm A, Volknandt W, Langer D, Nolte C, Kettenmann H, Zimmermann H. Localization of SNARE proteins and secretory organelle proteins in astrocytes in vitro and in situ. Neurosci Res 2004;48(3):249–57.*

of NPY-containing dense-core granules with cell membrane followed by peptide secretion in astrocytes [95].

Immunogold analysis of rat tissue demonstrated glutamate and D-serine accumulation in synaptic-like microvesicles in the perisynaptic processes of astrocytes [96]. In the rat cerebral cortex, D-serine colocalized with astrocytic processes [79]. The amino acid content of immunoisolated glial vesicles consisted of both glutamate and D-serine but lacked GABA [79]. Furthermore, vesicular D-serine and glutamate uptake is driven by an electrochemical potential generated by V-ATPase, coupled to chloride transport, and is stereoselective [79].

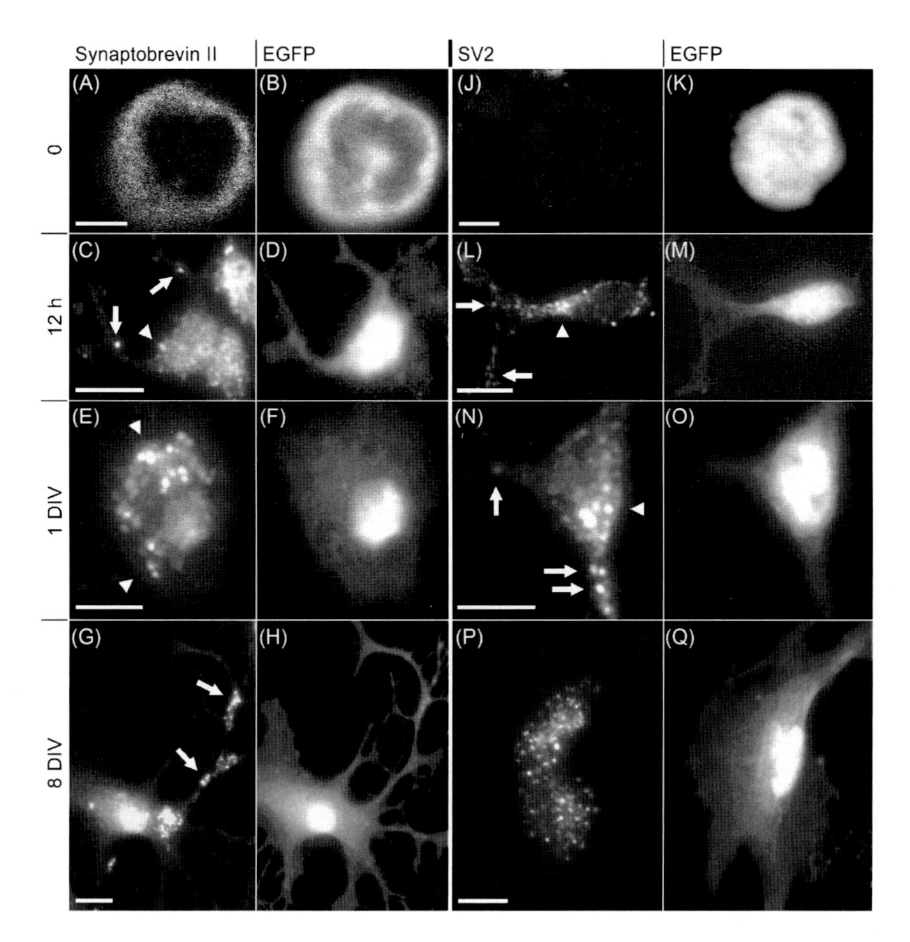

FIGURE 6.2 **Immunodetection of synaptobrevin II and SV2 in EGFP-expressing astrocytes isolated from 11-day-old animals.** Cells were fixed directly after isolation (A, B, J, K) or after culturing for 12 h (C, D, L, M), 1 DIV (E, F, N, O), and 8 DIV (G, H, P, Q). A, C, E, G, distribution of synaptobrevin II; J, L, N, P, distribution of SV2. Note the absence of indirect SV2 immunofluorescence in freshly isolated astrocytes. B, D, F, H, K, M, O, Q represent the corresponding distribution of EGFP fluorescence, representing astrocytes. Arrows, punctuate immunofluorescence in cellular extensions; arrow heads, punctuate staining in cell bodies. Images were adjusted to improve brightness and contrast. Scale bars: 10 μm. *Source: Reproduced with permission from Wilhelm A, Volknandt W, Langer D, Nolte C, Kettenmann H, Zimmermann H. Localization of SNARE proteins and secretory organelle proteins in astrocytes in vitro and in situ. Neurosci Res 2004;48(3):249–57.*

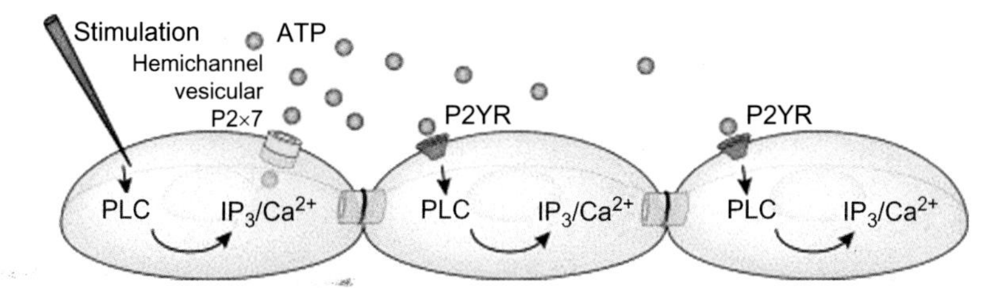

FIGURE 6.3 **Model of ATP-mediated astrocytic Ca^{2+} waves.** Stimulation of astrocytes results in release of ATP, which binds to P2YRs on neighboring astrocytes and mobilizes intracellular Ca^{2+}. Astrocytic Ca^{2+} waves are evoked by mechanical or electrical stimulation, agonist exposure, or photolysis of caged Ca^{2+}. *Source: Reproduced with permission from Tian GF, Takano T, Lin JH, Wang X, Bekar L, Nedergaard M. Imaging of cortical astrocytes using 2-photon laser scanning microscopy in the intact mouse brain. Adv Drug Deliv Rev 2006;58(7):773–87.*

Photostimulation of astrocytes led to transmitter release with a paired pulse index consistent with a presynaptic mechanism of action [49]. Pretreatment with tetanus neurotoxin abolished these flash-induced membrane capacitance increases [61]. Furthermore, reduction of synaptotagmin IV in astrocytes using RNA interference decreased calcium-dependent glutamate release in situ [64]. Electrical stimulation of hippocampal excitatory axons in rat hippocampal slices resulted in AMPA receptor-mediated SICs in oligodendrocyte precursor cells (OPCs) that were both quantal in nature and exhibited rapid kinetics [97]. These characteristics are consistent with exocytosis of glutamate-filled vesicles.

Time lapse confocal microscopy of astrocytes in vitro revealed punctate expression of clear vesicles that were heterogeneous in size [78]. While most of these vesicles remained immobile, subpopulations of highly mobile vesicles capable of fusing with the plasma membrane at astrocyte processes existed [78]. Furthermore, a separate study using cultured hippocampal astrocytes from rats estimated quantal release of only about 10% of total vesicle content during a physiologically stimulated release event via a "kiss-and-run" mechanism [98]. Full-collapse fusion has also been found to occur in astrocytes [99]. Rapid phase exocytosis was sustained almost entirely by vesicles undergoing kiss-and-run fusion whereas slow phase exocytosis was maintained by new vesicles undergoing full-collapse fusion. Intracellular calcium rises occurred near sites of exocytosis and were in strict temporal and spatial correlation with full-collapse fusion events [99] in a manner similar to neuronal exocytosis.

Modulation of Calcium-Dependent Exocytosis

Various chemical mediators have been found to modulate calcium-dependent exocytosis. Endocannabinoids activate CB1 receptors on astrocytes, which leads to phospholipase C-dependent release of intracellular calcium stores and release of glutamate [48,100,101]. This may lead to LTP if accompanied by postsynaptic NO production [100]. Prostaglandins mediate the activation of both AMPA/kainate receptors and mGluRs on astrocytes in acute slice preparations [102]. The application of prostaglandin E2 promoted calcium-dependent glutamate release whereas the inhibition of prostaglandin synthesis prevented the release [102].

Tumor necrosis factor α (TNFα) and its receptor TNF receptor 1 (TNFR1) play a role in the modulation of glutamate secretion from astrocytes. Using both wild-type and TNFα-deficient neurons and glia, Stellwagen and Malenka [103] demonstrated that synaptic scaling in response to activity blockade is mediated by TNFα released from astrocytes. In hemibrain horizontal slices from wild-type mice, TNFα activated astrocyte P2YR1 as well as induced astrocytic calcium elevations and glutamate release [104]. In mice lacking TNFα, however, intracellular calcium elevations could still be obtained, but glutamate release and neuromodulation could not. Cultured astrocytes from TNFα-deficient mice exhibited a defect in functional docking of VGLUT-expressing synaptic-like microvesicles [104]. Activation of the G-protein linked CXCR4 receptor by the chemokine stroma cell-derived factor 1 in hippocampal slices resulted in the rapid release of TNFα and subsequent stimulation of calcium-dependent glutamate exocytosis, a process that was absent in TNFα-deficient mice [105]. Interestingly, reactive microglia amplified this glutamate release.

Controversy Over Calcium-Induced Exocytosis

The above data collectively constitute convincing evidence for the vesicular release of gliotransmitters in response to Gq-linked GPCR-dependent calcium elevations within astrocytes. This release can be sensed by neighboring neurons and is thought to play a role in the modulation of neuronal activity. The calcium-dependent mechanism of gliotransmission, however, has been called into question for several reasons [106]. First, transgenic mice that express a Gq-coupled metabotropic receptor (MrgA1) only in astrocytes were created [107]. Activation of these receptors produced robust, widespread astrocyte calcium increases in acute mouse hippocampal slices but failed to increase neuronal calcium, produce SICs in CA1 pyramidal neurons, or have any effect on excitatory synapse activity [107]. Second, IP_3 receptor 2 (IP_3R2) knockout mice lack spontaneous Gq-linked GPCR-mediated calcium increases in astrocytes but neuronal Gq GPCR calcium increases remain intact. In these knockout mice, baseline excitatory neuronal synaptic activity and glutamate levels were not altered [108]. Developmental adaptations in these mice cannot be ruled out until conditional knockouts are created [109]. Third, the majority of studies have been conducted in vitro and may not be representative of the true brain state [74,109,110]. One study reported the lack of vesicular glutamate transporter and synaptic protein (VGLUT1, VGLUT2, synapsin 1, synaptotagmin) mRNA in the astrocyte transcriptome in vivo [111]. These proteins were found in neurons. Fourth, experimental design and pharmacological agents used in preparations often have unintended consequences. For example, slice preparations that remove magnesium from the bath relieve the voltage-dependent block on NMDA receptors, but also may initiate spontaneous calcium signaling [112]. "Specific" agonists and antagonists often have inadvertent targets [23,110]. These considerations should be taken into account when interpreting astrocyte calcium signaling studies.

One major problem with studying astrocytic calcium signaling in vivo has been the lack of available tools. Recent developments of new tools that enable real-time monitoring of calcium dynamics, however, may alleviate some of the difficulties. First, genetically encoded calcium indicator (GECI) technology will help visualize thin astrocytic processes with an exceptional signal-to-noise ratio [113]. Second, genetically altered mice, when fully characterized, can prove to be powerful tools [107]. For example, the reporter mouse PC::G5-tdT when crossed with GFAP-Cre drivers can reliably report astrocytic calcium transients [114]. Third, in utero electroporation techniques have allowed for the stable transfection of cells with plasmids featuring the GECI GCaMP [114]. Fourth, excellent calcium imaging has been accomplished using two-photon microscopy after loading of fluorescent indicators in vivo [91]. The creation of better pharmacological, astrocyte-specific tools and the creation of conditional knockout mice combined with powerful imaging strategies will be required to help elucidate the role of astrocyte calcium signaling in healthy and diseased brain.

ASTROCYTE CALCIUM SIGNALING IN EPILEPSY

Several lines of evidence have suggested that calcium elevations in astrocytes may result in the vesicular release of gliotransmitters, including glutamate and ATP. The role that these calcium spikes and subsequent gliotransmitter release play in epilepsy, however, has not

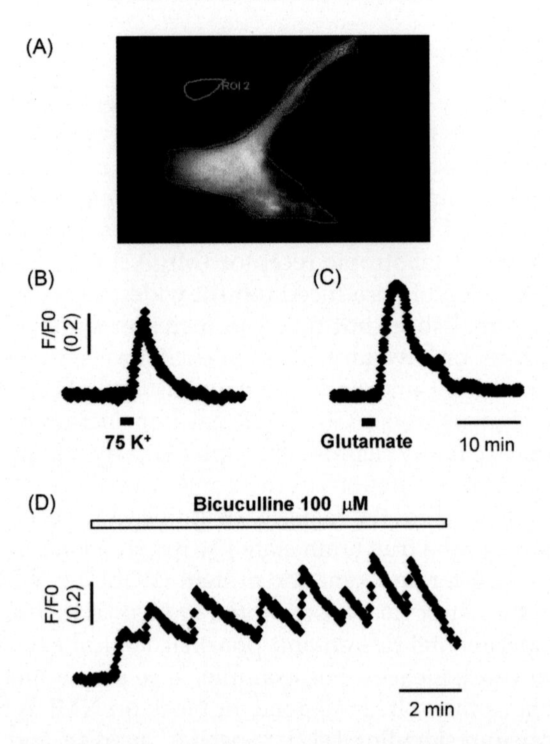

FIGURE 6.4 **Glutamate- and K^+-induced $[Ca^{2+}]_c$ transients in human neocortical astrocytes.** Panel (A) illustrates a typical astrocyte loaded with fura-2 in neurobasal medium. Original traces of $[Ca^{2+}]_c$ increases after (B) K^+ (75 mM, 30 s) or (C) glutamate (500 μM, 30 s) pulse applied as indicated by horizontal bars at the bottom of the traces. Panel (D) shows a representative image of $[Ca^{2+}]_c$ induced by 100 μM bicuculline for 15 min from a human neocortical astrocyte after 2 weeks in culture. *Source: Reproduced with permission from Cano-Abad MF, Herrera-Peco I, Sola RG, Pastor J, Garcia-Navarrete E, Moro RC, et al. New insights on culture and calcium signalling in neurons and astrocytes from epileptic patients. Int J Dev Neurosci 2011;29(2):121–9.*

been well studied. Thus far, few human tissue and animal model studies have examined this topic. Existing evidence suggests that dysregulation of calcium channels and increases in astrocytic calcium signaling may contribute to hyperexcitability and subsequent development of seizures [91,115–119].

Human Tissue Studies

Astrocytes cultured from human epileptic foci exhibited spatially and temporally coherent calcium waves [120]. In coculture with neurons, astrocyte cytosolic calcium initially increased in response to high K^+, glutamate, or bicuculline and slowly returned to baseline within 10 min (Fig. 6.4) [121]. Calcium imaging combined with electrophysiology revealed spontaneous calcium elevations in cortical and hippocampal slices from biopsies of epileptic patients [122]. These spikes were insensitive to the neuronal activity blocker tetrodotoxin (TTX), abolished by depleting internal calcium stores with thapsigargin, and propagated with the addition of ATP or glutamate. Electrical stimulation led to astrocytic calcium increases and glutamate release, which caused NMDA receptor-dependent SICs in neighboring neurons [122]. These SICs, however, were recorded in the absence of magnesium to maximize NMDA receptor activation, but it is important to remember that the lack of magnesium can have other unintended effects, such as the initiation of spontaneous calcium signaling [112].

Calcium signaling in astrocytes is altered in tissue from patients with temporal lobe epilepsy (TLE), the most common form of epilepsy. Astrocytes cultured from cortical tissue of patients with TLE exhibited decreased intracellular calcium after the addition of low concentrations of albumin whereas high concentrations of albumin increased intracellular calcium in an IP_3-dependent manner [123]. A separate study cultured astrocytes from brain regions that correlated with hyperexcitable EEG recordings and found that astrocytes in the "epileptic" cultures exhibited synchronization and larger oscillation intensities compared to control cultures [124]. Transverse slices of hippocampal tissue cultured from TLE patients exhibiting either hippocampal sclerosis (HS) or lesions in the temporal lobe not involving the hippocampus were studied [125]. A nondesensitizing response to kainic acid application suggested the presence of AMPA receptors on astrocytes in both cultures. The increased cyclothiazide-mediated potentiation in HS tissue compared to the lesioned group suggested an upregulation of the AMPA receptor flip splice variant in the HS tissue [125].

DNA microanalysis of human TLE tissue revealed increased expression of genes associated with astrocyte structure, blood–brain barrier function, and calcium regulation in the sclerotic hippocampus compared to non-HS tissue [126]. Immunohistochemical analysis of different voltage-dependent calcium channel $\alpha1$ subunits in hippocampal tissue revealed increased immunoreactivity for α_{1c} in the sclerotic hippocampus compared to autopsy controls [127]. Manning et al. [128] studied calcium responses in astrocytes cultured from cortical tissue of a 3-year-old patient with Rasmussen's encephalitis and found large spontaneous intracellular calcium oscillations. Comparable oscillations were not seen in control rat cortical astrocyte cultures. The oscillations in the epileptic cultures could be eliminated with calcium-free media, suggesting that the calcium oscillations were dependent on calcium flux across the membrane [128]. Human tissue studies are limited, however, and often lack appropriate controls.

In Vitro and In Situ Animal Models

Tashiro et al. [129] characterized calcium oscillations in neocortical slices from juvenile mice after application of either bicuculline-containing artificial cerebrospinal fluid (ACSF) or magnesium-free + high K^+ ACSF. Under both conditions, astrocytes exhibited slow calcium transients, a phenomenon that was absent in slices bathed in standard ACSF. Astrocytic calcium transients had no temporal relationship to the observed field potential spikes, suggesting a lack of correlation with epileptiform activity [129]. In the picrotoxin, magnesium-free perfusion of coronal neocortical-hippocampal slice model of epilepsy, seizure-like discharges evoked calcium increases in astrocyte endfeet followed by an arteriole response [130]. This effect was abolished by the pharmacological inhibition of calcium signals in astrocyte processes [130]. Incubation of astrocytes in rat slice preparations with IP_3 resulted in glutamate release, SICs, and epileptiform discharges in CA1 pyramidal neurons [16]. The SICs were a result of astrocytic glutamate release acting on ionotropic glutamate receptors in pyramidal neurons, which led to epileptiform discharges. Glutamate release also triggered a transport current due to the fusion of glutamate-containing vesicles in astrocytes. The fusion of vesicles and transport currents in astrocytes were blocked by tetanus toxin, suggesting glutamate release through a SNARE-dependent mechanism [16]. The potent convulsant 4-aminopyridine (4-AP) induced similar neuronal SICs and astrocytic transport currents, suggesting that astrocytic release of glutamate may play a role in seizure initiation [16]. Slices exposed to 4-AP also exhibited epileptiform bursting activity and paroxysmal neuronal depolarization events [131]. These events were insensitive to TTX, voltage-gated calcium channel blockers, and glutamate transporter inhibitors; reduced by APV (NMDA receptor antagonist) and CNQX (AMPA/kainate receptor antagonist); and initiated by photolysis of caged calcium and subsequent glutamate release in astrocytes. Bicuculline, penicillin, and the removal of either magnesium or calcium all triggered TTX-insensitive depolarization shifts resulting from extrasynaptic glutamate release [131]. Interestingly, blocking astrocyte CB1 receptors reduced the maintenance of 4-AP-induced epileptiform discharges in hippocampal slice cultures [132]. Chelation of astrocytic calcium abolished the CB1 receptor-mediated modulation of epileptiform activity [132].

Gómez-Gonzalo et al. [133] examined the cooperation of neurons and astrocytes in supporting ictal and interictal epileptiform events in rat entorhinal cortex (EC) slices treated with either picrotoxin in a magnesium-free solution or focal pulse application of NMDA. Using simultaneous patch-clamp recording and calcium imaging techniques, they showed that astrocytic calcium elevations correlated with both the initial development and the maintenance of focal seizure-like discharges. Interictal discharges, however, were not associated with calcium changes in astrocytes [130]. Similarly, seizure-like events, but not interictal events, regularly evoked astrocytic calcium spikes after bicuculline perfusion of in vitro isolated whole brain from young adult guinea pigs [133]. Perfusion of EC slices with either an mGluR or purinergic (P2) receptor antagonist reduced ictal discharge activation of astrocytes whereas interictal discharges were either unaffected or increased in frequency. Bath perfusion of slices with TFLLR, a PAR-1 thrombin receptor agonist known to selectively activate glutamate release in astrocytes, in the presence of TTX and D-AP5 (to block astrocyte-to-neuron signaling) blocked calcium changes in EC neurons but large elevations were still observed in astrocytes. Furthermore, TLFFR decreased the ictal discharge threshold in the picrotoxin in a magnesium-free solution model [133]. Therefore, astrocytes may play a role in the generation of ictal discharges.

In Vivo Animal Models

Stringer et al. [134] recorded reduced extracellular calcium levels followed by partial recovery in response to stimulus trains in the rat hippocampus during epileptiform activity. Local injection of the astrocyte-specific metabolic toxin fluorocitrate resulted in a small, but significant, decrease in extracellular calcium in CA1 but caused an increase in the DG. Focal application of fluorocitrate into the cortex caused seizures and decreased extracellular calcium levels [135]. Anesthetized, paralyzed rats given intravenous injections of picrotoxin exhibited decreased cortical and hippocampal calcium levels whereas kainic acid did not affect calcium levels [135]. Furthermore, decreases in extracellular calcium initiated

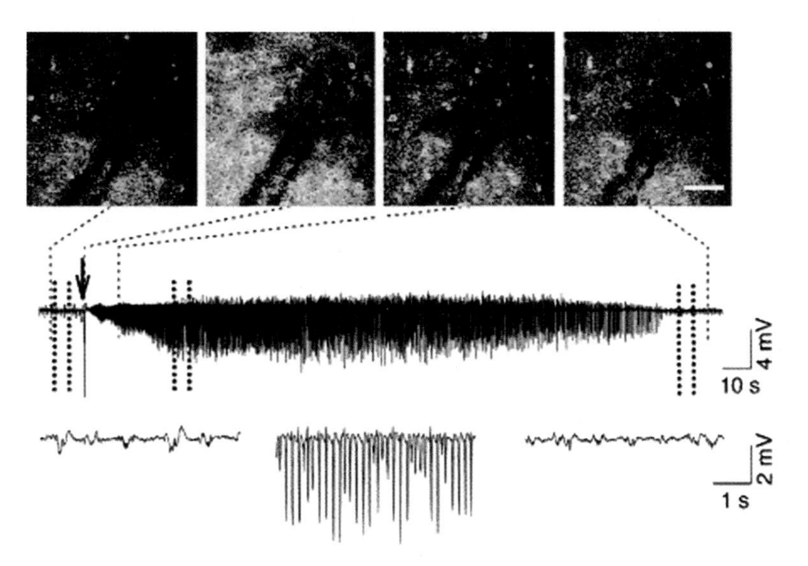

FIGURE 6.5 **4-AP induced seizure in vivo.** 4-AP (100 mM) was delivered locally (5–10 pulses of 5–10 ms at 10 psi, Picospritzer) by an electrode inserted into cortex at a distance of 100–150 μm from the pial surface for recording of the field potential, and triggered delayed spontaneous episodes of high frequency, large amplitude discharges preceded by astrocytic Ca^{2+} signaling. Upper panel: two-photon imaging showed abnormal Ca^{2+} signaling (second photo) preceding epileptiform discharges induced by 4-AP; Middle panel: recording of field potential showed an event of epileptiform high frequency, large amplitude discharges induced by 4-AP; Lower panel: expanded recordings of field potential during the periods indicated by dotted lines. *Source: Reproduced with permission from Tian GF, Takano T, Lin JH, Wang X, Bekar L, Nedergaard M. Imaging of cortical astrocytes using 2-photon laser scanning microscopy in the intact mouse brain. Adv Drug Deliv Rev 2006;58(7):773–87.*

astrocytic ATP release through hemichannels and subsequent increase in cytosolic calcium levels in surrounding astrocytes [136].

Two-photon microscopy is a powerful tool that has been used to study astrocytic calcium signaling in both the healthy and epileptic brain. Prolonged elevated intracellular astrocytic calcium levels were observed after pilocarpine-induced status epilepticus [137]. Elevated calcium was correlated with delayed neuronal death. Calcium chelation, mGluR5 antagonists, and NMDA receptor antagonists given after SE suppressed neuronal cell death [137]. Two-photon imaging of calcium signaling in the cortex of adult mice after local delivery of 4-AP revealed propagating calcium waves in conjunction with spontaneous seizure activity (Fig. 6.5) [91,131]. In a separate study, calcium dyes were bulk loaded into the neocortex using convection-enhanced delivery [138]. The local injection of 4-AP resulted on seizure-initiated, slowly evolving calcium waves of activity in astrocytes that started at the seizure focus and slowly spread, surpassing neuronal activity (Fig. 6.6) [138]. The antiepileptic drugs valproate, gabapentin, and phenytoin reduced the amplitude of neuronal discharges and the astrocytic calcium signaling response to 4-AP [131].

It is likely that differences in calcium responses between the healthy and epileptic brain are a result of dysregulated protein activity. Decreased levels of $CA_v2.2$, a nuclear protein

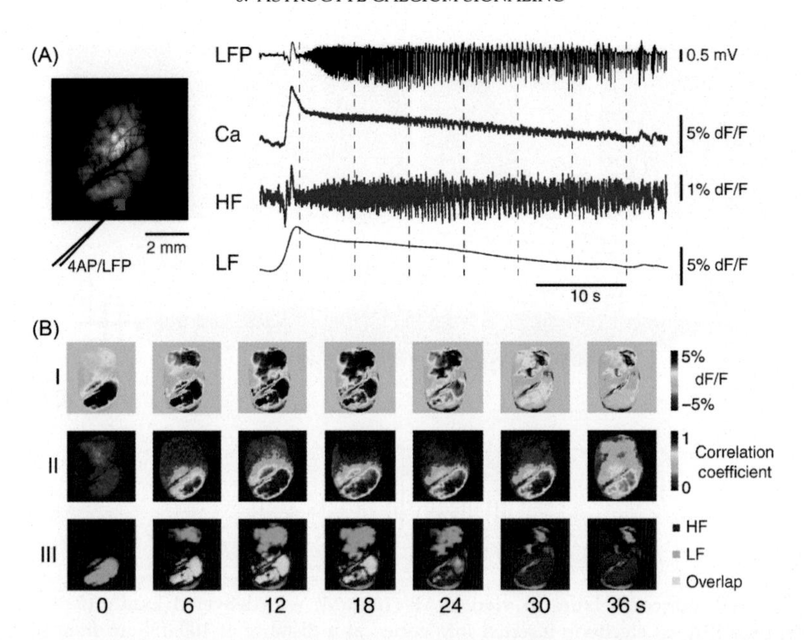

FIGURE 6.6 **The spatial distribution of HF (neuronal) and LF (glial) activity during a seizure.** (A) The top left shows the field of view and location of the 4-AP and LFP electrode. The blue box indicates the area from which the Ca trace and subsequent HF and LF are derived. (B) Spatial propagation of HF and LF activity throughout ictal event. The seizure spread is shown at 6-s intervals. (I) Glial propagation during seizures. The glial activity begins at the seizure focus and propagates to the rest of the hemisphere as a slowly propagating wave with onset in the focus and offset in the surrounding. The wave has spread several millimeters in only 6 s and reaches its maximum around 18 s. (II) The neuronal activity also begins at the seizure focus, but the activity, although consisting of rapidly propagating waves, does not progressively march across the cortex. Rather, the activity remains conserved to a relatively focal area, the majority of which is involved very early in seizure evolution. Occasionally, some waves propagate further laterally (as seen at 36 s), but the majority of the activity stays within the focus. (III) Spatial overlap of glial and neuronal activity during seizure evolution. The glial waves clearly propagate further into the surrounding areas as a slowly moving wave that starts in the focus and spreads laterally, whereas the neuronal activity stays within the focus. *Source: Reproduced with permission from Daniel AG, Laffont P, Zhao M, Ma H, Schwartz TH. Optical electrocorticogram (OECoG) using wide-field calcium imaging reveals the divergence of neuronal and glial activity during acute rodent seizures. Epilepsy Behav 2015;49:61–5.*

expressed in both neurons and astrocytes, were observed in the stratum pyramidale and in the stratum granulosum of the DG at acute stages in mice after pilocarpine-induced SE [139]. Increased expression of $Ca_v2.2$ was found in the stratum lucidum of CA3 and in the hilus of the DG. In the chronic phase, increased levels of $Ca_v2.2$ were seen in the stratum granulosum and molecular layer of DG [139]. The expression of $Ca_v1.3$ or $Ca_v2.1$ in reactive astrocytes was induced 1 week and 2 months after pilocarpine-induced SE [140]. An upregulation of L-type calcium channel immunoreactivity in reactive astrocytes was observed after kainic acid-induced epilepsy [141]. Stromal interaction molecule 1 and 2 (STIM1 and STIM2, respectively), calcium sensors primarily found on the endoplasmic reticulum, are expressed in mouse brain and spinal cord tissue, human tissue, and in both astrocyte and neuron cell cultures [142]. Resting calcium levels were much higher in neurons than astrocytes and the inhibition of store-operated calcium entry decreased steady-state intracellular calcium in primary neurons but not in astrocytes. These proteins were both highly expressed after pilocarpine-induced SE and in a hippocampal specimen from a patient with mesial temporal lobe epilepsy (MTLE), particularly in neurons [142].

The mGluRs expressed on astrocytes play a major role in astrocyte calcium signaling. Both mGluR2/3 and mGluR5 were upregulated during the chronic phase in reactive astrocytes in the electrical angular bundle stimulation model of epilepsy [143]. A loss of mGluR5, however, was seen in the ipsilateral hippocampus 3 days after intra-amygdala kainic acid-injections with recovery over time [144]. The discrepancy between these two studies may be partially because one study specifically examined reactive astrocyte expression in the chronic phase of epilepsy while the other looked at total expression (mGluR5 is found in both astrocytes and neurons) in the latent phase of epilepsy.

Intracellular calcium levels in astrocytes may serve as a novel therapeutic target. The addition of high K^+ to primary cultures of astrocytes led to the uptake of calcium [145]. Nimodipine, a calcium channel blocker, inhibited calcium uptake but had little effect on the unstimulated cells. Therefore, the therapeutic benefit of calcium blockers in epilepsy may be due to the prevention of astrocytic calcium uptake during a seizure. A novel non-L-type voltage sensitive calcium channel inhibitor (NP04636) reduced the number of rats that entered into kainic acid-induced SE, mortality rates, and gliosis as well as increased latency to SE and prevented the loss of hippocampal CA1 and CA3 pyramidal neurons [146]. Drugs that specifically target mechanisms involved in astrocytic calcium signaling may be promising for the treatment of epilepsy.

CONCLUSION

A large body of evidence supports the mechanism of calcium-dependent gliotransmitter release from astrocytes. In 1990, it was demonstrated that glutamate could induce calcium wave activity in astrocytes. It was then discovered that gliotransmitters could be released from astrocytes and these molecules could induce responses in nearby neurons. Finally, the existence of vesicular release proteins in astrocytes combined with pharmacological studies shed light onto the mechanisms of release. Calcium-dependent exocytosis of gliotransmitters, however, still remains controversial. Evidence using transgenic mice has demonstrated that calcium elevations within astrocytes have no effect on neighboring neurons. The role of astrocyte calcium signaling in epilepsy is not yet understood. Evidence thus far has suggested that hypersynchronization among neurons may be, in part, due to hyperactivity of astrocytes. Specific mechanisms involved in the generation of astrocytic calcium oscillations, therefore, may serve as novel therapeutic targets for the treatment of epilepsy. Additional in vivo studies examining astrocytic calcium signaling under physiological and pathological conditions are needed to fully understand the extent to which astrocytic calcium signaling shapes synaptic activity, hyperexcitability, epileptiform activity, and seizures.

References

[1] Charles AC, Merrill JE, Dirksen ER, Sanderson MJ. Intercellular signaling in glial cells: calcium waves and oscillations in response to mechanical stimulation and glutamate. Neuron 1991;6(6):983–92.

[2] Cornell-Bell AH, Finkbeiner SM, Cooper MS, Smith SJ. Glutamate induces calcium waves in cultured astrocytes: long-range glial signaling. Science 1990;247(4941):470–3.

[3] Fiacco TA, McCarthy KD. Intracellular astrocyte calcium waves in situ increase the frequency of spontaneous AMPA receptor currents in CA1 pyramidal neurons. J Neurosci 2004;24(3):722–32.

[4] Panatier A, Vallee J, Haber M, Murai KK, Lacaille JC, Robitaille R. Astrocytes are endogenous regulators of basal transmission at central synapses. Cell 2011;146(5):785–98.

[5] Perea G, Araque A. Properties of synaptically evoked astrocyte calcium signal reveal synaptic information processing by astrocytes. J Neurosci 2005;25(9):2192–203.

[6] Porter JT, McCarthy KD. Hippocampal astrocytes in situ respond to glutamate released from synaptic terminals. J Neurosci 1996;16(16):5073–81.

[7] Perea G, Araque A. Glial calcium signaling and neuron-glia communication. Cell Calcium 2005;38(3–4):375–82.

[8] Perea G, Araque A. Synaptic regulation of the astrocyte calcium signal. J Neural Transm 2005;112(1):127–35.

[9] Duffy S, MacVicar BA. Adrenergic calcium signaling in astrocyte networks within the hippocampal slice. J Neurosci 1995;15(8):5535–50.

[10] Pearce B, Cambray-Deakin M, Morrow C, Grimble J, Murphy S. Activation of muscarinic and of α_1-adrenergic receptors on astrocytes results in the accumulation of inositol phosphates. J Neurochem 1985;45(5): 1534–40.

[11] Shelton MK, McCarthy KD. Hippocampal astrocytes exhibit Ca^{2+}-elevating muscarinic cholinergic and histaminergic receptors in situ. J Neurochem 2000;74(2):555–63.

[12] Araque A, Martín ED, Perea G, Arellano JI, Buno W. Synaptically released acetylcholine evokes Ca^{2+} elevations in astrocytes in hippocampal slices. J Neurosci 2002;22(7):2443–50.

[13] Cotrina ML, Lin JH, Alves-Rodrigues A, Liu S, Li J, Azmi-Ghadimi H, et al. Connexins regulate calcium signaling by controlling ATP release. Proc Natl Acad Sci USA 1998;95(26):15735–40.

[14] Guthrie PB, Knappenberger J, Segal M, Bennett MV, Charles AC, Kater SB. ATP released from astrocytes mediates glial calcium waves. J Neurosci 1999;19(2):520–8.

[15] Wang Z, Haydon PG, Yeung ES. Direct observation of calcium-independent intercellular ATP signaling in astrocytes. Anal Chem 2000;72(9):2001–7.

[16] Kang J, Jiang L, Goldman SA, Nedergaard M. Astrocyte-mediated potentiation of inhibitory synaptic transmission. Nat Neurosci 1998;1(8):683–92.

[17] Navarrete M, Araque A. Endocannabinoids potentiate synaptic transmission through stimulation of astrocytes. Neuron 2010;68(1):113–26.

[18] Bal-Price A, Moneer Z, Brown GC. Nitric oxide induces rapid, calcium-dependent release of vesicular glutamate and ATP from cultured rat astrocytes. Glia 2002;40(3):312–23.

[19] Li N, Sul JY, Haydon PG. A calcium-induced calcium influx factor, nitric oxide, modulates the refilling of calcium stores in astrocytes. J Neurosci 2003;23(32):10302–10.

[20] Pascual M, Climent E, Guerri C. BDNF induces glutamate release in cerebrocortical nerve terminals and in cortical astrocytes. Neuroreport 2001;12(12):2673–7.

[21] Dani JW, Chernjavsky A, Smith SJ. Neuronal activity triggers calcium waves in hippocampal astrocyte networks. Neuron 1992;8(3):429–40.

[22] Pasti L, Volterra A, Pozzan T, Carmignoto G. Intracellular calcium oscillations in astrocytes: a highly plastic, bidirectional form of communication between neurons and astrocytes in situ. J Neurosci 1997;17(20):7817–30.

[23] Nedergaard M. Direct signaling from astrocytes to neurons in cultures of mammalian brain cells. Science 1994;263(5154):1768–71.

[24] Grosche J, Matyash V, Möller T, Verkhratsky A, Reichenbach A, Kettenmann H. Microdomains for neuron-glia interaction: parallel fiber signaling to Bergmann glial cells. Nat Neurosci 1999;2(2):139–43.

[25] Parri HR, Gould TM, Crunelli V. Spontaneous astrocytic Ca^{2+} oscillations in situ drive NMDAR-mediated neuronal excitation. Nat Neurosci 2001;4(8):803–12.

[26] Di Castro MA, Chuquet J, Liaudet N, Bhaukaurally K, Santello M, Bouvier D, et al. Local Ca^{2+} detection and modulation of synaptic release by astrocytes. Nat Neurosci 2011;14(10):1276–84.

[27] Hua X, Malarkey EB, Sunjara V, Rosenwald SE, Li WH, Parpura V. Ca^{2+}-dependent glutamate release involves two classes of endoplasmic reticulum Ca^{2+} stores in astrocytes. J Neurosci Res 2004;76(1):86–97.
[28] Hansson E. Regional heterogeneity among astrocytes in the central nervous system. Neurochem Int 1990;16(3):237–45.
[29] Fiacco TA, McCarthy KD. Astrocyte calcium elevations: properties, propagation, and effects on brain signaling. Glia 2006;54(7):676–90.
[30] Nimmerjahn A, Mukamel EA, Schnitzer MJ. Motor behavior activates Bergmann glial networks. Neuron 2009;62(3):400–12.
[31] Takata N, Hirase H. Cortical layer 1 and layer 2/3 astrocytes exhibit distinct calcium dynamics in vivo. PLoS One 2008;3(6):e2525.
[32] Santello M, Calì C, Bezzi P. Gliotransmission and the tripartite synapse Kreutz MR, Sala C, editors. Synaptic Plasticity. Wien: Springer; 2012.
[33] Angulo MC, Kozlov AS, Charpak S, Audinat E. Glutamate released from glial cells synchronizes neuronal activity in the hippocampus. J Neurosci 2004;24(31):6920–7.
[34] Bardoni R, Ghirri A, Zonta M, Betelli C, Vitale G, Ruggieri V, et al. Glutamate-mediated astrocyte-to-neuron signalling in the rat dorsal horn. J Physiol 2010;588(Pt 5):831–46.
[35] D'Ascenzo M, Fellin T, Terunuma M, Revilla-Sanchez R, Meaney DF, Auberson YP, et al. mGluR5 stimulates gliotransmission in the nucleus accumbens. Proc Natl Acad Sci USA 2007;104(6):1995–2000.
[36] Fellin T, Pascual O, Gobbo S, Pozzan T, Haydon PG, Carmignoto G. Neuronal synchrony mediated by astrocytic glutamate through activation of extrasynaptic NMDA receptors. Neuron 2004;43(5):729–43.
[37] Nie H, Zhang H, Weng HR. Bidirectional neuron-glia interactions triggered by deficiency of glutamate uptake at spinal sensory synapses. J Neurophysiol 2010;104(2):713–25.
[38] Parpura V, Haydon PG. Physiological astrocytic calcium levels stimulate glutamate release to modulate adjacent neurons. Proc Natl Acad Sci USA 2000;97(15):8629–34.
[39] Reyes-Haro D, Müller J, Boresch M, Pivneva T, Benedetti B, Scheller A, et al. Neuron-astrocyte interactions in the medial nucleus of the trapezoid body. J Gen Physiol 2010;135(6):583–94.
[40] Sasaki T, Kuga N, Namiki S, Matsuki N, Ikegaya Y. Locally synchronized astrocytes. Cereb Cortex 2011;21(8):1889–900.
[41] Shigetomi E, Bowser DN, Sofroniew MV, Khakh BS. Two forms of astrocyte calcium excitability have distinct effects on NMDA receptor-mediated slow inward currents in pyramidal neurons. J Neurosci 2008;28(26):6659–63.
[42] Dombeck DA, Khabbaz AN, Collman F, Adelman TL, Tank DW. Imaging large-scale neural activity with cellular resolution in awake, mobile mice. Neuron 2007;56(1):43–57.
[43] Hirase H, Qian L, Bartho P, Buzsaki G. Calcium dynamics of cortical astrocytic networks in vivo. PLoS Biol 2004;2(4):E96.
[44] Nimmerjahn A, Kirchhoff F, Kerr JN, Helmchen F. Sulforhodamine 101 as a specific marker of astroglia in the neocortex in vivo. Nat Methods 2004;1(1):31–7.
[45] Navarrete M, Perea G, Fernandez de Sevilla D, Gómez-Gonzalo M, Núñez A, Martín ED, et al. Astrocytes mediate in vivo cholinergic-induced synaptic plasticity. PLoS Biol 2012;10(2):e1001259.
[46] Wang X, Lou N, Xu Q, Tian GF, Peng WG, Han X, et al. Astrocytic Ca^{2+} signaling evoked by sensory stimulation in vivo. Nat Neurosci 2006;9(6):816–23.
[47] Bekar LK, He W, Nedergaard M. Locus coeruleus α-adrenergic-mediated activation of cortical astrocytes in vivo. Cereb Cortex 2008;18(12):2789–95.
[48] Navarrete M, Araque A. Endocannabinoids mediate neuron-astrocyte communication. Neuron 2008;57(6): 883–93.
[49] Perea G, Araque A. Astrocytes potentiate transmitter release at single hippocampal synapses. Science 2007;317(5841):1083–6.
[50] Bonansco C, Couve A, Perea G, Ferradas CÁ, Roncagliolo M, Fuenzalida M. Glutamate released spontaneously from astrocytes sets the threshold for synaptic plasticity. Eur J Neurosci 2011;33(8):1483–92.
[51] Honsek SD, Walz C, Kafitz KW, Rose CR. Astrocyte calcium signals at Schaffer collateral to CA1 pyramidal cell synapses correlate with the number of activated synapses but not with synaptic strength. Hippocampus 2012;22(1):29–42.
[52] Willmott NJ, Wong K, Strong AJ. Intercellular Ca^{2+} waves in rat hippocampal slice and dissociated glial-neuron cultures mediated by nitric oxide. FEBS Lett 2000;487(2):239–47.

[53] Willmott NJ, Wong K, Strong AJ. A fundamental role for the nitric oxide-G-kinase signaling pathway in mediating intercellular Ca^{2+} waves in glia. J Neurosci 2000;20(5):1767–79.

[54] Sergeeva M, Strokin M, Reiser G. Regulation of intracellular calcium levels by polyunsaturated fatty acids, arachidonic acid and docosahexaenoic acid, in astrocytes: possible involvement of phospholipase A_2. Reprod Nutr Dev 2005;45(5):633–46.

[55] Perea G, Yang A, Boyden ES, Sur M. Optogenetic astrocyte activation modulates response selectivity of visual cortex neurons in vivo. Nat Commun 2014;5:3262.

[56] Liu QS, Xu Q, Arcuino G, Kang J, Nedergaard M. Astrocyte-mediated activation of neuronal kainate receptors. Proc Natl Acad Sci USA 2004;101(9):3172–7.

[57] Liu QS, Xu Q, Kang J, Nedergaard M. Astrocyte activation of presynaptic metabotropic glutamate receptors modulates hippocampal inhibitory synaptic transmission. Neuron Glia Biol 2004;1(4):307–16.

[58] Jeftinija SD, Jeftinija KV, Stefanovic G. Cultured astrocytes express proteins involved in vesicular glutamate release. Brain Res 1997;750(1–2):41–7.

[59] Parpura V, Liu F, Brethorst S, Jeftinija K, Jeftinija S, Haydon PG. α-latrotoxin stimulates glutamate release from cortical astrocytes in cell culture. FEBS Lett 1995;360(3):266–70.

[60] Parpura V, Basarsky TA, Liu F, Jeftinija K, Jeftinija S, Haydon PG. Glutamate-mediated astrocyte-neuron signalling. Nature 1994;369(6483):744–7.

[61] Kreft M, Stenovec M, Rupnik M, Grilc S, Kržan M, Potokar M, et al. Properties of Ca^{2+}-dependent exocytosis in cultured astrocytes. Glia 2004;46(4):437–45.

[62] Innocenti B, Parpura V, Haydon PG. Imaging extracellular waves of glutamate during calcium signaling in cultured astrocytes. J Neurosci 2000;20(5):1800–8.

[63] Zhang Q, Pangršič T, Kreft M, Kržan M, Li N, Sul JY, et al. Fusion-related release of glutamate from astrocytes. J Biol Chem 2004;279(13):12724–33.

[64] Zhang Q, Fukuda M, Van Bockstaele E, Pascual O, Haydon PG. Synaptotagmin IV regulates glial glutamate release. Proc Natl Acad Sci USA 2004;101(25):9441–6.

[65] Mothet JP, Pollegioni L, Ouanounou G, Martineau M, Fossier P, Baux G. Glutamate receptor activation triggers a calcium-dependent and SNARE protein-dependent release of the gliotransmitter D-serine. Proc Natl Acad Sci USA 2005;102(15):5606–11.

[66] Coco S, Calegari F, Pravettoni E, Pozzi D, Taverna E, Rosa P, et al. Storage and release of ATP from astrocytes in culture. J Biol Chem 2003;278(2):1354–62.

[67] Gordon GR, Baimoukhametova DV, Hewitt SA, Rajapaksha WR, Fisher TE, Bains JS. Norepinephrine triggers release of glial ATP to increase postsynaptic efficacy. Nat Neurosci 2005;8(8):1078–86.

[68] Gordon GR, Iremonger KJ, Kantevari S, Ellis-Davies GC, MacVicar BA, Bains JS. Astrocyte-mediated distributed plasticity at hypothalamic glutamate synapses. Neuron 2009;64(3):391–403.

[69] Kozlov AS, Angulo MC, Audinat E, Charpak S. Target cell-specific modulation of neuronal activity by astrocytes. Proc Natl Acad Sci USA 2006;103(26):10058–63.

[70] Panatier A, Theodosis DT, Mothet JP, Touquet B, Pollegioni L, Poulain DA, et al. Glia-derived D-serine controls NMDA receptor activity and synaptic memory. Cell 2006;125(4):775–84.

[71] Henneberger C, Papouin T, Oliet SH, Rusakov DA. Long-term potentiation depends on release of D-serine from astrocytes. Nature 2010;463(7278):232–6.

[72] Takata N, Mishima T, Hisatsune C, Nagai T, Ebisui E, Mikoshiba K, et al. Astrocyte calcium signaling transforms cholinergic modulation to cortical plasticity in vivo. J Neurosci 2011;31(49):18155–65.

[73] Bergersen LH, Gundersen V. Morphological evidence for vesicular glutamate release from astrocytes. Neuroscience 2009;158(1):260–5.

[74] Hamilton NB, Attwell D. Do astrocytes really exocytose neurotransmitters? Nat Rev Neurosci 2010;11(4):227–38.

[75] Montana V, Ni Y, Sunjara V, Hua X, Parpura V. Vesicular glutamate transporter-dependent glutamate release from astrocytes. J Neurosci 2004;24(11):2633–42.

[76] Anlauf E, Derouiche A. Astrocytic exocytosis vesicles and glutamate: a high-resolution immunofluorescence study. Glia 2005;49(1):96–106.

[77] Bezzi P, Gundersen V, Galbete JL, Seifert G, Steinhäuser C, Pilati E, et al. Astrocytes contain a vesicular compartment that is competent for regulated exocytosis of glutamate. Nat Neurosci 2004;7(6):613–20.

[78] Crippa D, Schenk U, Francolini M, Rosa P, Verderio C, Zonta M, et al. Synaptobrevin2-expressing vesicles in rat astrocytes: insights into molecular characterization, dynamics and exocytosis. J Physiol 2006;570(Pt 3):567–82.

[79] Martineau M, Shi T, Puyal J, Knolhoff AM, Dulong J, Gasnier B, et al. Storage and uptake of D-serine into astrocytic synaptic-like vesicles specify gliotransmission. J Neurosci 2013;33(8):3413–23.
[80] Parpura V, Fang Y, Basarsky T, Jahn R, Haydon PG. Expression of synaptobrevin II, cellubrevin and syntaxin but not SNAP-25 in cultured astrocytes. FEBS Lett 1995;377(3):489–92.
[81] Pasti L, Zonta M, Pozzan T, Vicini S, Carmignoto G. Cytosolic calcium oscillations in astrocytes may regulate exocytotic release of glutamate. J Neurosci 2001;21(2):477–84.
[82] Wilhelm A, Volknandt W, Langer D, Nolte C, Kettenmann H, Zimmermann H. Localization of SNARE proteins and secretory organelle proteins in astrocytes in vitro and in situ. Neurosci Res 2004;48(3):249–57.
[83] Fremeau Jr. RT, Burman J, Qureshi T, Tran CH, Proctor J, Johnson J, et al. The identification of vesicular glutamate transporter 3 suggests novel modes of signaling by glutamate. Proc Natl Acad Sci USA 2002;99(22): 14488–93.
[84] Martineau M, Galli T, Baux G, Mothet JP. Confocal imaging and tracking of the exocytotic routes for D-serine-mediated gliotransmission. Glia 2008;56(12):1271–84.
[85] Maienschein V, Marxen M, Volknandt W, Zimmermann H. A plethora of presynaptic proteins associated with ATP-storing organelles in cultured astrocytes. Glia 1999;26(3):233–44.
[86] Hepp R, Perraut M, Chasserot-Golaz S, Galli T, Aunis D, Langley K, et al. Cultured glial cells express the SNAP-25 analogue SNAP-23. Glia 1999;27(2):181–7.
[87] Araque A, Li N, Doyle RT, Haydon PG. SNARE protein-dependent glutamate release from astrocytes. J Neurosci 2000;20(2):666–73.
[88] Jaiswal JK, Fix M, Takano T, Nedergaard M, Simon SM. Resolving vesicle fusion from lysis to monitor calcium-triggered lysosomal exocytosis in astrocytes. Proc Natl Acad Sci USA 2007;104(35):14151–6.
[89] Li D, Ropert N, Koulakoff A, Giaume C, Oheim M. Lysosomes are the major vesicular compartment undergoing Ca^{2+} -regulated exocytosis from cortical astrocytes. J Neurosci 2008;28(30):7648–58.
[90] Zhang Z, Chen G, Zhou W, Song A, Xu T, Luo Q, et al. Regulated ATP release from astrocytes through lysosome exocytosis. Nat Cell Biol 2007;9(8):945–53.
[91] Tian GF, Takano T, Lin JH, Wang X, Bekar L, Nedergaard M. Imaging of cortical astrocytes using 2-photon laser scanning microscopy in the intact mouse brain. Adv Drug Deliv Rev 2006;58(7):773–87.
[92] Bowser DN, Khakh BS. Vesicular ATP is the predominant cause of intercellular calcium waves in astrocytes. J Gen Physiol 2007;129(6):485–91.
[93] Calegari F, Coco S, Taverna E, Bassetti M, Verderio C, Corradi N, et al. A regulated secretory pathway in cultured hippocampal astrocytes. J Biol Chem 1999;274(32):22539–47.
[94] Prada I, Marchaland J, Podini P, Magrassi L, D'Alessandro R, Bezzi P, et al. REST/NRSF governs the expression of dense-core vesicle gliosecretion in astrocytes. J Cell Biol 2011;193(3):537–49.
[95] Ramamoorthy P, Whim MD. Trafficking and fusion of neuropeptide Y-containing dense-core granules in astrocytes. J Neurosci 2008;28(51):13815–27.
[96] Bergersen LH, Morland C, Ormel L, Rinholm JE, Larsson M, Wold JF, et al. Immunogold detection of L-glutamate and D-serine in small synaptic-like microvesicles in adult hippocampal astrocytes. Cereb Cortex 2012;22(7):1690–7.
[97] Bergles DE, Roberts JD, Somogyi P, Jahr CE. Glutamatergic synapses on oligodendrocyte precursor cells in the hippocampus. Nature 2000;405(6783):187–91.
[98] Chen X, Wang L, Zhou Y, Zheng LH, Zhou Z. "Kiss-and-run" glutamate secretion in cultured and freshly isolated rat hippocampal astrocytes. J Neurosci 2005;25(40):9236–43.
[99] Marchaland J, Calì C, Voglmaier SM, Li H, Regazzi R, Edwards RH, et al. Fast subplasma membrane Ca^{2+} transients control exo-endocytosis of synaptic-like microvesicles in astrocytes. J Neurosci 2008;28(37):9122–32.
[100] Gómez-Gonzalo M, Navarrete M, Perea G, Covelo A, Martín-Fernández M, Shigemoto R, et al. Endocannabinoids induce lateral long-term potentiation of transmitter release by stimulation of gliotransmission. Cereb Cortex 2015;25(10):3699–712.
[101] Min R, Nevian T. Astrocyte signaling controls spike timing-dependent depression at neocortical synapses. Nat Neurosci 2012;15(5):746–53.
[102] Bezzi P, Carmignoto G, Pasti L, Vesce S, Rossi D, Rizzini BL, et al. Prostaglandins stimulate calcium-dependent glutamate release in astrocytes. Nature 1998;391(6664):281–5.
[103] Stellwagen D, Malenka RC. Synaptic scaling mediated by glial TNFα. Nature 2006;440(7087):1054–9.
[104] Santello M, Bezzi P, Volterra A. TNFα controls glutamatergic gliotransmission in the hippocampal dentate gyrus. Neuron 2011;69(5):988–1001.

[105] Bezzi P, Domercq M, Brambilla L, Galli R, Schols D, De Clercq E, et al. CXCR4-activated astrocyte glutamate release via TNFα: amplification by microglia triggers neurotoxicity. Nat Neurosci 2001;4(7):702–10.
[106] Agulhon C, Fiacco TA, McCarthy KD. Hippocampal short- and long-term plasticity are not modulated by astrocyte Ca^{2+} signaling. Science 2010;327(5970):1250–4.
[107] Fiacco TA, Agulhon C, Taves SR, Petravicz J, Casper KB, Dong X, et al. Selective stimulation of astrocyte calcium in situ does not affect neuronal excitatory synaptic activity. Neuron 2007;54(4):611–26.
[108] Petravicz J, Fiacco TA, McCarthy KD. Loss of IP_3 receptor-dependent Ca^{2+} increases in hippocampal astrocytes does not affect baseline CA1 pyramidal neuron synaptic activity. J Neurosci 2008;28(19):4967–73.
[109] Nedergaard M, Verkhratsky A. Artifact versus reality—how astrocytes contribute to synaptic events. Glia 2012;60(7):1013–23.
[110] Sloan SA, Barres BA. Looks can be deceiving: reconsidering the evidence for gliotransmission. Neuron 2014;84(6):1112–5.
[111] Cahoy JD, Emery B, Kaushal A, Foo LC, Zamanian JL, Christopherson KS, et al. A transcriptome database for astrocytes, neurons, and oligodendrocytes: a new resource for understanding brain development and function. J Neurosci 2008;28(1):264–78.
[112] Stout C, Charles A. Modulation of intercellular calcium signaling in astrocytes by extracellular calcium and magnesium. Glia 2003;43(3):265–73.
[113] Srinivasan R, Huang BS, Venugopal S, Johnston AD, Chai H, Zeng H, et al. Ca^{2+} signaling in astrocytes from IP_3R_2-/- mice in brain slices and during startle responses in vivo. Nat Neurosci 2015;18(5):708–17.
[114] Wilcox KS, Gee JM, Gibbons MB, Tvrdik P, White JA. Altered structure and function of astrocytes following status epilepticus. Epilepsy Behav 2015;49:17–19.
[115] Carmignoto G, Haydon PG. Astrocyte calcium signaling and epilepsy. Glia 2012;60(8):1227–33.
[116] de Lanerolle NC, Lee TS, Spencer DD. Astrocytes and epilepsy. Neurotherapeutics 2010;7(4):424–38.
[117] Jabs R, Seifert G, Steinhäuser C. Astrocytic function and its alteration in the epileptic brain. Epilepsia 2008;49(Suppl. 2):3–12.
[118] McNamara JO, Huang YZ, Leonard AS. Molecular signaling mechanisms underlying epileptogenesis. Science's STKE Signal Transduction Knowledge Environment 2006;2006(356):re12.
[119] Steinhäuser C, Seifert G. Glial membrane channels and receptors in epilepsy: impact for generation and spread of seizure activity. Eur J Pharmacol 2002;447(2–3):227–37.
[120] Jung P, Cornell-Bell A, Madden KS, Moss F. Noise-induced spiral waves in astrocyte syncytia show evidence of self-organized criticality. J Neurophysiol 1998;79(2):1098–101.
[121] Cano-Abad MF, Herrera-Peco I, Sola RG, Pastor J, García-Navarrete E, Moro RC, et al. New insights on culture and calcium signalling in neurons and astrocytes from epileptic patients. Int J Dev Neurosci 2011;29(2):121–9.
[122] Navarrete M, Perea G, Maglio L, Pastor J, García de Sola R, Araque A. Astrocyte calcium signal and gliotransmission in human brain tissue. Cereb Cortex 2013;23(5):1240–6.
[123] Vega-Zelaya L, Ortega GJ, Sola RG, Pastor J. Plasma albumin induces cytosolic calcium oscillations and DNA synthesis in human cultured astrocytes. BioMed Res Int 2014;2014:539140.
[124] Balàzsi G, Cornell-Bell AH, Moss F. Increased phase synchronization of spontaneous calcium oscillations in epileptic human versus normal rat astrocyte cultures. Chaos 2003;13(2):515–8.
[125] Seifert G, Schröder W, Hinterkeuser S, Schumacher T, Schramm J, Steinhäuser C. Changes in flip/flop splicing of astroglial AMPA receptors in human temporal lobe epilepsy. Epilepsia 2002;43(Suppl. 5):162–7.
[126] Lee TS, Mane S, Eid T, Zhao H, Lin A, Guan Z, et al. Gene expression in temporal lobe epilepsy is consistent with increased release of glutamate by astrocytes. Mol Med 2007;13(1–2):1–13.
[127] Djamshidian A, Grassl R, Seltenhammer M, Czech T, Baumgartner C, Schmidbauer M, et al. Altered expression of voltage-dependent calcium channel α_1 subunits in temporal lobe epilepsy with Ammon's horn sclerosis. Neuroscience 2002;111(1):57–69.
[128] Manning Jr. TJ, Sontheimer H. Spontaneous intracellular calcium oscillations in cortical astrocytes from a patient with intractable childhood epilepsy (Rasmussen's encephalitis). Glia 1997;21(3):332–7.
[129] Tashiro A, Goldberg J, Yuste R. Calcium oscillations in neocortical astrocytes under epileptiform conditions. J Neurobiol 2002;50(1):45–55.
[130] Gómez-Gonzalo M, Losi G, Brondi M, Uva L, Sato SS, de Curtis M, et al. Ictal but not interictal epileptic discharges activate astrocyte endfeet and elicit cerebral arteriole responses. Front Cell Neurosci 2011;5:8.
[131] Tian GF, Azmi H, Takano T, Xu Q, Peng W, Lin J, et al. An astrocytic basis of epilepsy. Nat Med 2005;11(9):973–81.

[132] Coiret G, Ster J, Grewe B, Wendling F, Helmchen F, Gerber U, et al. Neuron to astrocyte communication via cannabinoid receptors is necessary for sustained epileptiform activity in rat hippocampus. PloS One 2012;7(5):e37320.
[133] Gómez-Gonzalo M, Losi G, Chiavegato A, Zonta M, Cammarota M, Brondi M, et al. An excitatory loop with astrocytes contributes to drive neurons to seizure threshold. PLoS Biol 2010;8(4):e1000352.
[134] Stringer JL, Mukherjee K, Xiang T, Xu K. Regulation of extracellular calcium in the hippocampus in vivo during epileptiform activity—role of astrocytes. Epilepsy Res 2007;74(2–3):155–62.
[135] Broberg M, Pope KJ, Lewis T, Olsson T, Nilsson M, Willoughby JO. Cell swelling precedes seizures induced by inhibition of astrocytic metabolism. Epilepsy Res 2008;80(2–3):132–41.
[136] Torres A, Wang F, Xu Q, Fujita T, Dobrowolski R, Willecke K, et al. Extracellular Ca^{2+} acts as a mediator of communication from neurons to glia. Sci Signal 2012;5(208):ra8.
[137] Ding S, Fellin T, Zhu Y, Lee SY, Auberson YP, Meaney DF, et al. Enhanced astrocytic Ca^{2+} signals contribute to neuronal excitotoxicity after status epilepticus. J Neurosci 2007;27(40):10674–84.
[138] Daniel AG, Laffont P, Zhao M, Ma H, Schwartz TH. Optical electrocorticogram (OECoG) using wide-field calcium imaging reveals the divergence of neuronal and glial activity during acute rodent seizures. Epilepsy Behav 2015;49:61–5.
[139] Xu JH, Long L, Wang J, Tang YC, Hu HT, Soong TW, et al. Nuclear localization of $Ca_v2.2$ and its distribution in the mouse central nervous system, and changes in the hippocampus during and after pilocarpine-induced status epilepticus. Neuropathol Appl Neurobiol 2010;36(1):71–85.
[140] Xu JH, Long L, Tang YC, Hu HT, Tang FR. $Ca_v1.2$, $Ca_v1.3$, and $Ca_v2.1$ in the mouse hippocampus during and after pilocarpine-induced status epilepticus. Hippocampus 2007;17(3):235–51.
[141] Westenbroek RE, Bausch SB, Lin RC, Franck JE, Noebels JL, Catterall WA. Upregulation of L-type Ca^{2+} channels in reactive astrocytes after brain injury, hypomyelination, and ischemia. J Neurosci 1998;18(7):2321–34.
[142] Steinbeck JA, Henke N, Opatz J, Gruszczynska-Biegala J, Schneider L, Theiss S, et al. Store-operated calcium entry modulates neuronal network activity in a model of chronic epilepsy. Exp Neurol 2011;232(2):185–94.
[143] Aronica E, van Vliet EA, Mayboroda OA, Troost D, da Silva FH, Gorter JA. Upregulation of metabotropic glutamate receptor subtype mGluR3 and mGluR5 in reactive astrocytes in a rat model of mesial temporal lobe epilepsy. Eur J Neurosci 2000;12(7):2333–44.
[144] Ulas J, Satou T, Ivins KJ, Kesslak JP, Cotman CW, Balazs R. Expression of metabotropic glutamate receptor 5 is increased in astrocytes after kainate-induced epileptic seizures. Glia 2000;30(4):352–61.
[145] Hertz L, Bender AS, Woodbury DM, White HS. Potassium-stimulated calcium uptake in astrocytes and its potent inhibition by nimodipine. J Neurosci Res 1989;22(2):209–15.
[146] Morales-Garcia JA, Luna-Medina R, Martinez A, Santos A, Perez-Castillo A. Anticonvulsant and neuroprotective effects of the novel calcium antagonist NP04634 on kainic acid-induced seizures in rats. J Neurosci Res 2009;87(16):3687–96.

CHAPTER

7

Potassium Channels

OUTLINE

OVERVIEW

Astrocytes express various types of ion channels, including K^+, Na^+, Ca^{2+}, Cl^-, and other anion channels and aquaporin water channels (Table 7.1). In this chapter, we focus on the role of astrocyte K^+ channels in extracellular K^+ homeostasis and alterations in epilepsy. In particular, alterations in the inwardly rectifying potassium channel $K_{ir}4.1$ have been found in both epilepsy and other neurological diseases [1,2]. Dysregulation of K^+ homeostasis may play a key role in the hyperexcitability of epilepsy.

Astrocyte K^+ Channels

Early physiological studies of astrocytes revealed that the resting membrane conductance of astrocytes is dominated by passive K^+ currents which maintain the membrane potential (~−80 to −90 mV) close to the potassium Nernst equilibrium potential (E_K) [3–6]. Astrocyte K^+ channels include voltage-independent K^+ channels and voltage-dependent K^+ channels (divided into "inwardly rectifying" and "outwardly rectifying" categories) (Table 7.1). We will briefly summarize the known characteristics of each type of channel, and then focus in

Astrocytes and Epilepsy
DOI: http://dx.doi.org/10.1016/B978-0-12-802401-0.00007-7

TABLE 7.1 Astrocyte Ion Channels

Ion Channel	Molecular Identity	Localization	Main Function
Potassium channels			
Voltage-independent K^+ channels	Two-pore domain K^+ channels: TREK1, TREK2 and TWIK1	Hippocampus	Contribute to resting membrane potential
Inward rectifier potassium channels	K_{ir} 4.1 (predominant)	Ubiquitous	Maintenance of resting membrane potential;
	K_{ir} 2.1, 2.2, 2.3		K^+ buffering
	K_{ir} 3.1		
	K_{ir} 6.1, 6.2		
Outward rectifying K^+ channels			
Delayed rectifier potassium channels	K_v1.1, K_v1.2, K_v1.5, and K_v1.6	Ubiquitous	Generally unknown; may be involved in regulation of proliferation
K_D			
Rapidly inactivating A-type potassium currents (K_A)	K_v1.4	Hippocampus; astrocytes in vitro	Unknown
Ca^{2+}-dependent K^+ channels	K_{Ca}3.1	Cortex	Unknown
Sodium channels	Na_v1.1, Na_v1.2, and Na_v1.3	Spinal cord cultured astrocytes;	Regulation of differentiation, proliferation and migration(?)
		only NA_v1.2 was detected in spinal cord astrocytes in situ	Can be upregulated in pathological conditions
Calcium channels	L- (Ca_v1.2), N- (Ca_v2.2), P/Q- (Ca_v2.1), R- (Ca_v2.3), and T (Ca_v3.1)	Astrocytes in vitro; functional expression in situ remains controversial	Unknown
Transient receptor potential, TRP, channels	TRPC1, TRPC4, TRPC5	Regional distribution is unknown	Store-operated Ca^{2+} entry
Chloride channels	CLC1, CLC2, CLC3	Ubiquitous	Chloride transport; regulation of cell volume
	Volume-regulated anion channels of unknown identity		
Aquaporins (water channels)	AQP4 (predominant)	AQP4—ubiquitous	Water transport
	AQP9	AQP9—astrocytes in brain stem; ependymal cells; tanycytes in hypothalamus and in subfornical organ	

Reproduced with permission from Verkhratsky A, Butt AM. Glial physiology and pathophysiology. Oxford, UK: Wiley-Blackwell; 2013 (Table 4.1).

more detail on the inwardly rectifying (K_{ir}) channels which have been shown to be altered in epilepsy and other disease states.

Voltage-Independent K^+ Channels

TREK1, TREK2, and TWIK1 are voltage-independent two-pore domain (K2P) K^+ channels found in cultured astrocytes and astrocytes in hippocampal slices [7,8]. TREK channels have been found to contribute to the "passive" current pattern in hippocampal astrocytes [7]. TREK channels have various properties including activation by physical (membrane stretch, acidosis, temperature) and chemical (lipids, volatile anesthetics) stimuli. The functional significance of changes in TREK channels in disease states and TREK channels as pharmacologic targets is beginning to be explored in detail in certain pathologies [9–11]. Interestingly, TREK-1 channels can mediate glutamate release on GPCR activation [12] which could affect excitability.

Inwardly Rectifying (K_{ir}) K^+ Channels

K_{ir} channels are inwardly rectifying potassium channels expressed in many cell types throughout the body. To date, 15 K_{ir} subunit genes have been identified and classified into seven subfamilies ($K_{ir}1.x$ to $K_{ir}7.x$). These subfamilies can be categorized into four functional groups: classical K_{ir} channels ($K_{ir}2.x$); G protein-gated K_{ir} channels ($K_{ir}3.x$); ATP-sensitive K^+ channels ($K_{ir}6.x$); and K^+-transport channels ($K_{ir}1.x$, $K_{ir}4.x$, $K_{ir}5.x$, and $K_{ir}7.x$) (Figs. 7.1 and 7.2) [13].

Their inward rectification refers to the voltage-dependence of channel activation: K_{ir} channels are closed when the membrane is depolarized and open at hyperpolarized potentials. This property is due to their sensitivity to Mg^{2+} and polyamines, which block the channel at depolarized potentials [14]. Increases in extracellular K^+ result in inward flux of K^+ ions through K_{ir} channels, critical for K^+ removal from the extracellular space (ECS). K_{ir} channels are inhibited by micromolar concentrations of Ba^{2+} and "barium-sensitive" potassium flux is therefore thought to be K_{ir}-mediated (Fig. 7.3).

In astrocytes, K_{ir} channels are abundantly expressed and are thought to be largely responsible for setting the resting membrane potential [15,16]. Thus, a primary mechanism for K^+ reuptake from the ECS is thought to be via K_{ir} channels. Glial K_{ir} contributes to K^+ reuptake and spatial K^+ buffering [4,17], which has been most clearly demonstrated in the retina [18–21]. In CNS astrocytes, the expression of $K_{ir}4.1$ and $K_{ir}5.1$ has been investigated most thoroughly and have the predominant functional role [22–24] (Fig. 7.4). $K_{ir}4.1$ is the principal K_{ir} channel expressed in astrocytes, appearing during embryonic development and gradually increasing between the early postnatal period and adulthood [7,25] (Figs. 7.5 and 7.6). Pharmacological or genetic inactivation of $K_{ir}4.1$ leads to impairment of extracellular K^+ regulation [26–28]. In addition, members of the strongly rectifying $K_{ir}2$ family may also contribute to astroglial K^+ buffering [15,29].

Outwardly Rectifying K^+ Channels

Outwardly rectifying K^+ channels (K_v channels) comprise over 40 subtypes [30]. These channels are closed at resting membrane potential but open in response to membrane depolarization. They are not blocked by micromolar Ba^{2+} but are sensitive to the voltage-dependent K^+ channel blockers tetraethylammonium (TEA) and 4-aminopyridine (4-AP). Members

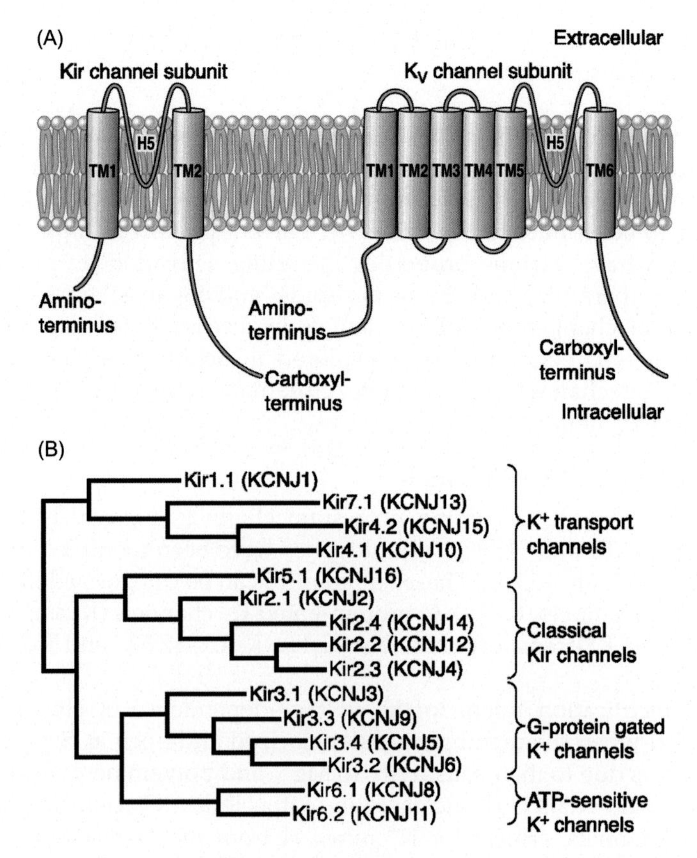

FIGURE 7.1 **Basic structure and K_{ir} channel phylogenetic tree.** (A) Primary structure of the K_{ir} channel subunit (*left*). Each K_{ir} subunit contains two transmembrane (TM1 and TM2) regions, a pore-forming (H5) loop, and cytosolic NH_2 and COOH termini. As a comparison, the structure of voltage-gated K^+ (K_v) channel subunit, which possesses six transmembrane (TM1-TM6) regions, is shown on the *right*. (B) Amino acid sequence alignment and phylogenetic analysis of the 15 known subunits of human K_{ir} channels. These subunits can be classified into four functional groups. *Source: Reproduced with permission from H. Hibino, A. Inanobe, K. Furutani, S. Murakami, I. Findlay, and Y. Kurachi. 2010. Inwardly rectifying potassium channels: their structure, function, and physiological roles. Physiol Rev 90:291–366 (Figure 2).*

of the K_v superfamily include $K_v1.4$ channels which mediate rapidly inactivating A-type potassium currents (K_A) and are responsible for fast hyperpolarization; and Ca^{2+}-dependent K^+ channels (K_{Ca}), which have a dual gating mechanism by both membrane voltage and cytosolic Ca^{2+}, and have been divided into three types (BK, IK, SK) differing in biophysical characteristics [31,32].

Astrocytes and K^+ Homeostasis

A critical role for astrocytes in maintaining extracellular K^+ homeostasis was first proposed by Leif Hertz in 1965 [33]. Neuronal activity is accompanied by influx of Na^+ and

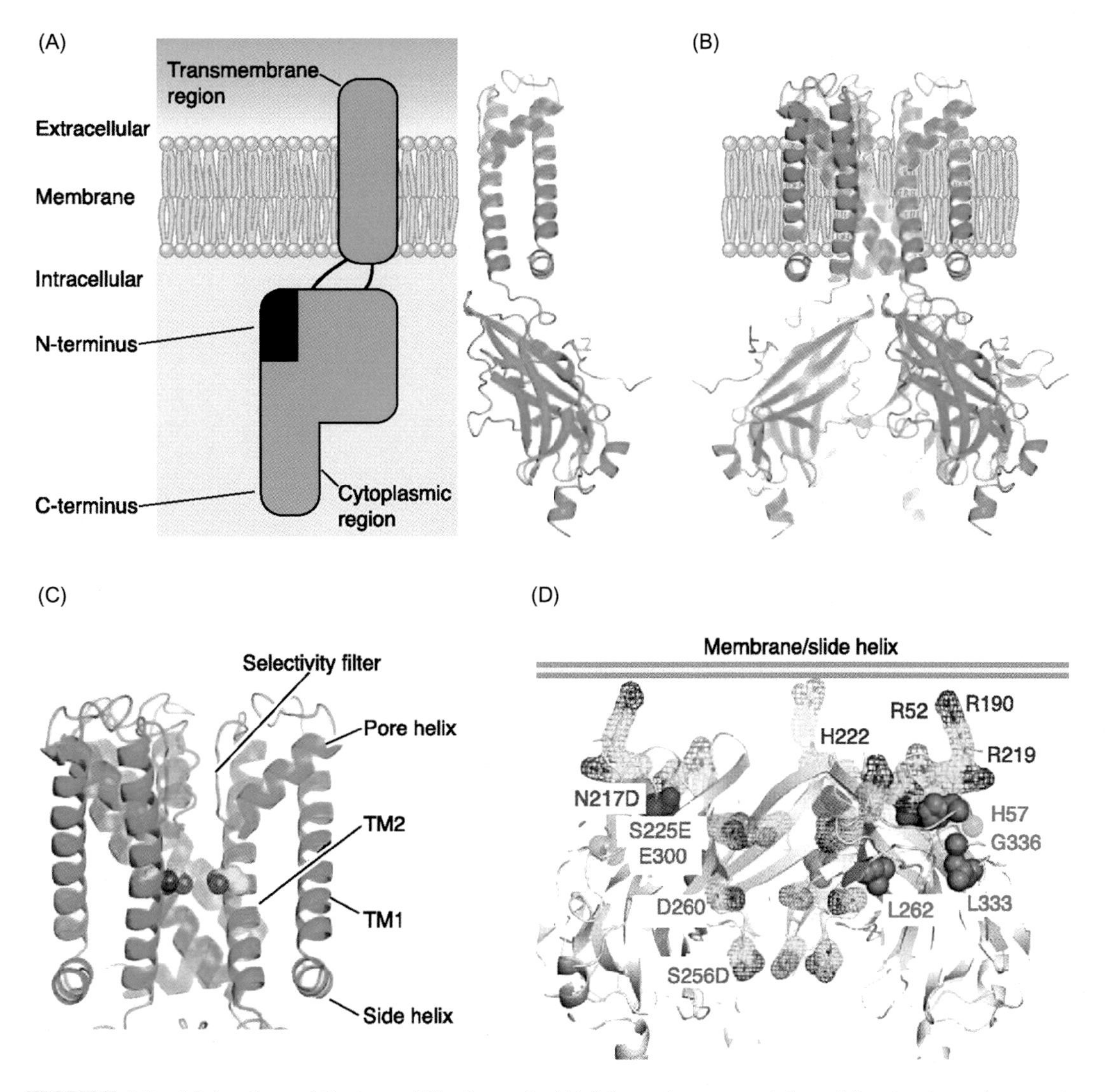

FIGURE 7.2 **Molecular architecture of K_{ir} channels.** (A) Schematic representation of the structure of a generic K_{ir} channel. The K_{ir} channel is divided into transmembrane and cytoplasmic domains. The NH_2 and COOH termini are cytosolic and together contribute to the formation of the cytoplasmic domain. (B) Tetrameric assembly of K_{ir} channels. The molecular architecture of a tetrameric K_{ir} channel (protein database ID 2QKS: K_{ir}3.1-K_{ir}Bac3.1 chimera) is represented as a cartoon model. The front subunit has been omitted for clarity. The organization of the tetramer of NH_2 and COOH termini leads to an extended pore for ion permeation. (C) Transmembrane domain. An enlarged view of the transmembrane domain indicates several important secondary structures for K_{ir} channel function. The transmembrane domain comprises three helices: TM1, pore, and TM2. At the membrane-cytoplasm interface, there is also an amphiphilic slide helix. The residue that is largely responsible for the interaction with polyamines and Mg^{2+} and thus inward rectification is indicated by the yellow/red spheres (D131 in K_{ir}3.1 which was mutated in the original coordinate). (D) Cytoplasmic domain. The opening of K_{ir} channels requires PtdIns(4,5)P_2. Those amino acid residues (*blue*) associated with the interaction with PtdIns(4,5)P_2 are distributed on the surface of the cytoplasmic domain toward the plasma membrane. The center of the cytoplasmic domain is a water-filled cavity that contributes to the ion permeation pathway. Some residues associated with inward rectification (*red*) map along this cavity. K_{ir}3.x channels are activated by direct interaction with a G protein βγ (Gβγ) subunit and channels containing K_{ir}3.2 and K_{ir}3.4 subunits can also be activated by Na^+. The amino acids responsible for these stimuli are indicated in green (Gβγ) and pink (Na^+). (C) and (D) have been derived from (B). *Source: Reproduced with permission from H. Hibino, A. Inanobe, K. Furutani, S. Murakami, I. Findlay, and Y. Kurachi. 2010. Inwardly rectifying potassium channels: their structure, function, and physiological roles. Physiol Rev 90:291–366 (Fig. 4).*

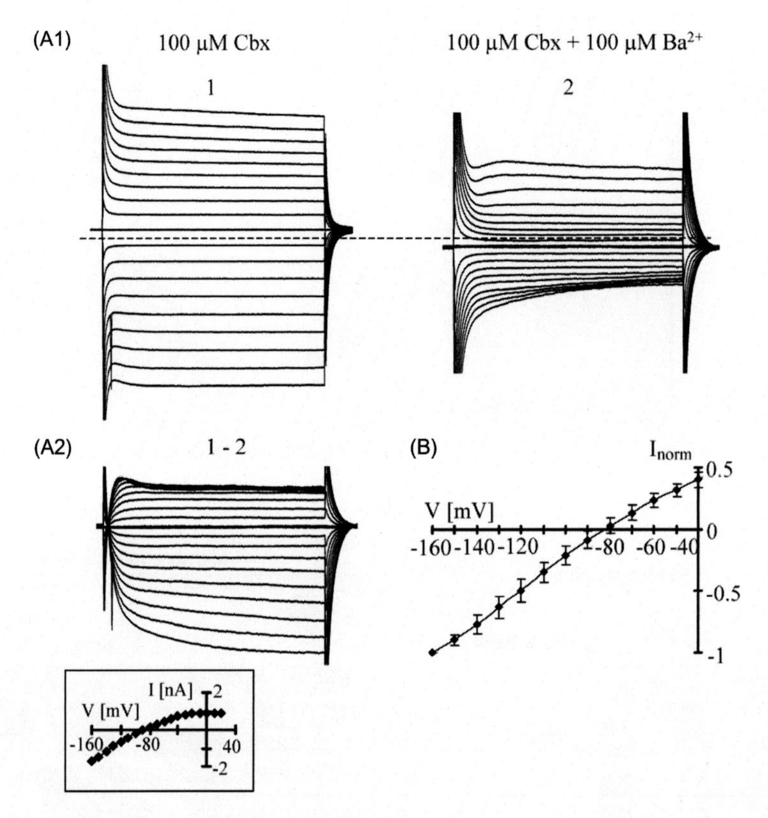

FIGURE 7.3 **Block of membrane currents of astrocytes in situ by Ba^{2+} in the presence of Cbx (100 μM).** (A1) Membrane currents were elicited in the whole-cell mode by depolarization and hyperpolarization between –160 and +20 mV (1; 10 mV increments; holding potential, –70 mV; dashed line, zero current). After adding 100 μM Ba^{2+} to the bath, currents were reduced (2). (A2) Ba^{2+}-sensitive currents were determined by subtracting respective current families (1–2). Ba^{2+}-sensitive currents (inset) reversed at –88.5 mV. (B) The averaged I/V relationship of Ba^{2+} (100 μM)-sensitive currents obtained in the presence of Cbx of six astrocytes (mean ± SD). In each cell, data were normalized to maximum inward currents. *Source: Reproduced with permission from G. Seifert, K. Hüttmann, D.K. Binder, C. Hartmann, A. Wyczynski, C. Neusch, and C. Steinhäuser. 2009. Analysis of astroglial K^+ channel expression in the developing hippocampus reveals a predominant role of the $K_{ir}4.1$ subunit. J Neurosci 29:7474–88.*

Ca^{2+} into neurons and efflux of K^+ from neurons into the ECS. Due to the limited ECS volume [34], small amounts of K^+ released by neurons can significantly affect extracellular potassium concentration. Prolonged increases in extracellular [K^+] could prolong neuronal depolarization and inhibit repolarization and therefore undermine neuronal and synaptic function [35].

Therefore, during normal activity extracellular [K^+] is tightly regulated by both *local K^+ uptake* and *K^+ spatial buffering*. Local K^+ uptake is accomplished by individual astrocytes through ion channels, exchangers, and transporters, notably $K_{ir}4.1$ but also importantly via Na^+/K^+/ATPase (sodium-potassium pump) and $Na^+/K^+/2Cl^-$ cotransporters [36,37]. Potassium spatial buffering [4,38,39] involves redistribution of K^+ ions throughout the gap-junction-connected glial "syncytium" from areas of high to low concentration, thereby

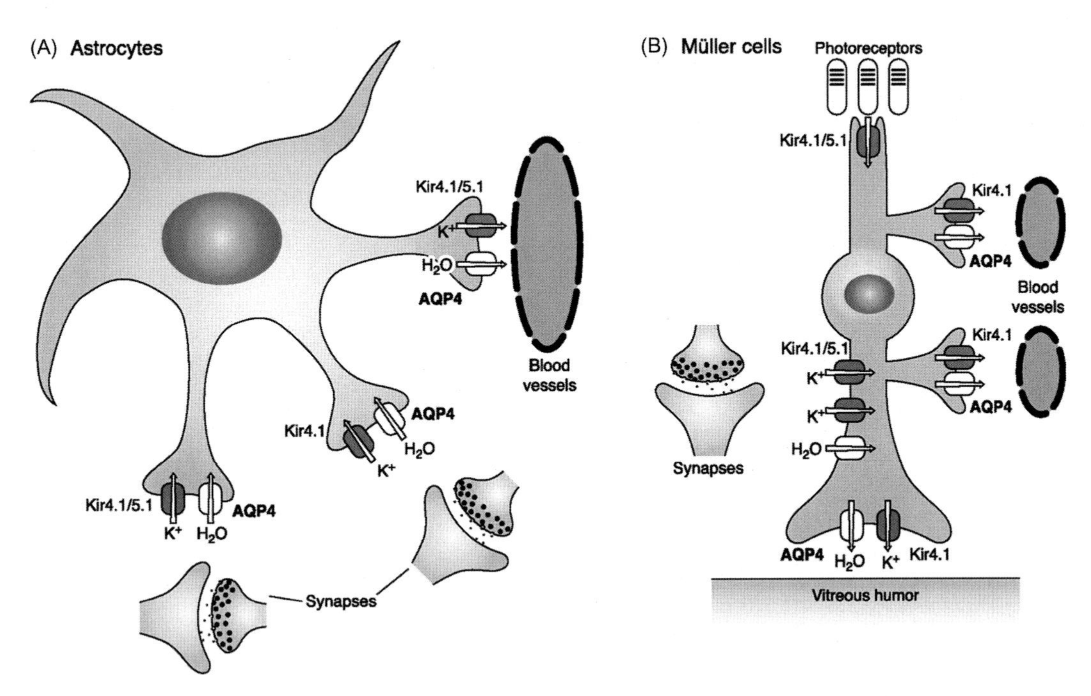

FIGURE 7.4 **Localization of astroglial K_{ir} channels.** (A) Brain astrocytes. Astrocytes harbor and localize homomeric $K_{ir}4.1$ channels and heteromeric $K_{ir}4.1/5.1$ channels in particular parts of their membrane: both the homomer and the heteromer are present in perisynaptic processes, and the heteromer is situated alone in the endfeet. (B) Retinal Müller cells. These cells express the heteromeric $K_{ir}4.1/5.1$ channel at perisynaptic sites and the homomeric $K_{ir}4.1$ channel at perivascular processes and in endfeet. In both the astrocytes and Müller cells, the K_{ir} channels coexist with the astroglial water channel aquaporin-4 (AQP4) in the same membrane domains (A and B). *Source: Reproduced with permission from H. Hibino, A. Inanobe, K. Furutani, S. Murakami, I. Findlay, and Y. Kurachi. 2010. Inwardly rectifying potassium channels: their structure, function, and physiological roles. Physiol Rev 90:291–366 (Fig. 16).*

helping to restore basal extracellular K^+ levels (Fig. 7.7). According to this model, the difference between glial syncytium membrane potential and local K^+ equilibrium potential at sites of high extracellular K^+ accumulation drives potassium ions into the astrocytic network. The inwardly rectifying K^+ channel $K_{ir}4.1$ is the primary K^+ channel responsible for astrocyte K^+ uptake [15,27,40]. Once taken up into astrocytes, K^+ can then propagate through the glial network to sites of lower $[K^+]_o$ [41]. Given the colocalization of $K_{ir}4.1$ channels and aquaporin-4 (AQP4) water channels, a similar process likely occurs for what might be called spatial "water" buffering, that is water distribution from sites of higher to lower to higher osmotic pressure within the astrocyte network (Fig. 7.8). These processes link water and K^+ reuptake into astrocytes leading to astrocyte swelling with intense neuronal activity (see also *Chapter 8: Water Channels*).

Genetic knockout mice have provided support for the roles of both astrocyte gap junctions and $K_{ir}4.1$ channels in regulating extracellular K^+ homeostasis. Mice completely deficient in astrocytic gap junction coupling (Cx30/43 double knockout mice) exhibited

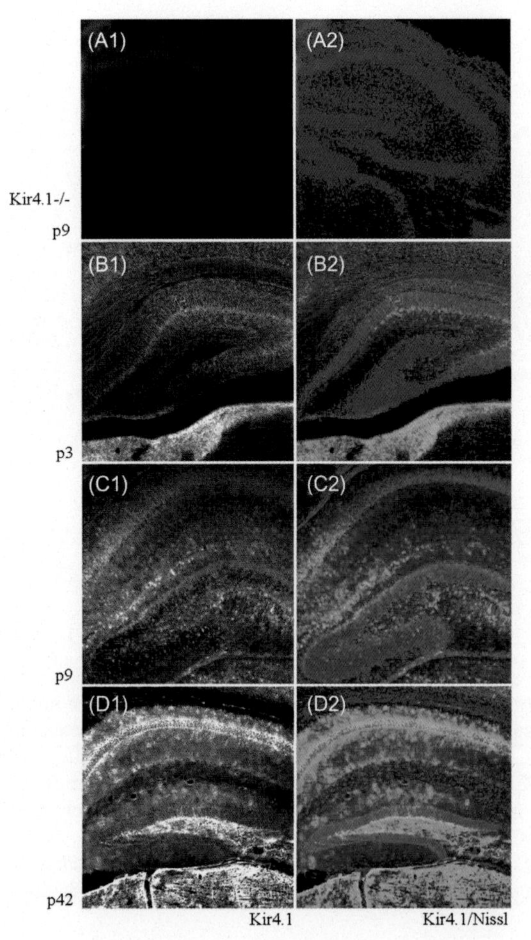

FIGURE 7.5 **$K_{ir}4.1$ immunoreactivity in the hippocampus.** (A1) $K_{ir}4.1$ immunoreactivity was lacking in $K_{ir}4.1^{-/-}$ mice. (A2) Perikarya of cells were visualized by Nissl staining (blue). (B1, C1, D1) $K_{ir}4.1$ immunoreactivity (gray levels) in hippocampal slices from mice at postnatal stages as indicated. (B2, C2, D2) Double labeling of $K_{ir}4.1$ (green) and Nissl (blue). Scale bar, 300 μm. *Source: Reproduced with permission from G. Seifert, K. Hüttmann, D.K. Binder, C. Hartmann, A. Wyczynski, C. Neusch, and C. Steinhäuser. 2009. Analysis of astroglial K^+ channel expression in the developing hippocampus reveals a predominant role of the $K_{ir}4.1$ subunit. J Neurosci 29:7474–88.*

impaired K^+ buffering, spontaneous epileptiform activity, and a decreased seizure threshold [42]. $K_{ir}4.1$ global knockout mice demonstrated the importance of $K_{ir}4.1$ in K^+ buffering by several cell types, including Müller glia [26], cochlear epithelium [43], and brainstem [28]. Similar results were obtained with RNAi knockdown of $K_{ir}4.1$ in cultured astrocytes [44]. Generation of a glial-conditional knockout mouse (cKO) of $K_{ir}4.1$ via Cre-lox technology was performed using the human GFAP astrocyte promoter (*gfa2*). This led to loss of $K_{ir}4.1$ throughout the CNS, depolarization of all glial types, impairment of astrocyte K^+ and glutamate uptake, ataxia, seizures, and early postnatal lethality [40]. A follow-up in vivo study in the same cKO mice demonstrated the role of $K_{ir}4.1$ in setting the membrane potential of glial cells and its contribution to astrocyte potassium permeability [45]. In another mouse model with glia-specific $K_{ir}4.1$ deletion, K^+ clearance was delayed after synaptic

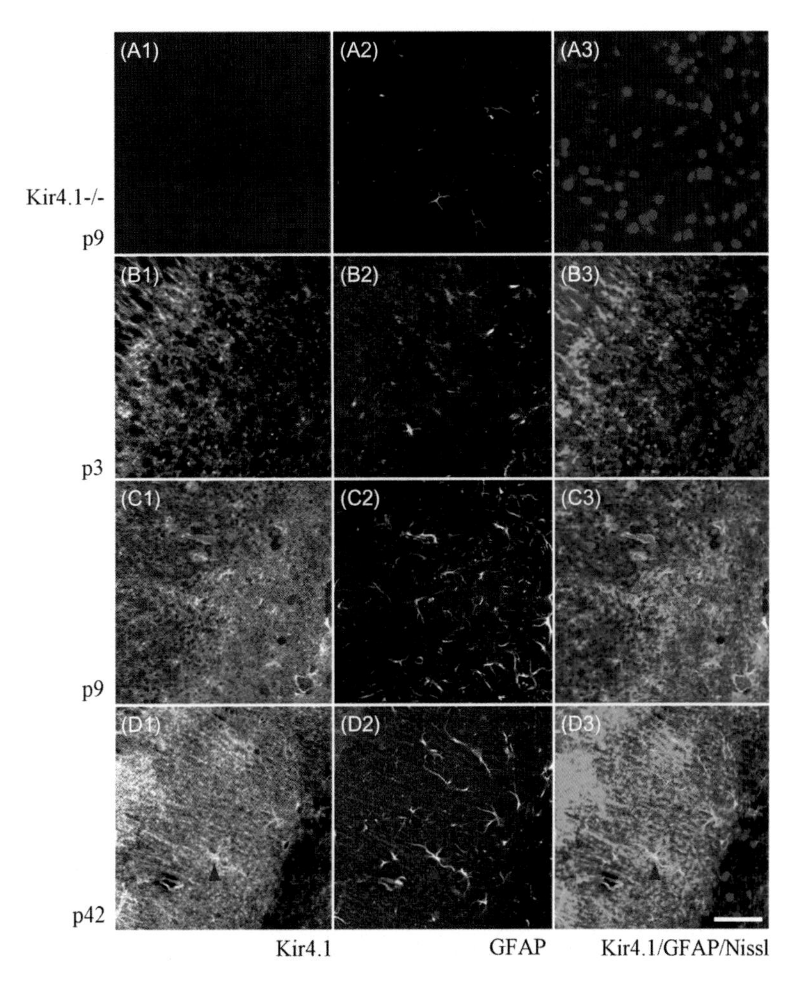

FIGURE 7.6 Confocal analysis of colabeling of $K_{ir}4.1$ and GFAP in strata radiatum and lacunosum-moleculare of the developing CA1 region. (A1) $K_{ir}4.1$ immunoreactivity was absent in $K_{ir}4.1^{-/-}$ mice. (A2) GFAP staining in the same section. (A3) The merge along with Nissl staining (blue; GFAP is in red). (B–D) $K_{ir}4.1$ immunoreactivity (B1, C1, D1), GFAP staining (B2, C2, D2), and the merge along with Nissl counterstaining (B3, C3, D3) at p3, p9, and p42 as indicated ($K_{ir}4.1$, green). Red arrowheads (D) indicate colocalization of $K_{ir}4.1$ and GFAP. Scale bar, 50 µm. In the top left, part of the pyramidal layer is visible. *Source: Reproduced with permission from G. Seifert, K. Hüttmann, D.K. Binder, C. Hartmann, A. Wyczynski, C. Neusch and, C. Steinhäuser. 2009. Analysis of astroglial K^+ channel expression in the developing hippocampus reveals a predominant role of the $K_{ir}4.1$ subunit. J Neurosci 29:7474–88.*

activation in stratum radiatum of hippocampal slices [46]. Alteration in $K_{ir}4.1$-mediated K^+ uptake may directly impact neuronal plasticity [47]. These studies have clearly indicated the critical importance of astrocytic $K_{ir}4.1$ in the maintenance of astrocyte membrane potential and overall CNS potassium homeostasis, and the potential epileptogenic effect of K^+ dysregulation.

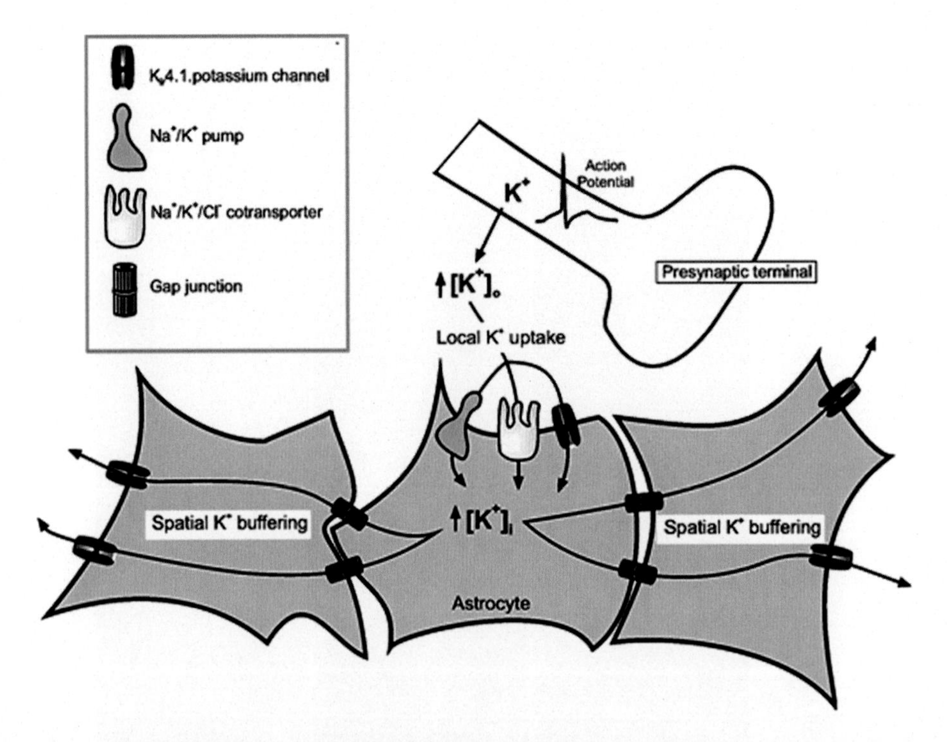

FIGURE 7.7 **Astrocytes control extracellular potassium homeostasis.** Potassium is released into the extracellular space during neuronal activity (K^+ efflux underlies the recovery phase—repolarization—of the action potential). Buffering of extracellular potassium occurs through astroglial inward rectifier potassium channels $K_{ir}4.1$, Na^+/K^+/ATPase and $Na^+/K^+/2Cl^-$ cotransporters, which all provide for local potassium buffering. Astrocytes take up excess K^+, redistribute the K^+ through the astroglial syncytium via gap junctions (spatial potassium buffering), and release K^+ through $K_{ir}4.1$ distantly. *Source: Reproduced with permission from Verkhratsky and Butt, Glial physiology and pathophysiology. Wiley-Blackwell, 2013 (Fig. 4.43).*

ALTERATIONS IN K^+ HOMEOSTASIS IN HUMAN EPILEPSY

Therefore, in the context of epilepsy it is important to determine whether there are changes in astrocyte K^+ regulation. During the intense neuronal activity of seizures, extracellular [K^+] may increase from ~3mM to a ceiling of 10–12mM [37,48–50]. (Under nonphysiological conditions such as hypoxia/ischemia and spreading depression, extracellular [K^+] may rise as high as 40–60mM [49,51,52].) Any impairment of glial K^+ uptake would be expected to be proconvulsant based on many previous studies. In the hippocampus, millimolar, and even submillimolar increases in extracellular K^+ concentration powerfully enhance epileptiform activity [53–56]. High-K^+ also reliably induces epileptiform activity in hippocampal slices from human patients with intractable temporal lobe epilepsy and hippocampal sclerosis [57]. Since both extracellular K^+ concentration and osmolarity have been shown to dramatically modulate neural excitability [58], it is plausible that changes in astrocytic K^+ or water channels could contribute to hyperexcitability in epilepsy.

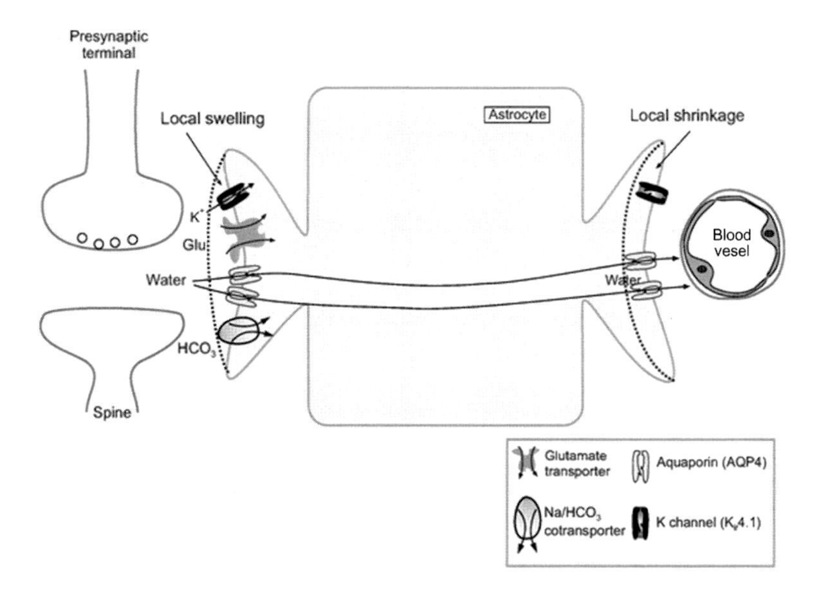

FIGURE 7.8 **Water transport and neuronal-activity-dependent regulation of extracellular volume by astrocytes.** Synaptic activity causes a local elevation in the concentrations of extracellular glutamate, K^+, and CO_2. Glutamate and K^+ are accumulated by astrocytes via glutamate transporters and $K_{ir}4.1$ channels, respectively, whereas CO_2 enters the cell in the form of HCO_3^- via Na^+/HCO_3^- cotransporter. These events increase osmotic pressure within the astrocyte, which favors water influx through aquaporins, resulting in local swelling of the astrocyte processes and shrinkage of the extracellular space. Locally accumulated water is redistributed through the astroglial network and can be extruded distantly, for example at blood vessels (also through aquaporins), resulting in local shrinkage of the astrocyte processes, and therefore increasing the extracellular volume near the site of efflux. *Source: Reproduced with permission from Verkhratsky and Butt, Glial physiology and pathophysiology. Wiley-Blackwell, 2013 (Fig. 4.46).*

Alterations in K_{ir} Currents and Channels in Epilepsy

Downregulation of astrocytic K_{ir} channels has indeed been found in the injured or diseased CNS. K_{ir} currents are reduced following injury-induced reactive gliosis in vitro [59], entorhinal cortex lesion [60], freeze lesion-induced cortical dysplasia [61,62], and traumatic [63] and ischemic [64] brain injury. In addition, human tissue studies have reported a significant loss of perivascular $K_{ir}4.1$ expression (~50%) and function in TLE patients, suggesting impaired spatial K^+ buffering in the epileptic hippocampus [65–71]. Reduction in K_{ir} conductance in astrocytes [69] and potassium clearance [71] were observed in human epilepsy specimens. Interestingly, two studies directly examined astrocytic K^+ currents in sclerotic versus nonsclerotic hippocampi resected from epileptic patients. Using ion-sensitive microelectrodes, Heinemann's group compared glial Ba^{2+}-sensitive K^+ uptake in the CA1 region of hippocampal slices obtained from patients with or without mesial temporal sclerosis [68,72]. Ba^{2+}, a blocker of K_{ir} channels, augmented stimulus-evoked K^+ elevation in nonsclerotic but not in sclerotic specimens, suggesting an impairment in K_{ir}-mediated K^+ buffering in sclerotic tissue. Direct evidence for downregulation of K_{ir} currents in the sclerotic CA1 region of hippocampus came from Steinhäuser's group who found, using comparative patch-clamp analysis, that there was a reduction in astroglial K_{ir} currents in sclerotic compared to non-sclerotic hippocampi [67]. These data indicate that dysfunction of astroglial K_{ir} channels could underlie impaired K^+ buffering and contribute to hyperexcitability in epileptic tissue [73]. More recently, Heuser et al. found that the loss of astrocytic $K_{ir}4.1$ expression was most pronounced around vessels in gliotic areas of the sclerotic hippocampus of mesial temporal lobe epilepsy (MTLE) patients [66]. Thus, increasing evidence suggests that the dysregulation of K^+ spatial buffering may play a role in hyperexcitability. Direct evidence for a potential causative role of reduced $K_{ir}4.1$ channel expression in epileptogenesis derives from an albumin model of BBB disruption in which $K_{ir}4.1$ downregulation occurs prior to onset of epilepsy; [74,75] however, other studies report no changes in astrocytic K_{ir} currents 7–16 days following systemic kainic acid injection [76] and increased $K_{ir}4.1$ expression in a rat pilocarpine model [77]. $K_{ir}4.1$ may also be regulated by inflammatory cytokines such as interleukin (IL)-1β [78] which are known to play a role in epilepsy (see *Chapter 13: Inflammation*).

Further evidence for a role of alteration in $K_{ir}4.1$ function in human epilepsy arises from genetic association studies. Quantitative trait loci mapping in mice led to identification of genetic variation in the $K_{ir}4.1$ gene Kcnj10; this led to human genetic association studies which identified an association between the human $K_{ir}4.1$ gene KCNJ10 mutations and seizure susceptibility in human populations [79,80]. Further evidence was adduced for an allelic association with variations in KCNJ10 and idiopathic generalized epilepsy [81,82]. Interestingly, human subjects were subsequently found with homozygous loss-of-function mutations in KCNJ10 leading to the definition of a specific autosomal recessive syndrome consisting of Epilepsy, Ataxia, Sensorineural deafness, and Tubulopathy (EAST syndrome) [83,84], a different version of which was also termed SeSAME syndrome (Seizures, Sensorineural deafness, Ataxia, Mental retardation, and Electrolyte imbalance) [85], both together termed SeSAME/EAST syndrome [86]. The combined phenotype of this syndrome parallels well the expression of $K_{ir}4.1$ in the brain, inner ear, and kidney. A subset of patients with autism and epilepsy were found to have gain-of-function mutations of KCNJ10 [87]. Single nucleotide variations in the KCNJ10 gene have also been found in TLE patients with hippocampal sclerosis [88].

Diverse K^+ Channelopathies in Animal and Human Epilepsies

It is important to recognize that aside from $K_{ir}4.1$/KCNJ10, a wide variety of mutations in K^+ channels have been described in distinct forms of animal and human epilepsies [89] (Tables 7.2 and 7.3). However, as most K^+ channels are not specific to astrocytes and are also expressed by neurons, a detailed discussion of each of these K^+ channel mutations and syndromes is beyond the scope of this chapter and is not pathophysiologically limited to astrocytes. $K_{ir}4.1$/KCNJ10 mutations represent a "purer" astrocytic contribution to epilepsy. Nevertheless, it is instructive to note the powerful impact of K^+ channel mutations—both neuronal and astrocytic—have on excitability and seizure susceptibility. In the case of $K_{ir}4.1$ loss-of-function mutations, seizure susceptibility likely arises from abrogation of the K^+ buffering mechanisms described above; whereas in the case of K_v channel mutations in neurons, seizure susceptibility is likely more related to loss of neuronal repolarization [90]. Interestingly, one of the newest and most promising antiepileptic drug mechanisms of

TABLE 7.2 K^+ Channelopathies in Human Epilepsies

Channel	Gene/Protein	Expression in Brain Regions Relevant to Epilepsy	Epilepsy Type	Gene Mutation/Channel Dysfunction	References
Voltage-gated K^+ channels	*KCNA1*/Kv1.1 (pore forming α subunit)	Axons/terminals of hippocampal neurons (*Shaffer* collateral axons and mossy fibers contacting CA3 neurons); hippocampal interneurons of hilus and CA1; neocortical pyramidal neurons	Generalized and partial seizures associated to EA1	*Loss-of-function* mutations generally associated with reduced current amplitudes; positive shift of the activation $V_{1/2}$; increased sensitivity to Zn^{2+} inhibition	Assaf and Chung (1984); Browne et al. (1994); Adelman et al. (1995); D'Adamo et al. (1998, 1999); Zuberi et al. (1999); Geiger and Jonas (2000); Cusimano et al. (2004); Imbrici et al. (2006, 2007, 2008); Guan et al. (2006)
	KCNAB2/Kvβ2 (β accessory subunit for Kv1 channels)	Widely expressed in cerebral cortex and hippocampus	Severe epilepsy including infantil spasms	Allele diletion/haploinsufficiency	Heilstedt et al. (2001)
	LGI1 (accessory protein for Kv1 channels)	Neocortex and hippocampus	Autosomal dominant lateral temporal lobe epilepsy	E383A, frameshift with protein truncation/Mutated LGI1 does not prevent Kvβ1-mediated Kv1 channel inactivation, a function performed by the WT protein	Kalachikov et al. (2002); Morante-Redolat et al. (2002)
	KCND2/Kv4.2 (I_A pore forming α subunit)	Dentrites of hippocampal neurons	Temporal lobe epilepsy	Truncated Kv4.2 subunit/attenuated I_A current density	Singh et al. (2006)
	KCNV2/Kv8.2 (silent subunit associating with Kv2 channels)	Pyramidal neurons and principal excitatory neurons of the pyramidal cell layers and the dentate gyrus; cortex, with high levels of transcript in layers 2/3 and 5	Febrile and afebrile partial seizures; epileptic encephalopathy	R7K, M285K/Reduction of Kv2.1 mediated current; M285K impairs the voltage-dependence of the channel	Jorge et al. (2011)

(Continued)

TABLE 7.2 (Continued)

Channel	Gene/Protein	Expression in Brain Regions Relevant to Epilepsy	Epilepsy Type	Gene Mutation/Channel Dysfunction	References
	KCNQ2-3/Kv7.2-3 (M current pore forming α subunit)	Widely expressed in brain at neuronal cell bodies	Benign familial neonatal convulsions	Five-base pair insertion deleting more than 300 amino acids from the *KCNQ2*; missense mutations in critical regions for *KCNQ3* channel function / Reduced *KCNQ* current amplitudes	Biervert et al. (1998); Charlier et al. (1998); Schroder et al. (1998); Singh et al. (1998)
	KCNH2/Kv11.2 (HERG channel pore forming α subunit)	Widely expressed in brain	Epilepsy associated with type 2 long QT syndrome	*Loss-of-function* mutations	Keller et al. (2009); Omichi et al. (2010); Tu et al. (2011); Zamorano-León et al. (2012)
Ca^{2+}-dependent K^+ channels	*KCNMA1*/$K_{Ca1.1}$ (BK channel pore forming α subunit)	Axons and presynaptic terminals of excitatory neurons of cortex and hippocampus	Generalized epilepsy and paroxysmal diskynesia	D434G/Increase of channel open probability and calcium dependence of $K_{Ca1.1}$ expressed alone or with $K_{Ca\beta1}$, $K_{Ca\beta2}$, or $K_{Ca\beta4}$; *loss-of-function* of $K_{Ca1.1}$/$K_{Ca\beta3}$-mediated currents	Du et al. (2005); Díez-Sampedro et al. (2006); Lee and Cui (2009); Yang et al. (2010)
	KCNMB3/$K_{Ca\beta3}$ (accessory protein for $K_{Ca1.1}$ channels)	Widely expressed at low levels in brain	Idiopathic generalized epilepsy	*Loss-of-function* of $K_{Ca1.1}$/$K_{Ca\beta3}$-mediated currents	Behrens et al. (2000); Hu et al. (2003); Lorenz et al. (2007)

Inwardly rectifying K^+ *channels*	*KCNJ2*/Kir2.1 (pore forming α subunit)	Hippocampus, caudate, putamen, nucleus accumbens; to lower levels in habenula and amygdale	Seizures associated to the Andersen Tawil Syndrome (ATS)	*Loss-of-function* mutations with dominant-negative effects	Haruna et al. (2007); Chan et al. (2010)
	KCNJ10/Kir4.1 (pore forming α subunit)	Oligodendrocytes and astrocytes surrounding synapses and blood vessels, mainly in the cortex, thalamus, hippocampus, and brainstem	Seizure susceptibility	R271C missense variation; no alteration in the biophysical properties of the channel when heterologously expressed	Buono et al. (2004); Connors et al. (2004); Shang et al. (2005)
			Epilepsy associated to EAST syndrome	*Loss-of-function* recessive mutations	Bockenhauer et al. (2009); Scholl et al. (2009)
			Epilepsy associated to autism spectrum disorders (ASDs)	R18Q, V84M, *gain-of-function* of Kir4.1 and Kir4.1/Kir5.1-mediated current	Sicca et al. (2011)
	KCNJ11/Kir6.2 (K_{ATP} channel pore forming α subunit) *ABCC8*/SUR1 (K_{ATP} channel accessory regulatory subunit)	Hippocampus (principal neurons, interneurons, and glial cells); neocortex, entorhinal, and piriform cortex	Developmental delay, epilepsy, and neonatal diabetes mellitus (DEND syndrome)	*Gain-of-function* mutations leading to decreased channel inhibition by ATP, or enhanced Mg^{2+}-nucleotide-induced activation	Karschin et al. (1997); Hattersley and Ashcroft (2005)

Reproduced with permission from D'Adamo M., Catacuzzeno L., DiGiovanni G., Franciolini F., and Pessia M. 2013. K^+ channelepsy: progress in the neurobiology of potassium channels and epilepsy. Frontiers in Cellular Neuroscience 7:1–21 (Table 1).

TABLE 7.3 K^+ Channelopathies in Animal Models of Epilepsy

Channel	Gene/ Protein	Animal Model/ Channel Disfunction	Epilepsy Phenotype	Functional Effects of Mutation on Neurons Relevant to Epilepsy	References
Voltage-gated K^+ channels	*Kcna1*/ Kv1.1 (pore forming α subunit)	Kv1.1$^{V408A/+}$ mice/ EA-1 mutation that alters the biophysical properties of the channel	Unknown	Cf. *Table 1* for WT expression; unknown	
		Kv1.1$^{-/-}$ knockout mice	Spontaneous seizures resembling human temporal lobe epilepsy	Hippocampus with neural loss, astrocytosis, and mossy fiber sprouting; mossy fiber stimulation mediates long-latency epileptiform burst discharges; mossy fibers and medial perforant path axons were hyperexcitable and produced greater pre- and postsynaptic responses with reduced paired-pulse ratios	Smart et al. (1998); Rho et al. (1999)
		Kv1.1$^{S309T/+}$ rats/80% smaller current amplitudes with dominant-negative effects	neuromyotonia and spontaneous convulsive seizures aggravated by stress	Cortical and hippocampal EEG with aberrant large spike activity associated with falling-down behavior, low-voltage fast wave discharges during the tonic stage, spike-and-wave discharges (2 Hz) during the clonic convulsive stage. Behavioral phenotypes and abnormal discharge patterns similar to other rodent models of temporal lobe epilepsy	Ishida et al. (2012)
		mceph/mceph mice, carrying a 11-basepair deletion in the *Kcna1* gene. The mutation leads to a frame shift and to a premature stop codon	Running seizures, complex partial seizures, and postanesthetic tonic-clonic seizures	Increased brain volume and hypertrophic brain cells; hippocampal hyperexcitability consistent with limbic status epilepticus	Donahue et al. (1996); Petersson et al. (2003)
		Adar2$^{-/-}$ mice	Increased susceptibility to epileptic seizure	Adenosine deaminase (*Adar2*) acting on Kv1.1 mRNA and leading to a *gain-of-function* of the resulting current (increased amplitude and faster recovery from inactivation). *Adar2* also edits the mRNA and AMPA glutamate receptors	Higuchi et al. (2000); Bhalla et al. (2004)

	Kcna2/ Kv1.2 (pore forming α subunit)	Kv1.2$^{-/-}$ knockout mice	Spontaneous generalized seizures	WT Expression: mainly overlapping with that of Kv1.1 channels; fibers and neuropil, but not somata, neocortex; synaptic terminals of entorhinal afferents. Strongly expressed in axon initial segment, where they participate to action potential generation	Brew et al. (2007)
	Kcnab2/ Kvβ2 (β subunit for Kv1 channels)	Kvβ2$^{-/-}$ knockout mice/β2 promotes the trafficking of Kv1.1 and Kv1.2 to the membrane surface	Increased neuronal excitability, occasional seizures	Cf. *Table 1* for WT expression; deficits in associative learning and memory; reduction in the slow afterhyperpolarization and concomitant increase in excitability of projection neurons in the lateral nucleus of the amygdale	McCormack et al. (2002); Connor et al. (2005); Perkowski and Murphy (2011)
	Kchip2 (Accessory subunit for Kv4 channels)	*Kchip2*$^{-/-}$ knockout mice/Reduced I_A current density and slowed recovery from inactivation in hyppocampal neurons	Increased susceptibility to seizure induced by kindling	WT expression; apical dendrites of hyppocampal pyramidal cells Chronic hyperexcitability of hyppocampal pyramidal neurons	Wang et al. (2013)
	Lgi1 (accessory protein for Kv1 channels)	*Lgi1*$^{-/-}$ knockout mice/Kvβ1-mediated Kv1 channel inactivation is not prevented, a function performed in WT mice	Lethal epilepsy; the heterozygous has a lower seizure thresholds	Cf. *Table 1* for WT expression A lack of *Lgi1* disrupts synaptic protein connection and selectively reduces AMPA receptor-mediated synaptic transmission in the hippocampus	Fukata et al. (2010)
	Kcnd2/ Kv4.2 (I_A pore forming α subunit)	Kv4.2$^{-/-}$ knockout mice	Enhanced susceptibility to kainite-induced seizure, but see Hu et al. (2006)	Cf. *Table 1* for WT expression. Increased epileptiform bursting in area CA1	Hu et al. (2006); Nerbonne et al. (2008); Barnwell et al. (2009)
Ca^{2+}-dependent K^+ channels	*Kcnb4*/ KCaβ4 (accessory subunit for K_{Ca}1.1 channels)	$K^{-/-}_{Ca\beta4}$ mice/ Gain-of-function of $K_{Ca1.1}$-mediated current	Temporal cortex seizures	WT expression: Axons and presynaptic terminals of Shaffer collaterals and CA3 hippocalpal neurons. Higher firing rate of dental gyrus neurons	Jin et al. (2000); Raffaelli et al. (2004); Brenner et al. (2005); Shruti et al. (2008); Sheehan et al. (2009)

(*Continued*)

TABLE 7.3 (Continued)

Channel	Gene/ Protein	Animal Model/ Channel Disfunction	Epilepsy Phenotype	Functional Effects of Mutation on Neurons Relevant to Epilepsy	References
Inwardlyrectifying K^+ channels	*Kcnj6*/ Kir3.2 (GIRK2 channel pore forming α subunit)	Kir3.2$^{-/-}$ knockout mice	Spontaneous convulsions and increased propensity for generalized seizures; more susceptible to pharmacologically induced seizure	WT expression: Cortex, hippocampus, weaker signal in thalamic nuclei and amygdaloid nuclei; reduced GIRK1 expression in brain	Signorini et al. (1997)
		Weaver (w/w) mice, G156S/alteration of the K^+ selectivity of the channel	Epileptic seizures	Neurodegeneration; calcium overload within cells, reduced GIRK1 expression in brain	Patil et al. (1995); Slesinger et al. (1996)
	Kcnj10/ Kir4.1 (pore forming α subunit)	Kir4.1$^{-/-}$ knockout mice	Stress-induced seizures	WT expression: Oligodendrocytes and astrocytes surrounding synapses and blood vessels in cortex, thalamus, hippocampus, braistem No membrane depolarization is observed in astrocytes following $[K^+]_o$ increase by neuronal activity	Neusch et al. (2001); Djukic et al. (2007)
		DBA/2 mouse strain, T262S missense variation resulting in barium-sensitive Kir currents in astrocytes substantially reduced; no alteration in the biophysical properties of the channel	Greater susceptibility to induced seizures compared to the C57BL/6 strain	Potassium and glutamate buffering by cortical astrocytes is impaired	Ferraro et al. (2004); Inyushin et al. (2010)
	Kcnj11/ Kir6.2 (K_{ATP} channel pore forming α subunit)	Kir6.2$^{-/-}$ knockout mice	High-voltage sharp-wave bursts EEG	Cf. *Table 1* for WT expression; Substantia nigra neurons are depolarized by hypoxia (WT neurons are instead hyperpolarized)	Yamada et al. (2001)

Reproduced with permission from D'Adamo M., Catacuzzeno L., DiGiovanni G., Franciolini F., and Pessia M. 2013. K^+ channelepsy: progress in the neurobiology of potassium channels and epilepsy. Frontiers in Cellular Neuroscience 7:1–21 (Table 2).

action is that of retigabine, which activates voltage-dependent K^+ channels of the K_v7 subfamily and is thus a "neuronal potassium channel opener" [91]. No similar drug yet exists to activate glial K_{ir} channels.

Loss of K^+ Homeostasis: A More General Astrocyte Mechanism in Neurological Diseases?

Clearance of the two most neuroactive and potentially neurotoxic substances, K^+ and glutamate, requires efficient astrocyte ion channel and transporter function, which may actually itself be regulated at low levels of stimulation to prevent hyperexcitability or excitotoxicity [92]. In view of the pivotal role of K^+ homeostasis in particular in normal CNS function, it is interesting that increasing evidence is emerging that loss of K^+ homeostasis and specifically loss of $K_{ir}4.1$ function has emerged not only in epilepsy but also may be a more widespread mechanism in diverse neurological diseases [2]. For example, two mouse models of Huntington's disease have been associated with $K_{ir}4.1$ dysfunction [93,94]. Symptom onset in these models was associated with decreased $K_{ir}4.1$ expression leading to elevated in vivo striatal extracellular $[K^+]$ and increased medium spiny neuron (striatal neuron) excitability; viral delivery of $K_{ir}4.1$ channels to striatal astrocytes restored $K_{ir}4.1$ function, normalized extracellular $[K^+]$, prolonged survival and improved motor phenotypes [94]. Reduced $K_{ir}4.1$ expression or function has also been associated with Alzheimer's disease [95], amyotrophic lateral sclerosis [96,97], pain [98], and Rett syndrome [2]. Furthermore, anti-$K_{ir}4.1$ antibodies have been detected in a subpopulation of patients with multiple sclerosis (MS) [99]. A recent immunohistochemical study of subcortical white matter MS lesions indicated loss of $K_{ir}4.1$ immunoreactivity in both acute and chronic active MS lesions [100].

Thus, restoration of K^+ and glutamate homeostasis may represent a new therapeutic option for diverse neurological diseases, including epilepsy. It is interesting that in the recent study of HD, restoration of $K_{ir}4.1$ channels by viral delivery not only restored K^+ homeostasis but also normalized levels of GLT1 [94], the primary astrocyte glutamate transporter which had already been implicated in HD pathology [101–103]. This is reminiscent of the finding in $K_{ir}4.1$ cKO mice of impaired glutamate uptake [40], together suggesting a cross-talk between $K_{ir}4.1$-mediated K^+ homeostasis and GLT1-mediated glutamate uptake. It will be interesting for the future to further understand the mechanistic interactions between $K_{ir}4.1$ channels, AQP4 channels (see *Chapter 8: Water Channels*), and glutamate transporters (see *Chapter 9: Glutamate Metabolism*) to derive therapeutic targets for selective or combined manipulation of K^+, glutamate, and water homeostasis tailored to particular translational requirements and disease states.

References

[1] Bedner P, Steinhäuser C. Altered K_{ir} and gap junction channels in temporal lobe epilepsy. Neurochem Int 2013;63(7):682–7.

[2] Olsen ML, Khakh BS, Skatchkov SN, Zhou M, Lee CJ, Rouach N. New insights on astrocyte ion channels: critical for homeostasis and neuron-glia signaling. J Neurosci 2015;35(41):13827–35.

[3] Kuffler SW, Nicholls JG, Orkand RK. Physiological properties of glial cells in the central nervous system of amphibia. J Neurophysiol 1966;29(4):768–87.

[4] Orkand RK, Nicholls JG, Kuffler SW. Effect of nerve impulses on the membrane potential of glial cells in the central nervous system of amphibia. J Neurophysiol 1966;29(4):788–806.
[5] Kuffler SW. Neuroglial cells: physiological properties and a potassium mediated effect of neuronal activity on the glial membrane potential. Proc R Soc Lond B Biol Sci 1967;168(1010):1–21.
[6] Ransom BR, Goldring S. Ionic determinants of membrane potential of cells presumed to be glia in cerebral cortex of cat. J Neurophysiol 1973;36(5):855–68.
[7] Seifert G, Hüttmann K, Binder DK, Hartmann C, Wyczynski A, Neusch C, et al. Analysis of astroglial K^+ channel expression in the developing hippocampus reveals a predominant role of the $K_{ir}4.1$ subunit. J Neurosci 2009;29(23):7474–88.
[8] Zhou M, Xu G, Xie M, Zhang X, Schools GP, Ma L, et al. TWIK-1 and TREK-1 are potassium channels contributing significantly to astrocyte passive conductance in rat hippocampal slices. J Neurosci 2009;29(26):8551–64.
[9] Alloui A, Zimmermann K, Mamet J, Duprat F, Noel J, Chemin J, et al. TREK-1, a K^+ channel involved in polymodal pain perception. EMBO J 2006;25(11):2368–76.
[10] Heurteaux C, Guy N, Laigle C, Blondeau N, Duprat F, Mazzuca M, et al. TREK-1, a K^+ channel involved in neuroprotection and general anesthesia. EMBO J 2004;23(13):2684–95.
[11] Vivier D, Bennis K, Lesage F, Ducki S. Perspectives on the two-pore domain potassium channel TREK-1 (TWIK-related K channel 1): a novel therapeutic target? J Med Chem 2015 Dec 14 [Epub ahead print]
[12] Woo DH, Han KS, Shim JW, Yoon BE, Kim E, Bae JY, et al. TREK-1 and Best1 channels mediate fast and slow glutamate release in astrocytes upon GPCR activation. Cell 2012;151(1):25–40.
[13] Hibino H, Inanobe A, Furutani K, Murakami S, Findlay I, Kurachi Y. Inwardly rectifying potassium channels: their structure, function, and physiological roles. Physiol Rev 2010;90(1):291–366.
[14] Lu Z. Mechanism of rectification in inward-rectifier K^+ channels. Annu Rev Physiol 2004;66:103–29.
[15] Butt AM, Kalsi A. Inwardly rectifying potassium channels (K_{ir}) in central nervous system glia: a special role for $K_{ir}4.1$ in glial functions. J Cell Mol Med 2006;10(1):33–44.
[16] Olsen ML, Sontheimer H. Functional implications for $K_{ir}4.1$ channels in glial biology: from K^+ buffering to cell differentiation. J Neurochem 2008;107(3):589–601.
[17] Ransom BR. Do glial gap junctions play a role in extracellular ion homeostasis? Dermietzel R, Spray DC, editors. Gap junctions in the nervous system. Georgetown, TX: Landes Bioscience; 1996. p. 159–73.
[18] Newman EA. High potassium conductance in astrocyte endfeet. Science 1986;233(4762):453–4.
[19] Newman EA. Inward-rectifying potassium channels in retinal glial (Müller) cells. J Neurosci 1993;13(8):3333–45.
[20] Newman EA, Frambach DA, Odette LL. Control of extracellular potassium levels by retinal glial cell K^+ siphoning. Science 1984;225(4667):1174–5.
[21] Newman EA, Karwoski CJ. Spatial buffering of light-evoked potassium increases by retinal glial (Müller) cells. Acta Physiol Scand Suppl 1989;582:51.
[22] Hibino H, Fujita A, Iwai K, Yamada M, Kurachi Y. Differential assembly of inwardly rectifying K^+ channel subunits, $K_{ir}4.1$ and $K_{ir}5.1$, in brain astrocytes. J Biol Chem 2004;279(42):44065–73.
[23] Higashi K, Fujita A, Inanobe A, Tanemoto M, Doi K, Kubo T, et al. An inwardly rectifying K^+ channel, $K_{ir}4.1$, expressed in astrocytes surrounds synapses and blood vessels in brain. Am J Physiol Cell Physiol 2001;281(3):C922–31.
[24] Olsen ML, Higashimori H, Campbell SL, Hablitz JJ, Sontheimer H. Functional expression of $K_{ir}4.1$ channels in spinal cord astrocytes. Glia 2006;53(5):516–28.
[25] Moroni RF, Inverardi F, Regondi MC, Pennacchio P, Frassoni C. Developmental expression of $K_{ir}4.1$ in astrocytes and oligodendrocytes of rat somatosensory cortex and hippocampus. Int J Dev Neurosci 2015;47(Pt B):198–205.
[26] Kofuji P, Ceelen P, Zahs KR, Surbeck LW, Lester HA, Newman EA. Genetic inactivation of an inwardly rectifying potassium channel ($K_{ir}4.1$ subunit) in mice: phenotypic impact in retina. J Neurosci 2000;20(15):5733–40.
[27] Kofuji P, Newman EA. Potassium buffering in the central nervous system. Neuroscience 2004;129(4):1045–56.
[28] Neusch C, Papadopoulos N, Müller M, Maletzki I, Winter SM, Hirrlinger J, et al. Lack of the $K_{ir}4.1$ channel subunit abolishes K^+ buffering properties of astrocytes in the ventral respiratory group: impact on extracellular K^+ regulation. J Neurophysiol 2006;95(3):1843–52.
[29] Neusch C, Weishaupt JH, Bähr M. K_{ir} channels in the CNS: emerging new roles and implications for neurological diseases. Cell Tissue Res 2003;311(2):131–8.
[30] Verkratsky A, Butt A. Glial physiology and pathophysiology. West Sussex, UK: Wiley-Blackwell; 2013.

[31] Longden TA, Dunn KM, Draheim HJ, Nelson MT, Weston AH, Edwards G. Intermediate-conductance calcium-activated potassium channels participate in neurovascular coupling. Br J Pharmacol 2011;164(3):922–33.
[32] Quandt FN, MacVicar BA. Calcium-activated potassium channels in cultured astrocytes. Neuroscience 1986;19(1):29–41.
[33] Hertz L. Possible role of neuroglia: a potassium-mediated neuronal-neuroglial-neuronal impulse transmission system. Nature 1965;206(989):1091–4.
[34] Sykova E, Nicholson C. Diffusion in brain extracellular space. Physiol Rev 2008;88(4):1277–340.
[35] Walz W. Role of astrocytes in the clearance of excess extracellular potassium. Neurochem Int 2000;36(4–5):291–300.
[36] D'Ambrosio R, Gordon DS, Winn HR. Differential role of K_{ir} channel and Na^+/K^+ pump in the regulation of extracellular K^+ in rat hippocampus. J Neurophysiol 2002;87(1):87–102.
[37] Xiong ZQ, Stringer JL. Astrocytic regulation of the recovery of extracellular potassium after seizures in vivo. Eur J Neurosci 1999;11(5):1677–84.
[38] Orkand RK. Extracellular potassium accumulation in the nervous system. Fed Proc 1980;39:1515–8.
[39] Walz W. Do neuronal signals regulate potassium flow in glial cells? Evidence from an invertebrate central nervous system. J Neurosci Res 1982;7(1):71–9.
[40] Djukic B, Casper KB, Philpot BD, Chin LS, McCarthy KD. Conditional knock-out of $K_{ir}4.1$ leads to glial membrane depolarization, inhibition of potassium and glutamate uptake, and enhanced short-term synaptic potentiation. J Neurosci 2007;27(42):11354–65.
[41] Seifert G, Carmignoto G, Steinhäuser C. Astrocyte dysfunction in epilepsy. Brain Res Rev 2010;63:212–21.
[42] Wallraff A, Köhling R, Heinemann U, Theis M, Willecke K, Steinhäuser C. The impact of astrocytic gap junctional coupling on potassium buffering in the hippocampus. J Neurosci 2006;26(20):5438–47.
[43] Marcus DC, Wu T, Wangemann P, Kofuji P. KCNJ10 ($K_{ir}4.1$) potassium channel knockout abolishes endocochlear potential. Am J Physiol Cell Physiol 2002;282(2):C403–7.
[44] Kucheryavykh YV, Kucheryavykh LY, Nichols CG, Maldonado HM, Baksi K, Reichenbach A, et al. Downregulation of $K_{ir}4.1$ inward rectifying potassium channel subunits by RNAi impairs potassium transfer and glutamate uptake by cultured cortical astrocytes. Glia 2007;55(3):274–81.
[45] Chever O, Djukic B, McCarthy KD, Amzica F. Implication of $K_{ir}4.1$ channel in excess potassium clearance: an in vivo study on anesthetized glial-conditional $K_{ir}4.1$ knock-out mice. J Neurosci 2010;30(47):15769–77.
[46] Haj-Yasein NN, Jense V, Vindedal GF, Gundersen GA, Klungland A, Otterson OP, et al. Evidence that compromised K^+ spatial buffering contributes to the epileptogenic effect of mutations in the human $K_{ir}4.1$ gene (KCNJ10). Glia 2011;59:1635–42.
[47] Sibille J, Pannasch U, Rouach N. Astroglial potassium clearance contributes to short-term plasticity of synaptically evoked currents at the tripartite synapse. J Physiol 2014;592(Pt 1):87–102.
[48] Heinemann U, Lux HD. Ceiling of stimulus induced rises in extracellular potassium concentration in the cerebral cortex of cat. Brain Res 1977;120(2):231–49.
[49] Somjen GG. Ion regulation in the brain: implications for pathophysiology. Neuroscientist 2002;8:254–67.
[50] Ballanyi K, Grafe P, ten Bruggencate G. Ion activities and potassium uptake mechanisms of glial cells in guinea-pig olfactory cortex slices. J Physiol 1987;382:159–74.
[51] Hansen AJ. Effect of anoxia on ion distribution in the brain. Physiol Rev 1985;65(1):101–48.
[52] Somjen GG. Mechanisms of spreading depression and hypoxic spreading depression-like depolarization. Physiol Rev 2001;81(3):1065–96.
[53] Rutecki PA, Lebeda FJ, Johnston D. Epileptiform activity induced by changes in extracellular potassium in hippocampus. J Neurophysiol 1985;54(5):1363–74.
[54] Yaari Y, Konnerth A, Heinemann U. Nonsynaptic epileptogenesis in the mammalian hippocampus in vitro. II. Role of extracellular potassium. J Neurophysiol 1986;56(2):424–38.
[55] Traynelis SF, Dingledine R. Potassium-induced spontaneous electrographic seizures in the rat hippocampal slice. J Neurophysiol 1988;59(1):259–76.
[56] Feng Z, Durand DM. Effects of potassium concentration on firing patterns of low-calcium epileptiform activity in anesthetized rat hippocampus: inducing of persistent spike activity. Epilepsia 2006;47:727–36.
[57] Gabriel S, Njunting M, Pomper JK, Merschhemke M, Sanabria ER, Eilers A, et al. Stimulus and potassium-induced epileptiform activity in the human dentate gyrus from patients with and without hippocampal sclerosis. J Neurosci 2004;24(46):10416–30.

[58] Schwartzkroin PA, Baraban SC, Hochman DW. Osmolarity, ionic flux, and changes in brain excitability. Epilepsy Res 1998;32(1–2):275–85.

[59] MacFarlane SN, Sontheimer H. Electrophysiological changes that accompany reactive gliosis in vitro. J Neurosci 1997;17(19):7316–29.

[60] Schröder W, Hager G, Kouprijanova E, Weber M, Schmitt AB, Seifert G, et al. Lesion-induced changes of electrophysiological properties in astrocytes of the rat dentate gyrus. Glia 1999;28(2):166–74.

[61] Bordey A, Lyons SA, Hablitz JJ, Sontheimer H. Electrophysiological characteristics of reactive astrocytes in experimental cortical dysplasia. J Neurophysiol 2001;85(4):1719–31.

[62] Bordey A, Hablitz JJ, Sontheimer H. Reactive astrocytes show enhanced inwardly rectifying K^+ currents in situ. Neuroreport 2000;11(14):3151–5.

[63] D'Ambrosio R, Maris DO, Grady MS, Winn HR, Janigro D. Impaired K^+ homeostasis and altered electrophysiological properties of post-traumatic hippocampal glia. J Neurosci 1999;19(18):8152–62.

[64] Köller H, Schroeter M, Jander S, Stoll G, Siebler M. Time course of inwardly rectifying K^+ current reduction in glial cells surrounding ischemic brain lesions. Brain Res 2000;872(1–2):194–8.

[65] Das A, Wallace GC, Holmes C, McDowell ML, Smith JA, Marshall JD, et al. Hippocampal tissue of patients with refractory temporal lobe epilepsy is associated with astrocyte activation, inflammation, and altered expression of channels and receptors. Neuroscience 2012;220:237–46.

[66] Heuser K, Eid T, Lauritzen F, Thoren AE, Vindedal GF, Tauboll E, et al. Loss of perivascular $K_{ir}4.1$ potassium channels in the sclerotic hippocampus of patients with mesial temporal lobe epilepsy. J Neuropathol Exp Neurol 2012;71(9):814–25.

[67] Hinterkeuser S, Schröder W, Hager G, Seifert G, Blümcke I, Elger CE, et al. Astrocytes in the hippocampus of patients with temporal lobe epilepsy display changes in potassium conductances. Eur J Neurosci 2000;12(6):2087–96.

[68] Kivi A, Lehmann TN, Kovacs R, Eilers A, Jauch R, Meencke HJ, et al. Effects of barium on stimulus-induced rises of $[K^+]_o$ in human epileptic non-sclerotic and sclerotic hippocampal area CA1. Eur J Neurosci 2000;12(6):2039–48.

[69] Bordey A, Sontheimer H. Properties of human glial cells associated with epileptic seizure foci. Epilepsy Res 1998;32:286–303.

[70] Schröder W, Hinterkeuser S, Seifert G, Schramm J, Jabs R, Wilkin GP, et al. Functional and molecular properties of human astrocytes in acute hippocampal slices obtained from patients with temporal lobe epilepsy. Epilepsia 2000;41(Suppl 6):S181–4.

[71] Jauch R, Windmuller O, Lehmann TN, Heinemann U, Gabriel S. Effects of barium, furosemide, ouabaine and 4,4′-diisothiocyanatostilbene-2,2′-disulfonic acid (DIDS) on ionophoretically-induced changes in extracellular potassium concentration in hippocampal slices from rats and from patients with epilepsy. Brain Res 2002;925(1):18–27.

[72] Heinemann U, Gabriel S, Jauch R, Schulze K, Kivi A, Eilers A, et al. Alterations of glial cell function in temporal lobe epilepsy. Epilepsia 2000;41(Suppl 6):S185–9.

[73] Steinhäuser C, Seifert G. Glial membrane channels and receptors in epilepsy: impact for generation and spread of seizure activity. Eur J Pharmacol 2002;447(2–3):227–37.

[74] David Y, Cacheaux LP, Ivens S, Lapilover E, Heinemann U, Kaufer D, et al. Astrocytic dysfunction in epileptogenesis: consequence of altered potassium and glutamate homeostasis? J Neurosci 2009;29(34):10588–99.

[75] Ivens S, Kaufer D, Flores LP, Bechmann I, Zumsteg D, Tomkins O, et al. TGF-β receptor-mediated albumin uptake into astrocytes is involved in neocortical epileptogenesis. Brain 2007;130(Pt 2):535–47.

[76] Takahashi DK, Vargas JR, Wilcox KS. Increased coupling and altered glutamate transport currents in astrocytes following kainic-acid-induced status epilepticus. Neurobiol Dis 2010;40(3):573–85.

[77] Nagao Y, Harada Y, Mukai T, Shimizu S, Okuda A, Fujimoto M, et al. Expressional analysis of the astrocytic $K_{ir}4.1$ channel in a pilocarpine-induced temporal lobe epilepsy model. Front Cell Neurosci 2013;7:104.

[78] Zurolo E, de Groot M, Iyer A, Anink J, van Vliet EA, Heimans JJ, et al. Regulation of $K_{ir}4.1$ expression in astrocytes and astrocytic tumors: a role for interleukin-1β. J Neuroinflammation 2012;9:280.

[79] Buono RJ, Lohoff FW, Sander T, Sperling MR, O'Connor MJ, Dlugos DJ, et al. Association between variation in the human KCNJ10 potassium ion channel gene and seizure susceptibility. Epilepsy Res 2004;58(2–3):175–83.

[80] Ferraro TN, Golden GT, Smith GG, Martin JF, Lohoff FW, Gieringer TA, et al. Fine mapping of a seizure susceptibility locus on mouse Chromosome 1: nomination of Kcnj10 as a causative gene. Mamm Genome 2004;15(4):239–51.

[81] Lenzen KP, Heils A, Lorenz S, Hempelmann A, Höfels S, Lohoff FW, et al. Supportive evidence for an allelic association of the human KCNJ10 potassium channel gene with idiopathic generalized epilepsy. Epilepsy Res 2005;63(2–3):113–8.
[82] Dai AI, Akcali A, Koska S, Oztuzcu S, Cengiz B, Demiryürek AT. Contribution of KCNJ10 gene polymorphisms in childhood epilepsy. J Child Neurol 2015;30(3):296–300.
[83] Bockenhauer D, Feather S, Stanescu HC, Bandulik S, Zdebik AA, Reichold M, et al. Epilepsy, ataxia, sensorineural deafness, tubulopathy, and KCNJ10 mutations. N Engl J Med 2009;360(19):1960–70.
[84] Reichold M, Zdebik AA, Lieberer E, Rapedius M, Schmidt K, Bandulik S, et al. KCNJ10 gene mutations causing EAST syndrome (epilepsy, ataxia, sensorineural deafness, and tubulopathy) disrupt channel function. Proc Natl Acad Sci USA 2010;107(32):14490–5.
[85] Scholl UI, Choi M, Liu T, Ramaekers VT, Hausler MG, Grimmer J, et al. Seizures, sensorineural deafness, ataxia, mental retardation, and electrolyte imbalance (SeSAME syndrome) caused by mutations in KCNJ10. Proc Natl Acad Sci USA 2009;106(14):5842–7.
[86] Sala-Rabanal M, Kucheryavykh LY, Skatchkov SN, Eaton MJ, Nichols CG. Molecular mechanisms of EAST/SeSAME syndrome mutations in $K_{ir}4.1$ (KCNJ10). J Biol Chem 2010;285(46):36040–8.
[87] Sicca F, Imbrici P, D'Adamo MC, Moro F, Bonatti F, Brovedani P, et al. Autism with seizures and intellectual disability: possible causative role of gain-of-function of the inwardly-rectifying K^+ channel $K_{ir}4.1$. Neurobiol Dis 2011;43(1):239–47.
[88] Heuser K, Nagelhus EA, Tauboll E, Indahl U, Berg PR, Lien S, et al. Variants of the genes encoding AQP4 and $K_{ir}4.1$ are associated with subgroups of patients with temporal lobe epilepsy. Epilepsy Res 2010;88(1):55–64.
[89] D'Adamo MC, Catacuzzeno L, Di Giovanni G, Franciolini F, Pessia M. K^+ channelepsy: progress in the neurobiology of potassium channels and epilepsy. Front Cell Neurosci. 2013;7:134.
[90] Smart SL, Lopantsev V, Zhang CL, Robbins CA, Wang H, Chiu SY, et al. Deletion of the $K_v1.1$ potassium channel causes epilepsy in mice. Neuron 1998;20(4):809–19.
[91] Barrese V, Miceli F, Soldovieri MV, Ambrosino P, Iannotti FA, Cilio MR, et al. Neuronal potassium channel openers in the management of epilepsy: role and potential of retigabine. Clin Pharmacol 2010;2:225–36.
[92] Cheung G, Sibille J, Zapata J, Rouach N. Activity-dependent plasticity of astroglial potassium and glutamate clearance. Neural Plast 2015;2015:109106.
[93] Khakh BS, Sofroniew MV. Astrocytes and Huntington's disease. ACS Chem Neurosci 2014;5(7):494–6.
[94] Tong X, Ao Y, Faas GC, Nwaobi SE, Xu J, Haustein MD, et al. Astrocyte $K_{ir}4.1$ ion channel deficits contribute to neuronal dysfunction in Huntington's disease model mice. Nat Neurosci 2014;17(5):694–703.
[95] Wilcock DM, Vitek MP, Colton CA. Vascular amyloid alters astrocytic water and potassium channels in mouse models and humans with Alzheimer's disease. Neuroscience 2009;159(3):1055–69.
[96] Kaiser M, Maletzki I, Hülsmann S, Holtmann B, Schulz-Schaeffer W, Kirchhoff F, et al. Progressive loss of a glial potassium channel (KCNJ10) in the spinal cord of the SOD1 (G93A) transgenic mouse model of amyotrophic lateral sclerosis. J Neurochem 2006;99(3):900–12.
[97] Bataveljic D, Nikolic L, Milosevic M, Todorovic N, Andjus PR. Changes in the astrocytic aquaporin-4 and inwardly rectifying potassium channel expression in the brain of the amyotrophic lateral sclerosis SOD1(G93A) rat model. Glia 2012;60(12):1991–2003.
[98] Vit JP, Ohara PT, Bhargava A, Kelley K, Jasmin L. Silencing the $K_{ir}4.1$ potassium channel subunit in satellite glial cells of the rat trigeminal ganglion results in pain-like behavior in the absence of nerve injury. J Neurosci 2008;28(16):4161–71.
[99] Brill L, Goldberg L, Karni A, Petrou P, Abramsky O, Ovadia H, et al. Increased anti-$K_{ir}4.1$ antibodies in multiple sclerosis: could it be a marker of disease relapse? Mult Scler 2015;21(5):572–9.
[100] Schirmer L, Srivastava R, Kalluri SR, Böttinger S, Herwerth M, Carassiti D, et al. Differential loss of $K_{ir}4.1$ immunoreactivity in multiple sclerosis lesions. Ann Neurol 2014;75(6):810–28.
[101] Liévens JC, Woodman B, Mahal A, Spasic-Boscovic O, Samuel D, Kerkerian-Le Goff L, et al. Impaired glutamate uptake in the R6 Huntington's disease transgenic mice. Neurobiol Dis 2001;8(5):807–21.
[102] Miller BR, Dorner JL, Shou M, Sari Y, Barton SJ, Sengelaub DR, et al. Up-regulation of GLT1 expression increases glutamate uptake and attenuates the Huntington's disease phenotype in the R6/2 mouse. Neuroscience 2008;153(1):329–37.
[103] Estrada-Sanchez AM, Rebec GV. Corticostriatal dysfunction and glutamate transporter-1 (GLT1) in Huntington's disease: interactions between neurons and astrocytes. Basal Ganglia 2012;2(2):57–66.

CHAPTER

8

Water Channels

OUTLINE

OVERVIEW

Recent studies have implicated glial cells in modulation of synaptic transmission, so it is plausible that glial cells may have a functional role in the hyperexcitability characteristic of epilepsy. Indeed, alterations in distinct astrocyte membrane channels, receptors, and transporters have all been associated with the epileptic state. This chapter focuses on the potential roles of the glial water channel aquaporin-4 (AQP4) in modulation of brain excitability and in epilepsy. We will review studies of mice lacking AQP4 ($AQP4^{-/-}$ mice) or α-syntrophin (an AQP4 anchoring protein) and discuss the available human studies demonstrating alterations of AQP4 in human epilepsy tissue specimens. We will conclude with new studies of AQP4 regulation and discuss the potential role(s) of AQP4 in the development of epilepsy (epileptogenesis). While many questions remain unanswered, the available data indicate that AQP4 and its molecular partners may represent important new therapeutic targets.

DOI: http://dx.doi.org/10.1016/B978-0-12-802401-0.00008-9

WATER AND ION HOMEOSTASIS AND EPILEPSY

Glial cells are involved in many important physiologic functions, such as sequestration and/or redistribution of K^+ during neural activity, neurotransmitter cycling, and provision of energy substrates to neurons [1]. Several recent lines of evidence strongly suggest that changes in glial cells potentially contribute to epilepsy [2–4]. First, many studies now link glial cells to modulation of synaptic transmission [5–9]. Second, functional alterations of specific glial membrane channels and receptors have been discovered in epileptic tissue [10–15]. Third, direct stimulation of astrocytes has been shown to be sufficient for neuronal synchronization in epilepsy models [9] (although see [16]). Thus, if the cellular and molecular mechanisms by which glial cells (especially astrocytes) modulate excitability are better understood, specific antiepileptic therapies based on modulation of glial receptors and channels can be contemplated [17]. It is likely that therapies directed to glial cells would have fewer deleterious side effects than current therapies targeting neurons.

Alteration of water and K^+ homeostasis could dramatically affect seizure susceptibility. First, brain tissue excitability is exquisitely sensitive to osmolarity and the size of the extracellular space (ECS) [18]. Decreasing ECS volume with hypoosmolar treatment produces hyperexcitability and enhanced epileptiform activity [19–22]. At least two mechanisms could be operative: decreasing ECS volume could increase extracellular neurotransmitter and ion concentrations and magnify ephaptic interactions among neurons due to increased ECS resistance [3]. Conversely, increasing ECS volume with hyperosmolar treatment powerfully attenuates epileptiform activity [20,22–24]. These experimental data parallel extensive clinical experience indicating that hypoosmolar states such as hyponatremia lower seizure threshold while hyperosmolar states elevate seizure threshold [25–27]. Second, millimolar and even submillimolar increases in extracellular K^+ concentration powerfully enhance epileptiform activity in the hippocampus [28–31]. High K^+ reliably induces epileptiform activity in hippocampal slices from human patients with intractable temporal lobe epilepsy [32].

AQUAPORINS

The aquaporins (AQPs) are a family of membrane proteins that function as water channels in many cell types and tissues in which fluid transport is crucial [33–35]. The AQPs are small hydrophobic integral membrane proteins (~30 kDa monomer) that facilitate bidirectional water transport in response to osmotic gradients [36]. Multiple mammalian AQPs have been identified including AQP0, AQP1, AQP2, AQP4, and AQP5 which transport water only (aquaporins), and AQP3, AQP7, and AQP9 which also transport glycerol (aquaglyceroporins) [35,36].

AQP4 [37–39] is of particular interest in neuroscience as it is expressed in brain and spinal cord by glial cells, especially at specialized membrane domains including astroglial endfeet in contact with blood vessels and astrocyte membranes that ensheathe glutamatergic synapses [40–44] (Figs. 8.1 and 8.2). Activity-induced radial water fluxes in neocortex have been demonstrated that may represent water movement via AQP channels in response to physiological activity [45,46]. Interestingly, AQP4 is a structural component of orthogonal arrays of particles (OAPs) seen in freeze-fracture electron micrographs [47–51].

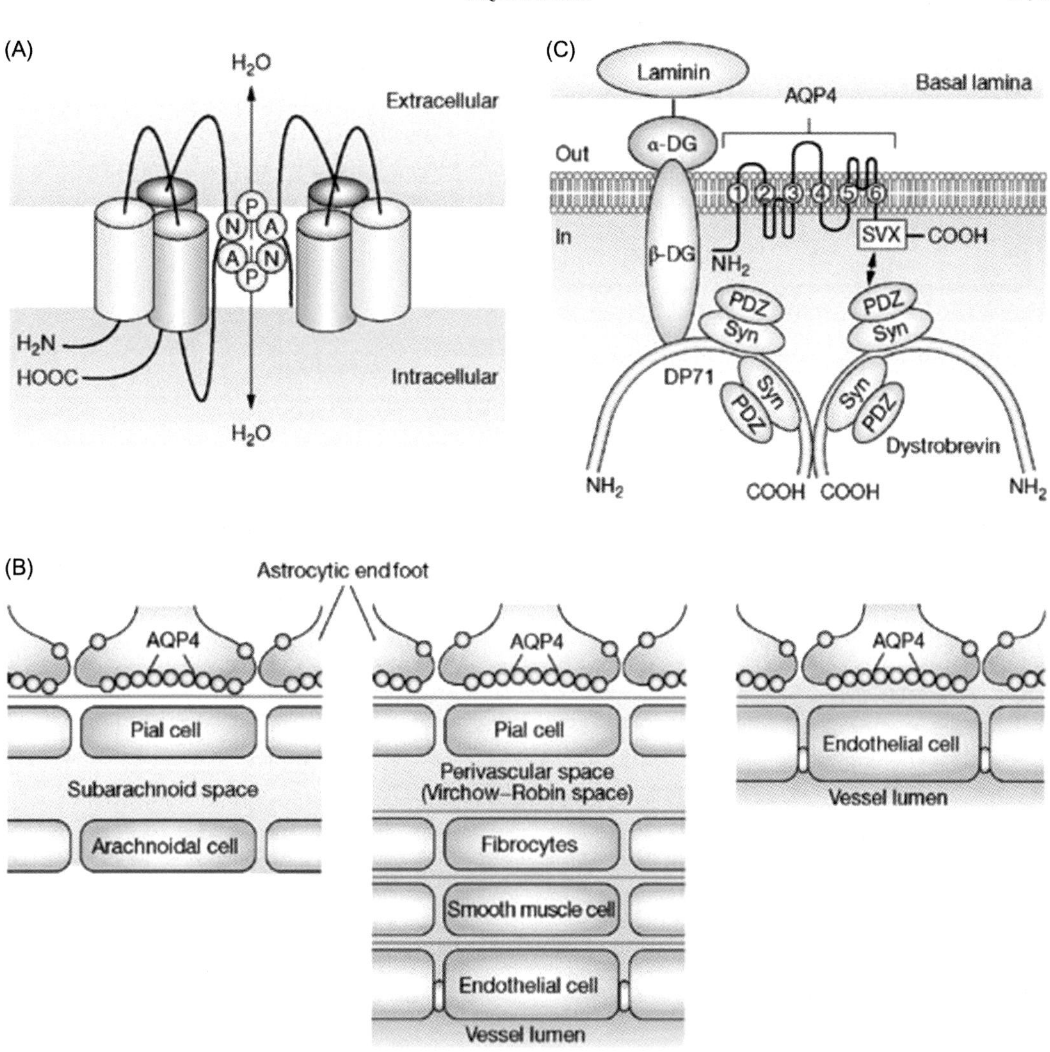

FIGURE 8.1 **The structure and localization of aquaporin-4.** (A) Each AQP4 monomer consists of six membrane-spanning α-helices with both termini located intracellularly. Water selectivity depends on two pore helices and their highly conserved Asn–Pro–Ala motifs. Like other aquaporins, AQP4 forms homotetramers (not shown). (B) Polarization of AQP4 within the astrocytic endfeet at the glial–pial interfaces of the subarachnoid space and the Virchow–Robin space, and the glial–endothelial interface of arteries and arterioles. Astrocytic processes, together with collagen fibers of the subpial space, form the glia limitans externa that seals the brain surface, and form a dense sheath around capillaries, thereby constituting an integral part of the blood–brain barrier. (C) AQP4 is connected to the dystrophin-associated protein complex, which connects the astrocytic cytoskeleton via α1-syntrophin with the basal lamina and is mainly responsible for the polarization of AQP4 at the astrocyte endfeet. *AQP4*, aquaporin-4; *α-DG*, α-dystroglycan; *β-DG*, β-dystroglycan; *DP71*, dystrophin variant DP71; *PDZ*, PDZ domain of α_1-syntrophin; *SXV*, Ser-X-Val domain of aquaporin-4; *Syn*, α_1-syntrophin. *Source: Reproduced with permission from Jarius S, Paul F, Franciotta D, Waters P, Zipp F, Hohlfeld R, et al. Mechanisms of disease: aquaporin-4 antibodies in neuromyelitis optica. Nat Clin Pract Neurol 2008;4(4):202–14.*

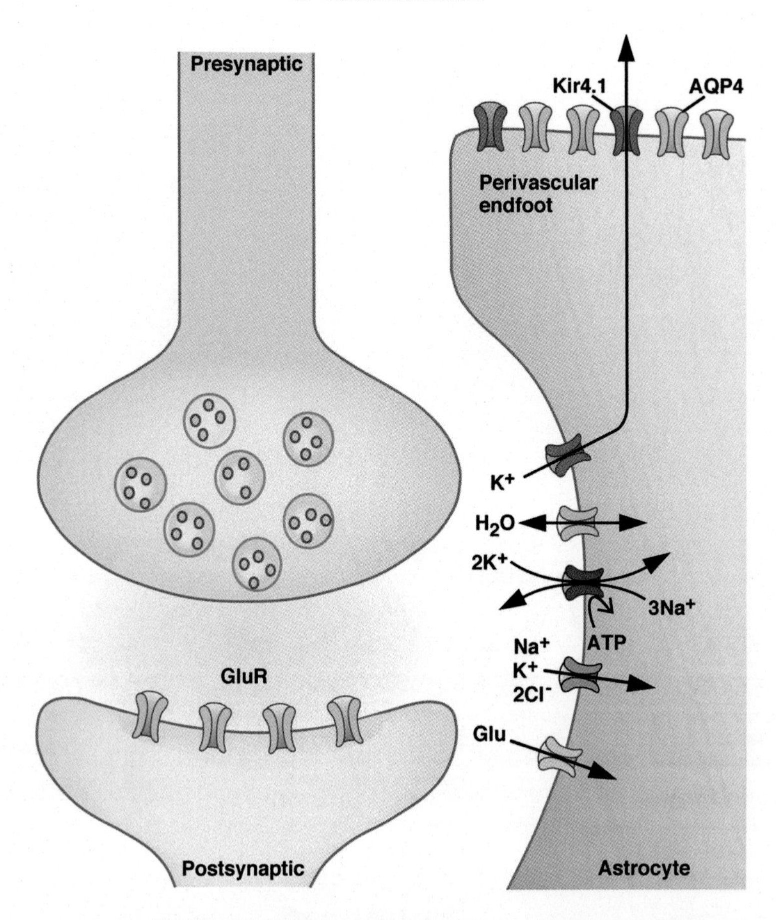

FIGURE 8.2 **Schematic of AQP4 distribution at the "tripartite" synapse.** In addition to its well-known perivascular location on astrocyte endfeet, AQP4 is also localized in perisynaptic astrocyte processes. Depicted here are the presynaptic (axon) and postsynaptic (dendrite) components of the synapse as well as the perisynaptic astrocyte process (together forming the "tripartite" synapse). Multiple channels and transporters (eg, AQP4, $K_{ir}4.1$, Na^+/K^+/ATPase, chloride cotransporters, glutamate transporters) are located on the perisynaptic astrocyte membrane. *Source: Reproduced with permission from Binder OK, Nagelhus EA, Ottersen OP. Aquaporin-4 and epilepsy. Glia 2012;60(8):1203–14.*

AQP4$^{-/-}$ Mice

AQP4$^{-/-}$ mice were originally generated by targeted gene disruption in 1997 [52], and recently a glial conditional deletion of AQP4 has been generated [53]. AQP4$^{-/-}$ mice are grossly normal phenotypically, do not manifest overt neurological abnormalities, altered blood–brain barrier properties, abnormal baseline intracranial pressure, impaired osmoregulation, or obvious brain dysmorphology [54–56] (Fig. 8.3). A mild hearing impairment has been recorded and the electroretinogram is perturbed [57,58]. Mice with AQP4 deletion have normal blood chemistries and hematologies, normal expression of other brain aquaporins (AQP1 and AQP9) and their only nonneural phenotype is a mild impairment in maximal urine concentrating ability [52].

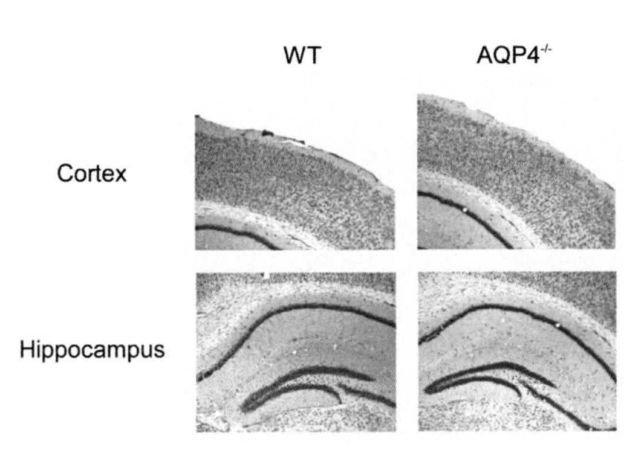

FIGURE 8.3 **Gross brain histology of wild-type (+/+) and AQP4 null (AQP4$^{-/-}$) mice.** Nissl stain of coronal section of cortex and dorsal hippocampus reveals no gross histologic abnormalities in AQP4$^{-/-}$ mice. *Source: Modified with permission from Binder OK, Oshio K, Ma T, Verkman AS, Manley GT. Increased seizure threshold in mice lacking aquaporin-4 water channels. Neuroreport 2004;15(2):259–62.*

Interestingly, more detailed studies of brain tissue from AQP4$^{-/-}$ mice demonstrated reduced osmotic water permeability as measured in isolated membrane vesicles [52], brain slices [59], and intact brain [60]. Also, water permeability was sevenfold reduced in primary astrocyte cultures from AQP4$^{-/-}$ mice as measured by a calcein fluorescence quenching method [61], and similar results were obtained with AQP4 RNAi knockdown experiments in wild-type astrocytes [62]. These data directly demonstrated that AQP4 provides the predominant pathway for transmembrane water movement in astrocytes.

In vivo studies of these mice demonstrated a functional role for AQP4 in brain water transport. AQP4$^{-/-}$ mice have markedly decreased accumulation of brain water (cerebral edema) following water intoxication and focal cerebral ischemia [63] and impaired clearance of brain water in models of vasogenic edema [56]. Clearance of seizure-induced edema may also be AQP4-dependent [64]. Impaired water flux into (in the case of cytotoxic edema) and out of (in the case of vasogenic edema) the brain makes sense based on the bidirectional nature of water flux across the AQP4 membrane channel at the blood–brain barrier. The recently generated glial-conditional AQP4 knockout mouse line demonstrates a 31% reduction in brain water uptake after systemic hypoosmotic stress [53]. Similarly, mice deficient in dystrophin or α-syntrophin, in which there is mislocalization of the AQP4 protein [65–67], show attenuated cerebral edema [66,68].

Altered ECS in AQP4$^{-/-}$ Mice

The ECS in brain comprises ~20% of brain tissue volume, consisting of a jelly-like matrix in which neurons, glia, and blood vessels are embedded [69]. The ECS contains ions, neurotransmitters, metabolites, peptides, and extracellular matrix molecules, forming the microenvironment for all cells in the brain and mediating glia–neuron communication via diffusible messengers, metabolites, and ions [70]. The size of the ECS is likely largely controlled by astrocytic mechanisms [71–73].

The ECS of AQP4$^{-/-}$ mice was assessed using a cortical fluorescence recovery after photobleaching (cFRAP) method to measure the diffusion of fluorescently-labeled macromolecules in the cortex [74]. ECS in mouse brain was labeled by exposure of the intact dura to fluorescein dextrans (M_r 4, 70, and 500 kD) after craniectomy, and fluorescein-dextran diffusion was detected by fluorescence recovery after laser-induced cortical photobleaching using confocal optics. The cFRAP method was applied to brain edema, seizure initiation, and AQP4 deficiency. In contrast to the slowed diffusion produced by brain edema and seizure activity, ECS diffusion was *faster* in $AQP4^{-/-}$ mice, indicating ECS expansion in AQP4 deficiency [74]. Similar results were obtained with follow-up studies using the TMA^+ method [75].

Seizure Phenotype of $AQP4^{-/-}$ Mice

Seizure susceptibility of $AQP4^{-/-}$ mice was initially examined using the convulsant ($GABA_A$ antagonist) pentylenetetrazol (PTZ) [55]. At 40 mg/kg PTZ (i.p.), all wild-type mice exhibited seizure activity, whereas six out of seven $AQP4^{-/-}$ mice did not exhibit seizure activity. At 50 mg/kg PTZ, both groups exhibited seizure activity; however, the latency to generalized (tonic-clonic) seizures was longer in $AQP4^{-/-}$ mice [55]. Since seizure propensity is exquisitely sensitive to ECS volume [18], the expanded ECS in AQP4 deficiency is consistent with the increased seizure threshold. Thus, more intense stimuli (eg, higher PTZ doses or a longer time after PTZ) may be required to overcome the expanded ECS of $AQP4^{-/-}$ mice in order to initiate a seizure.

In order to analyze the seizure phenotype of the $AQP4^{-/-}$ mice in greater detail, in vivo electroencephalographic (EEG) characterization with stimulation and recording was employed [76] (Fig. 8.4). $AQP4^{-/-}$ mice and wild-type controls were implanted in the right dorsal hippocampus with bipolar electrodes. Following postoperative recovery, electrical stimulations were given to assess electrographic seizure threshold and duration. $AQP4^{-/-}$ mice had a higher mean electrographic seizure threshold than wild-type controls, consistent with the prior PTZ studies [55]. However, $AQP4^{-/-}$ mice were also found to have remarkably prolonged stimulation-evoked seizures compared to wild-type controls [55].

Altered K^+ Homeostasis in $AQP4^{-/-}$ Mice

Because of the colocalization of AQP4 and inwardly-rectifying K^+ channels in glial endfeet, the hypothesis arose that AQP4 was indirectly involved in K^+ reuptake [40,77]. Impaired K^+ clearance from the ECS following the intense neuronal activity accompanying the seizure would lead to prolonged depolarization of neurons and inhibit seizure termination [10,28–30]. Indeed, in addition to modulation of brain water transport, AQP4 and its known molecular partners have been hypothesized to modulate ion homeostasis [78,79]. During rapid neuronal firing, extracellular $[K^+]$ increases from ~3 mM to a maximum of 10–12 mM; and K^+ released by active neurons is thought to be primarily taken up by glial cells [69,80–82]. Such K^+ reuptake into glial cells could be AQP4-dependent, as water influx coupled to K^+ influx is thought to underlie activity-induced glial cell swelling [83,84]. In support of this possibility was the known subcellular colocalization of AQP4 with the inwardly-rectifying K^+ channel $K_{ir}4.1$ in the retina [41,77,85]. $K_{ir}4.1^{-/-}$ mice, like $AQP4^{-/-}$ mice [57,58], demonstrate abnormal retinal and cochlear physiology presumably due to altered K^+ homeostasis [86–89]. $K_{ir}4.1$ is thought to contribute to K^+ reuptake and spatial K^+ buffering by glial cells [90–93], and pharmacological or genetic inactivation of $K_{ir}4.1$ leads to impairment of extracellular K^+ regulation [86,94–99].

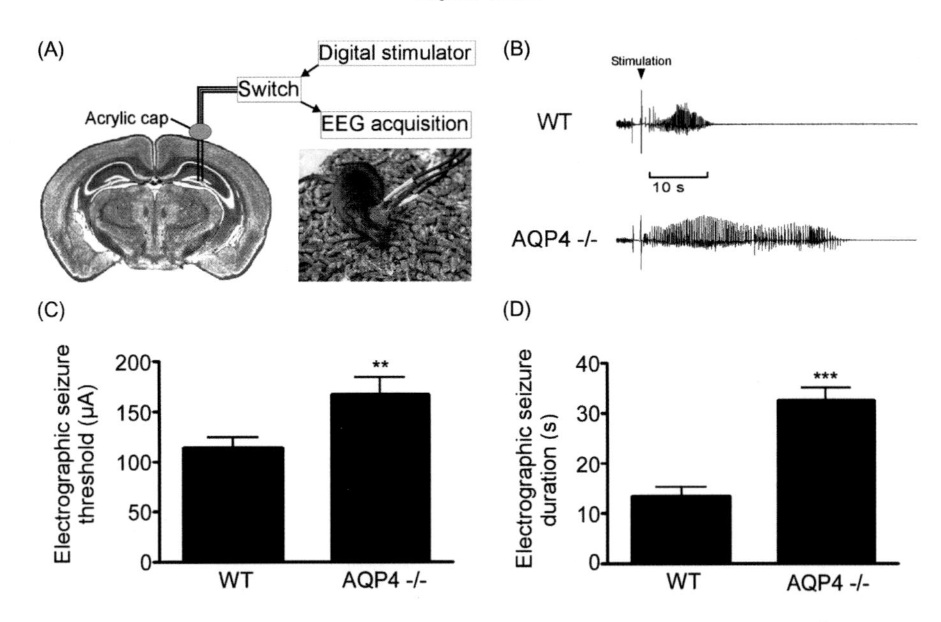

FIGURE 8.4 **Electrographic seizure threshold and duration in wild-type versus AQP4$^{-/-}$ mice.** (A) Bipolar electrodes implanted in the right hippocampus were connected to a stimulator and digital EEG acquisition system. Mice were awake and behaving normally at the onset of stimulation (inset). (B) Representative electroencephalograms from WT and AQP4$^{-/-}$ mice. Baseline EEG prior to stimulation is similar (left). Hippocampal stimulation-induced electrographic seizures are shown for a WT mouse (top) and an AQP4$^{-/-}$ mouse (bottom). The WT mouse had a 11-second seizure, whereas the AQP4$^{-/-}$ mouse had a much longer seizure (37 seconds). Behavioral arrest was observed in both animals during the seizure. Postictal depression is evident on the EEG in both cases. (C) Electrographic seizure threshold (μA) (mean ± SEM) in WT versus AQP4$^{-/-}$ mice. AQP4$^{-/-}$ mice had a higher electrographic seizure threshold than wild-type controls. (D) Electrographic seizure duration (seconds) (mean ± SEM) following hippocampal stimulation in WT versus AQP4$^{-/-}$ mice. AQP4$^{-/-}$ mice had remarkably longer stimulation-evoked seizures compared to wild-type controls. **, $P < 0.01$; ***, $P < 0.001$ compared to WT. *Source: Reproduced with permission from Binder DK, Yao X, Sick TJ, Verkman AS, Manley GT. Increased seizure duration and slowed potassium kinetics in mice lacking aquaporin-4 water channels. Glia 2006;53:631–6.*

To address the possibility that AQP4 deficiency was associated with a deficit in K^+ homeostasis, K^+ dynamics were examined in vivo in AQP4$^{-/-}$ mice [76] (Fig. 8.5). Neither baseline $[K^+]_o$ nor the "Lux-Heinemann ceiling" level of activity-induced $[K^+]_o$ elevation (~12 mM) [80,81] were altered in AQP4 deficiency, indicating that basic K^+ homeostasis was intact. Based on K^+ measurements made from cortex with double-barreled K^+-sensitive microelectrodes, stimulation-induced rises in $[K^+]_o$ were quite different in AQP4$^{-/-}$ compared to wild-type mice; in particular, there was a markedly slower rise and decay time for poststimulus changes in $[K^+]_o$ in AQP4$^{-/-}$ mice [76] (Fig. 8.5). A similar delay in K^+ kinetics was observed following cortical spreading depression in AQP4$^{-/-}$ mice using a fluorescent K^+ sensor [100].

Slowed $[K^+]_o$ rise time is consistent with increased ECS volume fraction in AQP4$^{-/-}$ mice [74]. Slowed $[K^+]_o$ decay is possibly due to impaired K^+ reuptake into AQP4$^{-/-}$ astrocytes. Interestingly, there is no difference in expression of $K_{ir}4.1$ protein [76] or $K_{ir}4.1$

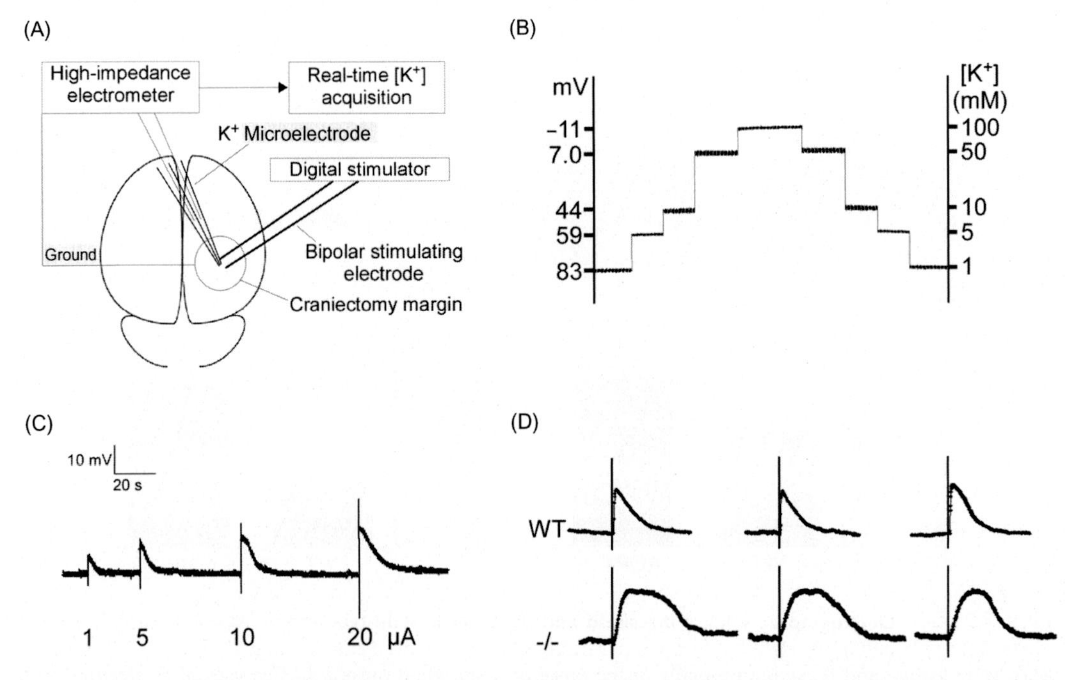

FIGURE 8.5 **In vivo K^+ measurements in WT and AQP4$^{-/-}$ mice.** (A) Double-barreled potassium-selective microelectrodes were introduced through a small craniectomy and connected to a high-impedance electrometer for real-time [K^+] acquisition. A bipolar electrode connected to a digital stimulator was positioned so as to provide direct current stimulation of the cortex at varying intensities. (B) Sample calibration curve for a K^+-sensitive microelectrode placed in serial K^+ standard solutions. (C) Representative family of K^+ curves in WT mouse following stimulation intensities of 1, 5, 10, and 20 μA. (D) Representative 20 μA stimulation curves from three different WT mice (top) and three different AQP4$^{-/-}$ mice (bottom). Calibration in (C) also applies to (D). *Source: Reproduced with permission from Binder DK, Yao X, Sick TJ, Verkman AS, Manley GT. Increased seizure duration and slowed potassium kinetics in mice lacking aquaporin-4 water channels. Glia 2006;53:631–6.*

immunoreactivity [101] in AQP4$^{-/-}$ mice nor AQP4 immunoreactivity in $K_{ir}4.1^{-/-}$ mice [101]. In addition, no alterations were observed in membrane potential, barium-sensitive $K_{ir}4.1$ K^+ current or current–voltage curves in AQP4$^{-/-}$ retinal Müller cells [102] or brain astrocytes [103]. Lack of alteration of K_{ir} channels in AQP4$^{-/-}$ mice suggests the interesting possibility that the slowed $[K^+]_o$ decay may be a secondary effect of slowed water extrusion ("deswelling") following stimulation, and this has been modeled carefully [104] but not directly demonstrated.

Stroschein et al. [105] investigated the impact of AQP4 on stimulus-induced alterations of $[K^+]_o$ in hippocampal slices. Antidromic stimulation evoked smaller increases and slower recovery of $[K^+]_o$ in the stratum pyramidale of AQP4$^{-/-}$ mice, consistent with the previous in vivo studies in cortex. Interestingly, astrocyte gap junction coupling as assessed with tracer filling during patch-clamp recording demonstrated enhanced tracer coupling in AQP4$^{-/-}$ mice, and laminar profiles indicated enhanced spatial redistribution of K^+ [105]. The functional consequences of alterations in gap junctional coupling are not completely clear; however, the complete absence of astrocytic gap junctions impairs K^+ homeostasis [106]. To further assess the direct link between AQP4 and K^+ homeostasis, Haj-Yasein et al. examined extracellular K^+ concentration in a distinct constitutive $AQP4^{-/-}$ mouse line [107] and found a layer-specific effect of AQP4 deletion on K^+ dynamics during synaptic stimulation [108].

Anchoring of AQP4 by the Dystrophin-Associated Protein Complex

AQP4 in endfeet membranes is anchored to the dystrophin-associated protein complex (DAPC) (Fig. 8.1C). For example, endfoot expression of AQP4 is dependent on α-syntrophin, a dystrophin-associated protein [67]. Accordingly, mice deficient in α-syntrophin show marked loss of AQP4 from perivascular and subpial membranes as judged by quantitative immunogold electron microscopy [68]. Similarly, dystrophin-deficient (*mdx*) mice exhibit a dramatic reduction of AQP4 in astroglial endfeet surrounding capillaries and at the glia limitans (cerebrospinal fluid–brain interface) despite no alteration in total AQP4 protein [66]. These studies clearly suggest that alterations in components of the DAPC may affect the subcellular targeting and function of AQP4. A recent study also demonstrates that dystrophin localization at the astrocytic endfoot is dependent on syntrophin [109]. Another study using the zero magnesium in vitro model of epilepsy demonstrated reduction in expression of a distinct member of the DAPC, β-dystroglycan, in cortical astrocytes following continuous seizures, which was accompanied by a decrease in AQP4 expression [110]. Astrocyte swelling is markedly impaired in response to hypoosmotic stress, oxygen-glucose deprivation or high K^+ in the cortex of α-syntrophin-negative GFAP/EGFP mice [111].

What are the functional consequences to seizure susceptibility of loss of DAPC components? Amiry-Moghaddam et al. [4] were the first to report a possible association between AQP4 and epilepsy by demonstrating an increased severity of hyperthermia-evoked seizures in α-syntrophin-deficient mice. These mice exhibited a deficit in extracellular K^+ clearance following evoked neuronal activity. Dystrophin-deficient mice were found to have altered seizure susceptibility in response to various chemical convulsants; in particular, *mdx* mice showed enhanced seizure severity and a shorter latency in the development of chemical kindling produced by administration of PTZ [112]. It is interesting to note that there is an increased incidence of epilepsy in forms of human muscular dystrophy in which the dystrophin complex is affected [113]. Taken together, these data suggest that AQP4 and its molecular partners together comprise a multifunctional "unit" responsible for clearance of K^+ and/or water following neural activity and that alterations in expression of this complex can lead to alterations in seizure susceptibility.

Human Tissue Studies

The most common pathology in patients with medically-intractable temporal lobe epilepsy is mesial temporal sclerosis (MTS), characterized by marked neuronal cell loss in specific hippocampal areas, gliosis, and microvascular proliferation [114]. Emerging work also demonstrates dysregulation of water and K^+ homeostasis in patients with mesial temporal lobe epilepsy. First, imaging studies demonstrate abnormal T2 prolongation by magnetic resonance imaging (MRI) in the epileptic hippocampus, possibly partially due to increased

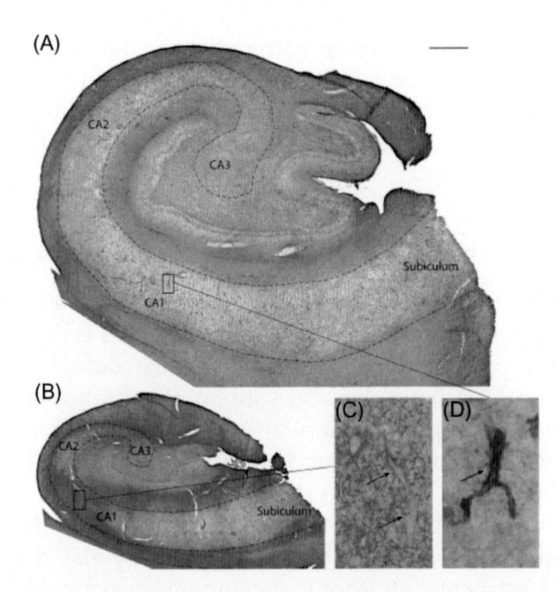

FIGURE 8.6 **Altered AQP4 distribution in mesial temporal lobe epilepsy (MTLE).** Although AQP4 is preferentially distributed around blood vessels in the non-MTLE hippocampus, this localization is lost in MTLE. AQP4 is demonstrated by preembedding immunohistochemistry on Vibratome sections of a representative non-MTLE (A and D) and MTLE (B and C) hippocampus. (A) In the non-MTLE hippocampus, immunoreactivity for AQP4 in the pyramidal layer of Ammon's horn (the area within the dashed line) is preferentially distributed around blood capillaries (scale bar, 1 mm). (D) This finding is demonstrated in the high-power field of CA1, where the arrow indicates a strongly immunopositive capillary amidst a weakly labeled neuropil (magnification, ×6 selected portion of A). (B) In the MTLE hippocampus, the preferential distribution of AQP4 around blood capillaries is lost in the pyramidal layer in areas of sclerosis (such as CA1). Scale is the same as in (A). (C) In the high-power field of CA1, moderate immunolabeling for AQP4 is present throughout the neuropil and also around blood capillaries, which are indicated by arrows (magnification, ×6 selected portion of B). *Source: Reproduced with permission from Eid T, Lee TS, Thomas MJ, Amiry-Moghaddam M, Bjornsen LP, Spencer DD, et al. Loss of perivascular aquaporin-4 may underlie deficient water and K^+ homeostasis in the human epileptogenic hippocampus. Proc Natl Acad Sci USA 2005;102(4):1193–8.*

water content [115]. This is accompanied by alterations in apparent diffusion coefficient (ADC) with diffusion-weighted MRI imaging [116]. Second, the expression and subcellular localization of AQP4 have been shown to be altered in sclerotic hippocampi obtained from patients with MTS. Using immunohistochemistry, rt-PCR and gene chip analysis, Lee et al. demonstrated an overall increase in AQP4 expression in sclerotic epilepsy tissue [117]. However, using quantitative immunogold electron microscopy, the same group found that there was mislocalization of AQP4 in the human epileptic hippocampus, with reduction in perivascular membrane expression [118] (Fig. 8.6). Thus, although there was an overall increase in AQP4 content by Western blot, rt-PCR, and gene chip analysis, the subcellular distribution of AQP4 in mesial temporal lobe epilepsy tissue had changed. Their hypothesis was that reduction in perivascular AQP4 expression would lead to water and K^+ dysregulation in the epileptic hippocampus, potentially contributing to hyperexcitability.

Medici et al. reported AQP4 expression in tissue from patients with focal cortical dysplasia (FCD) type IIB, normal-appearing (cryptogenic) epileptic cortex, and nonepileptic control tissue [119]. AQP4 expression and distribution in the cryptogenic cases were similar to control cases, that is, with intact perivascular immunoreactivity. In the patients with FCD type IIB, the pattern was different, with strong AQP4 immunoreactivity around dysplastic neurons but with very weak immunoreactivity for AQP4 around blood vessels [119].

Disruption of the DAPC has also been demonstrated in human epileptic tissue. In the study in which reduced perivascular membrane expression of AQP4 was noted [118], these authors also studied perivascular dystrophin expression. Like AQP4 expression, perivascular dystrophin expression was also reduced in tissue from the sclerotic epileptic hippocampus [118]. The more recent study also demonstrated reduced perivascular dystrophin expression in FCD type IIB [119]. Thus, subcellular alteration in AQP4 expression may result secondarily from alterations in dystrophin and/or other members of the DAPC.

Dysfunction of astroglial K_{ir} channels has also been found in specimens from patients with temporal lobe epilepsy [10,12]. First, using ion-sensitive microelectrodes, Kivi et al. demonstrated an impairment of glial barium-sensitive K^+ uptake in the CA1 region of MTS specimens [11]. Second, using patch-clamp analysis, Hinterkeuser et al. demonstrated a dramatic reduction in astroglial K_{ir} currents in MTS tissue [13]. Reduction in K_{ir} currents would be expected to contribute to hyperexcitability; evidence for this comes from the finding of stress-induced seizures in conditional $K_{ir}4.1^{-/-}$ mice [97]. For further details regarding the role of potassium channels in epilepsy please see *Chapter 7: Potassium Channels*.

Methodological Issues

In assessing the contribution of AQP4 and its molecular partners to epileptogenesis, it is critical to examine the methodology in the reported studies. First, studies of human tissue represent the "endpoint" of an already long-standing epileptogenic process. A sclerotic hippocampus with significant neuronal cell loss and gliosis may show many molecular changes in glial cells including the reported altered distribution of AQP4 [117,118], but what is unclear from these studies is whether AQP4 dysregulation represents a cause or consequence of epileptogenesis. Appropriate animal models of epilepsy permit dissection of the process of epileptogenesis in greater detail. In particular, in such models it is possible to examine the transition from a normal to an epileptic brain as assessed by in vivo electrophysiological recordings of spontaneous seizures following an initial epileptogenic insult. Description of molecular changes in glial cells during epileptogenesis [2] will be facilitated by studies in such in vivo models. Unlike in vivo models, studies with ex vivo slices enable the delineation of tissue and synaptic physiology in greater detail [4], but at the expense of isolating only a part of the in vivo network. Thus, a fruitful combination of methodologies for future study will include in vivo electrophysiologic recording for validation of epileptogenesis, ex vivo slice physiology, and examination of the cellular and molecular changes in glial cells leading up to the development of spontaneous seizures.

The studies discussed above suggest novel roles for AQP4 in control of seizure susceptibility [4,55,76], K^+ homeostasis [4,76,100], and ECS physiology [74]. These findings together with the changes in human epileptic tissue [117,118] lead to the unifying hypothesis that AQP4 and its molecular partners may play a functional role in epilepsy [3,120]. Thus, more recent studies have examined the pattern of expression, regulation and role(s) of AQP4 in various animal models of epileptogenesis (development of epilepsy).

Expression and Regulation of AQP4 During Epileptogenesis

Early studies demonstrated the perivascular localization of AQP4 [40,41,121,122] but did not examine region-specific expression of AQP4 in the brain in detail. A recent study examined subregional expression pattern of AQP4 within the mouse hippocampus, a structure critical to epilepsy [101]. AQP4 immunohistochemistry revealed a developmentally-regulated and laminar-specific pattern, with highest expression in the CA1 stratum lacunosum-moleculare (SLM) and the molecular layer (ML) of the dentate gyrus (DG) (Fig. 8.7). AQP4 was also ubiquitously expressed on astrocytic endfeet around blood vessels.

This study also addressed the cell type-specificity of AQP4 expression in the hippocampus (Fig. 8.8). The description of novel classes of hippocampal glial cells (classical astrocytes vs NG2 cells) with different morphology and functional properties [123–126] makes it particularly important to clearly identify which "glial cell" expresses AQP4. While AQP4 is thought to be expressed predominantly by "classical" astrocytes, some have reported expression in other cell types, such as activated microglia [127]. However, a recent study found no AQP4 expression in resting microglia of the neocortex [128] and conditional AQP4 deletion driven by the human glial fibrillary acidic protein (*GFAP*) promoter completely removed AQP4 from brain [53]. In line with the latter studies, AQP4 was found to colocalize with the astrocyte markers GFAP and S100β in the hippocampus (Fig. 8.8), and electrophysiological and postrecording RT-PCR analyses of individual cells revealed that AQP4 and

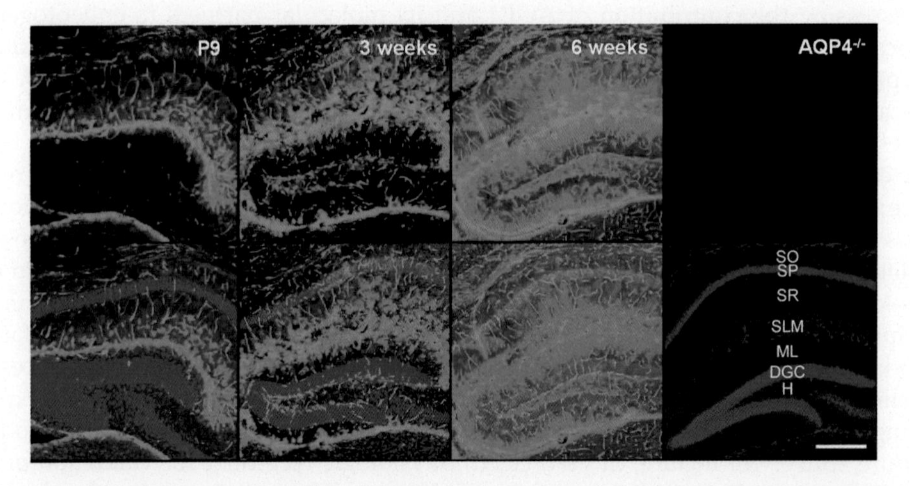

FIGURE 8.7 **Developmental regulation of AQP4 immunoreactivity in mouse hippocampus.** Low-power views of coronal mouse brain sections in P9 versus 3-week-old versus 6-week-old mice. While AQP4 immunoreactivity (green) is strongest in the CA1 SLM at P9, AQP4 immunoreactivity at 3 and 6 weeks is dramatically upregulated, particularly in CA1 SLM but also more diffusely throughout the hippocampus. Blood vessels are labeled well at all developmental stages. No specific immunoreactivity is present in sections from adult AQP4$^{-/-}$ mice (right). Nissl counterstains (blue) are shown in the bottom panels for each image. *SO*, stratum oriens; *SP*, stratum pyramidale; *SR*, stratum radiatum; *SLM*, stratum lacunosum-moleculare; *ML*, molecular layer; *DGC*, dentate granule cell layer; *H*, hilus. Scale bar, 300 μm. *Source: Reproduced with permission from Hsu MS, Seldin M, Lee DJ, Seifert G, Steinhäuser C, Binder DK. Laminar-specific and developmental expression of aquaporin-4 in the mouse hippocampus. Neuroscience 2011;178:21–32.*

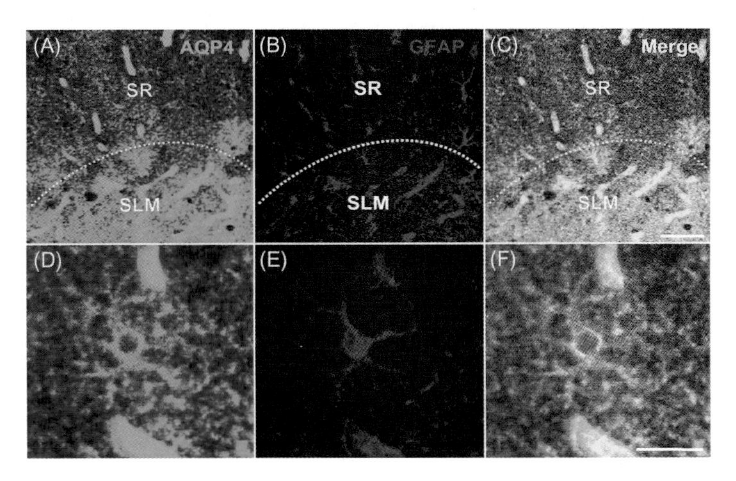

FIGURE 8.8 **Colocalization of AQP4 and GFAP.** Immunoreactivity for AQP4 (A), GFAP (B), and merged (C, F) is shown at the hippocampal CA1 SLM/SR border (dotted line) in a 3-week-old mouse. Note that in the SLM, there is significant colocalization of GFAP and AQP4 on cells with the morphology of *bona fide* astrocytes, whereas in the SR, GFAP-labeled astrocytes are seen that are AQP4-negative. AQP4 labels blood vessels in both layers. Higher-power images (D–F) demonstrate AQP4 and GFAP colocalization on an astrocyte in CA1 SR from a 6-week-old mouse. Scale bars, 50 μm (top), 20 μm (bottom). *Source: Reproduced with permission from Hsu MS, Seldin M, Lee DJ, Seifert G, Steinhäuser C, Binder DK. Laminar-specific and developmental expression of aquaporin-4 in the mouse hippocampus. Neuroscience 2011;178:21–32.*

$K_{ir}4.1$ were co-expressed in nearly all CA1 "classical astrocytes." In NG2 cells, AQP4 was detected at the transcript level but not with immunohistochemistry [101].

Seizure activity is accompanied by a variety of changes in gene expression, and repeated or prolonged seizure activity leads to glial changes, ultimately including the proliferation of "reactive" astrocytes (gliosis). The human studies indicated an upregulation and altered distribution of AQP4 protein in sclerotic tissue resected from patients with mesial temporal lobe epilepsy [117,118]. However, it is unclear whether AQP4 is upregulated by seizure activity per se and/or by other processes occurring during epileptogenesis (development of epilepsy), such as cell death, inflammation, or gliosis [129]. AQP4 is known to be dramatically upregulated in reactive astrocytes following injury [130–132] but may also be regulated by physiologic stimuli [133].

Of course, whether AQP4 dysregulation *precedes* or *follows* epileptogenesis impacts its functional role as cause or consequence. To address this question will require studies of AQP4 regulation with concurrent video-EEG monitoring during epileptogenesis to determine the exact timing of onset of spontaneous seizures relative to changes in AQP4 and its molecular partners. Three studies so far address this key issue. First, Kim et al. studied regulation of various AQPs in the rat pilocarpine model of epilepsy [134,135]. In control animals, AQP4 immunoreactivity was detected diffusely in the piriform cortex and hippocampus, with greatest expression at astrocyte endfeet. Following status epilepticus in this model the authors describe an "AQP4-deleted area" in the piriform cortex, associated with decreases in immunoreactivity for dystrophin and α-syntrophin, members of the DAPC [135]. However, EEG analysis for timing of spontaneous seizure onset was not investigated.

Second, Lee et al. [136] studied AQP4 distribution and regulation in the intrahippocampal model of epileptogenesis. In this model (Fig. 8.9), stereotactic microinjection of kainic acid into the mouse dorsal hippocampus leads to limbic status epilepticus followed by a "latent" period of several days (usually about 7 days) prior to the occurrence of spontaneous recurrent seizures (epilepsy) recorded by chronic video-EEG monitoring. Lee et al. found dramatic downregulation of AQP4 in multiple hippocampal subregions/laminae during epileptogenesis in this model (Fig. 8.10). Importantly, hippocampal AQP4 immunoreactivity was markedly downregulated during the latent phase prior to the occurrence of spontaneous seizures as well as during the chronic epileptic phase, with partial recovery in some hippocampal laminae. While marked reactive astrocytosis was observed, the reactive astrocytes in this model were largely AQP4-negative (Fig. 8.10A). Interestingly, in this study, $K_{ir}4.1$ expression on reactive astrocytes was preserved [136].

Third, Alvestad et al. [137] characterized electron microscopic (EM) localization of AQP4 in the latent and chronic phase in the intraperitoneal kainic acid rat model. Immunogold electron microscopic analysis revealed that adluminal AQP4 expression in astrocyte endfoot membranes was decreased in KA-treated rats during the latent phase. There was an accompanying reduction in adluminal α-syntrophin expression [137]. These data together with those of Kim et al. [134,135] and Lee et al. [136] suggest downregulation and/or mislocalization of AQP4 during early epileptogenesis. Thus, AQP4 dysregulation may contribute a critical role to loss of ion and water homeostasis leading to excitability and decreased seizure threshold during epileptogenesis.

Further studies will need to address the coordinate regulation of AQP4 and other relevant astrocyte molecules such as $K_{ir}4.1$ in these models. Finally, distinct epilepsy models in which epileptogenesis occurs in the *absence* of detectable cell death would prove useful in determining the threshold for AQP4 regulation [138,139].

Subcellular Alteration in AQP4 Distribution: A Common Disease Mechanism?

The hallmark of AQP4 expression in the CNS is its polarized expression at astrocyte endfeet ensheathing blood vessels [40,41,140]. Loss of such polarization is associated with pathology in mesial temporal lobe epilepsy as discussed above [118]. During seizures, there is focal swelling in the area of the seizure focus [23,74]; and the putative effect of loss of the perivascular pool of AQP4 would be to slow water egress from astrocyte to capillary leading to local astrocyte swelling, ECS constriction, and increased excitability [3].

Interestingly, a similar loss of AQP4 polarization has been observed in distinct models of neurological diseases. For example, in a mouse model of Alzheimer's disease, loss of AQP4 from endfoot membranes at sites of perivascular amyloid deposits was observed [141]. Such subcellular alterations could lead to perturbation of local water and K^+ homeostasis in affected brain regions, and thus contribute to cognitive decline and seizure susceptibility. Similarly, Badaut et al. found loss of polarization of AQP4 expression on astrocyte endfeet following subarachnoid hemorrhage (SAH) and in peritumoral tissue [130]; thus, posttraumatic epilepsy and tumor-associated epilepsy [142] could be associated with alterations in AQP4 expression and distribution. Further studies will require careful determination of perivascular/perisynaptic AQP4 expression ratios in a variety of conditions.

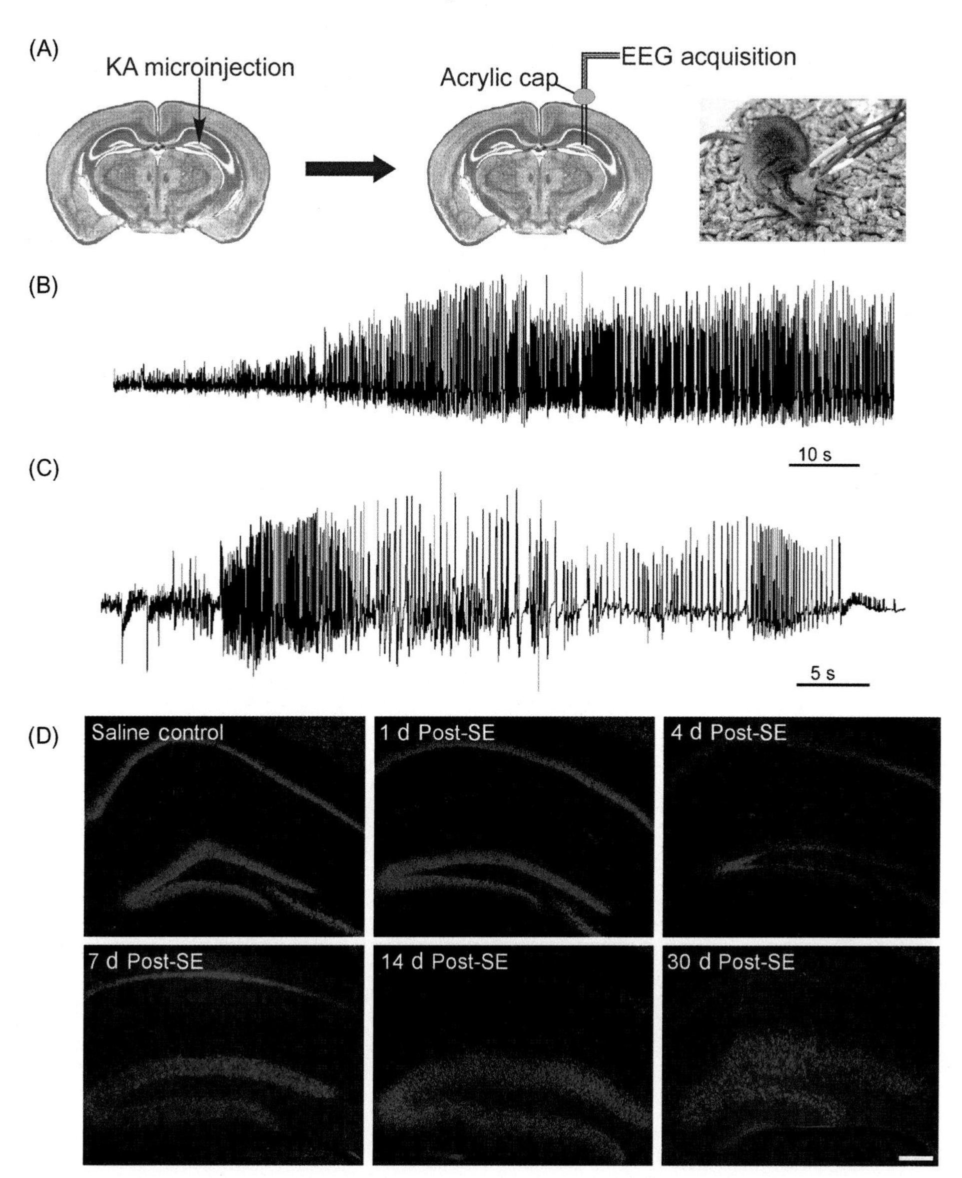

FIGURE 8.9 **Intrahippocampal kainic acid model of epilepsy.** (A) Intrahippocampal KA microinjection (left) is followed by bipolar hippocampal electrode placement (center) for chronic video-EEG recording in awake behaving mice (right). Within the first 30 minutes after IH KA injection, mice develop electrographic (B) and behavioral status epilepticus (SE). Approximately 1 week after SE, spontaneous recurrent seizures (epileptic seizures) are observed electrographically (C) and behaviorally. (D) Progressive histological changes in ipsilateral hippocampus (NeuN immunoreactivity), in particular progressive dentate granule cell dispersion, CA1 pyramidal cell loss and marked reactive astrogliosis (see next figure and book cover). Scale bar, 200 µm. *Source: Modified with permission from Lee DJ, Hsu MS, Seldin MM, Arellano JL, Binder DK. Decreased expression of the glial water channel aquaporin-4 in the intrahippocampal kainic acid model of epileptogenesis. Exp Neurol 2012;235(1):246–55.*

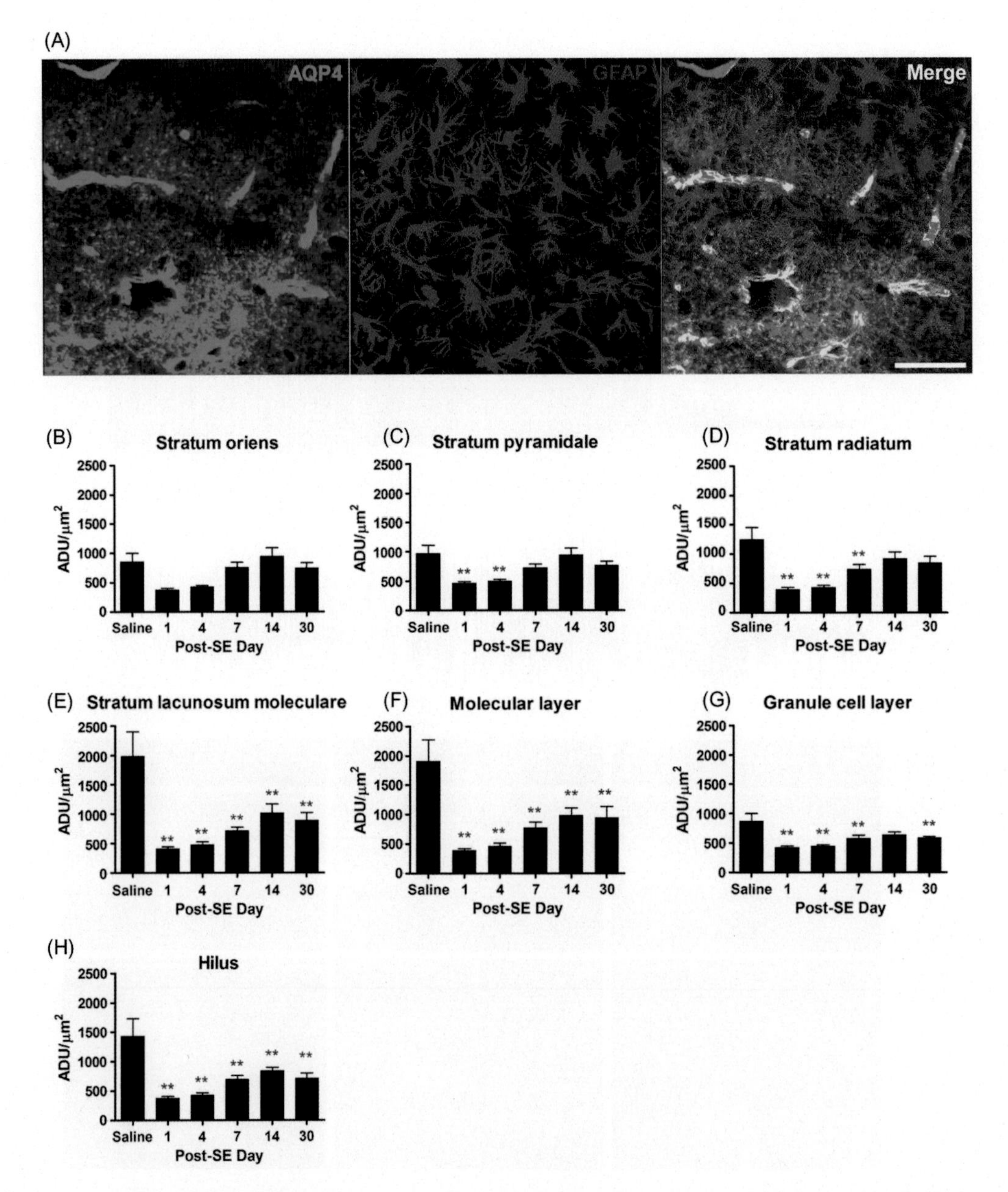

FIGURE 8.10 **Reduced AQP4 immunoreactivity following kainic acid status epilepticus.** Significant reduction in hippocampal AQP4 immunoreactivity was observed with delayed partial recovery. (A) Confocal 40× image in CA1 stratum lacunosum-moleculare 7 days post-SE. Note marked reactive phenotype of GFAP-positive astrocytes (center) but lack of strong AQP4 immunoreactivity. Scale bar, 50 μm. Laminar-specific analysis of AQP4 immunoreactivity after SE demonstrates decreased AQP4 immunoreactivity detected in various layers of the hippocampus throughout the study period (B–H). The initial decrease in AQP4 immunoreactivity is followed by a gradual increase. Persistent downregulation was observed in the stratum lacunosum-moleculare (SLM), molecular layer (ML), granule cell layer and hilus. *ADU*, arbitrary density units; $^{**}P < 0.01$ compared to saline control. *Source: Modified with permission from Lee DJ, Hsu MS, Seldin MM, Arellano JL, Binder DK. Decreased expression of the glial water channel aquaporin-4 in the intrahippocampal kainic acid model of epileptogenesis. Exp Neurol 2012;235(1):246–55.*

AQP4 and the Gliovascular Junction

Another area of astrocyte biology of potential importance for epilepsy is the "gliovascular junction" [78]. Recent studies have shown the close relationship between astrocyte endfeet ensheathing blood vessels, the targeted expression of AQP4 and $K_{ir}4.1$ on astrocyte endfeet, and the role of astrocytes in blood–brain barrier permeability [143] and control of cerebrovascular tone [144–148]. Molecular and cellular alterations in the gliovascular junction could perturb blood flow, water, and K^+ regulation, and therefore local tissue excitability. Interestingly, recent studies have shown that transient opening of the blood–brain barrier is sufficient for focal epileptogenesis [149,150]. For more details, please see *Chapter 12: Blood–Brain Barrier Disruption.*

Just as alterations in the perivascular pool of AQP4 on existing blood vessels could affect local water and K^+ regulation, it is also the case that many new blood vessels can be formed in epileptic tissue. MTS is characterized by neuronal cell loss in specific hippocampal areas, gliosis, synaptic reorganization, but also significant microvascular proliferation [114]. How does the regulation of AQP4 relate to angiogenesis during epileptogenesis? The increased number of microvessels in epileptic tissue may in part be due to upregulation of vascular endothelial growth factor (VEGF) following seizures [151–153]. VEGF administration itself has been shown to upregulate AQP4 [154], but it is unclear whether AQP4 distribution and/or function may be unique in "epileptic vasculature." Alterations in cerebrovascular tone in "epileptic vasculature" would be another interesting topic of investigation.

A Role for AQP4 at the Synapse?

It was clear from early immunogold studies of AQP4 localization that while the most abundant pool was localized at astrocyte endfeet, there was also a significant concentration in nonendfeet membranes, including those astrocyte membranes that ensheath glutamatergic synapses [40,41]. Given the role of AQP4 in rapid water fluxes and K^+ homeostasis, more specific roles of AQP4 in synaptic transmission, plasticity, and behavior have begun to be tested.

Skucas et al. evaluated long-term potentiation (LTP) and long-term depression (LTD) in hippocampal slices from AQP4$^{-/-}$ mice [155]. Interestingly, AQP4$^{-/-}$ mice exhibited a selective defect in LTP and LTD without a change in basal synaptic transmission or short-term plasticity. The impairment in LTP in AQP4$^{-/-}$ mice was specific for the type of LTP that depends on the neurotrophin brain-derived neurotrophic factor (BDNF), which is induced by stimulation at theta rhythm [theta-burst stimulation (TBS)-LTP], but there was no impairment in a form of LTP that is BDNF-independent, induced by high-frequency stimulation. LTD was also impaired in AQP4$^{-/-}$ mice, which was rescued by a scavenger of BDNF or blockade of Trk receptors. The AQP4$^{-/-}$ mice also exhibited a cognitive defect in location-specific object memory but not Morris water maze or contextual fear conditioning. These results suggest that AQP4 channels in astrocytes may play an unanticipated role in neurotrophin-dependent plasticity and influence behavior.

Based on these results, downregulation of AQP4 may not only lead to increased neural excitability due to abnormalities of water and potassium homeostasis but may also lead directly to abnormalities in synaptic plasticity (both LTP and LTD). This provides a potential explanation for the way that astrocytic changes in epilepsy may contribute not only to seizures but also to cognitive deficits. Cognitive impairment is very important because patients with temporal lobe epilepsy have many alterations in cognitive function and in particular hippocampal-dependent tasks such as spatial memory [156–159]. In addition, many other forms of plasticity are operative in the epileptic brain such as potentiation of synapses, reorganization of neuronal circuitry, and alteration in postnatal neurogenesis [160–162]. More experiments are necessary to elucidate the role of AQP4 in modulation of these processes underlying epileptogenesis.

Genetic Studies of AQP4 and Epilepsy

Based on the idea that AQP4 and $K_{ir}4.1$ act in concert to regulate water and K^+ homeostasis in the brain, the hypothesis arose that variants of the *AQP4* and *KCNJ10* (*$K_{ir}4.1$*) genes may be associated with forms of epilepsy. In a study of 218 Norwegian patients with TLE and 181 controls, Heuser et al. found several single nucleotide polymorphisms (SNPs) in the *KCNJ10* and *AQP4* genes associated with TLE [163]. In a mouse model with glial-specific deletion of *KCNJ10*, delayed stimulation-associated K^+ clearance in CA1 stratum radiatum was observed [99].

Restoration of Water and Ion Homeostasis

The above considerations lead to a new potential therapeutic opportunity: restoration of water and K^+ homeostasis by regulation of AQP4 targeting and distribution. Whether such a therapeutic approach in epilepsy may be reasonable is an open question. Might AQP4 upregulation be therapeutically feasible? So far, there is no known pharmacological method to upregulate AQP4. However, increasing understanding of the molecular partners involved (such as the DAPC) could lead to such approaches. For example, the extracellular matrix protein agrin, a heparan sulfate proteoglycan, may be a critical regulator of AQP4 membrane trafficking and polarity [164]. Astrocytes cultured with the neuronal agrin isoform A4B8 but not with the endothelial and meningeal isoform A0B0 demonstrated increases in the M23 splice variant of AQP4, increased water transport capacity, and increased membrane density of OAPs [164]. The glucocorticoid triamcinolone acetonide (TA) was shown to induce AQP4 downregulation in the normal retina but increased expression in the inflamed retina [165]. Targeting the DAPC for increased expression has the potential to be a novel therapeutic target, but a way to modulate its expression in the brain is still unknown. Current studies focus on the role of dystrophin in Duchenne muscular dystrophy and a few have found ways to upregulate the anchoring protein [166,167], but the expression changes in brain tissue and effects on AQP4 expression have not yet been examined.

CONCLUSIONS

Compelling evidence indicates that the glial water channel AQP4 plays a fundamental role in water transport in the brain. AQP4 is expressed in astrocytes, and along with the inwardly-rectifying K^+ channel $K_{ir}4.1$ is thought to be responsible for water and K^+

homeostasis during neural activity. Because osmolarity and K^+ have powerful effects on seizure susceptibility, AQP4 and its molecular partners may represent novel therapeutic targets for control of seizures. $AQP4^{-/-}$ mice have significantly prolonged seizure duration associated with alterations in extracellular K^+ clearance and gap junctional coupling. Dysfunctional K^+ homeostasis and upregulation and altered subcellular distribution of AQP4 have been observed in human epileptic tissue. The relevance of these findings to hippocampal epileptogenesis and also to human epilepsy requires further study. Restoration of water and K^+ homeostasis in epileptic tissue constitutes a novel therapeutic concept. AQP4 occupies a pivotal position: concentrated in both astrocytic endfeet and in astroglial processes investing synapses, AQP4 may be involved in glioneuronal interaction of water and ion fluxes and also in gliovascular interactions across the blood–brain barrier and thus play a critical role both during active tissue metabolism and in pathophysiologic states.

References

[1] Ransom B, Behar T, Nedergaard M. New roles for astrocytes (stars at last). Trends Neurosci 2003;26(10):520–2.
[2] Binder DK, Steinhäuser C. Functional changes in astroglial cells in epilepsy. Glia 2006;54(5):358–68.
[3] Wetherington J, Serrano G, Dingledine R. Astrocytes in the epileptic brain. Neuron 2008;58(2):168–78.
[4] Amiry-Moghaddam M, Williamson A, Palomba M, Eid T, de Lanerolle NC, Nagelhus EA, et al. Delayed K^+ clearance associated with aquaporin-4 mislocalization: phenotypic defects in brains of α-syntrophin-null mice. Proc Natl Acad Sci USA 2003;100(23):13615–20.
[5] Volterra A, Steinhäuser C. Glial modulation of synaptic transmission in the hippocampus. Glia 2004;47(3):249–57.
[6] Volterra A, Meldolesi J. Astrocytes, from brain glue to communication elements: the revolution continues. Nat Rev Neurosci 2005;6(8):626–40.
[7] Halassa MM, Haydon PG. Integrated brain circuits: astrocytic networks modulate neuronal activity and behavior. Annu Rev Physiol 2010;72:335–55.
[8] Halassa MM, Fellin T, Haydon PG. The tripartite synapse: roles for gliotransmission in health and disease. Trends Mol Med 2007;13(2):54–63.
[9] Tian GF, Azmi H, Takano T, Xu Q, Peng W, Lin J, et al. An astrocytic basis of epilepsy. Nat Med 2005;11:973–81.
[10] Steinhäuser C, Seifert G. Glial membrane channels and receptors in epilepsy: impact for generation and spread of seizure activity. Eur J Pharmacol 2002;447(2–3):227–37.
[11] Kivi A, Lehmann TN, Kovacs R, Eilers A, Jauch R, Meencke HJ, et al. Effects of barium on stimulus-induced rises of $[K^+]_o$ in human epileptic non-sclerotic and sclerotic hippocampal area CA1. Eur J Neurosci 2000;12(6):2039–48.
[12] Heinemann U, Gabriel S, Jauch R, Schulze K, Kivi A, Eilers A, et al. Alterations of glial cell function in temporal lobe epilepsy. Epilepsia 2000;41(Suppl 6):S185–9.
[13] Hinterkeuser S, Schröder W, Hager G, Seifert G, Blümcke I, Elger CE, et al. Astrocytes in the hippocampus of patients with temporal lobe epilepsy display changes in potassium conductances. Eur J Neurosci 2000;12(6):2087–96.
[14] de Lanerolle NC, Lee TS. New facets of the neuropathology and molecular profile of human temporal lobe epilepsy. Epilepsy Behav 2005;7(2):190–203.
[15] Seifert G, Schilling K, Steinhäuser C. Astrocyte dysfunction in neurological disorders: a molecular perspective. Nat Rev Neurosci 2006;7(3):194–206.
[16] Fiacco TA, Agulhon C, Taves SR, Petravicz J, Casper KB, Dong X, et al. Selective stimulation of astrocyte calcium in situ does not affect neuronal excitatory synaptic activity. Neuron 2007;54(4):611–26.
[17] Seifert G, Steinhäuser C. Neuron-astrocyte signaling and epilepsy. Exp Neurol 2013;244:4–10.
[18] Schwartzkroin PA, Baraban SC, Hochman DW. Osmolarity, ionic flux, and changes in brain excitability. Epilepsy Res 1998;32:275–85.
[19] Roper SN, Obenaus A, Dudek FE. Osmolality and nonsynaptic epileptiform bursts in rat CA1 and dentate gyrus. Ann Neurol 1992;31:81–5.

[20] Dudek FE, Obenhaus A, Tasker JG. Osmolality-induced changes in extracellular volume alter epileptiform bursts independent of chemical synapses in the rat: importance of non-synaptic mechanisms in hippocampal epileptogenesis. Neurosci Lett 1990;120:267–70.

[21] Chebabo SR, Hester MA, Aitken PG, Somjen GG. Hypotonic exposure enhances synaptic transmission and triggers spreading depression in rat hippocampal tissue slices. Brain Res 1995;695:203–16.

[22] Pan E, Stringer JL. Influence of osmolality on seizure amplitude and propagation in the rat dentate gyrus. Neurosci Lett 1996;207(1):9–12.

[23] Traynelis SF, Dingledine R. Role of extracellular space in hyperosmotic suppression of potassium-induced electrographic seizures. J Neurophysiol 1989;61(5):927–38.

[24] Haglund MM, Hochman DW. Furosemide and mannitol suppression of epileptic activity in the human brain. J Neurophysiol 2005;94(2):907–18.

[25] Andrew RD, Fagan M, Ballyk BA, Rosen AS. Seizure susceptibility and the osmotic state. Brain Res 1989;498:175–80.

[26] Carter CH. Status epilepticus treated by intravenous urea. Epilepsia 1962;3:198–200.

[27] Maa EH, Kahle KT, Walcott BP, Spitz MC, Staley KJ. Diuretics and epilepsy: will the past and present meet? Epilepsia 2011;52(9):1559–69.

[28] Rutecki PA, Lebeda FJ, Johnston D. Epileptiform activity induced by changes in extracellular potassium in hippocampus. J Neurophysiol 1985;54(5):1363–74.

[29] Yaari Y, Konnerth A, Heinemann U. Nonsynaptic epileptogenesis in the mammalian hippocampus in vitro. II. Role of extracellular potassium. J Neurophysiol 1986;56(2):424–38.

[30] Traynelis SF, Dingledine R. Potassium-induced spontaneous electrographic seizures in the rat hippocampal slice. J Neurophysiol 1988;59(1):259–76.

[31] Feng Z, Durand DM. Effects of potassium concentration on firing patterns of low-calcium epileptiform activity in anesthetized rat hippocampus: inducing of persistent spike activity. Epilepsia 2006;47:727–36.

[32] Gabriel S, Njunting M, Pomper JK, Merschhemke M, Sanabria ER, Eilers A, et al. Stimulus and potassium-induced epileptiform activity in the human dentate gyrus from patients with and without hippocampal sclerosis. J Neurosci 2004;24(46):10416–30.

[33] Verkman AS. More than just water channels: unexpected cellular roles of aquaporins. J Cell Sci 2005;118(Pt 15):3225–32.

[34] Amiry-Moghaddam M, Ottersen OP. The molecular basis of water transport in the brain. Nat Rev Neurosci 2003;4(12):991–1001.

[35] Agre P, King LS, Yasui M, Guggino WB, Ottersen OP, Fujiyoshi Y, et al. Aquaporin water channels—from atomic structure to clinical medicine. J Physiol 2002;542(Pt 1):3–16.

[36] Verkman AS. Physiological importance of aquaporin water channels. Ann Med 2002;34(3):192–200.

[37] Hasegawa H, Ma T, Skach W, Matthay MA, Verkman AS. Molecular cloning of a mercurial-insensitive water channel expressed in selected water-transporting tissues. J Biol Chem 1994;269(8):5497–500.

[38] Yang B, Ma T, Verkman AS. cDNA cloning, gene organization, and chromosomal localization of a human mercurial insensitive water channel. Evidence for distinct transcriptional units. J Biol Chem 1995;270(39):22907–13.

[39] Jung JS, Bhat RV, Preston GM, Guggino WB, Baraban JM, Agre P. Molecular characterization of an aquaporin cDNA from brain: candidate osmoreceptor and regulator of water balance. Proc Natl Acad Sci USA 1994;91(26):13052–6.

[40] Nielsen S, Nagelhus EA, Amiry-Moghaddam M, Bourque C, Agre P, Ottersen OP. Specialized membrane domains for water transport in glial cells: high-resolution immunogold cytochemistry of aquaporin-4 in rat brain. J Neurosci 1997;17(1):171–80.

[41] Nagelhus EA, Mathiisen TM, Ottersen OP. Aquaporin-4 in the central nervous system: cellular and subcellular distribution and coexpression with K_{ir}4.1. Neuroscience 2004;129(4):905–13.

[42] Rash JE, Yasumura T, Hudson CS, Agre P, Nielsen S. Direct immunogold labeling of aquaporin-4 in square arrays of astrocyte and ependymocyte plasma membranes in rat brain and spinal cord. Proc Natl Acad Sci USA 1998;95(20):11981–6.

[43] Oshio K, Binder DK, Yang B, Schecter S, Verkman AS, Manley GT. Expression of aquaporin water channels in mouse spinal cord. Neuroscience 2004;127(3):685–93.

[44] Jarius S, Paul F, Franciotta D, Waters P, Zipp F, Hohlfeld R, et al. Mechanisms of disease: aquaporin-4 antibodies in neuromyelitis optica. Nat Clin Pract Neurol 2008;4(4):202–14.

[45] Niermann H, Amiry-Moghaddam M, Holthoff K, Witte OW, Ottersen OP. A novel role of vasopressin in the brain: modulation of activity-dependent water flux in the neocortex. J Neurosci 2001;21(9):3045–51.
[46] Holthoff K, Witte OW. Directed spatial potassium redistribution in rat neocortex. Glia 2000;29(3):288–92.
[47] Verbavatz JM, Ma T, Gobin R, Verkman AS. Absence of orthogonal arrays in kidney, brain and muscle from transgenic knockout mice lacking water channel aquaporin-4. J Cell Sci 1997;110(Pt 22):2855–60.
[48] Yang B, Brown D, Verkman AS. The mercurial insensitive water channel AQP4 forms orthogonal arrays in stably transfected Chinese hamster ovary cells. J Biol Chem 1996;271(9):4577–80.
[49] Frigeri A, Gropper MA, Umenishi F, Kawashima M, Brown D, Verkman AS. Localization of MIWC and GLIP water channel homologs in neuromuscular, epithelial and glandular tissues. J Cell Sci 1995;108(Pt 9):2993–3002.
[50] Wolburg H, Berg K. Distribution of orthogonal arrays of particles in the Muller cell membrane of the mouse retina. Glia 1988;1(4):246–52.
[51] Wolburg H, Wolburg-Buchholz K, Fallier-Becker P, Noell S, Mack AF. Structure and functions of aquaporin-4-based orthogonal arrays of particles. Int Rev Cell Mol Biol 2011;287:1–41.
[52] Ma T, Yang B, Gillespie A, Carlson EJ, Epstein CJ, Verkman AS. Generation and phenotype of a transgenic knockout mouse lacking the mercurial-insensitive water channel aquaporin-4. J Clin Invest 1997;100(5):957–62.
[53] Haj-Yasein NN, Vindedal GF, Eilert-Olsen M, Gundersen GA, Skare O, Laake P, et al. Glial-conditional deletion of aquaporin-4 (AQP4) reduces blood-brain water uptake and confers barrier function on perivascular astrocyte endfeet. Proc Natl Acad Sci USA 2011;108:17815–20.
[54] Saadoun S, Tait MJ, Reza A, Davies DC, Bell BA, Verkman AS, et al. AQP4 gene deletion in mice does not alter blood-brain barrier integrity or brain morphology. Neuroscience 2009;161(3):764–72.
[55] Binder DK, Oshio K, Ma T, Verkman AS, Manley GT. Increased seizure threshold in mice lacking aquaporin-4 water channels. Neuroreport 2004;15(2):259–62.
[56] Papadopoulos MC, Manley GT, Krishna S, Verkman AS. Aquaporin-4 facilitates reabsorption of excess fluid in vasogenic brain edema. FASEB J 2004;18(11):1291–3.
[57] Li J, Patil RV, Verkman AS. Mildly abnormal retinal function in transgenic mice without Müller cell aquaporin-4 water channels. Invest Ophthalmol Vis Sci 2002;43(2):573–9.
[58] Li J, Verkman AS. Impaired hearing in mice lacking aquaporin-4 water channels. J Biol Chem 2001;276(33):31233–7.
[59] Solenov EI, Vetrivel L, Oshio K, Manley GT, Verkman AS. Optical measurement of swelling and water transport in spinal cord slices from aquaporin null mice. J Neurosci Methods 2002;113(1):85–90.
[60] Thiagarajah JR, Papadopoulos MC, Verkman AS. Noninvasive early detection of brain edema in mice by near-infrared light scattering. J Neurosci Res 2005;80(2):293–9.
[61] Solenov E, Watanabe H, Manley GT, Verkman AS. Sevenfold-reduced osmotic water permeability in primary astrocyte cultures from AQP-4-deficient mice, measured by a fluorescence quenching method. Am J Physiol Cell Physiol 2004;286(2):C426–32.
[62] Nicchia GP, Frigeri A, Liuzzi GM, Svelto M. Inhibition of aquaporin-4 expression in astrocytes by RNAi determines alteration in cell morphology, growth, and water transport and induces changes in ischemia-related genes. FASEB J 2003;17(11):1508–10.
[63] Manley GT, Fujimura M, Ma T, Noshita N, Filiz F, Bollen AW, et al. Aquaporin-4 deletion in mice reduces brain edema after acute water intoxication and ischemic stroke. Nat Med 2000;6(2):159–63.
[64] Lee DJ, Amini M, Hamamura MJ, Hsu MS, Seldin MM, Nalcioglu O, et al. Aquaporin-4-dependent edema clearance following status epilepticus. Epilepsy Res 2012;98(2–3):264–8.
[65] Frigeri A, Nicchia GP, Nico B, Quondamatteo F, Herken R, Roncali L, et al. Aquaporin-4 deficiency in skeletal muscle and brain of dystrophic mdx mice. FASEB J 2001;15(1):90–8.
[66] Vajda Z, Pedersen M, Füchtbauer EM, Wertz K, Stødkilde-Jørgensen H, Sulyok E, et al. Delayed onset of brain edema and mislocalization of aquaporin-4 in dystrophin-null transgenic mice. Proc Natl Acad Sci USA 2002;99(20):13131–6.
[67] Neely JD, Amiry-Moghaddam M, Ottersen OP, Froehner SC, Agre P, Adams ME. Syntrophin-dependent expression and localization of aquaporin-4 water channel protein. Proc Natl Acad Sci USA 2001;98(24):14108–13.
[68] Amiry-Moghaddam M, Otsuka T, Hurn PD, Traystman RJ, Haug FM, Froehner SC, et al. An α-syntrophin-dependent pool of AQP4 in astroglial end-feet confers bidirectional water flow between blood and brain. Proc Natl Acad Sci USA 2003;100(4):2106–11.
[69] Syková E. The extracellular space in the CNS: its regulation, volume and geometry in normal and pathological neuronal function. Neuroscientist 1997;3:28–41.

[70] Fields RD, Stevens-Graham B. New insights into neuron-glia communication. Science 2002;298(5593):556–62.
[71] Østby I, Øyehaug L, Einevoll GT, Nagelhus EA, Plahte E, Zeuthen T, et al. Astrocytic mechanisms explaining neural-activity-induced shrinkage of extraneuronal space. PLoS Comput Biol 2009;5(1):e1000272.
[72] Risher WC, Andrew RD, Kirov SA. Real-time passive volume responses of astrocytes to acute osmotic and ischemic stress in cortical slices and in vivo revealed by two-photon microscopy. Glia 2009;57(2):207–21.
[73] Andrew RD, Labron MW, Boehnke SE, Carnduff L, Kirov SA. Physiological evidence that pyramidal neurons lack functional water channels. Cereb Cortex 2007;17(4):787–802.
[74] Binder DK, Papadopoulos MC, Haggie PM, Verkman AS. In vivo measurement of brain extracellular space diffusion by cortical surface photobleaching. J Neurosci 2004;24(37):8049–56.
[75] Yao X, Hrabetova S, Nicholson C, Manley GT. Aquaporin-4-deficient mice have increased extracellular space without tortuosity change. J Neurosci 2008;28(21):5460–4.
[76] Binder DK, Yao X, Sick TJ, Verkman AS, Manley GT. Increased seizure duration and slowed potassium kinetics in mice lacking aquaporin-4 water channels. Glia 2006;53:631–6.
[77] Nagelhus EA, Horio Y, Inanobe A, Fujita A, Haug FM, Nielsen S, et al. Immunogold evidence suggests that coupling of K^+ siphoning and water transport in rat retinal Müller cells is mediated by a coenrichment of $K_{ir}4.1$ and AQP4 in specific membrane domains. Glia 1999;26(1):47–54.
[78] Simard M, Nedergaard M. The neurobiology of glia in the context of water and ion homeostasis. Neuroscience 2004;129(4):877–96.
[79] Manley GT, Binder DK, Papadopoulos MC, Verkman AS. New insights into water transport and edema in the central nervous system from phenotype analysis of aquaporin-4 null mice. Neuroscience 2004;129(4):983–91.
[80] Heinemann U, Lux HD. Ceiling of stimulus induced rises in extracellular potassium concentration in the cerebral cortex of cat. Brain Res 1977;120(2):231–49.
[81] Somjen GG. Ion regulation in the brain: implications for pathophysiology. Neuroscientist 2002;8:254–67.
[82] Xiong ZQ, Stringer JL. Astrocytic regulation of the recovery of extracellular potassium after seizures in vivo. Eur J Neurosci 1999;11(5):1677–84.
[83] Walz W. Swelling and potassium uptake in cultured astrocytes. Can J Physiol Pharmacol 1987;65(5):1051–7.
[84] Walz W. Mechanism of rapid K^+-induced swelling of mouse astrocytes. Neurosci Lett 1992;135(2):243–6.
[85] Connors NC, Adams ME, Froehner SC, Kofuji P. The potassium channel $K_{ir}4.1$ associates with the dystrophin-glycoprotein complex via alpha-syntrophin in glia. J Biol Chem 2004;279(27):28387–92.
[86] Kofuji P, Ceelen P, Zahs KR, Surbeck LW, Lester HA, Newman EA. Genetic inactivation of an inwardly rectifying potassium channel ($K_{ir}4.1$ subunit) in mice: phenotypic impact in retina. J Neurosci 2000;20(15):5733–40.
[87] Rozengurt N, Lopez I, Chiu CS, Kofuji P, Lester HA, Neusch C. Time course of inner ear degeneration and deafness in mice lacking the $K_{ir}4.1$ potassium channel subunit. Hear Res 2003;177(1–2):71–80.
[88] Marcus DC, Wu T, Wangemann P, Kofuji P. KCNJ10 ($K_{ir}4.1$) potassium channel knockout abolishes endocochlear potential. Am J Physiol Cell Physiol 2002;282(2):C403–7.
[89] Neusch C, Rozengurt N, Jacobs RE, Lester HA, Kofuji P. $K_{ir}4.1$ potassium channel subunit is crucial for oligodendrocyte development and in vivo myelination. J Neurosci 2001;21(15):5429–38.
[90] Newman EA. High potassium conductance in astrocyte endfeet. Science 1986;233(4762):453–4.
[91] Newman EA. Inward-rectifying potassium channels in retinal glial (Müller) cells. J Neurosci 1993;13(8):3333–45.
[92] Newman EA, Frambach DA, Odette LL. Control of extracellular potassium levels by retinal glial cell K^+ siphoning. Science 1984;225(4667):1174–5.
[93] Newman EA, Karwoski CJ. Spatial buffering of light-evoked potassium increases by retinal glial (Müller) cells. Acta Physiol Scand Suppl 1989;582:51.
[94] Ballanyi K, Grafe P, ten Bruggencate G. Ion activities and potassium uptake mechanisms of glial cells in guinea-pig olfactory cortex slices. J Physiol 1987;382:159–74.
[95] Kofuji P, Newman EA. Potassium buffering in the central nervous system. Neuroscience 2004;129(4):1045–56.
[96] Neusch C, Papadopoulos N, Müller M, Maletzki I, Winter SM, Hirrlinger J, et al. Lack of the $K_{ir}4.1$ channel subunit abolishes K^+ buffering properties of astrocytes in the ventral respiratory group: impact on extracellular K^+ regulation. J Neurophysiol 2006;95(3):1843–52.
[97] Djukic B, Casper KB, Philpot BD, Chin LS, McCarthy KD. Conditional knock-out of $K_{ir}4.1$ leads to glial membrane depolarization, inhibition of potassium and glutamate uptake, and enhanced short-term synaptic potentiation. J Neurosci 2007;27(42):11354–65.

[98] Seifert G, Hüttmann K, Binder DK, Hartmann C, Wyczynski A, Neusch C, et al. Analysis of astroglial K^+ channel expression in the developing hippocampus reveals a predominant role of the $K_{ir}4.1$ subunit. J Neurosci 2009;29(23):7474–88.
[99] Haj-Yasein NN, Jensen V, Vindedal GF, Gundersen GA, Klungland A, Ottersen OP, et al. Evidence that compromised K^+ spatial buffering contributes to the epileptogenic effect of mutations in the human $K_{ir}4.1$ gene (KCNJ10). Glia 2011;59(11):1635–42.
[100] Padmawar P, Yao X, Bloch O, Manley GT, Verkman AS. K^+ waves in brain cortex visualized using a long-wavelength K^+-sensing fluorescent indicator. Nat Methods 2005;2(11):825–7.
[101] Hsu MS, Seldin M, Lee DJ, Seifert G, Steinhäuser C, Binder DK. Laminar-specific and developmental expression of aquaporin-4 in the mouse hippocampus. Neuroscience 2011;178:21–32.
[102] Ruiz-Ederra J, Zhang H, Verkman AS. Evidence against functional interaction between aquaporin-4 water channels and $K_{ir}4.1$ potassium channels in retinal Müller cells. J Biol Chem 2007;282(30):21866–72.
[103] Zhang H, Verkman AS. Aquaporin-4 independent $K_{ir}4.1$ K^+ channel function in brain glial cells. Mol Cell Neurosci 2008;37(1):1–10.
[104] Jin BJ, Zhang H, Binder DK, Verkman AS. Aquaporin-4-dependent K^+ and water transport modeled in brain extracellular space following neuroexcitation. J Gen Physiol 2013;141(1):119–32.
[105] Strohschein S, Hüttmann K, Gabriel S, Binder DK, Heinemann U, Steinhäuser C. Impact of aquaporin-4 channels on K^+ buffering and gap junction coupling in the hippocampus. Glia 2011;59(6):973–80.
[106] Wallraff A, Kohling R, Heinemann U, Theis M, Willecke K, Steinhäuser C. The impact of astrocytic gap junctional coupling on potassium buffering in the hippocampus. J Neurosci 2006;26(20):5438–47.
[107] Thrane AS, Rappold PM, Fujita T, Torres A, Bekar LK, Takano T, et al. Critical role of aquaporin-4 (AQP4) in astrocytic Ca^{2+} signaling events elicited by cerebral edema. Proc Natl Acad Sci USA 2011;108(2):846–51.
[108] Haj-Yasein NN, Bugge CE, Jensen V, Ostby I, Ottersen OP, Hvalby O, et al. Deletion of aquaporin-4 increases extracellular K^+ concentration during synaptic stimulation in mouse hippocampus. Brain Struct Funct 2015;220(4):2469–74.
[109] Bragg AD, Amiry-Moghaddam M, Ottersen OP, Adams ME, Froehner SC. Assembly of a perivascular astrocyte protein scaffold at the mammalian blood-brain barrier is dependent on α-syntrophin. Glia 2006;53(8):879–90.
[110] Gondo A, Shinotsuka T, Morita A, Abe Y, Yasui M, Nuriya M. Sustained down-regulation of β-dystroglycan and associated dysfunctions of astrocytic endfeet in epileptic cerebral cortex. J Biol Chem 2014;289(44):30279–88.
[111] Anderova M, Benesova J, Mikesova M, Dzamba D, Honsa P, Kriska J, et al. Altered astrocytic swelling in the cortex of α-syntrophin-negative GFAP/EGFP mice. PLoS One 2014;9(11):e113444.
[112] De Sarro G, Ibbadu GF, Marra R, Rotiroti D, Loiacono A, Donato Di Paola E, et al. Seizure susceptibility to various convulsant stimuli in dystrophin-deficient mdx mice. Neurosci Res 2004;50(1):37–44.
[113] Tsao CY, Mendell JR. Coexisting muscular dystrophies and epilepsy in children. J Child Neurol 2006;21(2):148–50.
[114] Blümcke I, Beck H, Lie AA, Wiestler OD. Molecular neuropathology of human mesial temporal lobe epilepsy. Epilepsy Res 1999;36(2-3):205–23.
[115] Mitchell LA, Jackson GD, Kalnins RM, Saling MM, Fitt GJ, Ashpole RD, et al. Anterior temporal abnormality in temporal lobe epilepsy: a quantitative MRI and histopathologic study. Neurology 1999;52(2):327–36.
[116] Hugg JW, Butterworth EJ, Kuzniecky RI. Diffusion mapping applied to mesial temporal lobe epilepsy: preliminary observations. Neurology 1999;53(1):173–6.
[117] Lee TS, Eid T, Mane S, Kim JH, Spencer DD, Ottersen OP, et al. Aquaporin-4 is increased in the sclerotic hippocampus in human temporal lobe epilepsy. Acta Neuropathol (Berl) 2004;108(6):493–502.
[118] Eid T, Lee TS, Thomas MJ, Amiry-Moghaddam M, Bjornsen LP, Spencer DD, et al. Loss of perivascular aquaporin-4 may underlie deficient water and K^+ homeostasis in the human epileptogenic hippocampus. Proc Natl Acad Sci USA 2005;102(4):1193–8.
[119] Medici V, Frassoni C, Tassi L, Spreafico R, Garbelli R. Aquaporin-4 expression in control and epileptic human cerebral cortex. Brain Res 2011;1367:330–9.
[120] Dudek FE, Rogawski MA. Regulation of brain water: is there a role for aquaporins in epilepsy? Epilepsy Curr 2005;5(3):104–6.

[121] Badaut J, Lasbennes F, Magistretti PJ, Regli L. Aquaporins in brain: distribution, physiology, and pathophysiology. J Cereb Blood Flow Metab 2002;22(4):367–78.

[122] Badaut J, Verbavatz JM, Freund-Mercier MJ, Lasbennes F. Presence of aquaporin-4 and muscarinic receptors in astrocytes and ependymal cells in rat brain: a clue to a common function? Neurosci Lett 2000;292(2):75–8.

[123] Jabs R, Pivneva T, Hüttmann K, Wyczynski A, Nolte C, Kettenmann H, et al. Synaptic transmission onto hippocampal glial cells with hGFAP promoter activity. J Cell Sci 2005;118(Pt 16):3791–803.

[124] Matthias K, Kirchhoff F, Seifert G, Hüttmann K, Matyash M, Kettenmann H, et al. Segregated expression of AMPA-type glutamate receptors and glutamate transporters defines distinct astrocyte populations in the mouse hippocampus. J Neurosci 2003;23(5):1750–8.

[125] Wallraff A, Odermatt B, Willecke K, Steinhäuser C. Distinct types of astroglial cells in the hippocampus differ in gap junction coupling. Glia 2004;48(1):36–43.

[126] Nishiyama A, Yang Z, Butt A. Astrocytes and NG2-glia: what's in a name? J Anat. 2005;;207(6):687–93.

[127] Tomás-Camardiel M, Venero JL, de Pablos RM, Rite I, Machado A, Cano J. In vivo expression of aquaporin-4 by reactive microglia. J Neurochem 2004;91(4):891–9.

[128] Eilert-Olsen M, Haj-Yasein NN, Vindedal GF, Enger R, Gundersen GA, Hoddevik EH, et al. Deletion of aquaporin-4 changes the perivascular glial protein scaffold without disrupting the brain endothelial barrier. Glia 2012;60(3):432–40.

[129] Vezzani A, Granata T. Brain inflammation in epilepsy: experimental and clinical evidence. Epilepsia 2005;46(11):1724–43.

[130] Badaut J, Brunet JF, Grollimund L, Hamou MF, Magistretti PJ, Villemure JG, et al. Aquaporin-1 and aquaporin-4 expression in human brain after subarachnoid hemorrhage and in peritumoral tissue. Acta Neurochir Suppl 2003;86:495–8.

[131] Saadoun S, Papadopoulos MC, Davies DC, Krishna S, Bell BA. Aquaporin-4 expression is increased in oedematous human brain tumours. J Neurol Neurosurg Psychiatry 2002 Feb;72(2):262–5.

[132] Vizuete ML, Venero JL, Vargas C, Ilundáin AA, Echevarría M, Machado A, et al. Differential upregulation of aquaporin-4 mRNA expression in reactive astrocytes after brain injury: potential role in brain edema. Neurobiol Dis 1999;6(4):245–58.

[133] Saito N, Ikegami H, Shimada K. Effect of water deprivation on aquaporin-4 (AQP4) mRNA expression in chickens (Gallus domesticus). Brain Res Mol Brain Res 2005;141(2):193–7.

[134] Kim JE, Ryu HJ, Yeo SI, Seo CH, Lee BC, Choi IG, et al. Differential expressions of aquaporin subtypes in astroglia in the hippocampus of chronic epileptic rats. Neuroscience 2009;163(3):781–9.

[135] Kim JE, Yeo SI, Ryu HJ, Kim MJ, Kim DS, Jo SM, et al. Astroglial loss and edema formation in the rat piriform cortex and hippocampus following pilocarpine-induced status epilepticus. J Comp Neurol 2010;518(22):4612–28.

[136] Lee DJ, Hsu MS, Seldin MM, Arellano JL, Binder DK. Decreased expression of the glial water channel aquaporin-4 in the intrahippocampal kainic acid model of epileptogenesis. Exp Neurol 2012;235(1):246–55.

[137] Alvestad S, Hammer J, Hoddevik EH, Skare O, Sonnewald U, Amiry-Moghaddam M, et al. Mislocalization of AQP4 precedes chronic seizures in the kainate model of temporal lobe epilepsy. Epilepsy Res 2013;105(1-2):30–41.

[138] Bender RA, Dubé C, Baram TZ. Febrile seizures and mechanisms of epileptogenesis: insights from an animal model. Adv Exp Med Biol 2004;548:213–25.

[139] Dubé CM, Brewster AL, Richichi C, Zha Q, Baram TZ. Fever, febrile seizures and epilepsy. Trends Neurosci 2007;30(10):490–6.

[140] Nagelhus EA, Veruki ML, Torp R, Haug FM, Laake JH, Nielsen S, et al. Aquaporin-4 water channel protein in the rat retina and optic nerve: polarized expression in Müller cells and fibrous astrocytes. J Neurosci 1998;18(7):2506–19.

[141] Yang J, Lunde LK, Nuntagij P, Oguchi T, Camassa LM, Nilsson LN, et al. Loss of astrocyte polarization in the Tg-ArcSwe mouse model of Alzheimer's disease. J Alzheimers Dis 2011;27:711–22.

[142] Rajneesh KF, Binder DK. Tumor-associated epilepsy. Neurosurg Focus 2009;27(2):E4.

[143] Abbott NJ. Astrocyte-endothelial interactions and blood-brain barrier permeability. J Anat 2002;200(6):629–38.

[144] Metea MR, Newman EA. Glial cells dilate and constrict blood vessels: a mechanism of neurovascular coupling. J Neurosci 2006;26(11):2862–70.

[145] Mulligan SJ, MacVicar BA. Calcium transients in astrocyte endfeet cause cerebrovascular constrictions. Nature 2004;431(7005):195–9.

[146] Takano T, Tian GF, Peng W, Lou N, Libionka W, Han X, et al. Astrocyte-mediated control of cerebral blood flow. Nat Neurosci 2006;9(2):260–7.
[147] Zonta M, Angulo MC, Gobbo S, Rosengarten B, Hossmann KA, Pozzan T, et al. Neuron-to-astrocyte signaling is central to the dynamic control of brain microcirculation. Nat Neurosci 2003;6(1):43–50.
[148] Gordon GR, Mulligan SJ, MacVicar BA. Astrocyte control of the cerebrovasculature. Glia 2007;55(12):1214–21.
[149] Seiffert E, Dreier JP, Ivens S, Bechmann I, Tomkins O, Heinemann U, et al. Lasting blood-brain barrier disruption induces epileptic focus in the rat somatosensory cortex. J Neurosci 2004;24(36):7829–36.
[150] Ivens S, Kaufer D, Flores LP, Bechmann I, Zumsteg D, Tomkins O, et al. TGF-β receptor-mediated albumin uptake into astrocytes is involved in neocortical epileptogenesis. Brain 2007;130(Pt 2):535–47.
[151] Croll SD, Goodman JH, Scharfman HE. Vascular endothelial growth factor (VEGF) in seizures: a double-edged sword. Adv Exp Med Biol 2004;548:57–68.
[152] Nicoletti JN, Shah SK, McCloskey DP, Goodman JH, Elkady A, Atassi H, et al. Vascular endothelial growth factor is up-regulated after status epilepticus and protects against seizure-induced neuronal loss in hippocampus. Neuroscience 2008;151(1):232–41.
[153] Rigau V, Morin M, Rousset MC, de Bock F, Lebrun A, Coubes P, et al. Angiogenesis is associated with blood-brain barrier permeability in temporal lobe epilepsy. Brain 2007;130(Pt 7):1942–56.
[154] Rite I, Machado A, Cano J, Venero JL. Intracerebral VEGF injection highly upregulates AQP4 mRNA and protein in the perivascular space and glia limitans externa. Neurochem Int 2008;52:897–903.
[155] Skucas VA, Mathews IB, Yang J, Cheng Q, Treister A, Duffy AM, et al. Impairment of select forms of spatial memory and neurotrophin-dependent synaptic plasticity by deletion of glial aquaporin-4. J Neurosci 2011;31(17):6392–7.
[156] Amlerova J, Laczo J, Vlcek K, Javurkova A, Andel R, Marusic P. Risk factors for spatial memory impairment in patients with temporal lobe epilepsy. Epilepsy Behav 2013;26(1):57–60.
[157] Chin J, Scharfman HE. Shared cognitive and behavioral impairments in epilepsy and Alzheimer's disease and potential underlying mechanisms. Epilepsy Behav 2013;26:343–51.
[158] Brooks-Kayal A, Bath KG, Berg A, Galanopoulou A, Holmes G, Jensen F, et al. Issues related to symptomatic and disease modifying treatments affecting cognitive and neuropsychiatric comorbidities of epilepsy. Epilepsia 2013;54(Suppl 4):44–60.
[159] Bell B, Lin JJ, Seidenberg M, Hermann B. The neurobiology of cognitive disorders in temporal lobe epilepsy. Nat Rev Neurol 2011;7(3):154–64.
[160] Morimoto K, Fahnestock M, Racine RJ. Kindling and status epilepticus models of epilepsy: rewiring the brain. Prog Neurobiol 2004;73(1):1–60.
[161] Parent JM. The role of seizure-induced neurogenesis in epileptogenesis and brain repair. Epilepsy Res 2002;50(1-2):179–89.
[162] Scharfman HE. Epilepsy as an example of neural plasticity. Neuroscientist 2002;8(2):154–73.
[163] Heuser K, Nagelhus EA, Tauboll E, Indahl U, Berg PR, Lien S, et al. Variants of the genes encoding AQP4 and $K_{ir}4.1$ are associated with subgroups of patients with temporal lobe epilepsy. Epilepsy Res 2010;88(1):55–64.
[164] Noell S, Fallier-Becker P, Beyer C, Kroger S, Mack AF, Wolburg H. Effects of agrin on the expression and distribution of the water channel protein aquaporin-4 and volume regulation in cultured astrocytes. Eur J Neurosci 2007;26(8):2109–18.
[165] Zhao M, Bousquet E, Valamaneh F, Farman N, Jeanny J, Jaisser F, et al. Differential regulations of AQP4 and $K_{ir}4.1$ by triamcinolone acetonide and dexamethasone in the healthy and inflamed retina. Invest Ophthalmol Vis Sci 2011;52(9):6340–7.
[166] Kayali R, Ku J, Khitrov G, Jung ME, Prikhodko O, Bertoni C. Read-through compound 13 restores dystrophin expression and improves muscle function in the mdx mouse model for Duchenne muscular dystrophy. Hum Mol Genet 2012;21(18):4007–20.
[167] Malerba A, Boldrin L, Dickson G. Long-term systemic administration of unconjugated morpholino oligomers for therapeutic expression of dystrophin by exon skipping in skeletal muscle: implications for cardiac muscle integrity. Nucleic Acid Ther 2011;21(4):293–8.

CHAPTER

9

Glutamate Metabolism

OUTLINE

OVERVIEW

The majority of energy consumed by the brain is used to maintain membrane potentials of neurons after action potentials are fired. Glutamate, the main neurotransmitter in the central nervous system (CNS), is responsible for a large percentage of this energy expenditure. Accumulation of glutamate, however, can be toxic to neurons and, therefore, must be tightly regulated. The brain uses a combination of the blood–brain barrier (BBB), powerful glial transporters, metabolic enzymes, and glutamate cycling to maintain low extracellular levels of glutamate. These protective mechanisms, however, can become dysregulated in epilepsy. Processes involved in glutamate homeostasis and the implications of an impaired regulatory system in epilepsy will be discussed in detail. Both human and animal studies have shown that glutamate uptake into astrocytes is hindered, the metabolic processing of glutamate is slowed, and the cycling of glutamate to glutamine is downregulated in epileptic tissue.

Astrocytes and Epilepsy
DOI: http://dx.doi.org/10.1016/B978-0-12-802401-0.00009-0

METABOLISM IN THE BRAIN

Under basal conditions, glucose, a six-carbon sugar, is the primary metabolic substrate for the brain. Glucose is almost entirely oxidized to carbon dioxide and water, although other fates for glucose are possible. Along with O_2, it is supplied via the blood circulation. The brain has a high energy demand; despite representing only about 2% of total body mass, the brain requires about 20% of the oxygen and 25% of the glucose consumed by the human body [1]. Most of the energy consumed is used to maintain the membrane potential of neurons. Glucose can enter cells through glucose transporters (GLUTs) and it is immediately phosphorylated by hexokinase to produce glucose-6-phosphate (G6P). From there, G6P can be metabolized in a number of different ways: (1) It can continue through glycolysis, ultimately committing to the breakdown of G6P to produce ATP; (2) G6P may enter the pentose phosphate pathway to ultimately generate reducing equivalents in the form of NADPH; or (3) It may be stored as glycogen (only in astrocytes). In addition, glucose can be used to synthesize lipids, amino acids, or neurotransmitters, if necessary.

Glycolysis is the metabolic pathway that converts glucose into pyruvate, a three-carbon α-keto acid. The fate of pyruvate will depend on the ratio of reducing equivalents in the brain, namely the ratio of NADH to NAD^+ available in the cell. When the NADH:NAD^+ ratio is low, pyruvate is preferentially converted to acetyl coenzyme A (acetyl-CoA), enters the tricarbocylic acid (TCA) cycle and is oxidized to produce energy and reducing equivalents of NADH. When the ratio of NADH:NAD^+ is high, however, pyruvate may be converted to lactate via the enzyme lactate dehydrogenase (LDH). Subsequently, NADH is oxidized to NAD^+ in that same process. Over the last couple of decades, lactate has received a great deal of attention as an alternative metabolic substrate in the brain.

Astrocyte–Neuron Lactate Shuttle Model

It is widely accepted that glucose is the primary energy source for the brain and, under basal physiological conditions, glucose is almost entirely oxidized to provide the necessary energy to support brain function. Other sources of energy such as ketones, fatty acids, acetate, lactate, and certain amino acids, however, may be used when supplemental energy is needed. A popular alternative energy source to glucose is lactate. Several studies have provided evidence that under certain circumstances, lactate may be preferentially metabolized by neurons [2–8]. Initially proposed in 1994 by Pellerin and Magistretti [3], this idea became known as the astrocyte–neuron lactate shuttle (ANLS) model [4]. In this model, glutamate stimulates glycolysis and the subsequent release of lactate from astrocytes, which can then be taken up by neurons and metabolized via the tricarboxylic acid (TCA) cycle. Growing evidence over the years has helped expand the model to include glutamate-stimulated aerobic glycolysis and glycogenolysis (Figs. 9.1 and 9.2) [2].

In the absence of glutamatergic activation (Fig. 9.1), glucose is taken up by astrocytes via the glucose transporter GLUT1. Some of this glucose will undergo oxidative metabolism via glycolysis and the TCA cycle in the mitochondria to produce ATP and reducing equivalents. Some glucose, however, may be converted into lactate and released from the astrocyte through monocarboxylate transporters, MCT1 and MCT4. Once in the extracellular space, lactate is taken up by neurons via MCT2 in parallel with glucose. Pyruvate produced from

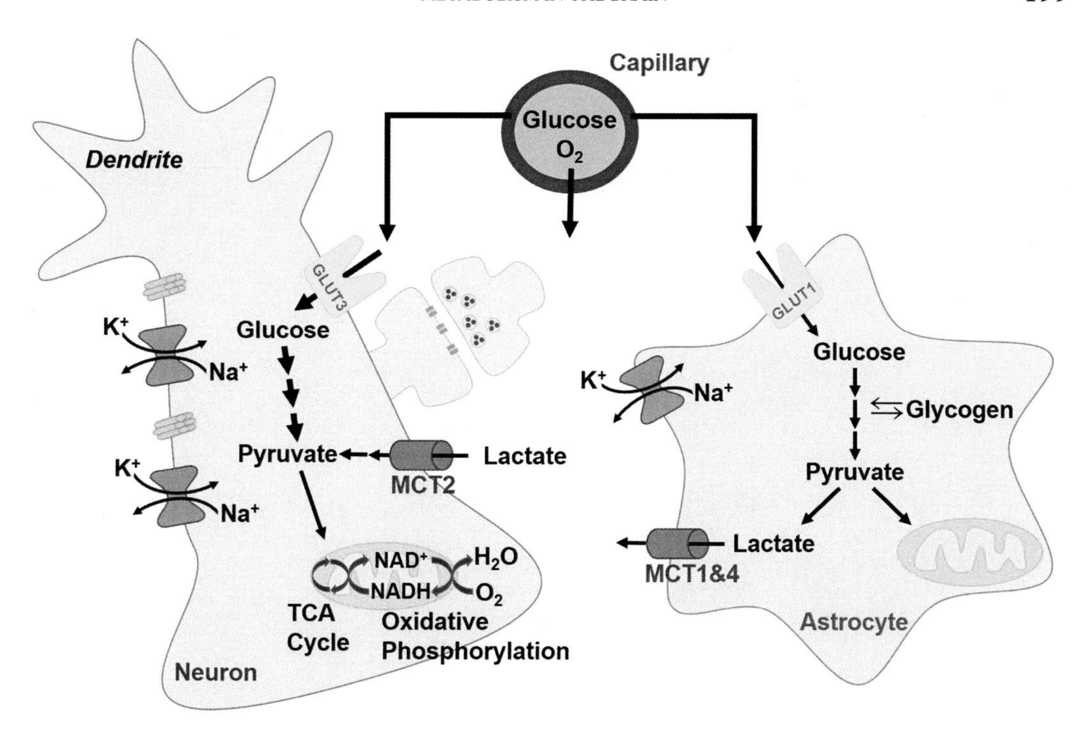

FIGURE 9.1 **Basal neuroenergetics in the absence of glutamatergic activation.** Glucose, the main cerebral energy substrate in the adult brain under normal conditions, and O_2 are supplied via the blood circulation to parenchymal cells. In astrocytes, glucose is taken up via the glucose transporter GLUT1. Part of it is processed oxidatively via the TCA cycle in mitochondria while the remaining part is converted to lactate, which is released via MCT1 and MCT4 in the extracellular space. In neurons, glucose is transported via the glucose transporter GLUT3. In parallel, lactate from the extracellular space is taken up via the monocarboxylate transporter MCT2. Pyruvate arising from both glucose and lactate is used oxidatively in mitochondria to satisfy neuronal resting energy needs. It cannot be ruled out that neurons could produce a certain amount of lactate but it is considered that they have a net lactate consumption. Overall, glucose is almost entirely oxidized to CO_2 and H_2O within the brain, yielding a net O_2/glucose consumption ratio close to six. *Source: Recreated with permission from Pellerin L, Bouzier-Sore AK, Aubert A, Serres S, Merle M, Costalat R, et al. Activity-dependent regulation of energy metabolism by astrocytes: an update. Glia 2007;55(12):1251–1262.*

both glucose and lactate can then be used oxidatively to provide neurons with the required resting energy needs.

Upon glutamatergic activation (Fig. 9.2), glutamate released from the presynaptic neuron binds to and activates AMPA and NMDA receptors on the postsynaptic neuron, allowing for the entry of sodium and other cations. After excitatory postsynaptic potentials are generated, the sodium-potassium ATPase (Na^+/K^+-ATPase) acts to reestablish the resting ion gradient, which requires substantial energy expenditure. In neurons, oxidative phosphorylation in the mitochondria is activated to replenish ATP, but this is done at the cost of NADH. To replenish NADH, TCA cycle activity is enhanced. As pyruvate (obtained from glucose through the process of glycolysis) enters the TCA cycle and the cytoplasmic levels drop, conditions inside

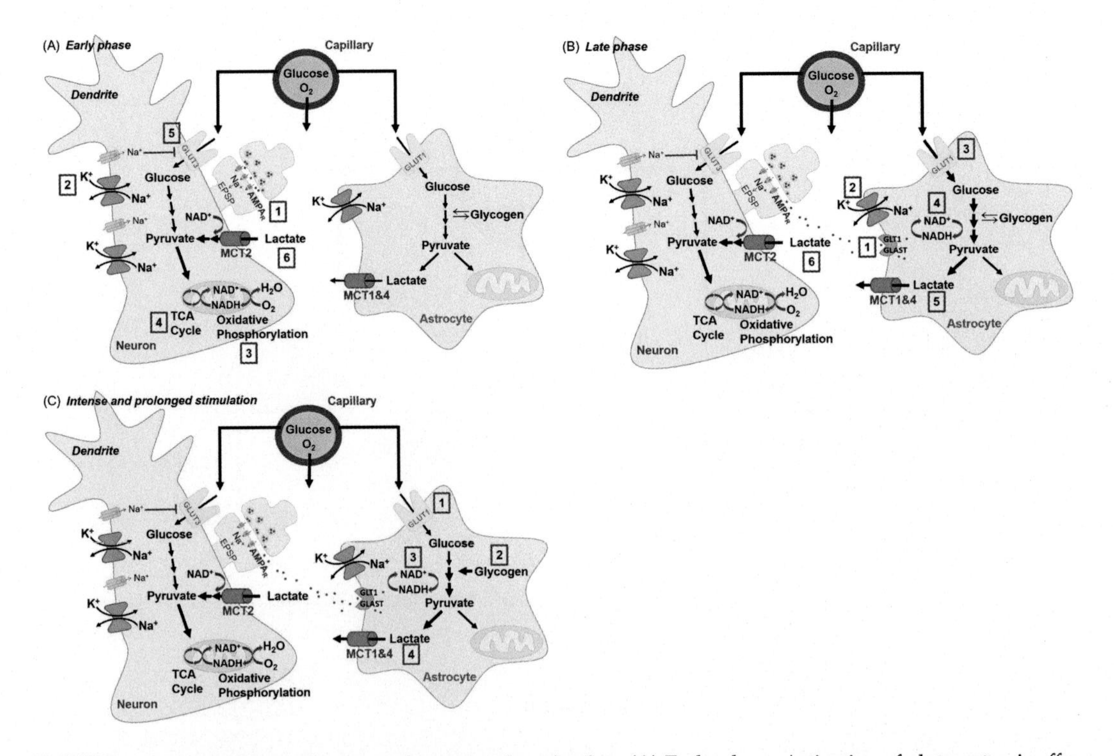

FIGURE 9.2 **Neuroenergetics upon glutamatergic activation.** (A) Early phase. Activation of glutamatergic afferents leads to synaptic release of glutamate, AMPA receptor (AMPAR) activation and generation of an excitatory postsynaptic potential (EPSP) caused by Na^+ entry within the postsynaptic spine (1). Depolarization propagates to the dendrite and causes opening of voltage-sensitive Na^+ channels, leading to further Na^+ entry. Reestablishment of ion gradients is accomplished by the Na^+/K^+-ATPase (2) that creates considerable energy expenditure. As a consequence, oxidative phosphorylation is activated (3) and mitochondrial NADH levels first decrease [9]. Then, enhanced TCA cycle activity will ensure (4) a supply of NADH for oxidative phosphorylation and support ATP production. As pyruvate utilization in the TCA cycle increases and its cytoplasmic levels decrease, the conditions become favorable for both enhanced glucose and lactate use. Surprisingly, activation of AMPARs and coupled Na^+ entry lead to a reduction in glucose uptake and utilization in neurons (5) [10], thus further favoring lactate utilization as preferential oxidative substrate (6) [11]. This would cause a transient drop in extracellular lactate levels as measured in vivo [12,13]. (B) Late phase. Glutamate released in the synaptic cleft is taken up by astrocytes to be recycled via the specific glutamate transporters GLAST and GLT1 (1). A large Na^+ influx caused by glutamate uptake takes place and activates the Na^+/K^+-ATPase (2) [14], glucose transport (3) [15], and (4) glucose utilization [3] in astrocytes. The enhancement of aerobic glycolysis in astrocytes first causes a large increase in cytosolic NADH [9] that normalizes with the conversion of pyruvate into lactate and its release via monocarboxylate transporters expressed on astrocytes (mainly MCT1 and MCT4) (5). Such a lactate release following glutamatergic activation corresponds to the increase in extracellular lactate levels measured in vivo [12,16]. Lactate produced by astrocytes during this later phase of activation not only replenishes the extracellular pool but also could help sustain neuronal energy needs as activation persists. Metabolic events occurring in the early and late phases described earlier constitute the so-called astrocyte–neuron lactate shuttle (ANLS) and its importance grows with the degree of glutamatergic activation. Such a view is supported by a series of experiments conducted in vivo [5–7]. (C) Intense and prolonged stimulation. Upon strong and long-lasting stimulation that occurs in certain conditions, glucose utilization becomes very important, in part, due to intense glutamate reuptake in astrocytes, that extracellular glucose levels are insufficient to sustain such uptake (1). In such a situation, glycogen present in the astrocyte is mobilized to provide the necessary glycosyl units (2) as previously demonstrated in vivo [17]. Glycolysis is the predominant pathway (3) and lactate is produced (4) to maintain the high glycolytic rate. Resynthesis of glycogen will cause additional glucose uptake that might contribute to create a mismatch between glucose utilization and oxygen consumption, a phenomenon known as "uncoupling" [18]. *Source: Recreated with permission from Pellerin L, Bouzier-Sore AK, Aubert A, Serres S, Merle M, Costalat R, et al. Activity-dependent regulation of energy metabolism by astrocytes: an update. Glia 2007;55(12):1251–2.*

the neuron become favorable for the use of both glucose and lactate. Interestingly, real-time epifluorescence microscopy studies show that glutamate may stimulate glucose uptake in astrocytes while simultaneously inhibiting its uptake in neurons [10]. Thus, it is possible that during glutamatergic activation, neurons prefer lactate as an oxidative substrate.

After glutamatergic activity, astrocytes can take up the glutamate from the synaptic cleft through glutamate transporters. This is accompanied by an influx of sodium ions, thus activating the Na^+/K^+-ATPase on astrocytes. In turn, aerobic glycolysis of glucose to pyruvate is increased, causing an initial rise in cytosolic NADH. This is normalized by the conversion of pyruvate to lactate by LDH, which can be released through MCT1 and MCT4. The now extracellular lactate can be taken up by neurons to meet additional energy demands, provide reducing equivalents to neurons [19], or used to replenish the extracellular pool. Further support of the ANLS model comes from the recent finding that inhibition of LDH hyperpolarized neurons and reduced seizures and epileptiform activity; neuron hyperpolarization could be reversed with pyruvate [20]. In fact, reducing circulating glucose concentrations by switching to a ketogenic diet (a diet consisting of mostly fat with minimal carbohydrates) has gained support as a therapeutic method to treat epilepsy [21].

Upon prolonged stimulation, glycogen breakdown and mobilization from astrocytes may be necessary to provide supplemental energy for neurons. Although glycogen storage is relatively low in the CNS compared to other peripheral tissues, it is still the largest energy reserve in the brain. Glycogen storage is more prevalent in nonneuronal cells [22], namely astrocytes [23], and its metabolism can yield lactate and ultimately ATP under anaerobic conditions to meet the high energy demands of the brain. Glycogenolysis can be triggered by various neuromodulators including noradrenaline (NA), vasoactive intestinal peptide (VIP), monoamines, and adenosine [24–26]. Glycolysis and glycogenolysis are thought to be activated successively after intense neuronal stimulation occurs. In this sense, glycogenolysis is sometimes viewed as an extension of the ANLS model.

The concept of the ANLS has gained a great amount of support over the years, but its acceptance is still controversial [27–31]. Although it is widely accepted that the brain can consume lactate during hypoglycemia and hyperlactemia [32], whether glucose or lactate is the primary energy source for neurons during neuronal activation is still debated. Criticisms of this model include evidence that both neurons and astrocytes have glucose transporters [29] and the importance of oxidative phosphorylation in energy production compared to the astrocyte-neuron lactate shuttle [28]. Recent evidence has demonstrated that the glucose oxidation in nerve terminals is substantial under both resting and activated conditions, suggesting that pyruvate from astrocytes would not be a major fuel source during neuronal activity [30]. It can be agreed, however, that glucose is and has always been the primary energy source in the brain but, under certain metabolic conditions, other energy sources such as lactate may be used.

Role of Energy Metabolism in Epilepsy

Seizures are the result of excessive synchronous neuronal activity in the brain and are associated with an increase in extracellular glutamate. The uptake of glutamate into glial cells induces a delayed, long-term stimulation of astrocytic glycolysis, which requires Na^+/K^+-ATPase activation [33]. Synaptic activity is accompanied by an acute fall in glucose concentration and is followed by a local surge in lactate concentration. During intense neuronal activity, the energy demand exceeds that of the glucose supplied from the bloodstream. Patients with epilepsy experience high levels of glucose uptake and hypermetabolism during seizures and low levels of glucose uptake and hypometabolism in between seizures [34]. Astrocytes contribute to the neuronal energy demand, potentially through the breakdown of glycogen storage.

Elevated levels of glycogen have been reported in the white matter, gray matter, and hippocampus of patients with temporal lobe epilepsy (TLE) [35]. Glycogen synthesis and storage occurs primarily in astrocytes because astrocytes have the glycogen synthesizing enzyme, glycogen synthase, whereas neurons do not. When energy is in high demand, astrocytes break down glycogen with the enzyme glycogen phosphorylase, and produce supplemental energy for neurons, potentially in the form of lactate to be shuttled through the ANLS. In mice injected with the glycogenic agent and convulsant methionine sulfoximine (MSO), brain glycogen increased in the preconvulsive state, was rapidly mobilized during seizures, and returned to high levels after seizures [36]. MSO is an inhibitor of glutamine synthetase (GS) [37], and a decrease in GS and consequently a decrease in the glutamate–glutamine–GABA shunt can result in decreased inhibitory transmission and reduced glutamate clearance from the extracellular space, perhaps accounting for the convulsant effects of MSO. Astrocytes play a major role in both brain metabolism and glutamate homeostasis, which may become dysregulated in epilepsy.

GLUTAMATE HOMEOSTASIS IN THE BRAIN

Glutamate is the main excitatory neurotransmitter in the CNS. It is an L-amino acid with a carboxylic acid side chain that exerts its signaling role by acting on glutamate receptors found on cell membranes. Extracellular glutamate levels determine the extent of receptor stimulation; consequently, it is important to maintain low levels of glutamate in the extracellular space to ensure a high signal-to-noise ratio. Extracellular glutamate concentrations have been estimated to be between 1 and 30 μM, with the majority of studies finding levels below 5 mM [38,39]. Intracellular glutamate levels are several thousand fold higher, with the highest concentrations of glutamate found within nerve terminals [38]. During synaptic transmission, extracellular glutamate within the synapse reached levels above 1 mM and those levels decayed within milliseconds [40].

Accumulation of glutamate in the extracellular space can lead to cell excitotoxicity. Therefore, the brain must have a powerful protective system to maintain glutamate homeostasis in the brain. First, the blood-brain barrier (BBB) helps create an isolated environment that prevents circulating glutamate from entering the brain. Second, at synaptic clefts, neurons are surrounded by astrocytes that have such a high glutamate uptake activity that the healthy, non-compromised brain is remarkably resistant to glutamate toxicity. Third, a number of metabolic enzymes exist that allow for the synthesis and metabolism of glutamate. Finally, a well-established system between neurons and astrocytes, called glutamate–glutamine cycling, allows for the rapid uptake and conversion of glutamate into a nontoxic compound in astrocytes followed by its release into the extracellular space for uptake by neurons. All of these mechanisms are required to prevent neuronal excitotoxicity and all are possible with the aid of astrocytes.

Blood–Brain Barrier

Astrocytes play a major role in protecting the brain. In addition to controlling the chemical composition and levels of ions and nutrients in the extracellular space, astrocytes also modulate synaptic activity, store glycogen, and contribute to synapse formation and remodeling. Importantly, astrocytes help form and maintain the BBB through their cell projections called astrocytic endfeet. This barrier helps limit the net influx of glutamate into the brain, therefore allowing the brain to maintain an autonomous microenvironment unique to that of the rest of the body.

The BBB is a highly selective permeable barrier that separates the circulating blood from the extracellular fluid in the brain. It is formed by a layer of cerebral capillary endothelial cells connected by tight junctions that surround the CNS (Fig. 9.3) [43]. These cells separate the BBB into two barriers: (1) the luminal domain (membrane of the endothelial cells facing the blood) and (2) abluminal domain (membrane of endothelial cells facing the brain). Therefore, any molecules entering or leaving the brain must pass through two membrane barriers, each with distinct properties. Facilitative carriers exist exclusively on the luminal membrane while excitatory amino acid transporters (EAATs) exist only on the abluminal membranes. The EAATs couple Na^+ transport with glutamate movement against the electrochemical gradient to move glutamate into the endothelial cells where it can diffuse into the blood via facilitative carriers [44]. This clever organization of the BBB allows for easy removal of glutamate from the brain extracellular fluid while preventing the net entry of serum glutamate into the brain. Thus, it promotes the maintenance of low extracellular glutamate concentrations in the brain. The role of the BBB in epilepsy will be discussed in *Chapter 12: Blood–Brain Barrier Disruption*.

Glutamate Transporters at the Synapse

In the CNS, a class of five excitatory amino acid transporters (EAAT1–5) are responsible for the synaptic reuptake of glutamate. The rodent homologs to EAAT1–5 are glutamate aspartate transporter (GLAST) glutamate transporter-1 (GLT1), excitatory amino acid carrier-1 (EAAC1), EAAT4, and EAAT5, respectively. GLAST/EAAT1 and GLT1/EAAT2 are predominantly expressed in astrocytes whereas EAAT3–5 are largely expressed in neurons. These transporters cotransport three Na^+ and one H^+ into the cell along with glutamate in exchange for the release of one K^+. EAAT1–3 are efficient glutamate transporters with small associated macroscopic anion currents whereas EAAT4 and EAAT5 are low-capacity transporters that have a predominant anion (Cl^-) conductance [45]. EAAT1 (GLAST) expression is most prominent in the cerebellum with only moderate levels in the forebrain whereas EAAT2 (GLT1) expression is most abundant in the forebrain and low within the cerebellum. EAAC (EAAT3) is expressed at low levels throughout the brain. EAAT4 is only found in the cerebellum whereas EAAT5 is only found in the retina.

Glutamate transporters play a major role in clearing glutamate from the extracellular space. At the tripartite synapse, the pre- and postsynaptic neurons are surrounded by astrocytes that abundantly express glutamate transporters. EAAT2 (GLT1) is responsible for the majority of glutamate clearance after synaptic activity and is essential in glutamate uptake. Although the cycle time of glutamate transport has been measured to be slow, ranging from

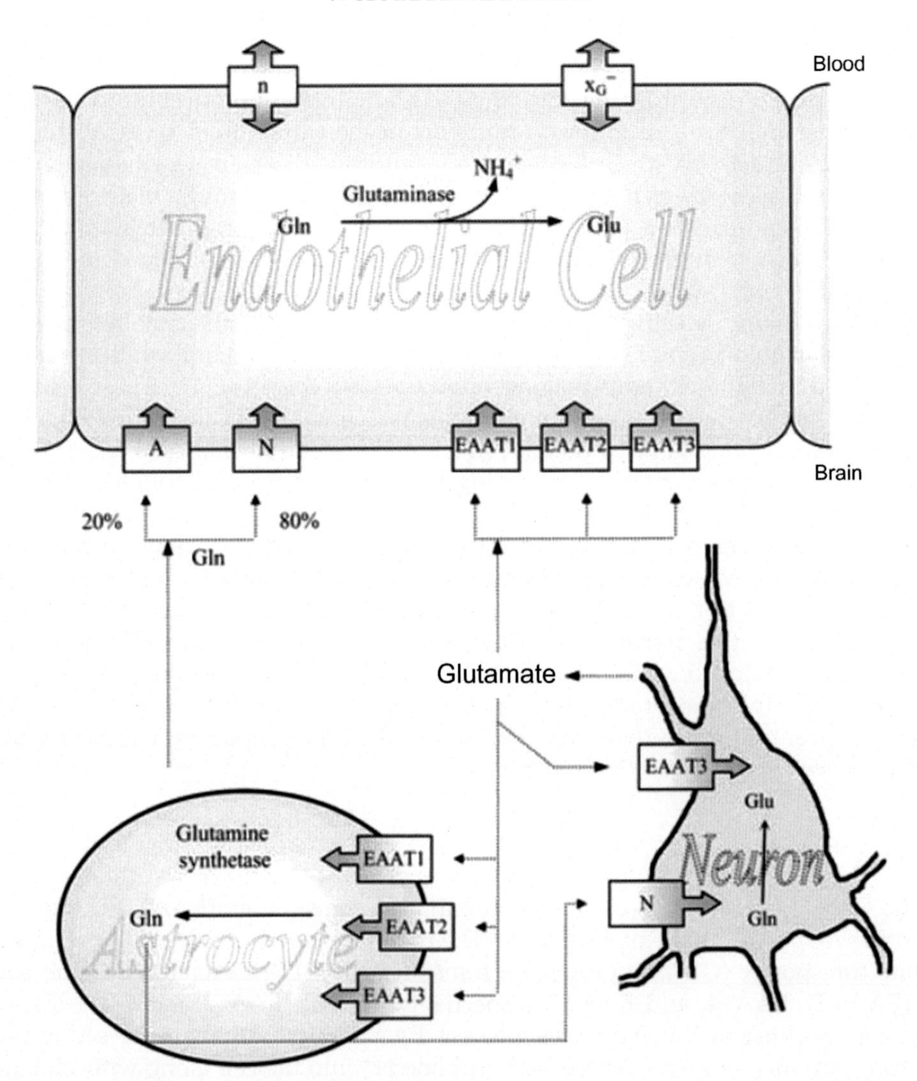

FIGURE 9.3 **Glutamate and glutamine transport between neurons, astrocytes, and endothelial cells.** The presence of Na^+-dependent carriers capable of pumping glutamine and glutamate from brain into endothelial cells, glutaminase within endothelial cells to hydrolyze glutamine to glutamate and NH_4^+, and facilitative carriers for glutamine and glutamate at the luminal membrane provides a mechanism for removing nitrogen and nitrogen-rich amino acids from brain [41]. EAAT1, EAAT2, and EAAT3 are present in endothelial cells [42], and astrocytes, EAAT3 is present in nerve cells. *A*, Na^+-dependent system A; *N*, Na^+-dependent system N; *EAAT*, Na^+-dependent glutamate transporter, x_G-, facilitative glutamate transporter; *n*, facilitative glutamine transporter. *Source: Reproduced with permission from Hawkins RA, O'Kane RL, Simpson IA, Vina JR. Structure of the blood-brain barrier and its role in the transport of amino acids. J Nutr 2006;136(1 Suppl):218S–26S.*

11–80 ms [38], it is possible that their highly efficient removal of glutamate from the extracellular space may be due to both their ability to quickly bind glutamate and their sheer abundance. Recently, it was discovered that even though GLT1 is anchored to the membrane at synapses, it can become untethered after glutamate is released into the extracellular space and can perform rapid, activity-regulated surface diffusion [46]. Therefore, GLT1 can readily move between synaptic and nonsynaptic sites to ensure glutamate clearance and effectively shape excitatory postsynaptic currents.

It is widely accepted that delayed clearance of extracellular glutamate is implicated in seizure development and spread [47–51]. During and Spencer [47] used microdialysis probes to record glutamate levels in the hippocampus before and during a seizure in human patients with complex partial epilepsy refractory to medical treatment. The dialysate concentration of glutamate rose before the occurrence of a seizure, which was followed by a sustained increase, potentially to neurotoxic levels, during a seizure in the epileptic hippocampus [47]. Elevated levels of extracellular glutamate have been reported in patients with various epilepsies, which could potentially be a consequence of altered expression or activity of glutamate transporters.

Glial Glutamate Transporters

Excitatory amino acid transporters, EAAT1 (GLAST) and EAAT2 (GLT1), are found throughout the CNS and are almost exclusively expressed on astrocytes. In fact, GLT1 protein is not expressed on neurons in the healthy adult CNS, however, mRNA encoding GLT1 has been found in some neurons [38]. GLT1 protein has been shown to be transiently expressed in several neuronal populations during development in the rodent and sheep brain, however, this expression disappears on maturation. GLAST is the dominant glutamate transporter in the cerebellum, expressed most densely on the Bergmann glia [52], and GLT1 is the predominant glutamate transporter in the forebrain, expressed most heavily in the hippocampus. GLAST levels are estimated to be 3200 and 18,000 per μm^3 tissue in the young adult rat stratum radiatum in the hippocampus (CA1) and the cerebellar molecular layer, respectively [53]. The number of GLT1 in the young adult rat stratum radiatum of the hippcampus (CA1) was measured to be 12,000 per μm^3 tissue and only 2800 per μm^3 in the cerebellar molecular layer [53]. Through both human tissue studies and the use of animal models of epilepsy, researchers have begun to elucidate the role of glial glutamate transporters in the pathology of epilepsy.

Human Tissue Studies

Several immunohistological studies on human tissue from patients with TLE have characterized localization and expression changes of EAAT1 and EAAT2 in hippocampal sclerosis (HS). In the sclerotic hippocampus, EAAT2 immunoreactivity (IR) was markedly reduced in subfields exhibiting neuron loss, such as CA1, hilus, and CA4 [54,55]. In addition, an increase of EAAT1 IR was observed in CA2/3 stratum radiatum of the sclerotic tissue [55]. A small decrease in EAAT1 IR has been seen in the CA4 and in the polymorphic and supragranular layer of the dentate gyrus in HS [56].

Although studies have shown a decrease in EAAT2 immunohistological expression in HS tissue from patients with TLE [56], either no change [55] or an increase in EAAT2 IR has been reported in non-HS tissue. One study, however, did observe a change in localization of EAAT2 IR in TLE human tissue, but it was determined that this occurred *postmortem* [57]. Non-HS cases also showed no change in EAAT1 IR compared to autopsy controls [55]. Western blotting analysis of tissue from patients with TLE have shown slightly reduced EAAT1 levels [57] but no change in EAAT2 protein expression [57–59].

A number of studies have used in situ hybridization or northern blotting determine mRNA levels in tissue resected from TLE patients. In the nonsclerotic tissue, no statistically significant difference in EAAT1 mRNA was observed compared to control. EAAT2 mRNA signal intensity was decreased in all subareas compared to control hippocampus, but was only found to be significant in CA3 [56]. No significant change in EAAT1 mRNA and a significant reduction in EAAT2 mRNA levels were observed in the HS tissue [56]. In the hippocampal dentate granule cells, no change in EAAT2 mRNA was found [60]. One study did notice, however, a relative increase in mRNA content per cell in TLE compared to control tissue [57].

In studies examining the dysplastic tissue of human patients with focal cortical dysplasia (FCD), decreased IR and more diffuse pattern of EAAT1 and EAAT2 expression were observed [61,62]. No changes were found in EAAT2 mRNA in the epileptic foci compared to the nonepileptic regions of human neocortical epilepsy tissue, but EAAT2 protein expression was reduced [63]. Support of these findings was found in a freeze-induced rat model of cortical dysplasia. The nonselective glutamate transporter antagonist DL-threo-β-benzylozyaspartic acid (TBOA) prolonged postsynaptic currents (PSCs) and decreased the threshold for evoking epileptiform activity in the lesioned cortex compared to controls [64]. This effect could be mimicked with the specific GLT1 antagonist dihydrokainate (DHK), suggesting a role of GLT1 in regulating network excitability.

When examining human tissue data, it is important to keep in mind that tissue is often resected after patients have had epilepsy for prolonged periods and have become medically refractory. Therefore, results found are often the endpoint of a disease and may not provide information on disease progression. Animal models must be used to determine whether glial glutamate transporters are involved in epileptogenesis or are a consequence of epilepsy.

Animal Models

Glial glutamate transporters are essential in maintaining low extracellular glutamate concentrations. GLT1-deficient mice suffered from lethal spontaneous seizure due to the elevated extracellular glutamate levels in the brain [49]. Deletion [49] or antisense oligonucleotide-mediate inhibition of synthesis [65] of GLT1 in rodents revealed that it is the major contributor to glutamate uptake from the extracellular space. Intraventricular administration of GLT1 antisense oligonucleotide in rats resulted in a 32-fold rise in extracellular glutamate, measured by a microdialysis probe in the striatum. In addition, this treatment resulted in a progressive motor syndrome that included mildly paretic hindlimbs and dystonic posture [65]. Overexpression of GLT1 in transgenic mice, on the other hand, attenuated epileptogenesis and reduced chronic seizure frequency in a pilocarpine-induced model of epilepsy [66]. Without GLT1, glutamate levels rise enough to cause neurotoxicity and seizures whereas overexpression of GLT1 can offer neuroprotective effects.

Mice lacking GLAST show increased cerebellar damage after brain injury and motor discoordination [67]. In the amygdala kindling model, duration of generalized seizures was prolonged by about 35% in GLAST-deficient mice [68]. There was no significant difference, however, in afterdischarge threshold or in seizure responses induced by the first stimulation. GLAST-deficient mice also showed a shorter latency period and more severe stages of pentylenetetrazol (PTZ)-induced seizures compared to wild-type mice [68]. The inhibition of synthesis of GLAST by antisense oligonucleotides in rats resulted in a 13-fold rise in extracellular glutamate dialysate in the striatum and a progressive motor syndrome [65]. Like GLT1, GLAST plays a role in modulating extracellular glutamate levels and deletion of it may result in neurotoxicity.

Kainic acid (KA) is a cyclic analog of L-glutamate and a potent agonist for kainate receptors that can lead to status epilepticus (SE) in animals. Injection of KA into animals is characterized by a latent period followed by refractory spontaneous seizures, regardless of administration route [69]. It is now one of the most widely used animal models of TLE pathology. Injection of KA into the amygdala of rats led to the development of spontaneous limbic seizures within 14–25 days. Sixty days postinjection, both mRNA and protein levels of GLT1 and GLAST were downregulated in the ipsi- and contralateral hippocampus to injection [70]. In 7-week-old rats treated with intraperitoneal (i.p.) injections of KA, neuronal death was associated with focal loss of GLT1 IR in reactive astrocytes and stronger IR for glutamate in the thalamus [71]. This loss was observed between 14 and 28 days postinjection and mimicked the pathology seen in human tissue studies. In a subcutaneous injection of KA model of epilepsy in adult rats, in situ hybridization revealed that GLT1 mRNA levels transiently increased in the stratum oriens and radiatum of the hippocampus, as well as the molecular and polymorphic layers of the dentate gyrus [72]. GLAST hippocampal mRNA was increased 12 hours after systemic KA administration in a rat model of limbic seizures, peaked at 48 hours, and returned to baseline by 7 days postinjection [73].

Administration of pilocarpine, a muscarinic receptor agonist, into animals is commonly used as a model of mesial temporal lobe epilepsy (MTLE). Intraperitoneal injection of pilocarpine into 90-day-old rats led to a period of SE, which was then followed by a latent period of several days, until rats developed spontaneous seizures (chronic phase). GLT1 and GLAST protein and mRNA levels were downregulated in the cortex during the latent period. In the chronic period, GLT1 mRNA and protein levels were decreased in the hippocampus. Immediately after SE, however, no change in either GLT1 or GLAST was observed [74].

The kindling model of epilepsy, in which repeated electrical stimulations are given to an animal to induce seizures, may leave long-lasting effects on the brain. GLAST protein was downregulated in the piriform cortex/amygdala in kindled rats as early as 24 hours after a stage 3 seizure and persisted through multiple stage 5 seizures. No change, however, was observed in the hippocampus or limbic forebrain. GLT1 protein remained unchanged in any region examined (hippocampus, piriform cortex/amygdala, and limbic forebrain) after single or multiple seizure stimulations [75]. In situ hybridization analysis revealed no changes in GLT1 mRNA in the hippocampus or cortex of kindled rats that were sacrificed 28 days after the final stimulation [76]. An increase of GLT1 mRNA was observed, however, bilaterally in the striatum in kindled animals. Western blot analysis revealed no change in GLAST or GLT1 protein.

A number of genetic models of epilepsy exist, including the spontaneous epileptic rat (SER), which is a double mutant obtained by crossing tremor rats with zitter rats. In the 9–12-week-old SER, GLAST mRNA was reduced and GLT1 mRNA was mildly increased in the hippocampus. The number of GLAST-positive cells in the hippocampus of the SER was reduced compared to control rat, particularly in CA3 and the dentate gyrus. GLAST protein levels were slightly downregulated whereas GLT1 protein levels were substantially upregulated in SER hippocampus [77]. Reduction of GLT1 mRNA expression, but no change in protein levels, were seen in the inferior colliculus, cortex, striatum, and CA1 of genetically epileptic-prone rats 7 days after a series of audiogenic stimulus exposures [78].

A genetic model of TLE can be created using an inbred mutant strain of DDY mice. These animals experience complex partial seizures with secondary generalization. GLT1 mRNA and protein in the parietal cortex remained unchanged but levels in the CA3 hippocampus were reduced. GLAST protein was reduced in the hippocampus, while no changes in mRNA or protein were observed in the parietal cortex [79].

The genetic absence epilepsy rat from Strasbourg (GAERS) is considered an animal model of inherited human absence epilepsy. The use of in situ hybridization revealed increases in GLT1 mRNA levels in the ventromedial nucleus of the thalamus and the subthalamic nucleus while increased GLAST mRNA was found in the primary somatosensory cortex and temporal cortex of GAERS [80]. In another study, epileptic discharges were recorded in the thalamo–cortical network began around 40 days after birth in GAERS. Although no GLAST or GLT1 protein changes were observed in adult rats, 30-day-old GAERS showed a reduction in cortical GLT1 and GLAST protein [81]. No change in GLT1 and GLAST protein expression was observed in the thalamus or hippocampus. The expression and activity of GLAST were decreased by 50% in newborn GAERS cortical astrocytes grown in primary culture, which was accompanied by a reduction in glutamate uptake [81].

In a model of posttraumatic epilepsy (PTE) that involves ferrous chloride injections into the rat cortex, GLT1 cortical protein levels were downregulated in the epileptic rat 1 and 4 days after induction, but not at 3 months. GLAST cortical protein expression, however, remained unchanged [82]. After injection of Fe^{3+} into the left amygdala, rats experienced spontaneous limbic behavioral seizures. In this model, GLT1 protein increased 5 and 15 days after iron injection on the contralateral side to injection and returned to basal levels by 30 days. GLAST protein showed an initial increase but the protein was persistently downregulated by 15 days postinjection [83].

A number of neurological disorders are accompanied by epilepsy. For example, in rats with cerebral ischemia-related epilepsy, GLT1 immunohistochemical staining was decreased in hippocampal CA1 and in the motor cortex compared to rats with cerebral ischemia without epilepsy [84]. In the tuberous sclerosis (TSC) epilepsy model in which the astrocyte-specific *Tsc1* gene is inactivated, mice exhibit decreased GLT1 and GLAST protein expression and decreased glutamate transport currents, suggesting that these glial glutamate transporters contribute to the glutamate dysregulation and seizures in TSC [85]. Sheepdogs with familial idiopathic epilepsy have reduced GLT1 immunostaining, particularly in the cerebral cortex and lateral nucleus of the thalamus, whereas there was no difference in GLAST immunolabeling compared to control dogs [86].

Taken together, these results strongly support the notion that astrocytes play a crucial role in protecting neurons from hyperexcitability and glutamate excitotoxicity. Elevated extracellular glutamate levels play a role in seizure generation and spread. Current literature on animal models and human tissue studies suggest a dysregulation of EAAT1/GLAST and EAAT2/GLT1 in certain epilepsies, however, whether dysregulation or dysfunction of glial transporters plays a causative role in epilepsy still remains unclear.

Neuronal Glutamate Transporters

EAAT3 (EAAC1) concentration is lower than that of the other EAATs and is mostly expressed in neurons. Unlike GLT1 and GLAST, EAAC1 is mostly intracellular and can transport cysteine. It is most abundantly expressed in the hippocampus, cerebellum, and basal ganglia [38] and is found in a variety of neurons. Although it is predominantly expressed by neurons in the healthy CNS, EAAC1 has also been found in oligodendrocytes and, in the injured brain, in microglia [87]. Administration of EAAC1 antisense oligonucleotide in rats has been shown to produce mild neurotoxicity and seizures in less than a week of treatment [65].

The expression of EAAT4 is predominantly found in the cerebellum, but low levels have been detected in the forebrain. It is most extensively found on the Purkinje cell plasma membrane and the highest levels are found perisynaptically [88]. The average concentration of EAAT4 in the molecular layer of adult rats is estimated to be 1900 molecules per μm^3 tissue. Even though EAAT4 shares a similar mechanism to EAAT1-3, it operates 5–10 times more slowly [89]. Unlike EAAT1-3, both EAAT4 and EAAT5 are highly associated with Cl^- conductance. EAAT5 is only found in the retina and is considered a low-capacity and low-affinity glutamate transporter [90]. Although reduced levels of EAAT4 mRNA and protein has been found in human epileptic neocortex tissue [63], EAAT5 has not been implicated in epilepsy and therefore will not be discussed further in this chapter.

Human Tissue Studies

Hippocampal neuronal loss is commonly observed in human TLE patients. Despite this, EAAT3 IR was increased in HS on pyramidal [60] and remaining granule cells [55,60]. In a separate study, fewer EAAT3 positive cells were found in the sclerotic hippocampus than in the non-HS in patients with pharmaco-resistant TLE. Individual neurons, however, showed an increase in EAAT3 IR in both HS and non-HS [56], although no differences between non-HS EAAT3 IR and autopsy control has also been reported [55]. The use of in situ hybridization revealed that EAAT3 mRNA did not change in non-HS tissue, but did decrease in the sclerotic hippocampus in CA4 compared to nonepileptic control tissue [56].

In neocortical tissue resected from patients with cortical dysplasia-induced medically intractable focal epilepsy, an increase of EAAT3 immunohistochemical staining was observed [60,62]. Northern blotting revealed elevated EAAT3 mRNA in dysplastic neurons in cortical dysplasia compared to postmortem control [60]. The localized epileptic foci showed reduced mRNA and protein levels of EAAT3 and EAAT4, assayed by real-time PCR and Western blot, respectively, compared to the nonepileptic regions in human neocortical epilepsy tissue [63].

Animal Models

The neuronal glutamate transporter EAAC1 is in lower abundance than GLT1 and GLAST and only plays a minor role in glutamate uptake. Despite this, EAAC1 has been implicated in epilepsy. Antisense oligonucleotide inhibition of EAAC1 in rats did not result in elevated extracellular glutamate in dialysate from the striatum, but it did produce mild neurotoxicity and resulted in spontaneous, recurrent seizures [65]. Mice deficient in EAAC1, however, were not more susceptible to PTZ-induced seizures compared to wild-type mice [91]. A variety of epilepsy models have been used to elucidate the role of EAAC1 in epilepsy.

Striking changes in EAAC1 IR have been observed in both the KA and pilocarpine models of epilepsy. An analysis of the various layers of the hippocampus revealed specific changes in EAAC1 IR in the subcutaneous KA-induced epilepsy model in rats. Reduced EAAC1 IR in the CA1 stratum lacunosum moleculare layer was observed within 4 hours after seizure onset [72] and continued to drop until it was almost completely eliminated by 5 days posttreatment. A transient increase of EAAC1 immunostaining in the dentate granule cells was observed 4 hours post-SE after KA treatment [72], while an increase in EAAC1 IR in the dentate granule cells was seen weeks after pilocarpine-induced seizures [60]. A decrease in EAAC1 IR in CA1 and CA3 pyramidal cells of the hippocampus was observed 5 days after KA administration. No immunohistological changes in EAAC1 IR were seen in the cerebral cortex or in the corpus callosum after KA-induced epilepsy [72].

Changes in EAAC1 IR often paralleled similar mRNA dysregulation. Decreased mRNA levels in CA1 and CA3 pyrimadal cells were seen as early as 4 hours after subcutaneous KA treatment and persisted out to 5 days. A consistent decrease in EAAC1 mRNA in the dentate gyrus up to 8 hours post KA-induced seizures was followed by a transient increase 16 hours after treatment before levels returned to baseline [72]. Similarly, an increase in EAAC1 mRNA levels in the dentate granule cells of rats was observed in the pilocarpine model of TLE 2 weeks after seizure induction [60]. In the amygdala KA model in rats, hippocampal EAAC1 mRNA levels were increased 60 days after injection, but protein levels were unchanged [70]. Western blot analysis of EAAC1 in the amygdala-kindled rat model showed increased levels of protein in the piriform cortex/amygdala and hippocampus after 24 hours in animals that reached stage 5 seizures, but no change was observed in the limbic forebrain [75].

SERs exhibit sporadic tonic convulsions and absence-like seizures, similar to those seen in humans. In both SER [77] and 4-month-old GAERS [81], hippocampal EAAC1 mRNA and protein levels showed no significant difference when compared to control mice. The 4-month-old GAERS also showed no change in EAAC1 mRNA and protein levels in the thalamus. In the cortex, however, mRNA levels were increased in 4-month-old GAERS, but no change in protein levels were found in either the 30-day- or 4-month-old animals [81]. In addition, no changes in hippocampal EAAC1 protein, but decreased levels of thalamic protein levels, were observed in the 30-day-old GAERS.

Various other epilepsy models showed significant increases in EAAC1 expression. Injections of Fe^{3+} into the rat amygdala causes chronic, spontaneous recurrent focal seizures with generalized limbic behaviors. In this model of epilepsy, rats experienced a bilateral increase in EAAC1 protein expression for up to 30 days after injection [83]. The peak of EAAC1 expression, however, was 5 days postinjection. In a rat model of global cerebral ischemia induced by chest compressions, 64% of rats developed epilepsy. Those who did exhibited increased EAAC1 IR in hippocampal CA1 and the motor cortex compared to ischemic mice who did not develop epilepsy [84].

Alterations in EAAT3/EAAC1 are clearly prominent in a variety of human tissue studies and in a number of different epilepsy models. Increased IR was commonly observed in epileptic human tissue, however, this was not always paralleled by animal studies. Looking at different time points and brain regions in animal epilepsy models revealed that changes in EAAC1 expression are not static but, in fact, are constantly changing. In comparison to EAAT1–3, EAAT4 and EAAT5 are not well studied in epilepsy. Functional assays of glutamate transporters will be helpful in future studies to determine what role these proteins play in epileptogenesis.

Metabolic Enzymes

Since little to no glutamate can enter the brain through the BBB, glutamate must be synthesized and metabolized within the CNS. A number of metabolic enzymes help regulate glutamate homeostasis in the brain. Overall mechanism, cellular distribution, and localization of a subset of these enzymes are listed in Table 9.1. Once glutamate is taken into the cell, it can be metabolized in a number of ways (Fig. 9.4) [92].

TABLE 9.1 Key Regulatory Enzymes in Glutamate Metabolism

Enzyme Name	Reaction Catalyzed	Predominant Cell Type	Intracellular Localization
Alanine aminotransferase (ALT)	Glutamate + pyruvate $\rightleftharpoons$ α-ketoglutarate + alanine	Astrocytes Neurons	Mitochondria Cytoplasm
Aspartate aminotransferase (AST)	Aspartate + α-ketoglutarate $\rightleftharpoons$ glutamate + oxaloacetate	Neurons	Cytoplasm Mitochondria
Glutamate Decarboxylase (GAD)	Glutamate + H_2O $\rightleftharpoons$ GABA+ CO_2	Neurons	Cytoplasmic vesicles
Glutamate dehydrogenase (GDH)	Glutamate + NAD(P) + $\rightleftharpoons$ α-ketoglutarate + NH_4 + NAD(P)H	Astrocytes Neurons	Mitochondria
Glutaminase	Glutamine + H_2O $\rightleftharpoons$ glutamate + NH_3	Astrocytes Neurons	Mitochondria
Glutamine Synthetase (GS)	Glutamate + NH_3 + ATP $\rightleftharpoons$ glutamine + ADP + P_i	Astrocytes	Cytoplasm
Cytosolic Malic Enzyme (ME)	Malate + $NADP^+$ $\rightleftharpoons$ Pyruvate + NADPH + CO_2	Astrocytes	Cytoplasm
Pyruvate Carboxylase (PC)	Pyruvate + CO_2 + ATP + H_2O $\rightleftharpoons$ oxaloacetate + ADP + P_i + $2H^+$	Astrocytes	Mitochondria

All amino acids listed refer to the L-stereoisomer.

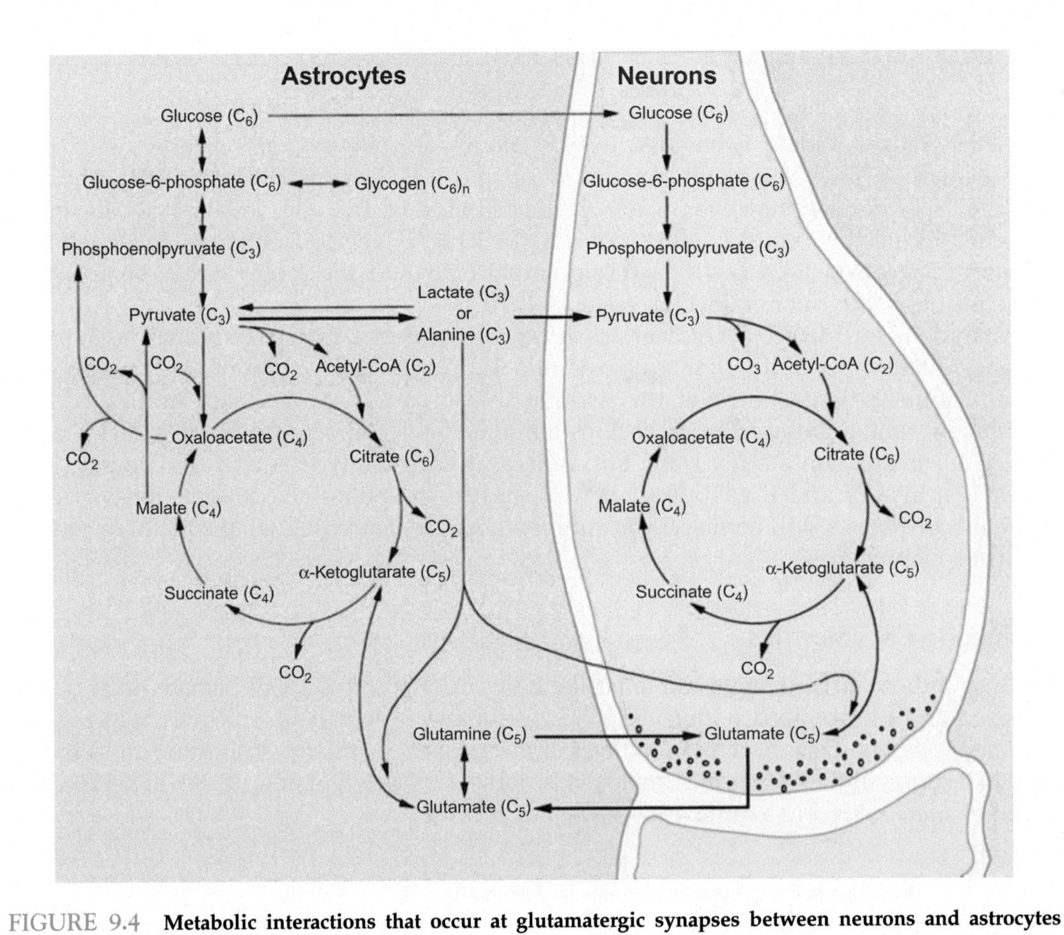

FIGURE 9.4 **Metabolic interactions that occur at glutamatergic synapses between neurons and astrocytes that ensheathe them.** The cartoon shows the cellular compartmentation in de novo formation of glutamate, its disposal by oxidative metabolism, and its return to neurons in the "glutamate–glutamine cycle." The number of carbon atoms in each intermediate is indicated in parentheses. Astrocytes and neurons are both capable of metabolizing glucose all the way to CO_2 and water, by initial glycolysis to pyruvate (not all intermediates are shown); subsequent formation of acetyl coenzyme A (acetyl-CoA) from pyruvate, releasing one molecule of CO_2; condensation of acetyl-CoA with oxaloacetate (OAA) in the tricarboxylic (TCA) cycle to form citrate; oxidation through different intermediates (which are not all shown) in the TCA cycle, producing two CO_2 molecules per acetyl-CoA, and regenerating one molecule of OAA, replacing the OAA molecule consumed during the initial formation of citrate, and readying it for condensation with another molecule of acetyl-CoA. Under some conditions, there may be an exchange of pyruvate or a pyruvate metabolite between cells, especially from astrocytes to neurons. During the oxidation of glucose a vast amount of energy is created (as ATP), but there is no net synthesis of any TCA cycle intermediate. Net synthesis of TCA intermediates is essential to replace leakage from the cycle and, especially, to synthesize transmitter glutamate. For this purpose, a molecule of OAA is generated de novo from pyruvate by addition of CO_2 (pyruvate carboxylation). This process cannot occur in neurons, which do not express pyruvate carboxylase (PC) activity, but it occurs readily in astrocytes. The newly generated molecule of OAA condenses with acetyl-CoA, generating one molecule of citrate from one molecule of glucose (two molecules of pyruvate) and, by further cycling, other TCA cycle intermediates. The TCA cycle intermediate α-ketoglutarate can be transaminated with alanine, formed from pyruvate to form glutamate. It is not known whether this transamination occurs in astrocytes or in neurons. If it occurs in astrocytes, the generated glutamate is amidated by the astrocyte-specific GS to glutamine, which can be safely released to the extracellular space, taken up in neurons and deamidated to glutamate. Since net amounts of glutamate are produced in the CNS and glutamate cannot leave the CNS across the blood–brain barrier, there must be mechanisms for disposal of excess glutamate. This is by formation of α-ketoglutarate from released transmitter glutamate after its accumulation in astrocytes; cycling in the astrocytic TCA cycle to malate, which can exit from mitochondria; formation of pyruvate or OAA, which is further converted to phosphoenolpyruvate. Pyruvate is oxidized via acetyl-CoA in either the cell where it was generated or in a different cell type. Phosphoenolpyruvate can be used for synthesis of glycogen, and possibly glucose, which is again oxidized. Although much more released transmitter glutamate is taken up by astrocytes than by neurons, it is not always oxidized. Another pathway is the formation of glutamine in astrocytes and return of glutamine to neurons as discussed above. *Source: Reproduced with permission from Hertz L, Dringen R, Schousboe A, Robinson SR. Astrocytes: glutamate producers for neurons. J Neurosci Res 1999;57(4):417–8.*

Synthesis of Glutamate

An important part of brain metabolism is the synthesis of amino acids, which can be done through a process called transamination. This chemical process involves the transfer of an α-amino group from an amino acid to the α-keto position of an α-keto acid. The enzyme that catalyzes this reaction is called an aminotransferase (formerly termed transaminase) and it requires pyridoxal phosphate (PLP) as a cofactor. These enzymes rely heavily on the availability of α-keto acids and are crucial in the production of nonessential amino acids, such as aspartate, glutamate, and alanine.

Two examples of important aminotransferases found in the brain are aspartate aminotransferase (AST) and alanine aminotransferase (ALT). AST catalyzes the reversible transfer of an amino group between aspartate and glutamate. It is found predominantly in neurons [93] and has two different isoforms; one associated with the cytoplasm and one associated with mitochondria [94]. ALT, on the other hand, catalyzes the reversible transfer of an amino group from alanine to α-ketoglutarate, producing pyruvate and glutamate. It is found in both the mitochondria and cytoplasm as well [95]. Although ALT is found in both neurons and astrocytes, its activity is significantly higher in astrocytes [96].

Alterations in aminotransferase enzymes have been reported in the epileptic brain. Increased ALT activity was observed in the frontal cortex, cerebellum, hippocampus, and pons-medulla in a rat model of electroconvulsive shock (ECS) induced generalized seizure activity 2 hour after ECS. Twenty-four hours after the fifth daily ECS, increased ALT activity was seen in the same regions with the exception of the hippocampus, where there was no change [97]. In contrast to this, ALT activity was not altered in the PTZ kindling of rats [98]. The administration of the convulsant isoniazid, a general pyridoxal enzyme inhibitor, into the adult rat caused the activity of ALT and AST to increase at the beginning of generalized tonic-clonic seizures [99]. Human tissue studies revealed that AST activity was higher in the cortex in MTLE compared to paradoxical temporal lobe epilepsy (PTLE) or mass lesion-associated temporal lobe epilepsy (MaTLE) [100].

Metabolic Fates of Glutamate

A pivotal role of astrocytes in glutamate metabolism is to prevent neurotoxicity. Once glutamate is brought into a cell, it can quickly be converted into other metabolic products. For example, glutamine synthetase (GS), primarily expressed on astrocytes, rapidly converts glutamate to glutamine inside the cell with the use of manganese as a cofactor. Since glutamine has no neurotransmitter action, it can safely be released into the extracellular space to be taken up by other cell types such as excitatory neurons. Glutamine is an effective precursor for glutamate and can be converted back to glutamate via the enzyme glutaminase (sometimes called phosphate-activated glutaminase, or PAG). This enzyme is strongly inhibited by its end product, glutamate, while it is only weakly inhibited by ammonia. Although glutaminase is present in both neurons and astrocytes [101], its activity is higher in neurons, namely due it its increased sensitivity to phosphate in neurons [102].

Increased IR of glutaminase in subpopulations of surviving neurons in the hippocampus, predominantly in CA1, CA3, the polymorphic layer, and the dentate gyrus, has been observed in patients with MTLE [103]. Glutaminase hippocampal protein levels and function, on the other hand, showed no significant difference between control and MTLE tissue. This lack of change may be explained by the significant reduction in neuronal density in patients with MTLE. In PTZ-kindled rats, however, glutaminase mRNA levels were not different from that of controls [104].

Inside an inhibitory neuron, L-glutamic acid can be converted to the chief inhibitory neurotransmitter γ-aminobutyric acid (GABA). This is accomplished by the enzyme glutamate decarboxylase (GAD) with the use of PLP as a cofactor. Two main isoforms of this enzyme exist in the brain: GAD_{67} and GAD_{65}. Encoded by the gene GAD1, GAD_{67} is found in both the cell body and nerve terminals. GAD_{65}, on the other hand, is encoded by the gene GAD2 and is restricted to the nerve terminals [105,106]. Although GAD is thought to be primarily expressed by GABA neurons, recent studies have suggested localization in astrocytes [107]. It can be hypothesized that reduced inhibitory transmission in the brain may play a role in epilepsy.

Reduced synthesis of GABA due to inhibition or decreased levels of GAD may result in seizures. In the PTZ-kindling model of 7-week-old rats, a downregulation of hippocampal GAD2 mRNA was observed. Twenty-four hours after generalized tonic-clonic seizures induced by PTZ, GAD1 hippocampal mRNA levels were decreased [104]. Interestingly, administration of the specific GAD inhibitor 3-mercaptopropionic acid (MPA) significantly increased the activity of both ALT and AST after the onset of MPA-induced seizures in both the adult rat and the 12-day-old animals [99]. Clearly, a dysregulation of glutamate metabolism in the brain may contribute to neuronal hyperexcitability.

Glutamate dehydrogenase (GDH) is a mitochondrial enzyme [95] that converts glutamate to its carbon skeleton α-ketoglutarate with the use of the cofactor $NAD(P)^+$. Although this enzyme has been detected in both astrocytes and nerve terminals [38], it is thought to favor astrocytic distribution [108]. Importantly, low glucose levels in the cell can activate this enzyme. Accordingly, the activity of GDH was negatively correlated with the duration of the first intractable seizure in patients with MTLE [100]. GDH activity was lower in the temporal cortex in MTLE compared to PTLE and MaTLE [100]. In contrast, Western blotting and immunogold analysis revealed that GDH activity did not change in the latent or chronic phase of intraperitoneal kainate-induced TLE in rats [109].

Pyruvate carboxylase (PC) is found exclusively on astrocytes [110] and catalyzes the condensation of pyruvate with CO_2 to form oxaloacetate (OAA) an important intermediate in the TCA cycle. OAA can then be condensed with acetyl-CoA to form citrate, which can be released by astrocytes to be taken up by neurons. Alternatively, OAA may continue through the TCA cycle to form reducing equivalents and the carbon skeleton for glutamate, α-ketoglutarate. PC deficiency has been linked to delayed development, seizures, and death [111]. Moreover, in the pilocarpine model of TLE in rats, the ratio of pyruvate carboxylation to pyruvate dehydrogenation for glutamate was reduced in the limbic structure [112]. No difference was observed, however, in the thalamus or hippocampus compared to control animals.

The glutamate precursor α-ketoglutarate may leave the mitochondrial TCA cycle as the dicarboxylic acid malate and enter the cytosol. Therefore, it can be converted to pyruvate by cytosolic malic enzyme (ME). This enzyme is primarily found on astrocytes, although colocalization with the oligodendrocyte marker myelin basic protein has been observed [113]. One possible metabolic fate of pyruvate, as discussed above, is conversion to lactate and shuttling to neurons. In idiopathic generalized epilepsy (IGE), a class of genetically determined, phenotypically related epilepsy syndromes, a single polymorphism (SNP) in the malic enzyme 2 (ME2) gene was linked to predisposition to IGE [114].

Glutamate–Glutamine Cycling

The healthy brain has a number of metabolic enzymes that play a role in glutamate synthesis and breakdown. Once entering the cell, glutamate can be converted to another neurotransmitter (GABA), cycled into an amino acid without neurotransmitter function (glutamine), or it can be metabolized into its carbon skeleton and enter the TCA cycle. This regulation of glutamate in the brain is essential to prevent neurotoxicity. Arguably, one of the most important enzymes in glutamate regulation is GS. Glutamate metabolism becomes altered in the epileptic brain, potentially contributing to the pathology of the disease.

Once glutamate is released by neurons, it is taken up by astrocytes through glial glutamate transporters and converted to glutamine by GS. A neurochemically inert molecule, glutamine can safely be released into the extracellular space where it can be taken up by neurons; it is converted to glutamate in glutamatergic neurons and to GABA by GABAergic neurons. Under physiological levels of inhibitory synaptic activity, blocking GS decreased both evoked inhibitory postsynaptic currents (eIPSC) and minimal stimulation-evoked quantal amplitudes by reducing the amount of GABA released at individual synapses [115]. This phenomenon could be reversed with the application of glutamine. Pharmacologically blocking the glutamate–glutamine cycle resulted in a reduced frequency of spontaneous epileptiform discharges (SEDs) in primary cultures of hippocampal neurons [116]. Together, these data suggest a role for the glutamate–glutamine cycle in supporting enhanced synaptic activity.

Human Tissue Studies

Glutamate–glutamine cycling between astrocytes and neurons have been shown to be decreased in human epileptic tissue (Fig. 9.5) [118,119]. Patients with mutations in the GS gene exhibit seizures, although this mutation is rare [120]. In tissue resected from patients with cortical dysplasia-induced focal epilepsy, a reduction of GS immunoreactivity was observed [62]. Similarly, GS protein levels in patients with malignant glioma and epilepsy were significantly lower than nonepileptic patients [121]. Tissue specimens obtained from drug-resistant epileptic patients, however, showed no change in GS activity when compared to nonepileptogenic brain tumor controls [122].

Deficiency in GS can lead to an accumulation of glutamate within both astrocytes and in the extracellular space in TLE [123]. GS immunoreactivity was prominently reduced in areas of the sclerotic hippocampus with neuron loss (CA1 and CA4) in patients with TLE [54]. Complimentary to this, decreased hippocampal expression and enzymatic activity of GS was observed in patients with MTLE [59]. Of note, GS was deficient in the most distal astrocyte processes in the subiculum in MTLE. This is similar to the redistribution of GS from the distal and thinner to proximal and thicker astrocytic branches seen in the rat pilocarpine model of TLE, even though the overall number of GS-positive cells did not change [124]. No changes in GS mRNA were observed in the HS or non-HS in MTLE tissue, except for an approximately 50% increase in CA3 of the nonsclerotic MTLE hippocampal tissue compared to autopsy control and sclerotic hippocampus [125].

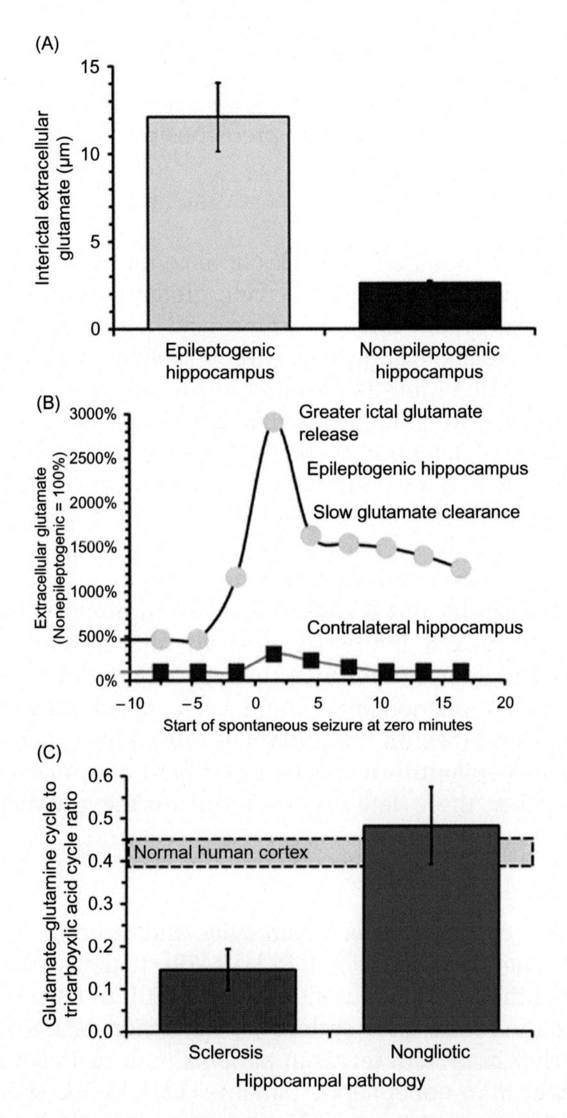

FIGURE 9.5 **Extracellular glutamate and glutamate–glutamine cycling in the human epileptogenic hippocampus.** (A) Interictal concentrations of extracellular glutamate measured by in vivo microdialysis are increased approximately fivefold in the epileptogenic hippocampus versus the contralateral nonepileptogenic hippocampus ($P < 0.0001$). (B) Extracellular glutamate measured by in vivo microdialysis is increased sixfold in the epileptogenic hippocampus during a seizure versus interictally ($P < 0.05$). Note the delayed clearance of extracellular glutamate in the epileptogenic hippocampus. (C) Magnetic resonance spectroscopy (MRS) of epilepsy patients prior to surgery reveal decreased cycling of glutamate to glutamine in the sclerotic (gliotic) hippocampus in MTLE. The glutamate–glutamine cycle remains directly proportional with energy metabolism over the entire range of cerebral electrical and metabolic activity. As a consequence of this linear relationship, the ratio of the glutamate–glutamine cycle to the tricarboxylic acid (TCA) cycle rates remains relatively constant (0.4–0.5) from deep anesthesia with marked slowing of the EEG through bicuculline-induced status epilepticus over a threefold variation in the rate of mitochondrial glucose oxidation. This ratio is the same in animal models and human subjects. The low values measured in mesial temporal sclerosis suggest significant impairment of the glutamate–glutamine cycle involving dysfunction of glutamine synthetase. *Source: Reproduced with permission from Eid T, Williamson A, Lee TS, Petroff OA, deLanerolle NC. Glutamate and astrocytes – key players in human mesial temporal lobe epilepsy? Epilepsia 2008;49(Suppl 2):42–52.*

Animal Models

After reporting that hippocampal GS is deficient in tissue resected from MTLE patients, Eid et al. continuously microinfused MSO into the right hippocampus of rats, which resulted in an 82–97% reduction in GS. Most animals exhibited recurrent seizures and some animals exhibited neuropathological features that resembled MTLE, thus supporting their hypothesis that GS deficiency may cause recurrent seizures [126]. Another study showed that chronic treatment with the H1-antihistamines diphenhydramine or pyrilamine increased seizure susceptibility in both the amygdala kindling and PTZ-induced seizure models in nonepileptic rats [127]. This effect persisted even after the antihistamines were withdrawn and the mechanism was shown to involve GS impairment.

The complete knockout of the GS gene resulted in early embryonic lethality [128]. Haploinsufficient mice, however, survived and did not experience spontaneous behavioral seizures. In the warm-air-induced hyperthermia febrile seizure (FS) model, 2-week-old haploinsufficient mice showed increased susceptibility to provoked seizures [129]. The structure, glial properties, and expression of GLT1 and vesicular glutamate transporter 1, other components of the glutamate–glutamine cycle, were not altered in these animals. This suggests that GS deficiency increases susceptibility to FS.

In the mouse intrahippocampal kainic acid (IHKA) model of epilepsy, transient elevations of GS hippocampal immunoreactivity were observed during the latent phase, but not the chronic phase, of the disease [130]. These results mirrored the increase of whole hippocampal GS protein expression during the latent phase, which returned to control levels during the chronic phase, in the IHKA model in rats [109]. GS levels did, however, remain increased in the CA3, CA1, and subiculum during the chronic phase. The activity of GS was significantly decreased 24 hours after KA injection in a rat model of epilepsy [131].

A number of additional animal models have been used to study GS in epilepsy. In the amygdala kindling model of rats, GS protein expression and activity did not change in the cortex or hippocampal CA1 and CA3 subregions during seizures. GS protein expression transiently increased by about 70% in the ipsilateral dentate gyrus during stage 4 seizures, along with a 37% increase in GS activity [132]. No significant changes in GS activity or protein expression were observed during stage 1–3 seizures. GS protein was slightly decreased in the thalamus of young GAERS, but not in adults. No difference was observed in protein or mRNA levels in the cortex or hippocampus when comparing the control rats to GAERS [133]. In the PTZ kindling model in rodents, GS immunoreactivity and activity were transiently upregulated in the cortex after stage 4 seizures [132]. Immunolabeling for GS in sheepdogs with familial idiopathic epilepsy, however, showed no difference compared to controls [86].

GS protein is sensitive to oxidative and nitrosative stress, which could result in GS deficiency. Nitric oxide inhibits GS and blocking nitric oxide synthase (NOS) increases the activity of GS [134]. Cytokines such as IL-1β and TNF-α administered to human fetal brain cell cultures generated NO and subsequently inhibited GS [135]. Neuroinflammation, including infiltration of microglial cells and upregulation of cytokines, is increased after seizures. Therefore, oxidative and nitrosative stress due to neuroinflammation may play a role in negatively regulating GS in epilepsy.

Epilepsy models have been used to determine the role of nitrogen on GS activity. After subcutaneous KA administration into rats, GS activity was decreased and was found to be modulated by NO in the epileptic rats [136]. In rats exposed to PTZ-induced repetitive epileptic seizures (PIRS), a regiment consisting of PTZ injections every other day for 2 weeks, GS activity was significantly reduced in the piriform–entorhinal cortex and the granular and molecular layers of the dentate gyrus [137]. In addition, PIRS in rats led to significantly higher levels of nitrated GS compared to controls, and this increase was directly correlated to seizure score. PIRS rats also showed no change in hippocampal or cortical protein levels, but did exhibit a nonsignificant reduction of total hippocampal GS activity.

Both human tissue and animal model studies have suggested a role for GS in the pathology of epilepsy. Human tissue studies have primarily shown decreased GS levels, particularly in areas with decreased neuronal density. Primary astrocyte cultures do not innately express GS unless they are cocultured with neurons. Taken together, one may propose that a reduction in astrocyte–neuron interaction may underlie part of the loss of GS in epilepsy [138]. Decreased glutamate–glutamine–GABA cycling in the brain may contribute to excitability and, therefore, may represent a novel therapeutic target.

CONCLUSION

This chapter demonstrated that astrocytes are essential in maintaining normal metabolic and glutamate homeostasis in the brain. Perturbations in glutamate metabolism can lead to increased excitability. Available evidence indicates that various epilepsy models and syndromes are associated with changes in brain glutamate metabolism. Although our understanding remains incomplete, virtually all aspects of the neuronal–glial synthesis, breakdown, and cycling of glutamate may be altered during epileptogenesis. Future studies should focus on determining the role of changes in glutamate metabolism in early epileptogenesis and the consequences for the epileptic brain. Ideally, such work would lead to the development of novel antiepileptic strategies targeted at properly regulating glutamate metabolism without dampening cognitive function.

References

[1] Hoyer S. The young-adult and normally aged brain. Its blood flow and oxidative metabolism. A review—part I. Arch Gerontol Geriatr 1982;1(2):101–16.

[2] Pellerin L, Bouzier-Sore AK, Aubert A, Serres S, Merle M, Costalat R, et al. Activity-dependent regulation of energy metabolism by astrocytes: an update. Glia 2007;55(12):1251–62.

[3] Pellerin L, Magistretti PJ. Glutamate uptake into astrocytes stimulates aerobic glycolysis: a mechanism coupling neuronal activity to glucose utilization. Proc Natl Acad Sci USA 1994;91(22):10625–9.

[4] Pellerin L, Magistretti PJ. Sweet sixteen for ANLS. J Cereb Blood Flow Metab 2012;32(7):1152–66.

[5] Serres S, Bezancon E, Franconi JM, Merle M. Ex vivo NMR study of lactate metabolism in rat brain under various depressed states. J Neurosci Res 2005;79(1–2):19–25.

[6] Serres S, Bezancon E, Franconi JM, Merle M. Ex vivo analysis of lactate and glucose metabolism in the rat brain under different states of depressed activity. J Biol chem 2004;279(46):47881–9.

[7] Serres S, Bouyer JJ, Bezancon E, Canioni P, Merle M. Involvement of brain lactate in neuronal metabolism. NMR Biomed 2003;16(6–7):430–9.

[8] Larrabee MG. Lactate metabolism and its effects on glucose metabolism in an excised neural tissue. J Neurochem 1995;64(4):1734–41.

[9] Kasischke KA, Vishwasrao HD, Fisher PJ, Zipfel WR, Webb WW. Neural activity triggers neuronal oxidative metabolism followed by astrocytic glycolysis. Science 2004;305(5680):99–103.

[10] Porras OH, Loaiza A, Barros LF. Glutamate mediates acute glucose transport inhibition in hippocampal neurons. J Neurosci 2004;24(43):9669–73.
[11] Bouzier-Sore AK, Voisin P, Canioni P, Magistretti PJ, Pellerin L. Lactate is a preferential oxidative energy substrate over glucose for neurons in culture. J Cereb Blood Flow Metab 2003;23(11):1298–306.
[12] Hu Y, Wilson GS. A temporary local energy pool coupled to neuronal activity: fluctuations of extracellular lactate levels in rat brain monitored with rapid-response enzyme-based sensor. J Neurochem 1997;69(4):1484–90.
[13] Mangia S, Garreffa G, Bianciardi M, Giove F, Di Salle F, Maraviglia B. The aerobic brain: lactate decrease at the onset of neural activity. Neuroscience 2003;118(1):7–10.
[14] Pellerin L, Magistretti PJ. Glutamate uptake stimulates Na^+/K^+-ATPase activity in astrocytes via activation of a distinct subunit highly sensitive to ouabain. J Neurochem 1997;69(5):2132–7.
[15] Loaiza A, Porras OH, Barros LF. Glutamate triggers rapid glucose transport stimulation in astrocytes as evidenced by real-time confocal microscopy. J Neurosci 2003;23(19):7337–42.
[16] Demestre M, Boutelle M, Fillenz M. Stimulated release of lactate in freely moving rats is dependent on the uptake of glutamate. J Physiol 1997;499(Pt 3):825–32.
[17] Swanson RA, Morton MM, Sagar SM, Sharp FR. Sensory stimulation induces local cerebral glycogenolysis: demonstration by autoradiography. Neuroscience 1992;51(2):451–61.
[18] Fox PT, Raichle ME. Focal physiological uncoupling of cerebral blood flow and oxidative metabolism during somatosensory stimulation in human subjects. Proc Natl Acad Sci USA 1986;83(4):1140–4.
[19] Cerdán S, Rodrigues TB, Sierra A, Benito M, Fonseca LL, Fonseca CP, et al. The redox switch/redox coupling hypothesis. Neurochem Int 2006;48(6-7):523–30.
[20] Sada N, Lee S, Katsu T, Otsuki T, Inoue T. Epilepsy treatment. Targeting LDH enzymes with a stiripentol analog to treat epilepsy. Science 2015;347(6228):1362–7.
[21] Scharfman HE. Neuroscience. Metabolic control of epilepsy. Science 2015;347(6228):1312–3.
[22] Cataldo AM, Broadwell RD. Cytochemical identification of cerebral glycogen and glucose-6-phosphatase activity under normal and experimental conditions. II. Choroid plexus and ependymal epithelia, endothelia and pericytes. J Neurocytol 1986;15(4):511–24.
[23] Magistretti PJ, Sorg O, Yu N, Martin JL, Pellerin L. Neurotransmitters regulate energy metabolism in astrocytes: implications for the metabolic trafficking between neural cells. Dev Neurosci 1993;15(3–5):306–12.
[24] Magistretti PJ, Hof PR, Martin JL. Adenosine stimulates glycogenolysis in mouse cerebral cortex: a possible coupling mechanism between neuronal activity and energy metabolism. J Neurosci 1986;6(9):2558–62.
[25] Magistretti PJ, Morrison JH, Shoemaker WJ, Sapin V, Bloom FE. Vasoactive intestinal polypeptide induces glycogenolysis in mouse cortical slices: a possible regulatory mechanism for the local control of energy metabolism. Proc Natl Acad Sci USA 1981;78(10):6535–9.
[26] Magistretti PJ, Dietl MM, Hof PR, Martin JL, Palacios JM, Schaad N, et al. Vasoactive intestinal peptide as a mediator of intercellular communication in the cerebral cortex. Release, receptors, actions, and interactions with norepinephrine. Ann N Y Acad Sci 1988;527:110–29.
[27] Chih CP, Roberts Jr EL. Energy substrates for neurons during neural activity: a critical review of the astrocyte-neuron lactate shuttle hypothesis. J Cereb Blood Flow Metab 2003;23(11):1263–81.
[28] Hall CN, Klein-Flügge MC, Howarth C, Attwell D. Oxidative phosphorylation, not glycolysis, powers presynaptic and postsynaptic mechanisms underlying brain information processing. J Neuroscience 2012;32(26):8940–51.
[29] Mangia S, Simpson IA, Vannucci SJ, Carruthers A. The in vivo neuron-to-astrocyte lactate shuttle in human brain: evidence from modeling of measured lactate levels during visual stimulation. J Neurochem 2009;109(Suppl 1): 55–62.
[30] Patel AB, Lai JC, Chowdhury GM, Hyder F, Rothman DL, Shulman RG, et al. Direct evidence for activity-dependent glucose phosphorylation in neurons with implications for the astrocyte-to-neuron lactate shuttle. Proc Natl Acad Sci USA 2014;111(14):5385–90.
[31] Lundgaard I, Li B, Xie L, Kang H, Sanggaard S, Haswell JD, et al. Direct neuronal glucose uptake Heralds activity-dependent increases in cerebral metabolism. Nat commun 2015;6:6807.
[32] Nemoto EM, Hoff JT, Severinghaus JW. Lactate uptake and metabolism by brain during hyperlactatemia and hypoglycemia. Stroke 1974;5(1):48–53.
[33] Bittner CX, Valdebenito R, Ruminot I, Loaiza A, Larenas V, Sotelo-Hitschfeld T, et al. Fast and reversible stimulation of astrocytic glycolysis by K^+ and a delayed and persistent effect of glutamate. J Neurosci 2011;31(12):4709–13.

[34] Engel Jr. J, Kuhl DE, Phelps ME, Rausch R, Nuwer M. Local cerebral metabolism during partial seizures. Neurology 1983;33(4):400–13.
[35] Dalsgaard MK, Madsen FF, Secher NH, Laursen H, Quistorff B. High glycogen levels in the hippocampus of patients with epilepsy. J Cereb Blood Flow Metab 2007;27(6):1137–41.
[36] Bernard-Helary K, Lapouble E, Ardourel M, Hévor T, Cloix JF. Correlation between brain glycogen and convulsive state in mice submitted to methionine sulfoximine. Life Sci 2000;67(14):1773–81.
[37] Ronzio RA, Meister A. Phosphorylation of methionine sulfoximine by glutamine synthetase. Proc Natl Acad Sci USA 1968;59(1):164–70.
[38] Danbolt NC. Glutamate uptake. Prog Neurobiol 2001;65:1–105.
[39] Moussawi K, Riegel A, Nair S, Kalivas PW. Extracellular glutamate: functional compartments operate in different concentration ranges. Front Syst Neurosci 2011;5:94.
[40] Clements JD, Lester RA, Tong G, Jahr CE, Westbrook GL. The time course of glutamate in the synaptic cleft. Science 1992;258(5087):1498–501.
[41] Lee WJ, Hawkins RA, Viña JR, Peterson DR. Glutamine transport by the blood-brain barrier: a possible mechanism for nitrogen removal. Am J physiol 1998;274(4 Pt 1):C1101–7.
[42] O'Kane RL, Martínez-López I, DeJoseph MR, Viña JR, Hawkins RA. Na^+-dependent glutamate transporters (EAAT1, EAAT2, and EAAT3) of the blood-brain barrier. A mechanism for glutamate removal. J Biol Chem 1999;274(45):31891–5.
[43] Hawkins RA, O'Kane RL, Simpson IA, Viña JR. Structure of the blood-brain barrier and its role in the transport of amino acids. J Nutr 2006;136(1 Suppl):218S–226SS.
[44] Hawkins RA. The blood-brain barrier and glutamate. Am J Clin Nutr 2009;90(3):867S–874SS.
[45] Jensen AA, Fahlke C, Bjørn-Yoshimoto WE, Bunch L. Excitatory amino acid transporters: recent insights into molecular mechanisms, novel modes of modulation and new therapeutic possibilities. Curr Opin Pharmacol 2015;20C:116–23.
[46] Murphy-Royal C, Dupuis JP, Varela JA, Panatier A, Pinson B, Baufreton J, et al. Surface diffusion of astrocytic glutamate transporters shapes synaptic transmission. Nat Neurosci 2015;18(2):219–26.
[47] During MJ, Spencer DD. Extracellular hippocampal glutamate and spontaneous seizure in the conscious human brain. Lancet 1993;341(8861):1607–10.
[48] Glass M, Dragunow M. Neurochemical and morphological changes associated with human epilepsy. Brain Res Rev 1995;21:29–41.
[49] Tanaka K, Watase K, Manabe T, Yamada K, Watanabe M, Takahashi K, et al. Epilepsy and exacerbation of brain injury in mice lacking the glutamate transporter GLT1. Science 1997;276:1699–702.
[50] Campbell SL, Hablitz JJ. Glutamate transporters regulate excitability in local networks in rat neocortex. Neuroscience 2004;127:625–35.
[51] Hubbard JA, Hsu MS, Fiacco TA, Binder DK. Glial cell changes in epilepsy: overview of the clinical problem and therapeutic opportunities. Neurochem Int 2013;63(7):638–51.
[52] Lehre KP, Levy LM, Ottersen OP, Storm-Mathisen J, Danbolt NC. Differential expression of two glial glutamate transporters in the rat brain: quantitative and immunocytochemical observations. J Neurosci 1995;15 (3 Pt 1):1835–53.
[53] Lehre KP, Danbolt NC. The number of glutamate transporter subtype molecules at glutamatergic synapses: chemical and stereological quantification in young adult rat brain. J Neurosci 1998;18(21):8751–7.
[54] van der Hel WS, Notenboom RG, Bos IW, van Rijen PC, van Veelen CW, de Graan PN. Reduced glutamine synthetase in hippocampal areas with neuron loss in temporal lobe epilepsy. Neurology 2005;64(2):326–33.
[55] Mathern GW, Mendoza D, Lozado A, Pretorius JK, Dehnes Y, Danbolt NC, et al. Hippocampal GABA and glutamate transporter immunoreactivity in patients with temporal lobe epilepsy. Neurology 1999;52(3):453–72.
[56] Proper EA, Hoogland G, Kappen SM, Jansen GH, Rensen MGA, Schrama LH, et al. Distribution of glutamate transporters in the hippocampus of patients with pharmaco-resistant temporal lobe epilepsy. Brain 2002;125:32–43.
[57] Tessler S, Danbolt NC, Faull RLM, Storm-Mathisen J, Emson PC. Expression of the glutamate transporters in human temporal lobe epilepsy. Neuroscience 1999;88(4):1083–91.
[58] Bjørnsen LP, Eid T, Holmseth S, Danbolt NC, Spencer DD, de Lanerolle NC. Changes in glial glutamate transporters in human epileptogenic hippocampus: inadequate explanation for high extracellular glutamate during seizures. Neurobiol Dis 2007;25(2):319–30.

[59] Eid T, Thomas MJ, Spencer DD, Rundén-Pran E, Lai JC, Malthankar GV, et al. Loss of glutamine synthetase in the human epileptogenic hippocampus: possible mechanisms for raised extracellular glutamate in mesial temporal lobe epilepsy. Lancet 2004;363:28–37.
[60] Crino PB, Jin H, Shumate MD, Robinson MB, Coulter DA, Brooks-Kayal AR. Increased expression of the neuronal glutamate transporter (EAAT3/EAAC1) in hippocampal and neocortical epilepsy. Epilepsia 2002;43(3):211–8.
[61] Ulu MO, Tanriverdi T, Oz B, Biceroglu H, Isler C, Eraslan BS, et al. The expression of astroglial glutamate transporters in patients with focal cortical dysplasia: an immunohistochemical study. Acta Neuropathologica 2010;152:845–53.
[62] González-Martínez JA, Ying Z, Prayson R, Bingaman W, Najm I. Glutamate clearance mechanism in resected cortical dysplasia. J Neurosurg 2011;114:1195–202.
[63] Rakhade SN, Loeb JA. Focal reduction of neuronal glutamate transporters in human neocortical epilepsy. Epilepsia 2008;49(2):226–36.
[64] Campbell SL, Hablitz JJ. Decreased glutamate transport enhances excitability in a rat model of cortical dysplasia. Neurobiol Dis 2008;32(2):254–61.
[65] Rothstein JD, Dykes-Hoberg M, Pardo CA, Bristol LA, Jin L, Kuncl RW, et al. Knockout of glutamate transporters reveals a major role for astroglial transport in excitotoxicity and clearance of glutamate. Neuron 1996;16:675–86.
[66] Kong Q, Takahashi K, Schulte D, Stouffer N, Lin Y, Lin CG. Increased glial glutamate transporter EAAT2 expression reduces epileptogenic processes following pilocarpine-induced status epilepticus. Neurobiol Dis 2012;47:145–54.
[67] Watase K, Hashimoto K, Kano M, Yamada K, Watanabe M, Inoue Y, et al. Motor discoordination and increased susceptibility to cerebellar injury in GLAST mutant mice. Eur J Neurosci 1998;10(3):976–88.
[68] Watanabe T, Morimoto K, Hirao T, Suwaki H, Watase K, Tanaka K. Amygdala-kindled and pentylenetetrazole-induced seizures in glutamate transporter GLAST-deficient mice. Brain Res 1999;845(1):92–6.
[69] Lévesque M, Avoli M. The kainic acid model of temporal lobe epilepsy. Neurosci Biobehav Rev 2013;37(10 Pt 2): 2887–99.
[70] Ueda Y, Doi T, Tokumaru J, Yokoyama H, Nakajima A, Mitsuyama Y, et al. Collapse of extracellular glutamate regulation during epileptogenesis: down-regulation and functional failure of glutamate transporter function in rats with chronic seizures induced by kainic acid. J Neurochem 2001;76(3):892–900.
[71] Sakurai M, Kurokawa H, Shimada A, Nakamura K, Miyata H, Morita T. Excitatory amino acid transporter 2 downregulation correlates with thalamic neuronal death following kainic acid-induced status epilepticus in rat. Neuropathology 2015;35(1):1–9.
[72] Simantov R, Crispino M, Hoe W, Broutman G, Tocco G, Rothstein JD, et al. Changes in expression of neuronal and glial glutamate transporters in rat hippocampus following kainate-induced seizure activity. Brain Res Mol Brain Res 1999;65(1):112–23.
[73] Nonaka M, Kohmura E, Yamashita T, Shimada S, Tanaka K, Yoshimine T, et al. Increased transcription of glutamate-aspartate transporter (GLAST/GluT-1) mRNA following kainic acid-induced limbic seizure. Brain Res Mol Brain Res 1998;55(1):54–60.
[74] Lopes MW, Soares FM, de Mello N, Nunes JC, Cajado AG, de Brito D, et al. Time-dependent modulation of AMPA receptor phosphorylation and mRNA expression of NMDA receptors and glial glutamate transporters in the rat hippocampus and cerebral cortex in a pilocarpine model of epilepsy. Exp Brain Res 2013;226(2):153–63.
[75] Miller HP, Levey AI, Rothstein JD, Tzingounis AV, Conn PJ. Alterations in glutamate transporter protein levels in kindling-induced epilepsy. J Neurochem 1997;68(4):1564–70.
[76] Akbar MT, Torp R, Danbolt NC, Levy LM, Meldrum BS, Ottersen OP. Expression of glial glutamate transporters GLT1 and GLAST is unchanged in the hippocampus in fully kindled rats. Neuroscience 1997;78(2):351–9.
[77] Guo F, Sun F, Yu JL, Wang QH, Tu DY, Mao XY, et al. Abnormal expressions of glutamate transporters and metabotropic glutamate receptor 1 in the spontaneously epileptic rat hippocampus. Brain Res Bull 2010;81(4–5):510–6.
[78] Akbar MT, Rattray M, Williams RJ, Chong NW, Meldrum BS. Reduction of GABA and glutamate transporter messenger RNAs in the severe-seizure genetically epilepsy-prone rat. Neuroscience 1998;85(4):1235–51.
[79] Ingram EM, Wiseman JW, Tessler S, Emson PC. Reduction of glial glutamate transporters in the parietal cortex and hippocampus of the EL mouse. J Neurochem 2001;79(3):564–75.

[80] Ingram EM, Tessler S, Bowery NG, Emson PC. Glial glutamate transporter mRNAs in the genetically absence epilepsy rat from Strasbourg. Brain Res Mol Brain Res 2000;75(1):96–104.

[81] Dutuit M, Touret M, Szymocha R, Nehlig A, Belin MF, Didier-Bazès M. Decreased expression of glutamate transporters in genetic absence epilepsy rats before seizure occurrence. J Neurochem 2002;80(6):1029–38.

[82] Samuelsson C, Kumlien E, Flink R, Lindholm D, Ronne-Engström E. Decreased cortical levels of astrocytic glutamate transport protein GLT-1 in a rat model of posttraumatic epilepsy. Neurosci Lett 2000;289(3):185–8.

[83] Ueda Y, Willmore LJ. Sequential changes in glutamate transporter protein levels during Fe^{3+}-induced epileptogenesis. Epilepsy Res 2000;39(3):201–9.

[84] Lu Z, Zhang W, Zhang N, Jiang J, Luo Q, Qiu Y. The expression of glutamate transporters in chest compression-induced audiogenic epilepsy: a comparative study. Neurol Res 2008;30(9):915–9.

[85] Wong M, Ess KC, Uhlmann EJ, Jansen LA, Li W, Crino PB, et al. Impaired glial glutamate transport in a mouse tuberous sclerosis epilepsy model. Ann Neurol 2003;54(2):251–6.

[86] Morita T, Takahashi M, Takeuchi T, Hikasa Y, Ikeda S, Sawada M, et al. Changes in extracellular neurotransmitters in the cerebrum of familial idiopathic epileptic shetland sheepdogs using an intracerebral microdialysis technique and immunohistochemical study for glutamate metabolism. J Vet Med Sci 2005;67(11):1119–26.

[87] Bianchi MG, Bardelli D, Chiu M, Bussolati O. Changes in the expression of the glutamate transporter EAAT3/EAAC1 in health and disease. Cell Mol Life Sci 2014;71(11):2001–15.

[88] Dehnes Y, Chaudhry FA, Ullensvang K, Lehre KP, Storm-Mathisen J, Danbolt NC. The glutamate transporter EAAT4 in rat cerebellar Purkinje cells: a glutamate-gated chloride channel concentrated near the synapse in parts of the dendritic membrane facing astroglia. J Neurosci 1998;18(10):3606–19.

[89] Mim C, Balani P, Rauen T, Grewer C. The glutamate transporter subtypes EAAT4 and EAATs 1-3 transport glutamate with dramatically different kinetics and voltage dependence but share a common uptake mechanism. J Gen Physiol 2005;126(6):571–89.

[90] Schneider N, Cordeiro S, Machtens JP, Braams S, Rauen T, Fahlke C. Functional properties of the retinal glutamate transporters GLT1c and EAAT5. J Biol Chem 2014;289(3):1815–24.

[91] Peghini P, Janzen J, Stoffel W. Glutamate transporter EAAC1-deficient mice develop dicarboxylic aminoaciduria and behavioral abnormalities but no neurodegeneration. EMBO J 1997;16(13):3822–32.

[92] Hertz L, Dringen R, Schousboe A, Robinson SR. Astrocytes: glutamate producers for neurons. J Neurosci 1999;57(4):417–28.

[93] Kugler P. Cytochemical demonstration of aspartate aminotransferase in the mossy-fibre system of the rat hippocampus. Histochemistry 1987;87(6):623–5.

[94] Fonnum F. The distribution of glutamate decarboxylase and aspartate transaminase in subcellular fractions of rat and guinea-pig brain. Biochem J 1968;106(2):401–12.

[95] Fonnum F. Regulation of the synthesis of the transmitter glutamate pool. Prog Biophys Mol Biol 1993;60(1):47–57.

[96] Westergaard N, Varming T, Peng L, Sonnewald U, Hertz L, Schousboe A. Uptake, release, and metabolism of alanine in neurons and astrocytes in primary cultures. J Neurosci Res 1993;35(5):540–5.

[97] Eraković V, Župan G, Varljen J, Laginja J, Simonić A. Altered activities of rat brain metabolic enzymes in electroconvulsive shock-induced seizures. Epilepsia 2001;42(2):181–9.

[98] Eraković V, Župan G, Varljen J, Laginja J, Simonić A. Altered activities of rat brain metabolic enzymes caused by pentylenetetrazol kindling and pentylenetetrazol-induced seizures. Epilepsy Res 2001;43(2):165–73.

[99] Netopilová M, Haugvicová R, Kubová H, Dršata J, Mareš P. Influence of convulsants on rat brain activities of alanine aminotransferase and aspartate aminotransferase. Neurochem Res 2001;26(12):1285–91.

[100] Malthankar-Phatak GH, de Lanerolle N, Eid T, Spencer DD, Behar KL, Spencer SS, et al. Differential glutamate dehydrogenase (GDH) activity profile in patients with temporal lobe epilepsy. Epilepsia 2006;47(8):1292–9.

[101] Kvamme E, Svenneby G, Hertz L, Schousboe A. Properties of phosphate activated glutaminase in astrocytes cultured from mouse brain. Neurochem Res 1982;7(6):761–70.

[102] Hogstad S, Svenneby G, Torgner IA, Kvamme E, Hertz L, Schousboe A. Glutaminase in neurons and astrocytes cultured from mouse brain: kinetic properties and effects of phosphate, glutamate, and ammonia. Neurochem Res 1988;13(4):383–8.

[103] Eid T, Hammer J, Rundén-Pran E, Roberg B, Thomas MJ, Osen K, et al. Increased expression of phosphate-activated glutaminase in hippocampal neurons in human mesial temporal lobe epilepsy. Acta Neuropathol 2007;113(2):137–52.

[104] Doi T, Ueda Y, Takaki M, Willmore LJ. Differential molecular regulation of glutamate in kindling resistant rats. Brain Res 2011;1375:1–6.
[105] Pinal CS, Tobin AJ. Uniqueness and redundancy in GABA production. Perspect Dev Neurobiol 1998;5(2-3):109–18.
[106] Esclapez M, Tillakaratne NJ, Kaufman DL, Tobin AJ, Houser CR. Comparative localization of two forms of glutamic acid decarboxylase and their mRNAs in rat brain supports the concept of functional differences between the forms. J Neurosci 1994;14(3 Pt 2):1834–55.
[107] Lee M, Schwab C, McGeer PL. Astrocytes are GABAergic cells that modulate microglial activity. Glia 2011;59(1):152–65.
[108] Aoki C, Milner TA, Sheu KF, Blass JP, Pickel VM. Regional distribution of astrocytes with intense immunoreactivity for glutamate dehydrogenase in rat brain: implications for neuron-glia interactions in glutamate transmission. J Neurosci 1987;7(7):2214–31.
[109] Hammer J, Alvestad S, Osen KK, Skare Ø, Sonnewald U, Ottersen OP. Expression of glutamine synthetase and glutamate dehydrogenase in the latent phase and chronic phase in the kainate model of temporal lobe epilepsy. Glia 2008;56(8):856–68.
[110] Yu AC, Drejer J, Hertz L, Schousboe A. Pyruvate carboxylase activity in primary cultures of astrocytes and neurons. J Neurochem 1983;41(5):1484–7.
[111] Rutledge SL, Snead III OC, Kelly DR, Kerr DS, Swann JW, Spink DL, et al. Pyruvate carboxylase deficiency: acute exacerbation after ACTH treatment of infantile spasms. Pediatr Neurol 1989;5(4):249–52.
[112] Hadera MG, Faure JB, Berggaard N, Tefera TW, Nehlig A, Sonnewald U. The anticonvulsant actions of carisbamate associate with alterations in astrocyte glutamine metabolism in the lithium-pilocarpine epilepsy model. J Neurochem 2014;132(5):532–45.
[113] Kurz GM, Wiesinger H, Hamprecht B. Purification of cytosolic malic enzyme from bovine brain, generation of monoclonal antibodies, and immunocytochemical localization of the enzyme in glial cells of neural primary cultures. J Neurochem 1993;60(4):1467–74.
[114] Greenberg DA, Cayanis E, Strug L, Marathe S, Durner M, Pal DK, et al. Malic enzyme 2 may underlie susceptibility to adolescent-onset idiopathic generalized epilepsy. Am J Hum Genet 2005;76(1):139–46.
[115] Liang SL, Carlson GC, Coulter DA. Dynamic regulation of synaptic GABA release by the glutamate-glutamine cycle in hippocampal area CA1. J Neurosci 2006;26(33):8537–48.
[116] Bacci A, Sancini G, Verderio C, Armano S, Pravettoni E, Fesce R, et al. Block of glutamate-glutamine cycle between astrocytes and neurons inhibits epileptiform activity in hippocampus. J Neurophysiol 2002;88(5):2302–10.
[117] Cavus I, Kasoff WS, Cassaday MP, Jacob R, Gueorguieva R, Sherwin RS, et al. Extracellular metabolites in the cortex and hippocampus of epileptic patients. Ann Neurol 2005;57(2):226–35.
[118] Petroff OAC, Errante LA, Rothman DL, Kim JH, Spencer DD. Glutamate-glutamine cycline in the epileptic human hippocampus. Epilepsia 2002;43(7):703–10.
[119] Eid T, Williamson A, Lee TS, Petroff OA, de Lanerolle NC. Glutamate and astrocytes--key players in human mesial temporal lobe epilepsy? Epilepsia 2008;49(Suppl 2):42–52.
[120] Häberle J, Görg B, Rutsch F, Schmidt E, Toutain A, Benoist JF, et al. Congenital glutamine deficiency with glutamine synthetase mutations. N Engl J Med 2005;353(18):1926–33.
[121] Rosati A, Marconi S, Pollo B, Tomassini A, Lovato L, Maderna E, et al. Epilepsy in glioblastoma multiforme: correlation with glutamine synthetase levels. J Neurooncol 2009;93(3):319–24.
[122] Steffens M, Huppertz HJ, Zentner J, Chauzit E, Feuerstein TJ. Unchanged glutamine synthetase activity and increased NMDA receptor density in epileptic human neocortex: implications for the pathophysiology of epilepsy. Neurochem Int 2005;47(6):379–84.
[123] Perez EL, Lauritzen F, Wang Y, Lee TS, Kang D, Zaveri HP, et al. Evidence for astrocytes as a potential source of the glutamate excess in temporal lobe epilepsy. Neurobiol Dis 2012;47(3):331–7.
[124] Papageorgiou IE, Gabriel S, Fetani AF, Kann O, Heinemann U. Redistribution of astrocytic glutamine synthetase in the hippocampus of chronic epileptic rats. Glia 2011;59(11):1706–18.
[125] Eid T, Lee TS, Wang Y, Perez E, Drummond J, Lauritzen F, et al. Gene expression of glutamate metabolizing enzymes in the hippocampal formation in human temporal lobe epilepsy. Epilepsia 2013;54(2):228–38.
[126] Eid T, Ghosh A, Wang Y, Beckström H, Zaveri HP, Lee TS, et al. Recurrent seizures and brain pathology after inhibition of glutamine synthetase in the hippocampus in rats. Brain 2008;131(Pt 8):2061–70.

[127] Hu WW, Fang Q, Xu ZH, Yan HJ, He P, Zhong K, et al. Chronic h1-antihistamine treatment increases seizure susceptibility after withdrawal by impairing glutamine synthetase. CNS Neurosci Ther 2012;18(8):683–90.
[128] He Y, Hakvoort TB, Vermeulen JL, Lamers WH, Van Roon MA. Glutamine synthetase is essential in early mouse embryogenesis. Dev Dyn 2007;236(7):1865–75.
[129] van Gassen KL, van der Hel WS, Hakvoort TB, Lamers WH, de Graan PN. Haploinsufficiency of glutamine synthetase increases susceptibility to experimental febrile seizures. Genes Brain Behav 2009;8(3):290–5.
[130] Lee DJ, Hsu MS, Seldin MM, Arellano JL, Binder DK. Decreased expression of the glial water channel aquaporin-4 in the intrahippocampal kainic acid model of epileptogenesis. Exp Neurol 2012;235(1):246–55.
[131] Waniewski RA, McFarland D. Intrahippocampal kainic acid reduces glutamine synthetase. Neuroscience 1990;34(2):305–10.
[132] Sun HL, Zhang SH, Zhong K, Xu ZH, Feng B, Yu J, et al. A transient upregulation of glutamine synthetase in the dentate gyrus is involved in epileptogenesis induced by amygdala kindling in the rat. PLoS ONE 2013;8(6):e66885.
[133] Dutuit M, Didier-Bazès M, Vergnes M, Mutin M, Conjard A, Akaoka H, et al. Specific alteration in the expression of glial fibrillary acidic protein, glutamate dehydrogenase, and glutamine synthetase in rats with genetic absence epilepsy. Glia 2000;32(1):15–24.
[134] Kosenko E, Llansola M, Montoliu C, Monfort P, Rodrigo R, Hernandez-Viadel M, et al. Glutamine synthetase activity and glutamine content in brain: modulation by NMDA receptors and nitric oxide. Neurochem Int 2003;43(4–5):493–9.
[135] Chao CC, Hu S, Ehrlich L, Peterson PK. Interleukin-1 and tumor necrosis factor-alpha synergistically mediate neurotoxicity: involvement of nitric oxide and of N-methyl-D-aspartate receptors. Brain Behav Immun 1995;9(4):355–65.
[136] Swamy M, Yusof WR, Sirajudeen KN, Mustapha Z, Govindasamy C. Decreased glutamine synthetase, increased citrulline-nitric oxide cycle activities, and oxidative stress in different regions of brain in epilepsy rat model. J Physiol Biochem 2011;67(1):105–13.
[137] Bidmon HJ, Görg B, Palomero-Gallagher N, Schleicher A, Häussinger D, Speckmann EJ, et al. Glutamine synthetase becomes nitrated and its activity is reduced during repetitive seizure activity in the pentylentetrazole model of epilepsy. Epilepsia 2008;49(10):1733–48.
[138] Eid T, Tu N, Lee TS, Lai JC. Regulation of astrocyte glutamine synthetase in epilepsy. Neurochem Int 2013;63(7):670–81.

CHAPTER

10

Adenosine Metabolism

OUTLINE

OVERVIEW

The purine nucleoside adenosine plays a neuromodulatory role in the brain. It is directly involved in a number of functions including metabolism, cellular communication, and DNA methylation. Purinergic signaling by adenosine and its metabolites can occur through a variety of receptors and have profound intracellular effects. Therefore, it is crucial to maintain homeostatic adenosine levels in the brain. This is accomplished by regulatory metabolic enzymes, including adenosine kinase, 5′-nucleotidase (5′-NT), and adenosine deaminase (ADA). Expression levels of these proteins, however, are altered in epileptic tissue. For example, ADK expression is upregulated, resulting in faster metabolism and clearance of adenosine. Adenosine can modulate neuronal activity and exhibits natural antiepileptic effects through its activation of A_1 receptors. A reduction in A_1 receptor expression, however, is commonly found in epileptic tissue. The combination of these changes diminish the natural anticonvulsant ability of adenosine. In this chapter, we will review changes in purinergic signaling and metabolism in epilepsy. We will also discuss the therapeutic

Astrocytes and Epilepsy
DOI: http://dx.doi.org/10.1016/B978-0-12-802401-0.00010-7

potential of targeting and altering adenosine metabolism, adenosine-regulated glutamate release, DNA hypermethylation, and the effects of a low glycemic diet.

PURINERGIC SIGNALING AND METABOLISM

Adenosine is a ubiquitous purine neuromodulator that is directly involved with energy metabolism and cellular communication. Purinergic signaling can be carried out by two different families of receptors: adenosine (P1) and nucleotide (P2X and P2Y) receptors (Table 3.1). P2X receptors (P2X1-P2X7) are ligand-gated ion channels whereas P1 and P2Y receptors are G-protein-coupled receptors (GPCRs, Table 10.1) [1–3]. Activation of P1 or P2Y receptors may inhibit cyclic adenosine monophosphate (cAMP, $G\alpha_i$), stimulate cAMP ($G\alpha_s$), or activate phospholipase C ($G\alpha_q$) and subsequently lead to both the activation of protein kinase C (PKC) and the release of internal calcium stores. Glial cellular distribution of the purinergic receptors is detailed in Table 10.2 [4]. A comprehensive overview of the cellular distribution of purinergic receptors can be found in Burnstock and Knight [5].

Various enzymes are involved in the metabolism of adenosine (Fig. 10.1) [6]. Adenosine kinase (ADK) phosphorylates adenosine to form AMP. Initially, ADK is predominantly expressed in neurons but by postnatal day 14 in rats, ADK expression migrates to an almost exclusively astrocytic distribution [7]. The major adenosine producing enzyme in the extracellular space is 5′-nucleotidase (5′-NT); it conducts the reverse reaction by hydrolyzing AMP to adenosine. Equilibrative nucleoside transporters (ENTs) can move nucleoside substrates, including adenosine, into the cell. Adenosine may be deaminated by the enzyme adenosine deaminase (ADA). More specifically, ADA catalyzes the deamination of

TABLE 10.1 Purinergic G-Protein-Coupled Receptor Classification

Receptor Family	Receptor Subtype	$G\alpha$ Subunit Subtype
P1	A_1	$G\alpha_i$
	A_{2A}	$G\alpha_s$
	A_{2B}	$G\alpha_s$
	A_3	$G\alpha_i$ and $G\alpha_q$
P2Y	Y_1	$G\alpha_q$
	Y_2	$G\alpha_q$
	Y_4	$G\alpha_q$ and $G\alpha_i$
	Y_6	$G\alpha_q$
	Y_{11}	$G\alpha_q$ and $G\alpha_s$
	Y_{12}	$G\alpha_i$
	Y_{13}	$G\alpha_i$
	Y_{14}	$G\alpha_i$

both adenosine and 2′-deoxyadenosine to inosine and 2′-deoxyinosine, respectively, while producing ammonia as a byproduct [8]. Altogether, these enzymes can regulate the total amount of available intracellular and extracellular adenosine in the brain.

Transmethylation is an enzymatic reaction characterized by the transfer of a methyl group. Adenosine is the end product of S-adenosylmethionine (SAM)-dependent transmethylation reactions. The final reaction in this process involves the hydrolysis of S-adenosylhomocysteine

TABLE 10.2 Purinergic Receptors in Glial Cells

				Schwann Cells			Oligodendrocytes		
Receptor	**Astrocyte**	**Müller (eye)**	**Enteric Glia**	**Myelin**	**Nonmyelin**	**Terminal**	**Progenitor**	**Myelin**	**Microglia**
ADENOSINE									
A_1	R, P, F					F	R	P, F	F
A_{2A}	F				R, P, F		R		F
A_{2B}	F				R		R		F
A_3	F						R		F
ATP (IONOTROPIC)									
$P2X_1$	R, P, F	R							
$P2X_2$	R, P, F								
$P2X_3$	R, P, F								
$P2X_4$	R, P, F	R							R, P, F
$P2X_5$	R, F	R							P
$P2X_6$	R, P								P
$P2X_7$	R, P	R, P, F	P	F	P, F		F		P, F
ATP (METABOTROPIC)									
$P2Y_1$	R, P, F				F		F		R, F
$P2Y_2$	R, F		F	F	F	F	F		F
$P2Y_4$	R, F		F				F		R, F
$P2Y_6$	R, F						F		R, F
$P2Y_{11}$							F		
$P2Y_{12}$	F								R, F
$P2Y_{13}$							F		R
$P2Y_{14}$	F								

R, mRNA evidence; *P*, protein evidence; *F*, functional evidence. Functional evidence includes calcium imaging, protein kinase activation, responses to selective agonists and antagonists, and electrophysiological studies.
Reproduced with permission from Fields RD, Burnstock G. Purinergic signalling in neuron-glia interactions. Nat Rev Neurosci 2006;7(6):423–36.

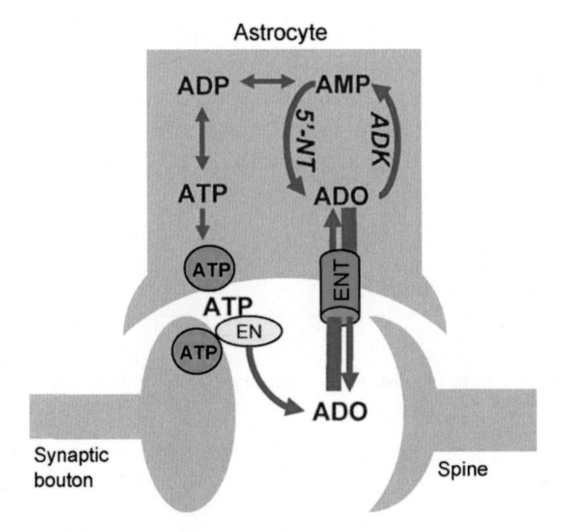

FIGURE 10.1 **Components of the astrocyte-based adenosine cycle.** Under physiological conditions, a major source of synaptic adenosine is vesicular release of ATP (orange circle) from astrocytes followed by its extracellular degradation to adenosine (ADO) by a cascade of ectonucleotidase (EN). Under conditions of high-frequency stimulation neurons contribute to synaptic ATP release. Extra- and intracellular levels of adenosine are rapidly equilibrated, mainly by equilibrative nucleoside transporters (ENTs). Thus, synaptic concentrations of adenosine are largely dependent on its intracellular metabolism as the driving force for the influx of adenosine into the cell. Intracellular metabolism of adenosine depends on the activity of the astrocytic enzyme adenosine kinase (ADK), which, together with 5′-nucleotidase (5′-NT), forms a substrate cycle between AMP and adenosine. This substrate cycle is directly linked to the energy pool of the brain involving ADP and ATP. *Source: Reproduced with permission from Boison D. The adenosine kinase hypothesis of epileptogenesis. Prog Neurobiol 2008;84(3):249–62.*

(SAH) to homocysteine and adenosine by the enzyme S-adenosylhomocysteinase. From there, homocysteine can be recycled back to methionine via a methyl transfer reaction and subsequently converted back to SAM. Interestingly, increased levels of ADK may lead to hypermethylated DNA by removing adenosine and driving the flux of methyl groups through the transmethylation pathway (Fig. 10.2, left panel) [9]. Both the reduction of ADK expression and the increase in adenosine levels shifts the equilibrium of S-adenosylhomocysteinase toward the formation of SAH, which blocks DNA methylation by DNA methyltransferase [9]. Therefore, adenosine is an epigenetic regulator of DNA methylation.

ADENOSINE METABOLISM AND P1 RECEPTORS IN EPILEPSY

Adenosine triphosphate (ATP) and adenosine are released during seizures [10] and the levels of both can be regulated through a number of metabolizing enzymes. Epileptic tissue has increased levels of 5′-NT, which leads to increases in adenosine [11]. Once in the extracellular space, adenosine acting through all four P1 adenosine receptors (A_1, A_{2A}, A_{2B}, and A_3) plays a role in the modulation of synaptic transmission by controlling neurotransmitter release and fine-tuning synaptic activity [12,13]. Adenosine can be neuroprotective through the activation of inhibitory A_1 receptors, although the expression levels of A_1 receptors may be altered in

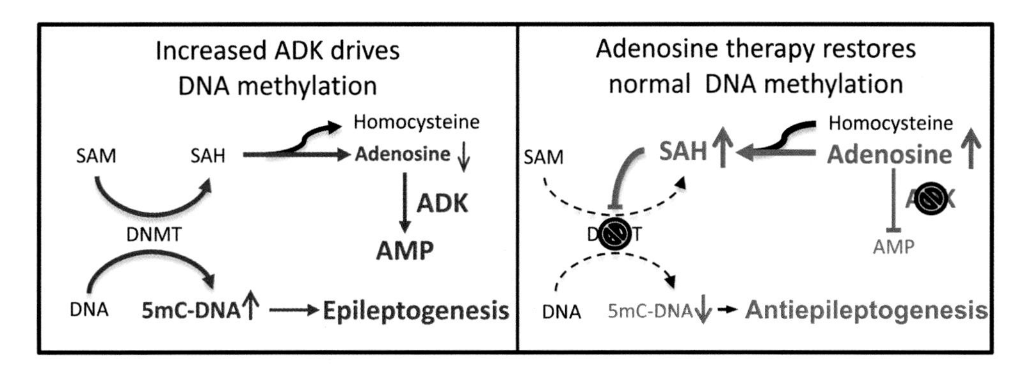

FIGURE 10.2 **The epigenetics of epileptogenesis.** Increased adenosine kinase (ADK) expression drives increased DNA methylation as a prerequisite for progressive epileptogenesis. Conversely, adenosine therapy restores normal DNA methylation and thereby prevents epileptogenesis. *SAM*, S-adenosylmethionine; *SAH*, S-adenosylhomocysteine; *DNMT*, DNA methyltransferase; *5mC*, 5-methyl-cytidine. *Source: Reproduced with permission from Boison D. Adenosinergic signaling in epilepsy. Neuropharmacology 2016;104:131–9.*

epileptic tissue [9,12,14]. Adenosine can be cleared from the extracellular space by equilbrative nucleotide transporters (ENTs) or metabolically by ADK, levels of which are upregulated in epileptic tissue (Fig. 10.3) [14]. Evidence for alterations in adenosine metabolism and P1 receptor expression in both human epileptic tissue and animal models of epilepsy are discussed.

Human Tissue Studies

Adenosine levels are elevated during a seizure and remain elevated during the postictal period [15]. ADK expression was increased in peritumoral tissue resected from patients with glial or glioneuronal tumors accompanied by epilepsy compared to tissue from seizure-free tumor patients [16]. Increased levels of ADK [17,18] and 5′-NT [19] as well as decreased expression of the adenosine A_1 receptor [20] were reported in tissue from patients with temporal lobe epilepsy (TLE). Proinflammatory molecules such as lipopolysaccharide (LPS) and interleukin-1β (IL-1β) increased ADK expression in cultured astrocytes from patients with TLE [17]. Membranes prepared from tissue resected from the temporal neocortex of patients with TLE or with epileptic cortical dysplasia were injected into *Xenopus laevis* oocytes [21,22]. In these oocytes and in neocortical slices from either human TLE tissue or pilocarpine-treated rats, GABA elicited inward currents [21]. Treatment with either ADA or nonselective adenosine receptor inhibitors reduced the GABA run-down current, a marker for $GABA_A$-receptor instability [21]. The use of transgenic mice and selective antagonists revealed that A_{2A}, A_{2B}, and A_3 receptors, but not A_1 receptors, were involved in maintaining GABA current stability and thus fine-tuning neuronal excitability [21,22]. Altered GABA run-down currents were also observed in human neocortical slices of periglioma epileptic tissue [22].

Acute encephalopathy with biphasic seizures and late reduced diffusion (AESD) is a childhood encephalopathy characterized by a biphasic clinical course of severe febrile seizures and restricted diffusion in the subcortical white matter. In an analysis of single nucleotide polymorphisms (SNPs) in patients with AESD, genetic variation in the A_{2A} receptor gene

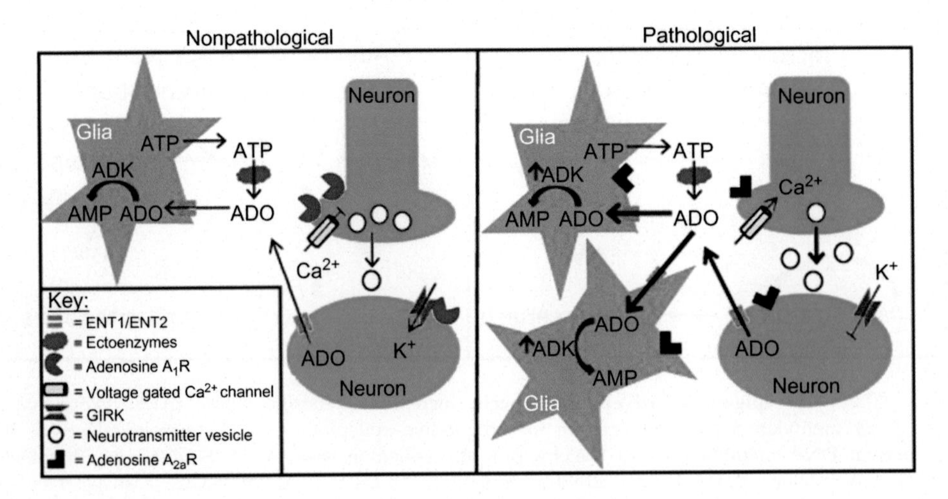

FIGURE 10.3 **Astrocyte-based ADK modulates network excitation by controlling the extracellular adenosine tone.** Nonpathological state (left panel): Extracellular adenosine (ADO) arises from two sources including: (1) Ca^{2+}-mediated release of ATP from astrocytes followed by catabolism to ADO through a series of ectoenzymes that include nucleoside triphosphate diphosphohydrolases, ectonucleotide pyrophosphatase/phosphodiesterases and 5′-nucleotidases; and (2) Direct ADO release into the extracellular space upon postsynaptic stimulation of neurons. Once in the extracellular space ADO inhibits neuron excitation through activation of A_1 receptors (A_1Rs) localized to the pre- and postsynaptic neuron membrane. On the presynaptic neuron, A_1R activation inhibits Ca^{2+}-dependent vesicular release of excitatory neurotransmitters by inhibiting P/Q- and N-type voltage gated Ca^{2+} channels; while on the postsynaptic neuron A_1R hyperpolarizes the cell by activating G-protein-coupled inwardly rectifying K^+ channels (GIRKS). ADO is cleared from the extracellular space by passive propagation through concentrative and equilibrative transporters (ENT1/ENT2) on the astrocyte membrane. Once in the astrocyte cytoplasm ADO is metabolized into AMP by ADK, which sets the ambient ADO tone. Pathological state (right panel): Sustained neuronal excitation, as observed in epilepsy, induces a shift in adenosine receptor expression levels with A_1R being superseded by A_{2A} receptors ($A_{2A}Rs$). As a consequence, there is a loss of A_1R activity, which translates to increased network excitability due to increased Ca^{2+}-dependent vesicular release of excitatory neurotransmitters and attenuated GIRK-mediated hyperpolarization. Furthermore, $A_{2A}R$ activation causes an increase in astrogliosis that is accompanied by increased ADK expression and activity. Pathological levels of ADK will drive ADO influx and metabolism; thereby, decreasing the extracellular ADO tone. *Source: Reproduced with permission from Aronica E, Sandau US, Iyer A, Boison D. Glial adenosine kinase—a neuropathological marker of the epileptic brain. Neurochem Int 2013;63(7):688–95.*

was a risk factor for AESD [23]. Similarly, certain SNPs in the A_1 receptor [24], ADK [25], and 5′-NT [25] genes were associated with developing seizures after traumatic brain injury (TBI).

Animal Models

The phenotype of astrogliosis (proliferation of and morphological alterations in "reactive" astrocytes), a common pathological feature of epilepsy, may be a result of adenosine largely acting through A_{2A} receptors [6]. Both A_{2A} receptors and ADK are upregulated in reactive astrocytes [9]. Selective A_{2A} receptor antagonists prevented astrogliosis in rat primary astrocytes [26,27]. The use of in vitro systems determined that A_{2A} receptor-dependent astrogliosis may occur through the basic fibroblast growth factor (bFGF) [26] or through the Akt/NF-κB [27] pathways.

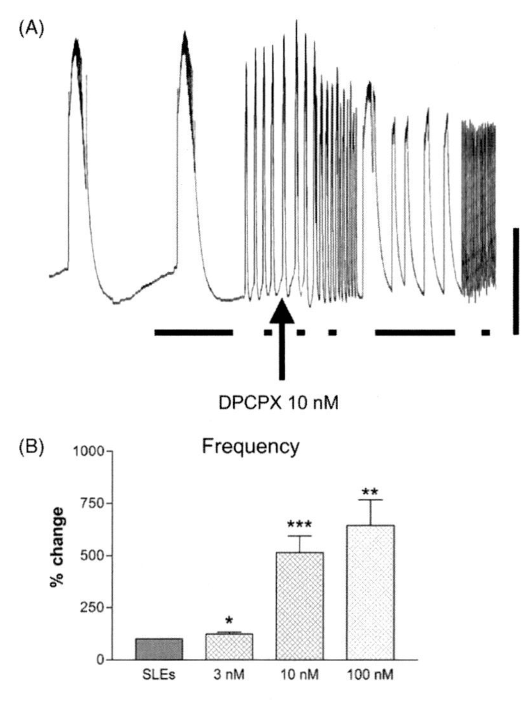

FIGURE 10.4 **Low concentrations of 8-cyclopentyl-1,3-dipropylxanthine (DPCPX), the selective adenosine A_1 receptor antagonist, caused seizure-like events (SLEs) to switch to status epilepticus-like late recurrent discharges (LRDs).** In (A), a slice that exhibited a stable, low frequency of SLEs during the previous 2 hours was exposed to a low concentration of DPCPX, shown by the arrow. Over the course of the 5–10 minutes wash in, a conversion of the epileptiform activity into LRD-like behavior occurred. The transition was associated with a gradual increase in frequency that reaches a maximum. A recovery of the DPCPX-induced switch was never observed, even following wash back to low Mg^{2+} alone for several hours. (B) The concentration dependence of the effects of DPCPX on the frequency of the events; the values are shown as a percentage change compared with control. Error bars represent SEMs. *represents a significance value of $P < 0.05$, **represents a significance value of $P < 0.01$ and ***represents a significance value of $P < 0.001$ all compared with predrug values. All time scale bars in (A) represent 2 minutes; note the faster and slower chart speeds used. Vertical scale bar represents 0.5 mV. *Source: Reproduced with permission from Avsar E, Empson RM. Adenosine acting via A_1 receptors, controls the transition to status epilepticus-like behaviour in an in vitro model of epilepsy. Neuropharmacology 2004;47(3):427–37.*

It is well known that adenosine has powerful anticonvulsant effects [28–33]. Rat horizontal entorhinal cortex brain slices at least 2 months after electrical stimulation-induced status epilepticus (SE) exhibited increased sensitivity to adenosine compared to control slices [31]. Pharmacological evidence suggested that adenosine acts through the A_1 receptor, but not the A_{2A} receptor, to inhibit membrane excitability [7,31,33–36]. In fact, antagonist of the A_1 receptor can accelerate seizure progression [7]. In the low magnesium in vitro entorhinal cortex slice preparation model of epilepsy, an adenosine A_1 receptor antagonist accelerated the progression of seizure-like events (SLEs) into late recurrent discharges (LRDs, epileptiform activity that resembles SE) (Fig. 10.4) [28]; this could be reversed with an A_1 receptor agonist (Fig. 10.5) [28]. Similarly, an inhibitor of 5′-NT, which effectively depleted extracellular adenosine, increased the frequency of SLEs whereas an inhibitor of ADA, which raised

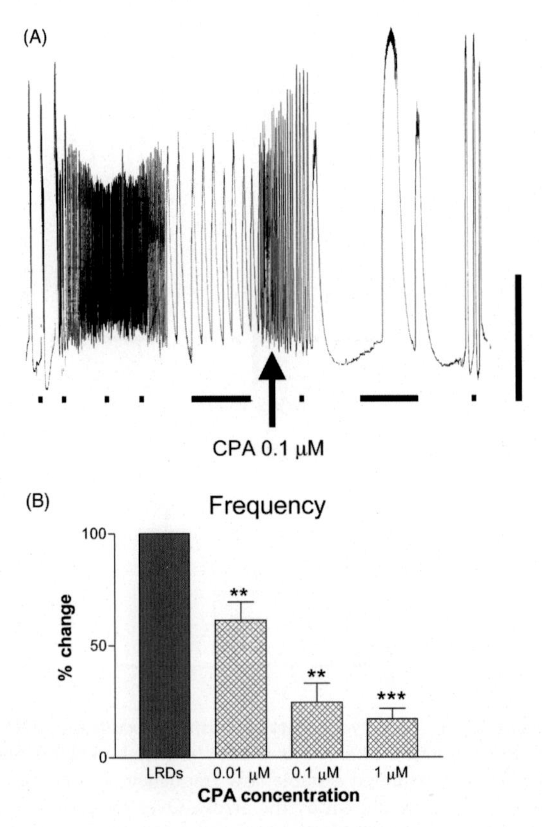

FIGURE 10.5 **The selective adenosine A1 receptor agonist N6-cyclopentyladenosine (CPA) was able to effectively reverse the LRDs back to SLEs.** This slice (shown in (A)) had spontaneously switched its electrographic behavior to produce LRDs. Application of 0.1 μM CPA, the A1 receptor agonist, shown by the arrow, effectively converted the high-frequency activity back to the SLEs, note also the recovery in the amplitude of the events following CPA. This slice, despite washing back to low Mg^{2+} alone, showed no reversal of the SLEs back into LRDs even 60 minutes after wash back, suggesting a permanent recovery by CPA. The concentration dependence of the ability of CPA to reverse the activity is shown in (B), where even 0.01 μM CPA was effective. Error bars represent SEMs. **represents a significance value of $P < 0.01$ and ***represents a significance value of $P < 0.001$ all compared with predrug values. All time scale bars in (A) represent 2 minutes; note the faster and slower chart speeds used. Vertical scale bar represents 0.5 mV. *Source: Reproduced with permission from Avsar E, Empson RM. Adenosine acting via A_1 receptors, controls the transition to status epilepticus-like behaviour in an in vitro model of epilepsy. Neuropharmacology 2004;47(3):427–37.*

extracellular adenosine, reversed these events [28]. A separate study, however, found that both ADA and S-adenosylhomocysteinase had no effect on neuronal activity in rat hippocampal slices [35]. ADK inhibitors also raised extracellular adenosine levels and reduced hyperpolarization and seizure-like activity in hippocampal slices from rats (Fig. 10.6) [7,35].

Pretreatment with the adenosine A_3 receptor agonist IB-MECA increased seizure latency and decreased neurological impairment in N-methyl-D-aspartate (NMDA) and pentylenetetrazol (PTZ), but not electrically-induced seizures [37]. The mortality rate in all three models, however, was improved with IB-MECA treatment [37]. In immature rat hippocampal

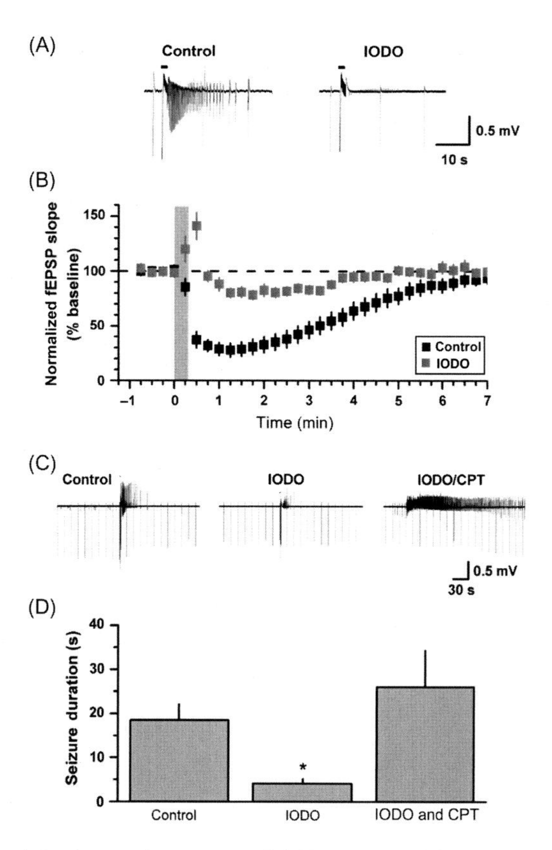

FIGURE 10.6 **Seizure activity is greatly attenuated following inhibition of ADK.** (A) Epileptiform activity induced by high-frequency stimulation (HFS, small black bar) in nominally Mg^{2+}-free artificial cerebrospinal fluid (aCSF) was greatly attenuated by the adenosine kinase (ADK) inhibitor iodotuberocidin (IODO, 5 mM; right panel) compared to control conditions in the absence of IODO (control; left panel). (B) The reversible depression of synaptic transmission induced by seizures under control conditions (filled squares) was also greatly attenuated in the presence of IODO (gray squares). Average duration of seizures (~18 seconds) in control conditions is indicated by gray bar, but too short to depict (~4 seconds) in the presence of IODO. (C) The IODO-induced depression of seizure activity (middle panel; IODO) is reversed by the A_1 receptor antagonist CPT (1 mM; IODO/CPT; right panel). Note: Seizure in IODO/CPT is a spontaneous seizure, seen in 6/9 slices exposed to this combination of drugs. (D) Pooled data showing the inhibition of HFS-evoked seizure duration by IODO ($n = 9$) and its reversal by CPT in the 3 slices not showing spontaneous seizure activity. *Source: Reproduced with permission from Etherington LA, Patterson GE, Meechan L, Boison D, Irving AJ, Dale N, et al. Astrocytic adenosine kinase regulates basal synaptic adenosine levels and seizure activity but not activity-dependent adenosine release in the hippocampus. Neuropharmacology 2009;56(2):429–37.*

slices perfused with the $GABA_A$-receptor antagonist bicuculline, however, the A_3 receptor agonist 2-Cl-IB-MECA increased excitatory effects, including the frequency of spontaneous discharges [38]. These effects could be blocked by A_3 receptor antagonists, but not by A_1 or A_{2A} antagonists [38]. The contrasting results could be explained by the excitatory actions of GABA in the immature brain.

Adenosine can decrease the firing rate of neurons [39]. Dialysate hippocampal purine levels were measured before and after electrically kindled seizures in rats in vivo [40].

While only a small increase in adenosine was observed, a two- to threefold increase in its metabolites (inosine, hypoxanthine, and xanthine) was found [40]. Decreased A_1 receptor density was reported in the amygdala kindling model in rats [36] but no alterations in A_1 receptor expression were reported in the rat perforant path stimulation model [41]. Application of the selective A_1 receptor agonist CPA reduced the severity of seizures whereas the selective antagonist DPCX had no effect on severity or progression to SE [41].

A persistent upregulation in ADK protein expression in reactive astrocytes was observed for several months in the electrical stimulation [17] and kainic acid [42–44] models of TLE. An initial loss of ADK expression was reported after intrahippocampal kainic acid injections, but this gradually increased over time until it was both overexpressed and exhibited increased enzymatic activity compared to control mice [42]. Furthermore, seizures and astrogliosis remained focal and restricted to ADK overexpressing areas in the intraamygdala kainic acid model [43,44].

Increased levels of astrocytic A_{2A} receptors [45] and 5′-NT [46] were reported within 1 week of kainic acid injections. This increase was not observed, however, in amygdala kindled rats [46]. In both the hippocampus and cerebral cortex, increased levels of ATP, adenosine diphosphate (ADP), and adenosine monophosphate (AMP) were found anywhere from 2 to 50 days after either intraperitoneal kainic acid or pilocarpine injections [47]. Treatment with an A_1 receptor antagonist reduced the latency to SE whereas an A_1 agonist provided both anticonvulsant and neuroprotective effects in the rat pilocarpine model of epilepsy [48]. Activation of the adenosine A_1 receptor also suppressed seizures in the intrahippocampal kainic acid model of epilepsy [49].

A single injection of the convulsant PTZ significantly increased adenosine uptake in the hippocampus, cerebellum, and cortex but decreased uptake in the striatum of mice [50]. Rats with increased ATP hydrolysis in synaptosomes from the hippocampus and cerebral cortex were more resistant to PTZ-induced seizures [51]. PTZ kindling led to decreased expression of the A_1 receptor binding sites [52]. A decrease in the expression of A_1 receptors would severely limit the anticonvulsant effects of adenosine.

Wistar Albino Glaxo/Rijswijk (WAG/Rij) rats develop spontaneous nonconvulsive seizures after 2 months of age and are therefore a genetic model of human absence epilepsy. Compared to age-matched Copenhagen Irish (ACI) control rats, cerebral A_{2A} receptor expression was lower in the 1.5-month-old (preepileptic) WAG/Rij rats but was elevated in the epileptic 6-month-old WAG/Rij rats [53]. Exposure of a specific A_{2A} receptor agonist to cortical and thalamic slices from the epileptic rats led to modulation of cAMP formation and stimulation of the MAPK pathway, an effect that was absent in the preepileptic WAG/Rij rats [53].

It is clear that reactive astrocytes have different properties than healthy ones, including upregulated ADK and A_{2A} receptor expression. One explanation for these changes may be that these systems undergo "interdependent maladaptive changes" during the development of epilepsy [9]. If the adenosine A_{2A} receptor becomes upregulated (and responds more efficiently to extracellular adenosine), then the upregulation of ADK may be triggered to limit adenosine availability. The now limited source of adenosine may then trigger increased A_{2A} receptor expression, creating a self-sustaining cycle. This, however, does not explain the overall reduction in A_1 receptor expression observed in epileptic tissue.

Transgenic Mice Studies

Purine inhibition of synaptic transmission is almost exclusively mediated by A_1 receptor [54]. Intrahippocampal kainic acid injections led to increased neuronal cell loss, seizure severity, and seizure-induced fatality in A_1 receptor knockout mice compared to wild-type mice [55]. After controlled cortical impact (CCI), seizures were observed in 83% of male and 100% of female A_1 receptor knockout mice, 0% of male and 14% of female heterozygotes, and in 33% of male and 25% of female wild-type mice [56]. Seizure scores were higher in the A_1 receptor knockout mice after CCI and lethal SE was only observed in knockout mice [56]. Adenosine A_{2A} receptor knockout mice, on the other hand, were only partially resistant to PTZ- and pilocarpine-induced seizures but not electroshock-induced seizures [57].

Although ATP can be broken down by multiple ectoenzymes to degradation products (ADP, AMP, adenosine, and inosine), the primary source of adenosine during seizures may be from the breakdown of AMP by 5′-NT [58]. Adenosine formation was almost completely absent in slices from 5′-NT knockout mice. Furthermore, A_1 receptor activation suppressed the spread of penicillin-induced seizures nearly twice as fast as in A_1 receptor knockout mice or wild-type mice treated with A_1 receptor antagonists [58].

Transgenic mice overexpressing ADK were more susceptible to brain injury. ADK was upregulated after seizures in the intraamygdala kainic acid model of epilepsy. Spontaneous seizures could be mimicked by simply overexpressing ADK or knocking out A_1 receptors in mice [44,59]. Mice that overexpressed ADK died within 3 days of kainic acid-induced SE whereas mice with reduced expression of ADK in the forebrain were resistant to epileptogenesis (prevented astrogliosis and spontaneous seizures) [44,59]. Transgenic mice with overexpressed ADK in hippocampal neurons showed injury within 15 minutes of transient middle cerebral artery occlusion (MCAO); wild-type mice were spared from injury even after 1 hour of MCAO [60]. Wild-type mice had reduced hippocampal ADK expression after MCAO and reperfusion [60].

Adeno-associated virus 8 (AAV8) gene therapy vectors were used to selectively modulate ADK expression. ADK cDNA was put under the control of the astrocyte-specific promotor gfaABC1D [61] and viral vectors were injected into the CA3 area of wild-type or spontaneously epileptic ADK transgenic mice to selectively overexpress ADK in astrocytes. This led to focal overexpression of ADK and spontaneous seizures [62,63]. ADK downregulation with AAV8-mediated RNA interference (RNAi) almost completely abolished the spontaneous seizures in ADK transgenic mice [63].

THERAPEUTIC MODULATION OF ADENOSINE

The regulation of adenosine metabolism offers ample therapeutic opportunities because adenosine, acting through the A_1 receptor, is a natural anticonvulsant [12,64–66]. A number of different enzymes play a role in controlling adenosine levels (Fig. 10.1) and all have the potential to serve as therapeutic targets. ADK, however, is a major negative regulator of adenosine levels and is found predominantly on astrocytes. Other membrane proteins, including the A_{2A} receptor and ENT, have been linked to controlling extracellular glutamate levels through inhibition of glutamate transporter-1 (GLT1). Glutamate has powerful effects on neuronal excitability and, therefore, needs to be tightly regulated. Furthermore,

DNA hypermethylation is linked to epilepsy and adenosine is known to play a major role in epigenetics by altering DNA methylation states. Finally, adenosine levels can potentially be regulated by a change in diet. Each of these therapeutic options will be discussed below.

Adenosine Metabolism

The known effects of adenosine in regulation of neuronal excitability [29,30,32] have been tested for efficacy in various animal models of epilepsy. Adenosine analogs had anticonvulsant effects, including decreased seizure occurrence and shortened seizure duration, in the amygdala kindling [67], 4-aminopyridine (4-AP), PTZ, picrotoxin, kainic acid, and strychnine models of epilepsy [68]. Similarly, adenosine injected into the rat prepiriform cortex provided dose-dependent protection against bicuculline-induced seizures, an effect that was potentiated by treatment with an ADK inhibitor, adenosine transport blockers, or an ADA inhibitor [69].

Several enzymes involved in adenosine metabolism could be targeted to modify the levels of adenosine during epilepsy. Adenosine transporter inhibitors, such as the potent inhibitor of equilibrative transporter 1 (ENT1) cannabidiol, may increase extracellular adenosine and suppress seizures [70,71]. ADK is upregulated in reactive astrocytes in epilepsy and ADK inhibition increases adenosine levels [12]. Homozygous ADK knockout mutant mice, however, displayed microvesicular hepatic steatosis 4 days after birth and died within 14 days with a fatty liver [72]. Adenosine receptor agonists, particularly A_1 receptor agonists, have also demonstrated anticonvulsant effects, but failed to undergo successful clinical development due to the severity of side effects at effective doses [73–75]. Furthermore, global application of either A_1 receptor agonists or ADK inhibitors induces side effects, including cardiovascular problems [12,66]. Therefore, focal adenosine augmentation therapies (AATs) are a promising approach for the suppression or even prevention of seizures [64–66,76].

ADK inhibitors increase adenosine concentrations in vivo and in vitro [73,77–79]. Fibroblasts engineered to release adenosine by inactivating both ADK and ADA in culture were then encapsulated into semipermeable polymers and grafted into brain ventricles [80]. This treatment paradigm offered nearly total protection from behavioral seizures and afterdischarges; control rats had full tonic-clonic convulsions. One major drawback, however, was that the encapsulated cells had a limited longevity and, therefore, seizure suppression was limited to 2 weeks [80].

A promising focal AAT was intraventricular grafting of adenosine-releasing cells encapsulated in a synthetic polymer. This therapy significantly reduced the number of severe seizures as well as the amplitude and duration of afterdischarges [81–83]. Shortly thereafter, embryonic stem cells (ESCs) were engineered to release adenosine by genetic disruption of ADK and were differentiated into neural precursor cells [84]. The ESCs were then injected into the rat hippocampus and prevented the development of seizures in the electrical kindling model of epilepsy [84].

Lentiviral RNAi was used to mediate the downregulation of ADK in human mesenchymal stem cells (hMSCs). These cells were then transplanted into the hippocampi of mice 1 week prior to intraamygdala kainic acid-induced SE [85]. Mice receiving ADK knockdown implants displayed a 35% reduction in seizure duration and a 65% reduction in neuronal cell loss compared to sham-treated and scrambled miRNA control mice (Fig. 10.7) [85].

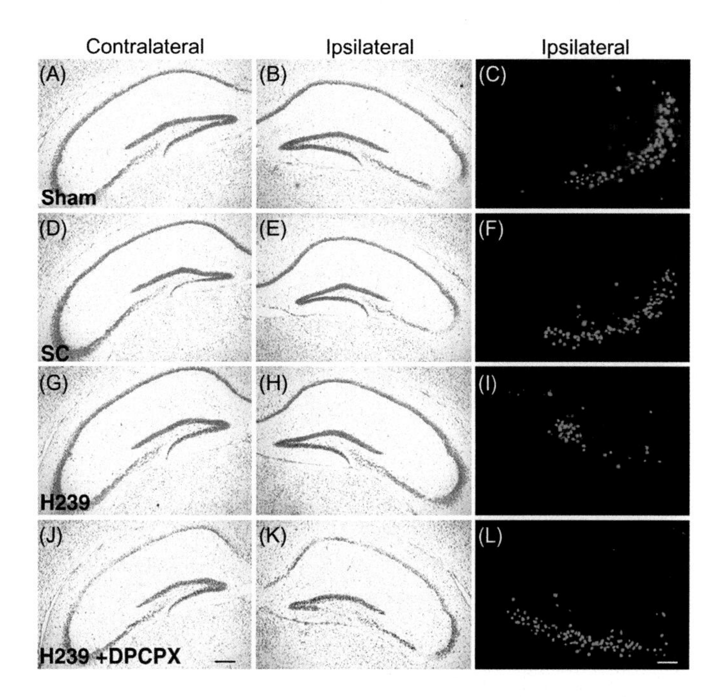

FIGURE 10.7 **Epilepsy-associated cell loss.** Representative micrographs of the hippocampal formation from coronal brain sections taken from sham-treated control mice (SHAM), recipients of scrambled control miRNA expressing hMSCs (SC), or H239 human mesenchymal stem cell (hMSC) graft recipients 24 hours after intraamygdaloid injection of 0.3 μg KA. Sections were stained either with cresyl violet (A, B, D, E, G, H, J, L) or with TUNEL (C, F, I, L green). (A–C) Typical apoptotic cell death in the CA3 region of the hippocampus of sham controls ipsilateral (ipsi) to the KA-injected amygdala. (D–F) CA3-selective apoptotic cell death in scrambled control-graft recipients was comparable to the sham-treated control animals. (G–I) H239 hMSC graft recipients show markedly decreased ipsilateral cell loss, which becomes evident by a hardly discernible lesion within the stratum pyramidale and by a prominent reduction in the number of TUNEL positive cells (I). (J–L) Sham- and scrambled control like CA3 injury in H239 graft recipients in which KA was paired with selective A_1 receptor antagonist DPCPX (1 mg/kg, intraperitoneal). Scale bars: A, B, D, E, G, H, J, K: 100 μm; C, F, I, L: 25 μm. *Source: Reproduced with permission from Ren G, Li T, Lan JQ, Wilz A, Simon RP, Boison D. Lentiviral RNAi-induced downregulation of adenosine kinase in human mesenchymal stem cell grafts: a novel perspective for seizure control. Exp Neurol 2007;208(1):26–37.*

Silk-based brain implants containing these same hMSCs also significantly reduced the number and duration of kainic acid-induced seizure (Fig. 10.8) [86].

Syzbala et al. [87] designed and implanted adenosine-loaded polymers into the hippocampal fissure of rats ipsilateral to the site of hippocampal electrical kindling. These were filled with adenosine-containing microspheres, coated with multiple macroscale adenosine-loaded silk films, and covered with a silk-based capping layer. The polymers released 1 μg adenosine per day for 10 days. Animals implanted with the adenosine-releasing implant were completely protected from seizures over the 10 days of treatment while rats receiving a control implant developed severe seizures (Fig. 10.9) [87]. After the initial 10 days of treatment, the adenosine polymers still offered protection and a significant reduction in seizure development [87].

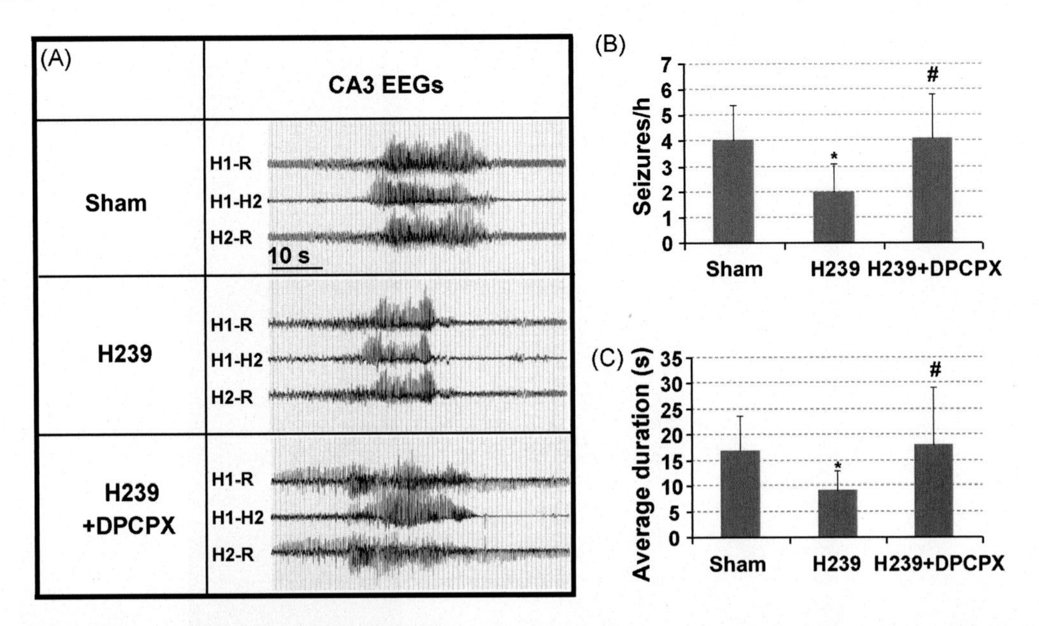

FIGURE 10.8 **Representative EEGs recorded from the ipsilateral CA3 3 weeks after kainic acid-induced seizures.** (A) Recordings from a montage using a bipolar electrode inserted into injured CA3 (H1; H2) and a cortical ground/reference electrode (R). H1-R and H2-R traces show hippocampus/cortical recordings whereas trace H1-H2 shows intrahippocampal recordings (ie, between the two tips of the bipolar electrode). Sham-treated controls were compared with recipients of H239 hMSCs before (middle) or after (bottom) injection of the A_1 receptor antagonist DPCPX (1 mg/kg, intraperitoneal). (B and C) Average number of seizures per hour and average seizure duration in respective treatment groups. Data are based on $n=6$ animals for each group and 16 hours of continuous EEG monitoring. Data are presented as ±SD and were analyzed using one-way ANOVA compared to sham. $*P < 0.001$, $\#P > 0.05$ compared to sham control. *Source: Reproduced with permission from Li T, Ren G, Kaplan DL, Boison D. Human mesenchymal stem cell grafts engineered to release adenosine reduce chronic seizures in a mouse model of CA3-selective epileptogenesis. Epilepsy Res 2009;84(2–3):238–41.*

The risk of cognitive impairment and other adverse effects are a major problem with current antiepileptic drugs and, therefore, should be a concern when testing new therapeutics. Adenosine, however, has been proven to be relatively safe. Local inhibition of ADK and subsequent augmentation of adenosine prevented psychosis and improved cognitive function without any known side effects in mice [88]. Rats and dogs treated with adenosine for 5 and 28 days, respectively, exhibited no disturbance in neurological function or histopathology [89]. A transient sedation in rats and an increase in muscle tone in dogs, however, were noted [89]. More importantly, in an open-label, double blinded, placebo-controlled trial of adenosine, blood pressure, heart rate, end-tidal carbon dioxide, and neurological function were all unaffected [90,91]. Some participants did report headaches and back pains [90,91]. In any drug treatment paradigm, however, cognitive function should be monitored.

Glutamate Regulation

Adenosine, acting through the A_{2A} receptor, stimulated astrocytic glutamate release and inhibited glutamate uptake through a cAMP/protein kinase A-dependent reduction of

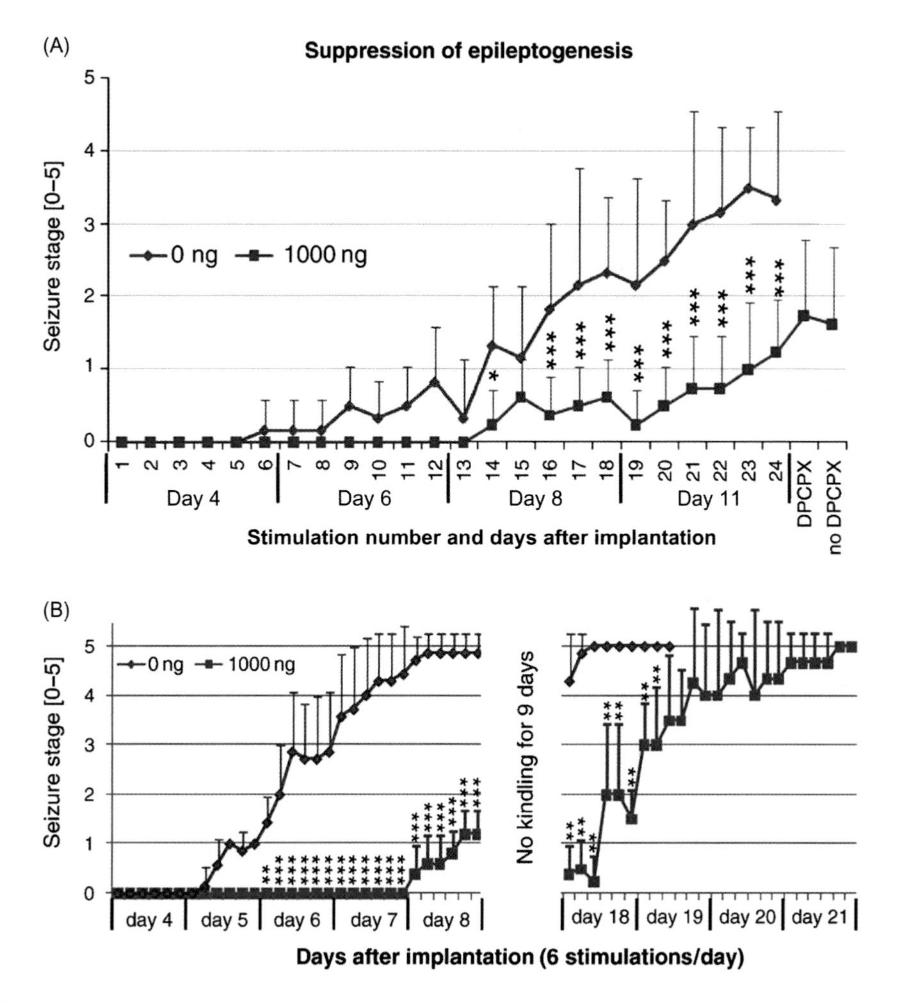

FIGURE 10.9 **Influence on epileptogenesis by adenosine-releasing polymers.** (A) Four days after infrahippocampal implantation of silk-based polymers with daily target release rates for adenosine of 0ng ($N = 5$, red), or 1000ng ($N = 8$, blue) kindling stimulations were delivered at a rate of 6 stimulations per day on days 4, 6, 8, and 11 following implantation. A total of 24 kindling stimulations were delivered. On day 12, the A_1 receptor antagonist DPCPX (1mg/kg, i.p.) was injected 30minutes prior to stimulation. Each animal was tested again on day 13 (no DPCPX). Seizure stages were averaged across animals from each group for each individual stimulus. Note that recipients of a target dose of 1000ng adenosine per display significant protection from kindling development, while DPCPX does not increase the seizure score. Errors are given as ± SD. Data were analyzed by two way ANOVA followed by a Bonferroni test; the significance of interaction between groups was determined as $F = 6.704$, $P < 0.0001$; significance levels of individual tests are indicated: $^{*}P < 0.05$, $^{**}P < 0.01$, $^{***}P < 0.001$. (B) 4 days after infrahippocampal implantation of silk-based polymers with daily target release rates for adenosine of 0ng ($N = 7$, red), or 1000ng ($N = 5$, blue) kindling stimulations were delivered at a rate of six stimulations per day on days 4, 5, 6, 7, and 8 following implantation. A total of 30 kindling stimulations were delivered. Please note the increased kindling frequency compared to (A). After the 30th kindling stimulation, kindling was discontinued for 9 days. Kindling stimulations were resumed at day 18. Seizure stages were averaged across animals from each group for each individual stimulus. Note that recipients of a target dose of 1000ng adenosine per day resumed kindling at day 18 at a level at which kindling was discontinued at day 8. After 7 consecutive stage 5 seizures kindling was discontinued in control animals due to animal welfare considerations. Errors are given as ± SD. Data were analyzed by two way ANOVA followed by a Bonferroni test; the significance of interaction between groups was determined as $F = 19.36$, $P < 0.0001$; significance levels of individual tests are indicated: $^{**}P < 0.01$, $^{***}P < 0.001$. *Source: Reproduced with permission from Szybala C, Pritchard EM, Lusardi TA, Li T, Wilz A, Kaplan DL, et al. Antiepileptic effects of silk-polymer based adenosine release in kindled rats. Exp Neurol 2009;219(1):126–35.*

glutamate transporter 1 (GLT1) [92–95]. Glutamate aspartate transporter (GLAST) expression was also upregulated in response to A_{2A} receptor activation [92,93]. Interestingly, both adenosine and A_{2A} receptor agonist enhanced glutamate release from hippocampal cultured rat glial cells in a bell-shaped manner; low concentrations enhanced glutamate release while higher concentrations had no effect [96]. Exposure of hippocampal slices from 15 to 20 days old rats to 1 mM ATP caused an increase in GLT1 expression, but also resulted in neuronal cell death [97].

ENT1-specific antagonists or siRNA knockdown of ENT1 significantly reduced GLT1 expression and glutamate uptake in cultured astrocytes [98]. Overexpression of ENT1 upregulated GLT1 expression and glutamate uptake [98]. Similarly, A_1 receptor antagonists or knockdown in astrocytes decreased GLT1 expression and function whereas A_1 receptor overexpression upregulated GLT1 [99]. The activation of adenylate cyclase also decreased GLT1 mRNA [99]. Therefore, the $G\alpha_i$-coupled A_1 receptor is another likely candidate for regulating GLT1 expression in astrocytes. A_1 and A_{2A} receptors as well as ENT1 can be targeted to modulate astrocytic glutamate uptake and release.

Epigenetic Roles

Epileptogenic areas of the brain are often characterized by DNA hypermethylation [9,100–104]. Methionine sulfoximine (MSO)-induced seizures were accompanied by several methylation changes, including increased methylation flux and a change in the ratio of SAM to SAH [105–107]. These effects could be diminished with the administration of adenosine and homocysteine thiolactone (Fig. 10.2, right panel) [9,106]. Therefore, DNA methylation is thought to play a pivotal role in the pathogenesis of epilepsy [108].

In the kainic acid model, recurrent spontaneous seizures, increased hippocampal DNA methylation, and disrupted adenosine homeostasis were observed [103]. The local delivery of adenosine using silk-based biodegradable brain implants reversed the DNA hypermethylation (reduced by ~51% in the hippocampus), inhibited mossy fiber sprouting in the hippocampus, and prevented the progression of epilepsy for at least 3 months [103]. DNA methyltransferase inhibitors also prevented epileptogenesis in the PTZ kindling model of epilepsy [103]. By restoring DNA methylation levels to normal, local AAT may serve as a promising antiepileptogenic treatment (Fig. 10.2). Although this therapy is very promising, the risk of mutagenesis and cell death should be carefully monitored in any treatment paradigm that involves epigenetic modulation.

Low Glycemic Diets

Low glycemic diets, such as the ketogenic diet or modified Atkins diet, have been used for nearly a century [109] and have been reported to have both antiepileptogenic and neuroprotective effects [110,111]. It has, however, been difficult to have a controlled, blinded trial of the ketogenic diet [109,112]. One successful trial was completed in 2008 and included 145 children with daily seizures who were medically refractory [113]. Participants were enrolled in the trial from December 2001 to July 2006 and had reduced seizures in response to the diet [113]. In December 2006, the Charlie Foundation commissioned an expert panel of pediatric epileptologists and dietitians from around the world to create a consensus regarding the clinical management of the ketogenic diet. For their detailed conclusions, see Kossoff et al. [112].

Although the anticonvulsant and antiepileptogenic effects are not entirely understood, it is known that the ketogenic diet forces the brain to use ketones instead of glucose as the primary energy source [114]. Other contributing properties may include elevated fatty acid levels, increased brain-derived neurotrophic factor (BDNF) expression, and attenuated neuroinflammation [115]. This diet may also have direct neuronal effects involving ATP-sensitive potassium channel modulation and enhanced purinergic and GABAergic transmission [115].

The ketogenic diet increases levels of ATP and adenosine in the brain [116–118]. As a result, adenosine, acting through the A_1 receptor, may exert powerful inhibitory effects on excitatory synaptic transmission [18,117,119]. Transgenic mice with an overexpression of ADK and mice lacking one (heterozygote) or both (knockout) A_1 receptor alleles on a control diet experienced electrographic seizures. After 3 weeks on the ketogenic diet, seizures were nearly abolished in transgenic ADK mice compared to wild-type mice [18]. Seizures were reduced in the A_1 receptor heterozygote mice and unaffected in knockout mice on the ketogenic diet. Injections of glucose or an A_1 receptor antagonist restored seizures [18]. These results provide further evidence that the ketogenic diet most likely exerts its antiseizure effects through adenosine activation of the A_1 receptor. Although the ketogenic diet offers promising antiseizure effects, the consequences of long-term dietary restrictions should be considered.

P2 RECEPTORS

P2X receptors are expressed on both neurons and glial cells [120] and are involved in a variety of functions including inflammation [121], transient increases in cytosolic calcium [122], and communication between glial and neuronal cells [123]. $P2X_1$–$P2X_4$ receptors are expressed in glutamatergic neurons [124] and $P2X_2$ receptors enhance excitatory synaptic transmission onto interneurons [125] and influence GABA release [126]. All P2X receptors have been reported to be expressed in astrocytes (Table 10.2) [4]. Their expression distribution in nerve terminals can be found in Fig. 10.10 [126] and their presence on GABAergic boutons is shown in Fig. 10.11 [126].

$P2X_7$ receptor activation has been linked to modulation of neuronal excitability, activation of microglia, and neuroinflammation [127]. Both $P2X_7$ receptor protein and mRNA expression has been reported on astrocytes [4], but other studies found a lack of $P2X_7$ receptor expression on astrocytes [128–130]. The $P2X_7$ receptor is abundant on glutamatergic and nonglutamatergic nerve terminals [131,132], including presynaptic terminals of mossy fiber synapses [133]. Activation of $P2X_7$ receptors led to long-lasting synaptic inhibition at mossy fiber-CA3 synapses [133]. $P2X_7$ receptors also play a role in ATP-elicited glutamate and GABA release [132,134–136].

ATP has been reported to evoke glutamate release in purified rat hippocampal nerve terminals, an effect that was mediated by several different P2X and P2Y receptors [124]. Activated astrocytes and microglia released ATP that stimulated astrocytic $P2Y_1$ receptors and increased excitatory postsynaptic current frequencies in an mGluR-dependent manner [137]. In general, however, P2Y receptors are considered inhibitory. Various P2Y receptors,

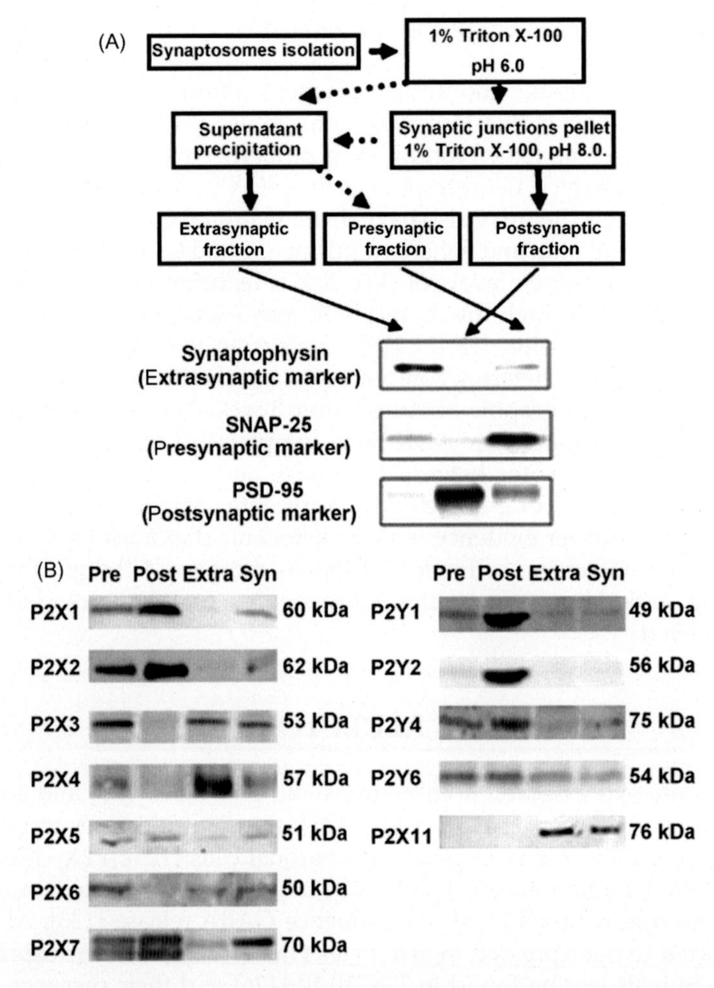

FIGURE 10.10 **Subsynaptic distribution of P2X and P2Y receptors in the cerebellum.** (A) Selective antibodies for each P2X and P2Y receptor were tested by Western blot analysis in a fraction enriched in the presynaptic active zone (pre), in the postsynaptic density (post), in nerve terminals outside the active zone (extra), and in the initial synaptosomal fraction (syn) from which fractionation began. These fractions were over 90% pure, as illustrated by the ability to recover the immunoreactivity for SNAP25 in the presynaptic active zone fraction, PSD95 in the postsynaptic density fraction and synaptophysin (a protein located in synaptic vesicles) in the extrasynaptic fraction. (B) Western blots of these fractions evaluating the subsynaptic distribution of the immunoreactivity of the antibodies selective for each P2X subunit and each P2Y receptor tested (20–80 μg of protein of each fraction were applied to SDS-PAGE gels). Each blot is representative of at least three blots from different groups of animals with similar results. For each fractionation procedure, Western blot analysis for the markers of each fraction was performed as illustrated in (A), in order to assess the efficiency of each fractionation. *Source: Reproduced with permission from Donato R, Rodrigues RJ, Takahashi M, Tsai MC, Soto D, Miyagi K, et al. GABA release by basket cells onto Purkinje cells, in rat cerebellar slices, is directly controlled by presynaptic purinergic receptors, modulating* Ca^{2+} *influx. Cell Calcium 2008;44(6):521–32.*

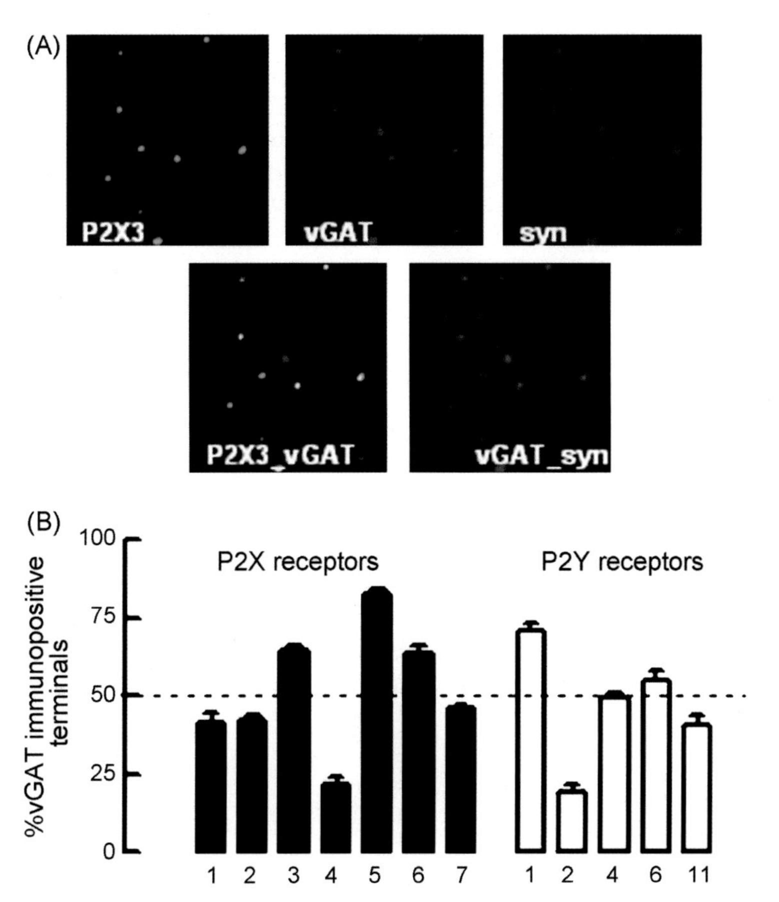

FIGURE 10.11 **Identification of P2X and P2Y receptors present in GABAergic boutons by triple immunocytochemical analysis of rat cerebellar single nerve terminals.** (A) An example of the immunocytochemical identification of $P2X_3$ receptors subunit (left panel, first row), GABAergic nerve terminals identified as immunopositive for vesicular GABA transporters (central panel, first row), and the total synaptosomal population identified as immunopositive for the synaptic marker synaptophysin (right panel, first row). The proportion of GABAergic terminals was measured after merging vGAT and synaptophysin (syn) images (right panel, second row) and comprised 54.2 ± 3.1% ($n = 4$). Merging images of vGAT and P2 antibody labeling (left panel, second row) allowed the percentage of rat cerebellar GABAergic nerve terminals endowed with each P2X subunit and P2Y receptor to be identified. (B) Summary of all the proportion of GABAergic terminals found to be positive for each of the subunits tested. The data are mean ± S.E.M. of 3–4 experiments and in each experiment, using different synaptosomal preparations from different animals, four different fields acquired from two different coverslips were analyzed. *Source: Reproduced with permission from Donato R, Rodrigues RJ, Takahashi M, Tsai MC, Soto D, Miyagi K, et al. GABA release by basket cells onto Purkinje cells, in rat cerebellar slices, is directly controlled by presynaptic purinergic receptors, modulating* Ca^{2+} *influx. Cell Calcium 2008;44(6):521–32.*

including $P2Y_1$ and $P2Y_4$, are found on presynaptic glutamatergic terminals [124] and presynaptic GABAergic synaptosomes [126]. P2Y receptor distribution on nerve terminals and GABAergic boutons is shown in Figs. 10.10 and 10.11, respectively [126]. Nucleotides, including ATP, inhibited transmitter release (glutamate, noradrenaline) through presynaptic P2Y receptors [13,138–141]. P2Y receptor activation inhibited calcium entry and depolarization-induced membrane capacitance [142,143] and excited interneurons, leading to increased GABAergic synaptic transmission [67,144]. Stimulation of $P2Y_1$ receptors inhibited GABA transport [145] and activation of $P2Y_4$ receptors inhibited GABA release [126].

P2 RECEPTORS IN EPILEPSY

ATP can be released from neurons and glia, particularly after excessive neuronal firing or chronic inflammation [10,146]. Extracellular ATP levels are kept low under normal physiological conditions, but can reach millimolar levels under pathological conditions. In fact, seizures can be induced by microinjection of ATP analogs into the prepiriform cortex [1]. Elevated extracellular ATP levels, acting through various purinergic receptors, increase susceptibility to neuronal damage, astrogliosis, microglia recruitment and activation, and the neuroinflammatory response [120,147–149]. These receptors can have conflicting roles in the brain. For example, $P2X_7$ and $P2X_4$ receptors are mainly proconvulsive [120,150] whereas $P2Y_{12}$ receptors are generally considered anticonvulsive [151]. A summary of P2 receptor changes in epilepsy can be found in Table 10.3 [150] and a schematic of their potential roles in SE and epilepsy is shown in Fig. 10.12 [150].

Human Tissue Studies

$P2X_7$ receptors were found in neurons, astrocytes, and microglia in human patients [159]. Surgically resected tissue from patients with TLE [129] or FCD [159] had elevated $P2X_7$ receptor expression. $P2Y_1$, $P2Y_2$, and $P2Y_4$ receptor expression levels were also elevated in resected epileptic tissue from patients with FCD, tuberous sclerosis complex (TSC), or low-grade astrocytoma compared to autopsy control brains of either patients without epilepsy or patients with gliosis but no seizures [158]. A missense single nucleotide polymorphism (SNP) was commonly found in the $P2X_7$ receptor gene in Caucasian children with febrile seizures compared to febrile cases without seizures [160]. In addition, SNPs were found in toll-like receptor 4 (TLR4), interleukin-6 (IL-6) receptor, and prostaglandin E receptor 3 subtype EP3 (PTGER3) genes, suggesting a link between $P2X_7$ receptors and inflammation [160].

Animal Models

The application of ATP depressed epileptiform activity in both hippocampal slices perfused in magnesium-free medium containing 4-AP [33] and horizontal entorhinal cortex/hippocampus slices from pilocarpine-treated rats perfused with both bicuculline and elevated potassium [161]. Furthermore, the effect of ATP was inhibited by the A_1 receptor antagonists whereas P2 receptor antagonists failed to inhibit ATP-induced depression

TABLE 10.3 P2 Receptors in Status Epilepticus and Epilepsy

Receptor	Cell Type Expression	Agonist	Receptor Expression in		Effect of Agonists/ Antagonists or Knockout on *status epilepticus*
			Status Epilepticus	**Epilepsy**	
$P2X_1$	Neurons (in vitro, RT-PCR) [174] (pre- and postsynaptic: ex vivo, W) [124], astrocytes (ex vivo, VC, qRT-PCR) [191] (in vitro, RT-PCR) [185], oligodendrocytes (in vitro, W. CI, PC) [173], microglia (in vitro, IH, RT-PCR) [198] (in vitro, CI, RT-PCR, W) [176]	ATP [4]	*Hippocampus:* No change (W, i.a. KA, mouse) [128] Upregulated (qPCR, i.p. KA, mouse) [152] *Cell type expression changes:* Not analyzed	Not studied	Not studied
$P2X_2$	Neurons (in vitro, RT-PCR) [174] (pre- and postsynaptic: ex vivo, W) [124], astrocytes (ex vivo, VC, qRT-PCR) [191] (in vitro, RT-PCR) [185], oligodendrocytes (in vitro, W, CI, PC) [173], microglia (in vitro, CI, RT-PCR, W) [176]	ATP [4]	*Hippocampus:* Decreased (W, i.a. KA, mouse) [128] No change (W, Pilo, rat) [165] *Cell-type expression changes:* Not analyzed	*Hippocampus:* Decreased (ISH) (seizure-sensitive strain, gerbil) [153] No change (W) (Pilo, rat) [165] *Cell type expression changes:* Not analyzed	Not studied
$P2X_3$	Neurons (in vitro, RT-PCR) [174] (pre- and postsynaptic: ex vivo, W) [124], astrocytes (in vitro, RT-PCR) [185], oligodendrocytes (in vitro, W, Cl, PC) [173], microglia (in vitro, CI, RT-PCR, W) [176]	ATP [4]	*Hippocampus:* No change (W, i.a. KA, mouse) [128]	Not studied	Not studied

(*Continued*)

TABLE 10.3 (Continued)

Receptor	Cell Type Expression	Agonist	Receptor Expression in: Status Epilepticus	Receptor Expression in: Epilepsy	Effect of Agonists/Antagonists or Knockout on *status epilepticus*
$P2X_4$	Neurons (in vitro, RT-PCR) [174] (pre- and postsynaptic: ex vivo, W) [124], astrocytes (in vitro, RT-PCR) [185], oligodendrocytes (in vitro, W, CI, PC) [173], microglia (in vitro, IH, RT-PCR) [198]; [176], endothelial cells (in vitro, CI, RT-PCR) [177]	ATP [4]	*Hippocampus:*	*Hippocampus:*	*$P2X_4R$ KO mice:*
			Increased (W/IH, i.p. KA, mouse) [154]	Decreased (ISH, W) (seizure-sensitive strain, gerbil) [153] (Pilo, rat) [165]	Decreased seizure-induced cell death (i.p. KA) [154]
			No change (W) (i.a. KA, mouse) [128] (Pilo, rat) [165]		No effect on seizures (i.p. KA) [154]
			Upregulated (qPCR, i.p. KA, mouse) [152]	*Cell type expression changes:*	Decreased inflammation and microglia density (i.p. KA) [154]
			Cell type expression changes:	Decreased in neurons (IH) [165]	No change in IL-1β levels (i.p. KA) [154]
			Increased in microglia, not astrocytes (GFP-reporter mice) [154]		
$P2X_5$	Neurons (in vitro, RT-PCR) [174] (pre- and postsynaptic: ex vivo, W) [124], astrocytes (ex vivo, VC, qRT-PCR) [191], endothelial cells (in vitro, CI, RT-PCR) [177]	ATP [4]	*Hippocampus:*	Not studied	Not studied
			No change (W, i.a. KA, mouse) [128]		
$P2X_6$	Neurons (in stiu, ISH) [181] (pre- and postsynaptic: ex vivo, W) [124], astrocytes (in vitro, RT-PCR) [185], microglia (in vitro, CI, RT-PCR, W) [176]	ATP [181]	Not studied	Not studied	Not studied

$P2X_7$	Neurons (in vitro, RT-PCR) [174] (in vivo) [128] (presynaptic: in situ, RT-PCR, ISH, W, IH, PC) [184] (pre- and postsynaptic: ex vivo, W) [124], astrocytes (in vitro, RT-PCR) [185] (in situ, IH) [190] (in vitro, CI, PC) [178], oligodendrocytes (in vitro, W, CI, PC) [173], microglia (in vitro, IH, RT-PCR) [198] (in vitro, CI, RT-PCR, W) [176], endothelial cells (in vitro, CI, RT-PCR) [177]	ATP [4]	*Hippocampus:*	*Hippocampus:*	*BzATP (Agonist):*
			Increased (W) (i.a. KA, GFP-reporter mouse) [128] (Pilo, rat) [165]	Increased (W) (Pilo, rat) [165]	Increased seizures (i.a. KA, mouse) [128]
			Upregulated (qPCR, i.p. KA, mouse) [152]	*Cortex:*	No effect on seizures (Pilo, mouse) [155]
			Increased (W/IH, i.a. KA, rat) [156]	Increased (W. human) and (W/IH, i.a. KA mouse) [129]	Increased microglia activation (Pilo, rat) [180]
			Increased (ICC, i.p. KA, rat) [153]	*Cell type expression changes:*	Increase in astrocyte loss (Pilo, rat) [155]
			Cortex:	Increased in neurons (IH) [165]	Increased TNFα immunoreactivity (Pilo, rat) [155]
			Increased (W, i.a. KA, GFP-reporter mouse) [129]	Increased in neurons and microglia, not astrocytes (GFP-reporter mice) [129]	Decreased seizure-induced cell death (Pilo, rat) [155]
			Cell type expression changes:		*$P2X_7R$ KO mice:*
			Increased in neurons, not astrocytes/microglia (GFP-reporter mice) [128]		Decreased seizures (i.a. KA) [128]
			Increased in neurons, astrocytes, and microglia (IH) [165]		Increased seizures (Pilo) [155]
			Increased in neurons, not astrocytes/microglia (GFP-reporter mice) [129]		No effect on seizures (i.p. KA and i.p. Pic) [155]
					A-43, A-74, BBG, OxATP, IgG-$P2X_7$ (Antagonists):

(*Continued*)

TABLE 10.3 (Continued)

Receptor Cell Type Expression	Agonist	Receptor Expression in		Effect of Agonists/ Antagonists or Knockout on *status epilepticus*
		Status Epilepticus	Epilepsy	
		Increased in neurons (IH) [156], Increased in microglia, not astrocytes (ICC) [153]		Decreased responses in activated microglia (i.p. KA, mouse) [152]
				Decreased seizures (i.a. KA: mouse [128], rat pups [156], mouse [129])
				Decreased kindling score (PTZ, rat) [157]
				Improved motor and cognitive functions (PTZ, rat) [157]
				Increased seizures (Pilo, mouse) [155]
				Decreased seizure-induced cell death (i.a. KA: mouse [128], rat pups [156])
				Increased seizure-induced cell death (Pilo, rat) [155]
				Decreased microglia activation (i.a. KA, mouse [128]; Pilo, rat [180])
				Decreased IL-1β levels (i.a. KA, mouse) [128]
				Decreased astrocyte loss (Pilo, rat) [155]

$P2Y_1$	Neurons (in vitro, RT-PCR) [174] (pre- & postsynaptic: ex vivo, W) [124], astrocytes (in vitro, RT-PCR) [185] (in vitro, ICC, CMF) [186], oligodendrocytes (in vitro, W, CI, PC) [173], microglia (in vitro, CI, RT-PCR, W) [176] (in situ, IH) [187], endothelial cells (in vitro, RI.BA, RT-PCR) [199]	ADP > ATP [197]	Not studied	*Not specified:* Increased (W, human) [158] *Cell type expression changes:* Increased in astrocytes (IH) [158]	Not studied
$P2Y_2$	Neurons (in vitro, RT-PCR) [174] (pre- and postsynaptic: ex vivo, W) [124], astrocytes (in vitro, RT-PCR) [185] (in vitro, ICC, CMF) [186], oligodendrocytes (in vitro, W, CI, PC) [173], microglia (in vitro, CI, RT-PCR, W) [176] (in situ, IH) [187].	UTP/ATP [194]	Not studied	*Not specified:* Increased (W, human) [158] *Cell type expression changes:* Increased in astrocytes (IH) [158]	Not studied
$P2Y_4$	Neurons (in vitro, RT-PCR) [174] (pre- and postsynaptic: ex vivo, W) [124], astrocytes (in vitro, RT-PCR) [185] (in vitro, ICC, CMF) [186], oligodendrocytes (in vitro, W, CI, PC) [173], microglia (in vitro, CI, RT-PCR, W) [176] (in situ, IH) [187]	UTP > ATP [194]	Not studied	*Not specified:* Increased (W, human) [158] *Cell type expression changes:* Increased in astrocytes (IH) [158]	Not studied
$P2Y_6$	Neurons (in vitro, RT-PCR) [174] (postsynaptic: ex vivo, W) [124], astrocytes (in vitro, RT-PCR) [185] (in vitro, ICC, CMF) [186], microglia (in vitro, CI, RT-PCR, W) [176] (in situ, IH) [187], endothelial cells (in vitro, CI, RT-PCR) [177]	UDP > UTP > > ADP [194]	*Hippocampus:* Upregulated (qPCR, i.p. KA, mouse) [152] *Cell type expression changes:* Not analyzed	Not studied	*UDP (agonist):* Increased microglia currents (i.p. KA, mouse) [152]

(*Continued*)

TABLE 10.3 (Continued)

Receptor	Cell Type Expression	Agonist	Receptor Expression in		Effect of Agonists/ Antagonists or Knockout on *status epilepticus*
			Status Epilepticus	**Epilepsy**	
$P2Y_{11}$	Neurons (in vitro and ex vivo, qPCR) [193], endothelial cells (in vitro, CI, RT-PCR) [177]	ATP [182]	Not studied	Not studied	Not studied
$P2Y_{12}$	Neurons (in vitro, CI, RT-PCR) [189] (postsynaptic: ex vivo, W) [124], astrocytes (in vitro, RT-PCR, CI) [188], oligodendrocytes (in situ, IH, W) [175], microglia (in vitro, CI, RT-PCR, W) [176] (in situ, IH) [187], endothelial cells (in vitro, RT-PCR) [195]	ADP > ATP [196]	*Hippocampus:* Upregulated (qPCR, i.p. KA, mouse) [152] *Cell type expression changes:* Increased in microglia (GFP-reporter mice) [152]	Not studied	*2-MeSADP (agonist):* Increased microglia currents i.p. KA, mouse [152] Higher motility of microglia processes i.p. KA, mouse [152] *P2Y12R KO mice:* Increased seizure phenotype (i.c.v KA) [151] Reduced hippocampal microglial processes (i.p/i.c.v. KA) (Eyo et al. [151])
$P2Y_{13}$	Neurons (ex vivo, RT-PCR, EFS) [138], astrocytes (in vitro, ICC, CMF) [186], microglia (in vitro, CI, RT-PCR, W) [176].	ADP > > ATP [183]	*Hippocampus:* Decreased shortly after SE and Upregulated at later time-points (qPCR, i.p. KA, mouse) [152] *Cell type expression changes:* Not analyzed	Not studied	Not studied
$P2Y_{14}$	Neurons (in situ, qPCR, IH, CI) [192], astrocytes (in vitro, ICC, CMF) [186] (in vitro, RT-PCR, CI) [188], microglia (in vitro, CI, RT-PCR, W) [176]	UDP-sugars [179]	Not studied	Not studied	Not studied

A-43, A-438079; *A-74*, A-740003; *BzATP*, 2′(3′)-O-(4-benzoylbenzoyl)adenosine 5′-triphosphate; *BBG*, brilliant blue G; *CI*, calcium imaging; *CMF*, calcium microfluorimetry; *EFS*, electrical field stimulation; *GFP*, P2rx7 GFP-reporter mouse; *i.a. KA*, intraamygdala kainic acid; *ICC*, immunocytochemistry; *i.c.v. KA*, intracerebroventricular kainic acid; *IgG-P2X₇*, anti-$P2X_7R$ antibody; *IH*, immunohistochemistry; *IL-1β*, Interleukin-1β; *i.p. KA*, intraperitoneal kainic acid; *i.p. Pic*, intraperitoneal picrotoxin; *ISH*, in situ hybridization; *OxATP*, oxidized ATP; *PC*, patch clamp; *Pilo*, pilocarpine; *PTZ*, pentylenetetrazol; *qPCR*, quantitative polymerase chain reaction; *qRT-PCR*, real-time reverse transcription polymerase chain reaction; *RLBA*, radioligand binding assay; *RT-PCR*, reverse transcription polymerase chain reaction; *TNFα*, Tumor necrosis factor alpha; *VC*, voltage clamp; *W*, Western blot. *Reproduced with permission from Engel T, Alves M, Sheedy C, Henshall DC. ATPergic signalling during seizures and epilepsy. Neuropharmacology 2016;104:140–53.*

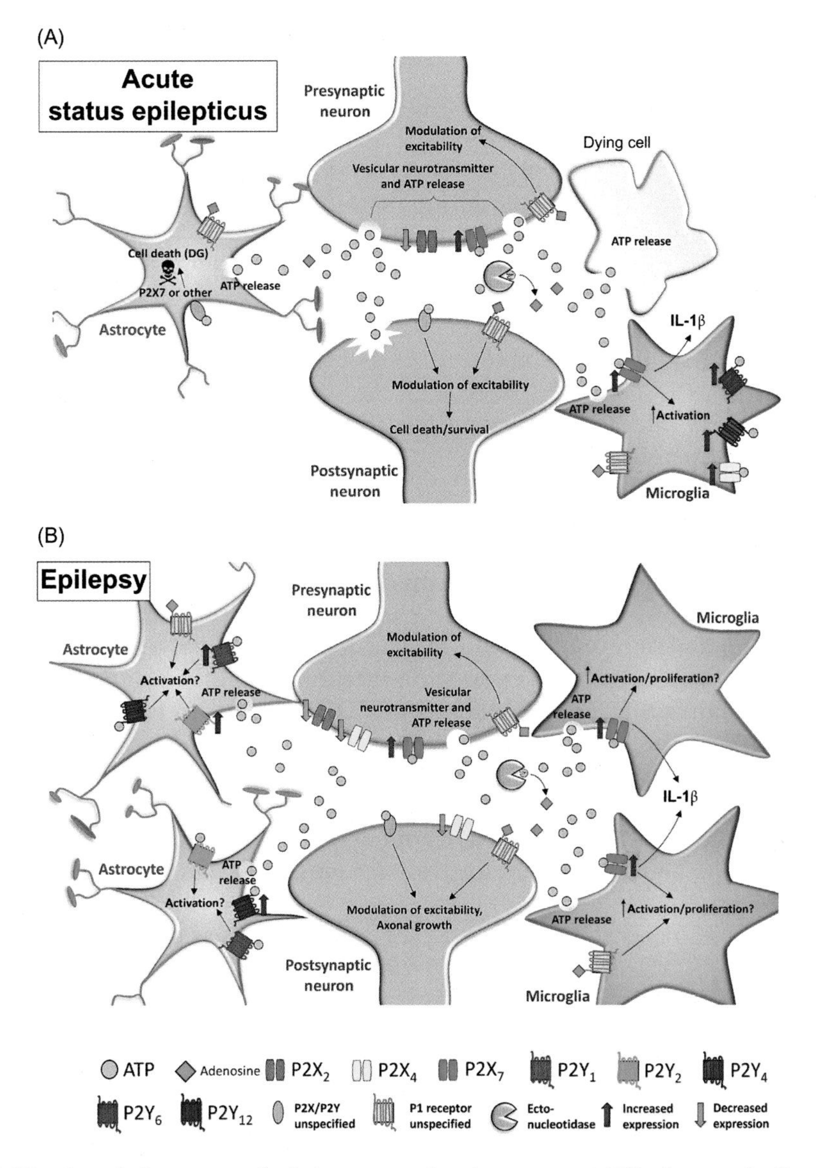

FIGURE 10.12 **Speculative cartoon depicting proposed main sources of ATP release and cell types reported to express P2X and P2Y receptors during acute prolonged seizures (status epilepticus) and during epilepsy.** (A) During acute, prolonged seizures, the main contributors to extracellular ATP release are dying neurons and increased neuronal activity. Extracellular ATP then activates P2 receptors particularly at the presynapse modulating neurotransmitter release. ATP also binds to P2 receptors on microglial cells which leads to their activation and the release of IL-1β. The activation of P2 receptors during seizures contributes also to seizure-induced cell death and increased cellular excitability. After its release, ATP is rapidly broken down into different breakdown products including adenosine which in turn activates P1 receptors further modulating neuronal excitability and glial activation. (B) The main sources of ATP release during epilepsy are activated astrocytes and microglia. Released ATP binds pre- and postsynaptically to P2 receptors modulating neurotransmitter release and cellular excitability. P1 (not shown) and P2 receptor activation on glial cells leads to their activation and proliferation and increased release of inflammatory cytokines such as IL-1β. Again, ATP is rapidly broken down in the extracellular space leading to the activation of P1 receptors. *Source: Reproduced with permission from Engel T, Alves M, Sheedy C, Henshall DC. ATPergic signalling during seizures and epilepsy. Neuropharmacology 2016;104:140–53.*

[33,161]. The specific $P2X_7$ receptor antagonist A-740003, however, did reduce the amplitude of slow field potentials [161]. The observed anticonvulsant effects of ATP, therefore, could be due to its degradation into adenosine.

Inbred DBA/2 (D2) mice are inherently susceptible to audiogenic seizures and exhibit increased extracellular ATP compared to C57BL/6 mice [162]. Electrical stimulation of the wild-type rat cortex also resulted in increased extracellular ATP levels [163]. In rat hippocampal slices bathed in magnesium-free artificial cerebrospinal fluid (aCSF), nonspecific P2 antagonists, a $P2X_{1,2/3,3}$ antagonist, and an ecto-ATPase inhibitor each caused significant decreases in electrically evoked electrographic seizure-like events (eSLEs) [164]. $P2Y_1$ antagonists, on the other hand, had no significant effect on eSLEs [164].

Changes in $P2X_2$, $P2X_4$, and $P2X_7$ receptor expression were examined in the acute, latent, and chronic phase of pilocarpine-injected rats (Fig. 10.13) [165]. Compared to control animals, $P2X_2$ expression was unaltered whereas $P2X_4$ receptor expression was downregulated in the chronic phase of epilepsy [165]. $P2X_7$ receptor expression was upregulated in both the acute and chronic phases. Immunostaining revealed that $P2X_7$ receptor expression was increased in both glial cells and glutamatergic nerve terminals [165].

The seizure-sensitive gerbil, a genetic model of epilepsy, exhibited decreased hippocampal expression of $P2X_2$ and $P2X_4$ receptors compared to the seizure-resistant gerbil [153]. Downregulation in hippocampal $P2X_2$ receptor protein expression but no change in $P2X_1$, $P2X_3$, $P2X_4$, or $P2X_5$ receptor expression was found within 24 hours after intraamygdala kainic acid-induced SE [128]. In contrast, $P2X_4$ receptor protein levels were upregulated in activated microglia 24–48 hours after intraperitoneal kainic acid-induced seizures [154]. Upregulation of hippocampal $P2X_1$, $P2X_4$, $P2X_7$, $P2Y_6$, $P2Y_{12}$, and $P2Y_{13}$ mRNA were observed within 48 hours after intraperitoneal kainic acid-induced SE, although a transient downregulation of $P2Y_{13}$ mRNA at 3 hours post-SE was reported [152]. Furthermore, larger $P2Y_6$, $P2Y_{12}$, and P2X receptor-mediated responses in reactive microglia were observed [152].

Hippocampal and neocortical $P2X_7$ receptor expression was upregulated 24 hours after intraamygdala kainic acid-induced SE [128–130,156,166]. Within the first 8 hours after SE, neurons were the major cell type expressing $P2X_7$ receptors but in the epileptic mouse (14 day post-SE), increased $P2X_7$ receptor expression was noted in both neurons and reactive microglia [129]. Treatment with $P2X_7$ receptor antagonists reduced kainic acid-induced seizure severity, seizure duration, and seizure-induced neuronal cell death [128,129,146,156] whereas ATP injection exacerbated seizures [128]. Similarly, the $P2X_7$ receptor antagonist brilliant blue G (BBG) decreased mean kindling score as well as improved motor performance, spatial learning, and memory in the rat PTZ kindling model of epilepsy [157]. In rat hippocampal slices bathed in magnesium-free aCSF, however, BBG had no significant effect on eSLEs [164]. The discrepancy involving the role of the $P2X_7$ receptor in seizure development may be due to the difference in seizure models (electrical stimulation vs drug-induced seizures) or the difference in approach (in vivo vs in situ).

$P2X_7$ receptors play a role in glial modulation during epilepsy. BzATP antagonism of the $P2X_7$ receptor exacerbated pilocarpine-induced astroglial cell loss in the frontoparietal cortex and the molecular layer of the dentate gyrus (DG) but alleviated SE-induced astroglial swelling in the CA1 region of the hippocampus [167,168]. The $P2X_7$ receptor antagonists

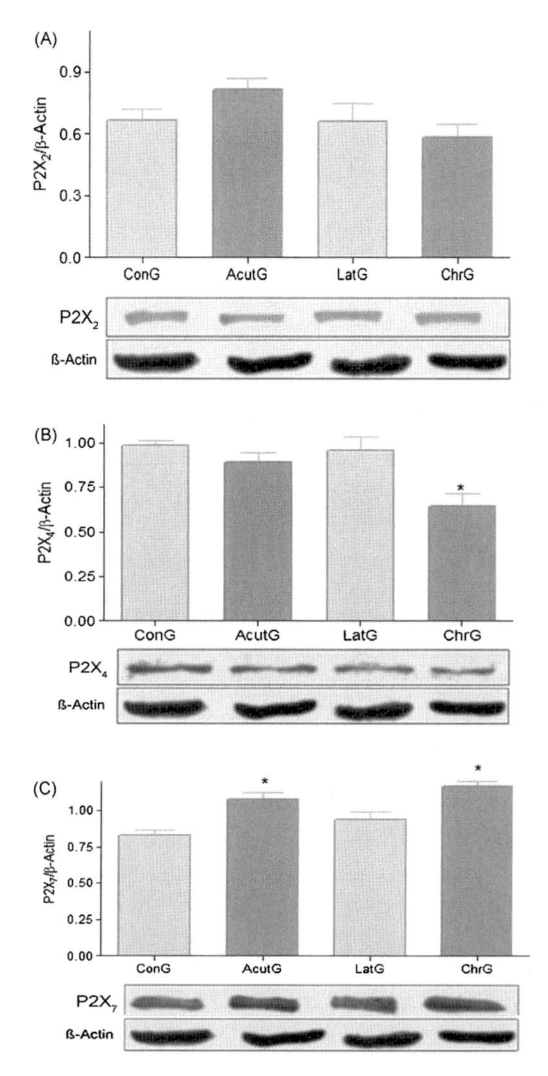

FIGURE 10.13 **Western blot study of $P2X_2$, $P2X_4$, and $P2X_7$ receptor expression in the hippocampus of control (ConG) and pilocarpine-injected rats in the acute (AcuG), latent (LatG), and chronic (ChrG) phase of epilepsy.** Cell extracts containing 40 μg protein from hippocampi of control and temporal lobe epilepsy (TLE)-suffering rats were separated by SDS-PAGE and submitted to Western blot analysis for detection of P2X receptor expression. β-actin protein expression was determined as an internal control for normalization and relative quantification of purinergic receptor gene expression was calculated. (A) Statistical analysis revealed no significant difference in $P2X_2$ receptor expression in control and TLE animals. (B) $P2X_4$ receptor expression in hippocampi of ChrG was decreased when compared to the ConG. (C) Expression of $P2X_7$ receptors was increased in hippocampi of AcuG and ChrG when compared to ConG (*$P < 0.05$). *Source: Reproduced with permission from Doná F, Ulrich H, Persike DS, Conceição IM, Blini JP, Cavalheiro EA, et al. Alteration of purinergic $P2X_4$ and $P2X_7$ receptor expression in rats with temporal lobe epilepsy induced by pilocarpine. Epilepsy Res 2009;83(2–3):157–67.*

oxidized ATP (OxATP) and BBG prevented astroglial apoptosis in the molecular layer of the DG as well as attenuated microglial activation in both the CA1 hippocampal region and frontoparietal cortex, but had no effect in the piriform cortex [167–169]. Furthermore, intracerebroventricular infusion of BzATP accelerated neutrophil and monocyte infiltration into the frontoparietal cortex whereas OxATP decreased the number of monocytes and neutrophils [170]. Infiltration of these leukocytes into the piriform cortex was unaffected by either compound [170]. Taken together, these studies suggest that $P2X_7$ receptors differentially modulate SE-induced astroglial cell death, microgliosis, and inflammation in a brain region-dependent manner.

Transgenic Mice Studies

Pannexin-1 (Panx1), a hemichannel protein, can be opened by $P2X_7$ receptor activation [155]. Activation of Panx1 in juvenile mice contributed to neuronal hyperexcitability in seizures [171], potentially through its involvement in ATP and glutamate release from both neurons and astrocytes [155]. The use of pharmacological tools and Panx1 knockout mice determined that antagonism of Panx1 ameliorated kainic acid-induced SE and attenuates ATP release [171]. Engel et al. [128] demonstrated that treatment of wild-type mice with $P2X_7$ receptor antagonists during SE reduced seizure duration and hippocampal damage in the intraamygdala kainic acid model of epilepsy. Kim and Kang [155], however, found that administration of $P2X_7$ receptor antagonists and gene silencing of either the $P2X_7$ receptor or Panx1 in wild-type mice led to increased pilocarpine-induced seizure susceptibility. In addition, $P2X_7$ receptor knockout mice exhibited greater sensitivity to pilocarpine-induced seizures [155]. These discrepancies may possibly be attributed to differences in seizure severity between the two epilepsy models [150].

The $P2X_7$ receptor also may be involved in inflammatory pathways. $P2X_7$ receptor inhibition resulted in diminished microglial proliferation and IL-1β release [128,146]. LPS and subsequent ATP injections in wild-type mice resulted in increases in IL-1 production and successive IL-6 production; this effect was lacking in $P2X_7$ receptor knockout mice [172]. Therefore, cytokine signaling cascades may be impaired in $P2X_7$ receptor knockout mice. A schematic of the potential sites of action of ATP released during SE is shown in Fig. 10.14 [146].

Reduced neuronal death and microglial activation were observed in $P2X_4$ knockout mice [154]. Susceptibility to kainic acid-induced seizures, however, was unaltered in $P2X_4$ knockout mice [154]. P2Y receptor knockout mice, on the other hand, exhibited increased kainic acid-induced seizure susceptibility but reduced microgliosis [151]. Future studies using transgenic mice should focus on the relative contributions and roles of other P2X and P2Y receptors in the development of epilepsy.

CONCLUSION

Adenosine is a ubiquitous anticonvulsant purine nucleoside found throughout the brain. Purinergic signaling, carried out by P1 and P2 receptors, has a significant impact on neuronal excitability. Levels of adenosine can be regulated by a variety of metabolic enzymes.

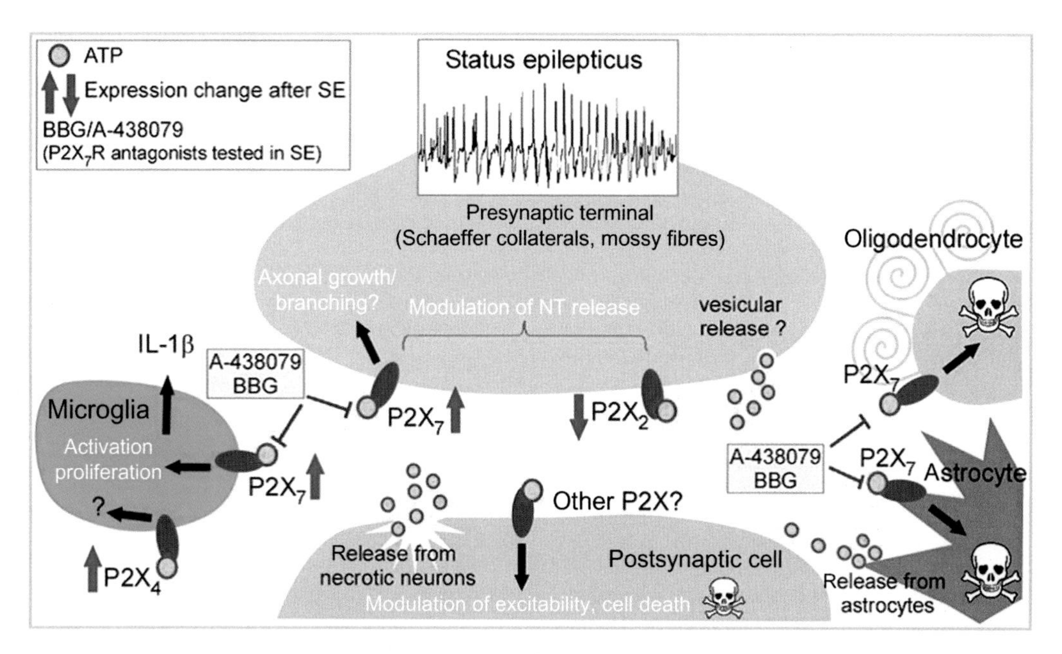

FIGURE 10.14 **Potential sites of action of ATP released during status epilepticus (SE), expressional responses of individual P2X receptors, and consequences of receptor activation.** Cartoon depicts the various different cell types reported to express P2X receptors and their presumed cellular locations. ATP is released during sustained neuronal activity and from damaged neurons to act presynaptically and postsynaptically on neurons, particularly targeting presynaptic receptors to modulate neurotransmitter release. ATP may also act on receptors of microglia to promote activation and release of IL-1β and act on astrocytes and oligodendrocytes to trigger cell death. Drugs such the $P2X_7$ receptor antagonists A-438079 and brilliant blue G (BBG) have been reported to reduce seizures and gliosis after SE. *SE*, status epilepticus; *IL-1β*, interleukin-1β. *Source: Reproduced with permission from Henshall DC, Engel T. P2X purinoceptors as a link between hyperexcitability and neuroinflammation in status epilepticus. Epilepsy Behav 2015;49:8–12.*

In epileptic tissue, however, the levels of both receptors and metabolic enzymes are altered and consequently modify available adenosine levels. Therefore, both purinergic receptors and adenosine metabolic enzymes serve as promising therapeutic targets for the treatment of epilepsy. Drug intervention, however, is best applied locally as to not disturb adenosine metabolism throughout the rest of the body. In particular, manipulation of astrocyte-specific ADK expression is a promising option because (1) astrocytes become reactive and more abundant in the epileptic brain; (2) ADK expression is elevated in human tissue and animal model studies of epilepsy; and (3) astrocyte-specific enzyme (ADK) inhibition potentially avoids side effects on neurons and could limit cognitive impairment, a common adverse effect seen with many antiepileptic drugs today. Future studies should focus on temporally and anatomically selective pharmacological manipulation of adenosine metabolism, ideally at the seizure focus.

References

[1] Burnstock G. Physiology and pathophysiology of purinergic neurotransmission. Physiol Rev 2007;87(2):659–797.
[2] Burnstock G. Introduction to purinergic signalling in the brain. Adv Exp Med Biol 2013;986:1–12.
[3] Burnstock G. An introduction to the roles of purinergic signalling in neurodegeneration, neuroprotection and neuroregeneration. Neuropharmacology 2015;10(12):1919.
[4] Fields RD, Burnstock G. Purinergic signalling in neuron-glia interactions. Nat Rev Neurosci 2006;7(6):423–36.
[5] Burnstock G, Knight GE. Cellular distribution and functions of P2 receptor subtypes in different systems. Int Rev Cytol 2004;240:31–304.
[6] Boison D. The adenosine kinase hypothesis of epileptogenesis. Prog Neurobiol 2008;84(3):249–62.
[7] Etherington LA, Patterson GE, Meechan L, Boison D, Irving AJ, Dale N, et al. Astrocytic adenosine kinase regulates basal synaptic adenosine levels and seizure activity but not activity-dependent adenosine release in the hippocampus. Neuropharmacology 2009;56(2):429–37.
[8] Cortés A, Gracia E, Moreno E, Mallol J, Lluís C, Canela EI, et al. Moonlighting adenosine deaminase: a target protein for drug development. Medicinal Res Rev 2015;35(1):85–125.
[9] Boison D. Adenosinergic signaling in epilepsy. Neuropharmacology 2016;104:131–9.
[10] Dale N, Frenguelli BG. Release of adenosine and ATP during ischemia and epilepsy. Curr Neuropharmacol 2009;7(3):160–79.
[11] Schetinger MR, Morsch VM, Bonan CD, Wyse AT. NTPDase and 5′-nucleotidase activities in physiological and disease conditions: new perspectives for human health. Biofactors 2007;31(2):77–98.
[12] Boison D. Adenosine dysfunction in epilepsy. Glia 2012;60(8):1234–43.
[13] Gonçalves J, Queiroz G. Presynaptic adenosine and P2Y receptors. Handb Exp Pharmacol 2008;184:339–72.
[14] Aronica E, Sandau US, Iyer A, Boison D. Glial adenosine kinase—a neuropathological marker of the epileptic brain. Neurochem Int 2013;63(7):688–95.
[15] During MJ, Spencer DD. Adenosine: a potential mediator of seizure arrest and postictal refractoriness. Ann Neurol 1992;32(5):618–24.
[16] de Groot M, Iyer A, Zurolo E, Anink J, Heimans JJ, Boison D, et al. Overexpression of ADK in human astrocytic tumors and peritumoral tissue is related to tumor-associated epilepsy. Epilepsia 2012;53(1):58–66.
[17] Aronica E, Zurolo E, Iyer A, de Groot M, Anink J, Carbonell C, et al. Upregulation of adenosine kinase in astrocytes in experimental and human temporal lobe epilepsy. Epilepsia 2011;52(9):1645–55.
[18] Masino SA, Li T, Theofilas P, Sandau US, Ruskin DN, Fredholm BB, et al. A ketogenic diet suppresses seizures in mice through adenosine A_1 receptors. J Clin Invest 2011;121(7):2679–83.
[19] Lie AA, Blümcke I, Beck H, Wiestler OD, Elger CE, Schoen SW. 5′-Nucleotidase activity indicates sites of synaptic plasticity and reactive synaptogenesis in the human brain. J Neuropathol Exp Neurol 1999;58(5):451–8.
[20] Glass M, Faull RL, Bullock JY, Jansen K, Mee EW, Walker EB, et al. Loss of A_1 adenosine receptors in human temporal lobe epilepsy. Brain Res 1996;710(1–2):56–68.
[21] Roseti C, Martinello K, Fucile S, Piccari V, Mascia A, Di Gennaro G, et al. Adenosine receptor antagonists alter the stability of human epileptic $GABA_A$ receptors. Proc Natl Acad Sci USA 2008;105(39):15118–23.
[22] Roseti C, Palma E, Martinello K, Fucile S, Morace R, Esposito V, et al. Blockage of A_{2A} and A_3 adenosine receptors decreases the desensitization of human $GABA_A$ receptors microtransplanted to *Xenopus* oocytes. Proc Natl Acad Sci USA 2009;106(37):15927–31.
[23] Shinohara M, Saitoh M, Nishizawa D, Ikeda K, Hirose S, Takanashi J, et al. ADORA2A polymorphism predisposes children to encephalopathy with febrile status epilepticus. Neurology 2013;80(17):1571–6.
[24] Wagner AK, Miller MA, Scanlon J, Ren D, Kochanek PM, Conley YP. Adenosine A_1 receptor gene variants associated with post-traumatic seizures after severe TBI. Epilepsy Res 2010;90(3):259–72.
[25] Diamond ML, Ritter AC, Jackson EK, Conley YP, Kochanek PM, Boison D, et al. Genetic variation in the adenosine regulatory cycle is associated with posttraumatic epilepsy development. Epilepsia 2015;56(8):1198–206.
[26] Brambilla R, Cottini L, Fumagalli M, Ceruti S, Abbracchio MP. Blockade of A_{2A} adenosine receptors prevents basic fibroblast growth factor-induced reactive astrogliosis in rat striatal primary astrocytes. Glia 2003;43(2):190–4.
[27] Ke RH, Xiong J, Liu Y, Ye ZR. Adenosine A_{2A} receptor induced gliosis via Akt/NF-κB pathway in vitro. Neurosci Res 2009;65(3):280–5.
[28] Avsar E, Empson RM. Adenosine acting via A_1 receptors, controls the transition to status epilepticus-like behaviour in an in vitro model of epilepsy. Neuropharmacology 2004;47(3):427–37.

[29] Bedner P, Steinhäuser C. Crucial role for astrocytes in epilepsy Parpura V, Verkhratsky A, editors. Pathological potential of neuroglia. New York: Springer; 2014. pp. 155–86.
[30] Dunwiddie TV. Endogenously released adenosine regulates excitability in the in vitro hippocampus. Epilepsia 1980;21(5):541–8.
[31] Hargus NJ, Jennings C, Perez-Reyes E, Bertram EH, Patel MK. Enhanced actions of adenosine in medial entorhinal cortex layer II stellate neurons in temporal lobe epilepsy are mediated via A_1-receptor activation. Epilepsia 2012;53(1):168–76.
[32] Lee KS, Schubert P, Heinemann U. The anticonvulsive action of adenosine: a postsynaptic, dendritic action by a possible endogenous anticonvulsant. Brain Res 1984;321(1):160–4.
[33] Ross FM, Brodie MJ, Stone TW. Modulation by adenine nucleotides of epileptiform activity in the CA3 region of rat hippocampal slices. Br J Pharmacol 1998;123(1):71–80.
[34] Arrigoni E, Chamberlin NL, Saper CB, McCarley RW. Adenosine inhibits basal forebrain cholinergic and noncholinergic neurons in vitro. Neuroscience 2006;140(2):403–13.
[35] Pak MA, Haas HL, Decking UK, Schrader J. Inhibition of adenosine kinase increases endogenous adenosine and depresses neuronal activity in hippocampal slices. Neuropharmacology 1994;33(9):1049–53.
[36] Rebola N, Coelho JE, Costenla AR, Lopes LV, Parada A, Oliveira CR, et al. Decrease of adenosine A_1 receptor density and of adenosine neuromodulation in the hippocampus of kindled rats. Eur J Neurosci 2003;18(4): 820–8.
[37] Von Lubitz DK, Carter MF, Deutsch SI, Lin RC, Mastropaolo J, Meshulam Y, et al. The effects of adenosine A_3 receptor stimulation on seizures in mice. Eur J Pharmacol 1995;275(1):23–9.
[38] Laudadio MA, Psarropoulou C. The A_3 adenosine receptor agonist 2-Cl-IB-MECA facilitates epileptiform discharges in the CA3 area of immature rat hippocampal slices. Epilepsy Res 2004;59(2-3):83–94.
[39] Kovács Z, Slézia A, Bali ZK, Kovács P, Dobolyi A, Szikra T, et al. Uridine modulates neuronal activity and inhibits spike-wave discharges of absence epileptic Long Evans and Wistar Albino Glaxo/Rijswijk rats. Brain Res Bull 2013;97:16–23.
[40] Ådén U, O'Connor WT, Berman RF. Changes in purine levels and adenosine receptors in kindled seizures in the rat. Neuroreport 2004;15(10):1585–9.
[41] Hamil NE, Cock HR, Walker MC. Acute down-regulation of adenosine A_1 receptor activity in status epilepticus. Epilepsia 2012;53(1):177–88.
[42] Gouder N, Scheurer L, Fritschy JM, Boison D. Overexpression of adenosine kinase in epileptic hippocampus contributes to epileptogenesis. J Neurosci 2004;24(3):692–701.
[43] Li T, Lytle N, Lan JQ, Sandau US, Boison D. Local disruption of glial adenosine homeostasis in mice associates with focal electrographic seizures: a first step in epileptogenesis? Glia 2012;60(1):83–95.
[44] Li T, Ren G, Lusardi T, Wilz A, Lan JQ, Iwasato T, et al. Adenosine kinase is a target for the prediction and prevention of epileptogenesis in mice. J Clin Invest 2008;118(2):571–82.
[45] Orr AG, Hsiao EC, Wang MM, Ho K, Kim DH, Wang X, et al. Astrocytic adenosine receptor A_{2A} and G_s-coupled signaling regulate memory. Nature Neurosci 2015;18(3):423–34.
[46] Schoen SW, Ebert U, Löscher W. 5′-Nucleotidase activity of mossy fibers in the dentate gyrus of normal and epileptic rats. Neuroscience 1999;93(2):519–26.
[47] Bonan CD, Walz R, Pereira GS, Worm PV, Battastini AM, Cavalheiro EA, et al. Changes in synaptosomal ectonucleotidase activities in two rat models of temporal lobe epilepsy. Epilepsy Res 2000;39(3):229–38.
[48] Vianna EP, Ferreira AT, Doná F, Cavalheiro EA, da Silva Fernandes MJ. Modulation of seizures and synaptic plasticity by adenosinergic receptors in an experimental model of temporal lobe epilepsy induced by pilocarpine in rats. Epilepsia 2005;46(Suppl 5):166–73.
[49] Gouder N, Fritschy JM, Boison D. Seizure suppression by adenosine A_1 receptor activation in a mouse model of pharmacoresistant epilepsy. Epilepsia 2003;44(7):877–85.
[50] Pagonopoulou O, Angelatou F. Time development and regional distribution of $[^3H]$nitrobenzylthioinosine adenosine uptake site binding in the mouse brain after acute Pentylenetetrazol-induced seizures. J Neurosci Res 1998;53(4):433–42.
[51] Bonan CD, Amaral OB, Rockenbach IC, Walz R, Battastini AM, Izquierdo I, et al. Altered ATP hydrolysis induced by pentylenetetrazol kindling in rat brain synaptosomes. Neurochem Res 2000;25(6):775–9.
[52] Cremer CM, Palomero-Gallagher N, Bidmon HJ, Schleicher A, Speckmann EJ, Zilles K. Pentylenetetrazole-induced seizures affect binding site densities for GABA, glutamate and adenosine receptors in the rat brain. Neuroscience 2009;163(1):490–9.

[53] D'Alimonte I, D'Auro M, Citraro R, Biagioni F, Jiang S, Nargi E, et al. Altered distribution and function of A_{2A} adenosine receptors in the brain of WAG/Rij rats with genetic absence epilepsy, before and after appearance of the disease. Eur J Neurosci 2009;30(6):1023–35.
[54] Masino SA, Diao L, Illes P, Zahniser NR, Larson GA, Johansson B, et al. Modulation of hippocampal glutamatergic transmission by ATP is dependent on adenosine A_1 receptors. J Pharmacol Exp Ther 2002;303(1):356–63.
[55] Fedele DE, Li T, Lan JQ, Fredholm BB, Boison D. Adenosine A_1 receptors are crucial in keeping an epileptic focus localized. Exp Neurol 2006;200(1):184–90.
[56] Kochanek PM, Vagni VA, Janesko KL, Washington CB, Crumrine PK, Garman RH, et al. Adenosine A_1 receptor knockout mice develop lethal status epilepticus after experimental traumatic brain injury. J Cereb Blood Flow Metab 2006;26(4):565–75.
[57] El Yacoubi M, Ledent C, Parmentier M, Costentin J, Vaugeois JM. Adenosine A_{2A} receptor deficient mice are partially resistant to limbic seizures. Naunyn-Schmiedeberg's Arch Pharmacol 2009;380(3):223–32.
[58] Lovatt D, Xu Q, Liu W, Takano T, Smith NA, Schnermann J, et al. Neuronal adenosine release, and not astrocytic ATP release, mediates feedback inhibition of excitatory activity. Proc Natl Acad Sci USA 2012;109(16):6265–70.
[59] Li T, Quan Lan J, Fredholm BB, Simon RP, Boison D. Adenosine dysfunction in astrogliosis: cause for seizure generation? Neuron Glia Biol 2007;3(4):353–66.
[60] Pignataro G, Maysami S, Studer FE, Wilz A, Simon RP, Boison D. Downregulation of hippocampal adenosine kinase after focal ischemia as potential endogenous neuroprotective mechanism. J Cerebral Blood Flow Metab 2008;28(1):17–23.
[61] Lee Y, Messing A, Su M, Brenner M. GFAP promoter elements required for region-specific and astrocyte-specific expression. Glia 2008;56(5):481–93.
[62] Shen HY, Sun H, Hanthorn MM, Zhi Z, Lan JQ, Poulsen DJ, et al. Overexpression of adenosine kinase in cortical astrocytes and focal neocortical epilepsy in mice. J Neurosurg 2014;120(3):628–38.
[63] Theofilas P, Brar S, Stewart KA, Shen HY, Sandau US, Poulsen D, et al. Adenosine kinase as a target for therapeutic antisense strategies in epilepsy. Epilepsia 2011;52(3):589–601.
[64] Boison D. Adenosine augmentation therapies (AATs) for epilepsy: prospect of cell and gene therapies. Epilepsy Res 2009;85(2–3):131–41.
[65] Boison D. Inhibitory RNA in epilepsy: research tools and therapeutic perspectives. Epilepsia 2010;51(9):1659–68.
[66] Boison D. Adenosine Augmentation Therapy Noebels JL, Avoli M, Rogawski MA, Olsen RW, Delgado-Escueta AV, editors. Jasper's Basic Mechanisms of the Epilepsies (4th ed). Bethesda (MD): National Center for Biotechnology Information (US); 2012.
[67] Barraco RA, Swanson TH, Phillis JW, Berman RF. Anticonvulsant effects of adenosine analogues on amygdaloid-kindled seizures in rats. Neurosci Lett 1984;46(3):317–22.
[68] Li M, Kang R, Shi J, Liu G, Zhang J. Anticonvulsant activity of B2, an adenosine analog, on chemical convulsant-induced seizures. PloS one 2013;8(6):e67060.
[69] Zhang G, Franklin PH, Murray TF. Manipulation of endogenous adenosine in the rat prepiriform cortex modulates seizure susceptibility. J Pharmacol Exp Ther 1993;264(3):1415–24.
[70] Boison D. Adenosine and epilepsy: from therapeutic rationale to new therapeutic strategies. Neuroscientist 2005;11(1):25–36.
[71] Carrier EJ, Auchampach JA, Hillard CJ. Inhibition of an equilibrative nucleoside transporter by cannabidiol: a mechanism of cannabinoid immunosuppression. Proc Natl Acad Sci USA 2006;103(20):7895–900.
[72] Boison D, Scheurer L, Zumsteg V, Rülicke T, Litynski P, Fowler B, et al. Neonatal hepatic steatosis by disruption of the adenosine kinase gene. Proc Natl Acad Sci USA 2002;99(10):6985–90.
[73] McGaraughty S, Cowart M, Jarvis MF, Berman RF. Anticonvulsant and antinociceptive actions of novel adenosine kinase inhibitors. Curr Top Med Chem 2005;5(1):43–58.
[74] Ugarkar BG, Castellino AJ, DaRe JM, Kopcho JJ, Wiesner JB, Schanzer JM, et al. Adenosine kinase inhibitors. 2. Synthesis, enzyme inhibition, and antiseizure activity of diaryltubercidin analogues. J Med Chem 2000;43(15):2894–905.
[75] Ugarkar BG, DaRe JM, Kopcho JJ, Browne III CE, Schanzer JM, Wiesner JB, et al. Adenosine kinase inhibitors. 1. Synthesis, enzyme inhibition, and antiseizure activity of 5-iodotubercidin analogues. J Med Chem 2000;43(15):2883–93.
[76] Boison D. Adenosine kinase: exploitation for therapeutic gain. Pharmacol Rev 2013;65(3):906–43.
[77] Kowaluk EA, Bhagwat SS, Jarvis MF. Adenosine kinase inhibitors. Curr Pharm Design 1998;4(5):403–16.

[78] Kowaluk EA, Jarvis MF. Therapeutic potential of adenosine kinase inhibitors. Expert Opin Investig Drugs 2000;9(3):551–64.
[79] McGaraughty S, Cowart M, Jarvis MF. Recent developments in the discovery of novel adenosine kinase inhibitors: mechanism of action and therapeutic potential. CNS Drug Rev 2001;7(4):415–32.
[80] Huber A, Padrun V, Deglon N, Aebischer P, Möhler H, Boison D. Grafts of adenosine-releasing cells suppress seizures in kindling epilepsy. Proc Natl Acad Sci USA 2001;98(13):7611–6.
[81] Boison D, Huber A, Padrun V, Deglon N, Aebischer P, Möhler H. Seizure suppression by adenosine-releasing cells is independent of seizure frequency. Epilepsia 2002;43(8):788–96.
[82] Boison D, Scheurer L, Tseng JL, Aebischer P, Möhler H. Seizure suppression in kindled rats by intraventricular grafting of an adenosine releasing synthetic polymer. Exp Neurol 1999;160(1):164–74.
[83] Fedele DE, Koch P, Scheurer L, Simpson EM, Möhler H, Brüstle O, et al. Engineering embryonic stem cell derived glia for adenosine delivery. Neurosci Lett 2004;370(2–3):160–5.
[84] Li T, Steinbeck JA, Lusardi T, Koch P, Lan JQ, Wilz A, et al. Suppression of kindling epileptogenesis by adenosine releasing stem cell-derived brain implants. Brain J Neurol 2007;130(Pt 5):1276–88.
[85] Ren G, Li T, Lan JQ, Wilz A, Simon RP, Boison D. Lentiviral RNAi-induced downregulation of adenosine kinase in human mesenchymal stem cell grafts: a novel perspective for seizure control. Exp Neurol 2007;208(1):26–37.
[86] Li T, Ren G, Kaplan DL, Boison D. Human mesenchymal stem cell grafts engineered to release adenosine reduce chronic seizures in a mouse model of CA3-selective epileptogenesis. Epilepsy Res 2009;84(2–3):238–41.
[87] Szybala C, Pritchard EM, Lusardi TA, Li T, Wilz A, Kaplan DL, et al. Antiepileptic effects of silk-polymer based adenosine release in kindled rats. Exp Neurol 2009;219(1):126–35.
[88] Shen HY, Singer P, Lytle N, Wei CJ, Lan JQ, Williams-Karnesky RL, et al. Adenosine augmentation ameliorates psychotic and cognitive endophenotypes of schizophrenia. J Clin Invest 2012;122(7):2567–77.
[89] Chiari A, Yaksh TL, Myers RR, Provencher J, Moore L, Lee CS, et al. Preclinical toxicity screening of intrathecal adenosine in rats and dogs. Anesthesiology 1999;91(3):824–32.
[90] Eisenach JC, Hood DD, Curry R. Preliminary efficacy assessment of intrathecal injection of an American formulation of adenosine in humans. Anesthesiology 2002;96(1):29–34.
[91] Eisenach JC, Hood DD, Curry R. Phase I safety assessment of intrathecal injection of an American formulation of adenosine in humans. Anesthesiology 2002;96(1):24–8.
[92] Matos M, Augusto E, Machado NJ, dos Santos-Rodrigues A, Cunha RA, Agostinho P. Astrocytic adenosine A_{2A} receptors control the amyloid-beta peptide-induced decrease of glutamate uptake. J Alzheimer's Dis 2012;31(3):555–67.
[93] Matos M, Augusto E, Santos-Rodrigues AD, Schwarzschild MA, Chen JF, Cunha RA, et al. Adenosine A_{2A} receptors modulate glutamate uptake in cultured astrocytes and gliosomes. Glia 2012;60(5):702–16.
[94] Nishizaki T. ATP- and adenosine-mediated signaling in the central nervous system: adenosine stimulates glutamate release from astrocytes via A_{2A} adenosine receptors. J Pharmacol Sci 2004;94(2):100–2.
[95] Nishizaki T, Nagai K, Nomura T, Tada H, Kanno T, Tozaki H, et al. A new neuromodulatory pathway with a glial contribution mediated via A_{2A} adenosine receptors. Glia 2002;39(2):133–47.
[96] Li XX, Nomura T, Aihara H, Nishizaki T. Adenosine enhances glial glutamate efflux via A_{2a} adenosine receptors. Life Sci 2001;68(12):1343–50.
[97] Frizzo ME, Frizzo JK, Amadio S, Rodrigues JM, Perry ML, Bernardi G, et al. Extracellular adenosine triphosphate induces glutamate transporter-1 expression in hippocampus. Hippocampus 2007;17(4):305–15.
[98] Wu J, Lee MR, Choi S, Kim T, Choi DS. ENT1 regulates ethanol-sensitive EAAT2 expression and function in astrocytes. Alcohol Clin Exp Res 2010;34(6):1110–7.
[99] Wu J, Lee MR, Kim T, Johng S, Rohrback S, Kang N, et al. Regulation of ethanol-sensitive EAAT2 expression through adenosine A_1 receptor in astrocytes. Biochem Biophys Res Commun 2011;406(1):47–52.
[100] Kobow K, El-Osta A, Blümcke I. The methylation hypothesis of pharmacoresistance in epilepsy. Epilepsia 2013;54(Suppl 2):41–7.
[101] Kobow K, Kaspi A, Harikrishnan KN, Kiese K, Ziemann M, Khurana I, et al. Deep sequencing reveals increased DNA methylation in chronic rat epilepsy. Acta Neuropathol 2013;126(5):741–56.
[102] Miller-Delaney SF, Bryan K, Das S, McKiernan RC, Bray IM, Reynolds JP, et al. Differential DNA methylation profiles of coding and non-coding genes define hippocampal sclerosis in human temporal lobe epilepsy. Brain J Neurol 2015;138(Pt 3):616–31.

[103] Williams-Karnesky RL, Sandau US, Lusardi TA, Lytle NK, Farrell JM, Pritchard EM, et al. Epigenetic changes induced by adenosine augmentation therapy prevent epileptogenesis. J Clin Invest 2013;123(8):3552–63.

[104] Kobow K, Blümcke I. The methylation hypothesis: do epigenetic chromatin modifications play a role in epileptogenesis? Epilepsia 2011;52(Suppl 4):15–19.

[105] Gill MW, Schatz RA. The effect of diazepam on brain levels of S-adenosyl-L-methionine and S-adenosyl-L-homocysteine: possible correlation with protection from methionine sulfoximine seizures. Res Commun Chem Pathol Pharmacol 1985;50(3):349–63.

[106] Schatz RA, Wilens TE, Tatter SB, Gregor P, Sellinger OZ. Possible role of increased brain methylation in methionine sulfoximine epileptogenesis: effects of administration of adenosine and homocysteine thiolactone. J Neurosci Res 1983;10(4):437–47.

[107] Sellinger OZ, Schatz RA, Porta R, Wilens TE. Brain methylation and epileptogenesis: the case of methionine sulfoximine. Ann Neurol 1984;16(Suppl):S115–20.

[108] Qureshi IA, Mehler MF. Epigenetic mechanisms underlying human epileptic disorders and the process of epileptogenesis. Neurobiol Dis 2010;39(1):53–60.

[109] Freeman JM. Seizures, EEG events, and the ketogenic diet. Epilepsia 2009;50(2):329–30.

[110] Kossoff EH. The ketogenic diet: it's about "time." Dev Med Child Neurol 2009;51(4):252–3.

[111] Kossoff EH, Rho JM. Ketogenic diets: evidence for short- and long-term efficacy. Neurotherapeutics 2009;6(2):406–14.

[112] Kossoff EH, Zupec-Kania BA, Amark PE, Ballaban-Gil KR, Christina Bergqvist AG, Blackford R, et al. Optimal clinical management of children receiving the ketogenic diet: recommendations of the International Ketogenic Diet Study Group. Epilepsia 2009;50(2):304–17.

[113] Neal EG, Chaffe H, Schwartz RH, Lawson MS, Edwards N, Fitzsimmons G, et al. The ketogenic diet for the treatment of childhood epilepsy: a randomised controlled trial. Lancet Neurol 2008;7(6):500–6.

[114] Yellen G. Ketone bodies, glycolysis, and KATP channels in the mechanism of the ketogenic diet. Epilepsia 2008;49(Suppl 8):80–2.

[115] Masino SA, Rho JM. Mechanisms of Ketogenic Diet Action Noebels JL, Avoli M, Rogawski MA, Olsen RW, Delgado-Escueta AV, editors. Jasper's Basic Mechanisms of the Epilepsies (4th ed). Bethesda (MD): National Center for Biotechnology Information (US); 2012.

[116] Masino SA, Geiger JD. Are purines mediators of the anticonvulsant/neuroprotective effects of ketogenic diets? Trends Neurosci 2008;31(6):273–8.

[117] Masino SA, Geiger JD. The ketogenic diet and epilepsy: is adenosine the missing link? Epilepsia 2009;50(2):332–3.

[118] Masino SA, Kawamura M, Wasser CD, Pomeroy LT, Ruskin DN. Adenosine, ketogenic diet and epilepsy: the emerging therapeutic relationship between metabolism and brain activity. Curr Neuropharmacol 2009;7(3):257–68.

[119] Masino SA, Kawamura Jr. M, Ruskin DN, Geiger JD, Boison D. Purines and neuronal excitability: links to the ketogenic diet. Epilepsy Res 2012;100(3):229–38.

[120] Henshall DC, Diaz-Hernandez M, Miras-Portugal MT, Engel T. P2X receptors as targets for the treatment of status epilepticus. Front Cell Neurosci 2013;7:237.

[121] Khakh BS, North RA. P2X receptors as cell-surface ATP sensors in health and disease. Nature 2006;442(7102):527–32.

[122] Sebastián-Serrano Á, de Diego-García L, Martínez-Frailes C, Ávila J, Zimmermann H, Millán JL, et al. Tissue-nonspecific alkaline phosphatase regulates purinergic transmission in the central nervous system during development and disease. Comput Struct Biotechnol J 2015;13:95–100.

[123] Sperlágh B, Vizi ES, Wirkner K, Illes P. $P2X_7$ receptors in the nervous system. Prog Neurobiol 2006;78(6):327–46.

[124] Rodrigues RJ, Almeida T, Richardson PJ, Oliveira CR, Cunha RA. Dual presynaptic control by ATP of glutamate release via facilitatory $P2X_1$, $P2X_{2/3}$, and $P2X_3$ and inhibitory $P2Y_1$, $P2Y_2$, and/or $P2Y_4$ receptors in the rat hippocampus. J Neurosci 2005;25(27):6286–95.

[125] Khakh BS, Gittermann D, Cockayne DA, Jones A. ATP modulation of excitatory synapses onto interneurons. J Neurosci 2003;23(19):7426–37.

[126] Donato R, Rodrigues RJ, Takahashi M, Tsai MC, Soto D, Miyagi K, et al. GABA release by basket cells onto Purkinje cells, in rat cerebellar slices, is directly controlled by presynaptic purinergic receptors, modulating Ca^{2+} influx. Cell Calcium 2008;44(6):521–32.

[127] Engel T, Jimenez-Pacheco A, Miras-Portugal MT, Diaz-Hernandez M, Henshall DC. $P2X_7$ receptor in epilepsy; role in pathophysiology and potential targeting for seizure control. Int J Physiol Pathophysiol Pharmacol 2012;4(4):174–87.

[128] Engel T, Gomez-Villafuertes R, Tanaka K, Mesuret G, Sanz-Rodriguez A, Garcia-Huerta P, et al. Seizure suppression and neuroprotection by targeting the purinergic $P2X_7$ receptor during status epilepticus in mice. FASEB J 2012;26(4):1616–28.
[129] Jimenez-Pacheco A, Mesuret G, Sanz-Rodriguez A, Tanaka K, Mooney C, Conroy R, et al. Increased neocortical expression of the $P2X_7$ receptor after status epilepticus and anticonvulsant effect of $P2X_7$ receptor antagonist A-438079. Epilepsia 2013;54(9):1551–61.
[130] Rappold PM, Lynd-Balta E, Joseph SA. $P2X_7$ receptor immunoreactive profile confined to resting and activated microglia in the epileptic brain. Brain Res 2006;1089(1):171–8.
[131] Alloisio S, Cervetto C, Passalacqua M, Barbieri R, Maura G, Nobile M, et al. Functional evidence for presynaptic $P2X_7$ receptors in adult rat cerebrocortical nerve terminals. FEBS Lett 2008;582(28):3948–53.
[132] Marcoli M, Cervetto C, Paluzzi P, Guarnieri S, Alloisio S, Thellung S, et al. $P2X_7$ pre-synaptic receptors in adult rat cerebrocortical nerve terminals: a role in ATP-induced glutamate release. J Neurochem 2008;105(6):2330–42.
[133] Armstrong JN, Brust TB, Lewis RG, MacVicar BA. Activation of presynaptic $P2X_7$-like receptors depresses mossy fiber-CA3 synaptic transmission through p38 mitogen-activated protein kinase. J Neurosci 2002; 22(14):5938–45.
[134] Duan S, Anderson CM, Keung EC, Chen Y, Swanson RA. $P2X_7$ receptor-mediated release of excitatory amino acids from astrocytes. J Neurosci 2003;23(4):1320–8.
[135] Papp L, Vizi ES, Sperlagh B. Lack of ATP-evoked GABA and glutamate release in the hippocampus of $P2X_7$ receptor-/- mice. Neuroreport 2004;15(15):2387–91.
[136] Sperlágh B, Kofalvi A, Deuchars J, Atkinson L, Milligan CJ, Buckley NJ, et al. Involvement of $P2X_7$ receptors in the regulation of neurotransmitter release in the rat hippocampus. J Neurochem 2002;81(6):1196–211.
[137] Pascual O, Ben Achour S, Rostaing P, Triller A, Bessis A. Microglia activation triggers astrocyte-mediated modulation of excitatory neurotransmission. Proc Natl Acad Sci USA 2012;109(4):E197–205.
[138] Csölle C, Heinrich A, Kittel Á, Sperlágh B. P2Y receptor mediated inhibitory modulation of noradrenaline release in response to electrical field stimulation and ischemic conditions in superfused rat hippocampus slices. J Neurochem 2008;106(1):347–60.
[139] Heinrich A, Kittel Á, Csölle C, Sylvester Vizi E, Sperlágh B. Modulation of neurotransmitter release by P2X and P2Y receptors in the rat spinal cord. Neuropharmacology 2008;54(2):375–86.
[140] Lechner SG, Dorostkar MM, Mayer M, Edelbauer H, Pankevych H, Boehm S. Autoinhibition of transmitter release from PC12 cells and sympathetic neurons through a P2Y receptor-mediated inhibition of voltage-gated Ca^{2+} channels. Eur J Neurosci 2004;20(11):2917–28.
[141] Queiroz G, Talaia C, Gonçalves J. ATP modulates noradrenaline release by activation of inhibitory P2Y receptors and facilitatory P2X receptors in the rat vas deferens. J Pharmacol Exp Ther 2003;307(2):809–15.
[142] Kulick MB, von Kügelgen I. P2Y-receptors mediating an inhibition of the evoked entry of calcium through N-type calcium channels at neuronal processes. J Pharmacol Exp Ther 2002;303(2):520–6.
[143] Powell AD, Teschemacher AG, Seward EP. P2Y purinoceptors inhibit exocytosis in adrenal chromaffin cells via modulation of voltage-operated calcium channels. J Neurosci 2000;20(2):606–16.
[144] Bowser DN, Khakh BS. ATP excites interneurons and astrocytes to increase synaptic inhibition in neuronal networks. J Neurosci 2004;24(39):8606–20.
[145] Jacob PF, Vaz SH, Ribeiro JA, Sebastião AM. $P2Y_1$ receptor inhibits GABA transport through a calcium signalling-dependent mechanism in rat cortical astrocytes. Glia 2014;62(8):1211–26.
[146] Henshall DC, Engel T. P2X purinoceptors as a link between hyperexcitability and neuroinflammation in status epilepticus. Epilepsy Behav 2015;49:8–12.
[147] Franke H, Illes P. Nucleotide signaling in astrogliosis. Neurosci Lett 2014;565:14–22.
[148] Monif M, Burnstock G, Williams DA. Microglia: proliferation and activation driven by the $P2X_7$ receptor. Int J Biochem Cell Biol 2010;42(11):1753–6.
[149] Rodrigues RJ, Tomé AR, Cunha RA. ATP as a multi-target danger signal in the brain. Front Neurosci 2015;9:148.
[150] Engel T, Alves M, Sheedy C, Henshall DC. ATPergic signalling during seizures and epilepsy. Neuropharmacology 2016;104:140–53.
[151] Eyo UB, Peng J, Swiatkowski P, Mukherjee A, Bispo A, Wu LJ. Neuronal hyperactivity recruits microglial processes via neuronal NMDA receptors and microglial $P2Y_{12}$ receptors after status epilepticus. J Neurosci 2014;34(32):10528–40.
[152] Avignone E, Ulmann L, Levavasseur F, Rassendren F, Audinat E. Status epilepticus induces a particular microglial activation state characterized by enhanced purinergic signaling. J Neurosci 2008;28(37):9133–44.

[153] Kang TC, An SJ, Park SK, Hwang IK, Won MH. $P2X_2$ and $P2X_4$ receptor expression is regulated by a $GABA_A$ receptor-mediated mechanism in the gerbil hippocampus. Brain Res Mol Brain Res 2003;116(1–2):168–75.

[154] Ulmann L, Levavasseur F, Avignone E, Peyroutou R, Hirbec H, Audinat E, et al. Involvement of $P2X_4$ receptors in hippocampal microglial activation after status epilepticus. Glia 2013;61(8):1306–19.

[155] Kim JE, Kang TC. The $P2X_7$ receptor-pannexin-1 complex decreases muscarinic acetylcholine receptor-mediated seizure susceptibility in mice. J Clin Invest 2011;121(5):2037–47.

[156] Mesuret G, Engel T, Hessel EV, Sanz-Rodriguez A, Jimenez-Pacheco A, Miras-Portugal MT, et al. $P2X_7$ receptor inhibition interrupts the progression of seizures in immature rats and reduces hippocampal damage. CNS Neurosci Ther 2014;20(6):556–64.

[157] Soni N, Koushal P, Reddy BV, Deshmukh R, Kumar P. Effect of GLT1 modulator and $P2X_7$ antagonists alone and in combination in the kindling model of epilepsy in rats. Epilepsy Behav 2015;48:4–14.

[158] Sukigara S, Dai H, Nabatame S, Otsuki T, Hanai S, Honda R, et al. Expression of astrocyte-related receptors in cortical dysplasia with intractable epilepsy. J Neuropathol Exp Neurol 2014;73(8):798–806.

[159] Wei YJ, Guo W, Sun FJ, Fu WL, Zheng DH, Chen X, et al. Increased expression and cellular localization of $P2X_7R$ in cortical lesions of patients with focal cortical dysplasia. J Neuropathol Exp Neurol 2016;75(1):61–8.

[160] Emsley HC, Appleton RE, Whitmore CL, Jury F, Lamb JA, Martin JE, et al. Variations in inflammation-related genes may be associated with childhood febrile seizure susceptibility. Seizure 2014;23(6):457–61.

[161] Klaft ZJ, Schulz SB, Maslarova A, Gabriel S, Heinemann U, Gerevich Z. Extracellular ATP differentially affects epileptiform activity via purinergic $P2X_7$ and adenosine A_1 receptors in naive and chronic epileptic rats. Epilepsia 2012;53(11):1978–86.

[162] Wieraszko A, Seyfried TN. Increased amount of extracellular ATP in stimulated hippocampal slices of seizure prone mice. Neurosci Lett 1989;106(3):287–93.

[163] Wu PH, Phillis JW. Distribution and release of adenosine triphosphate in rat brain. Neurochem Res 1978;3(5):563–71.

[164] Lopatář J, Dale N, Frenguelli BG. Minor contribution of ATP P2 receptors to electrically-evoked electrographic seizure activity in hippocampal slices: evidence from purine biosensors and P2 receptor agonists and antagonists. Neuropharmacology 2011;61(1–2):25–34.

[165] Doná F, Ulrich H, Persike DS, Conceicão IM, Blini JP, Cavalheiro EA, et al. Alteration of purinergic $P2X_4$ and $P2X_7$ receptor expression in rats with temporal lobe epilepsy induced by pilocarpine. Epilepsy Res 2009;83(2–3):157–67.

[166] Kim JE, Kwak SE, Jo SM, Kang TC. Blockade of P2X receptor prevents astroglial death in the dentate gyrus following pilocarpine-induced status epilepticus. Neurol Res 2009;31(9):982–8.

[167] Choi HK, Ryu HJ, Kim JE, Jo SM, Choi HC, Song HK, et al. The roles of $P2X_7$ receptor in regional-specific microglial responses in the rat brain following status epilepticus. Neurological Sci 2012;33(3):515–25.

[168] Kim JE, Ryu HJ, Yeo SI, Kang TC. $P2X_7$ receptor differentially modulates astroglial apoptosis and clasmatodendrosis in the rat brain following status epilepticus. Hippocampus 2011;21(12):1318–33.

[169] Kim JY, Ko AR, Kim JE. $P2X_7$ receptor-mediated PARP1 activity regulates astroglial death in the rat hippocampus following status epilepticus. Front Cell Neurosci 2015;9:352.

[170] Kim JE, Ryu HJ, Yeo SI, Kang TC. $P2X_7$ receptor regulates leukocyte infiltrations in rat frontoparietal cortex following status epilepticus. J Neuroinflammation 2010;7:65.

[171] Santiago MF, Veliskova J, Patel NK, Lutz SE, Caille D, Charollais A, et al. Targeting pannexin1 improves seizure outcome. PloS One 2011;6(9):e25178.

[172] Solle M, Labasi J, Perregaux DG, Stam E, Petrushova N, Koller BH, et al. Altered cytokine production in mice lacking $P2X_7$ receptors. J Biol Chem 2001;276(1):125–32.

[173] Agresti C, Meomartini ME, Amadio S, Ambrosini E, Serafini B, Franchini L, et al. Metabotropic P2 receptor activation regulates oligodendrocyte progenitor migration and development. Glia 2005;50(2):132–44.

[174] Amadio S, D'Ambrosi N, Cavaliere F, Murra B, Sancesario G, Bernardi G, et al. P2 receptor modulation and cytotoxic function in cultured CNS neurons. Neuropharmacology 2002;42(4):489–501.

[175] Amadio S, Tramini G, Martorana A, Viscomi MT, Sancesario G, Bernardi G, et al. Oligodendrocytes express $P2Y_{12}$ metabotropic receptor in adult rat brain. Neuroscience 2006;141(3):1171–80.

[176] Bianco F, Fumagalli M, Pravettoni E, D'Ambrosi N, Volonte C, Matteoli M, et al. Pathophysiological roles of extracellular nucleotides in glial cells: differential expression of purinergic receptors in resting and activated microglia. Brain Res Rev 2005;48(2):144–56.

[177] Bintig W, Begandt D, Schlingmann B, Gerhard L, Pangalos M, Dreyer L, et al. Purine receptors and Ca^{2+} signalling in the human blood-brain barrier endothelial cell line hCMEC/D3. Purinergic Signal 2012;8(1):71–80.

[178] Carrasquero LM, Delicado EG, Bustillo D, Gutiérrez-Martin Y, Artalejo AR, Miras-Portugal MT. $P2X_7$ and $P2Y_{13}$ purinergic receptors mediate intracellular calcium responses to BzATP in rat cerebellar astrocytes. J Neurochem 2009;110(3):879–89.

[179] Chambers JK, Macdonald LE, Sarau HM, Ames RS, Freeman K, Foley JJ, et al. A G protein-coupled receptor for UDP-glucose. J Biol Chem 2000;275(15):10767–71.

[180] Choi J, Ifuku M, Noda M, Guilarte TR. Translocator protein (18 kDa, TSPO)/peripheral benzodiazepine receptor specific ligands induce microglia functions consistent with an activated state. Glia 2011;59(2):219–30.

[181] Collo G, North RA, Kawashima E, Merlo-Pich E, Neidhart S, Surprenant A, et al. Cloning of $P2X_5$ and $P2X_6$ receptors and the distribution and properties of an extended family of ATP-gated ion channels. J Neurosci 1996;16(8):2495–507.

[182] Communi D, Robaye B, Boeynaems JM. Pharmacological characterization of the human $P2Y_{11}$ receptor. Br J Pharmacol 1999;128(6):1199–206.

[183] Communi D, Gonzalez NS, Detheux M, Brezillon S, Lannoy V, Parmentier M, et al. Identification of a novel human ADP receptor coupled to G_i. J Biol Chem 2001;276(44):41479–85.

[184] Deuchars SA, Atkinson L, Brooke RE, Musa H, Milligan CJ, Batten TF, et al. Neuronal $P2X_7$ receptors are targeted to presynaptic terminals in the central and peripheral nervous systems. J Neurosci 2001;21(18):7143–52.

[185] Dixon SJ, Yu R, Panupinthu N, Wilson JX. Activation of P2 nucleotide receptors stimulates acid efflux from astrocytes. Glia 2004;47(4):367–76.

[186] Fischer W, Appelt K, Grohmann M, Franke H, Norenberg W, Illes P. Increase of intracellular Ca^{2+} by P2X and P2Y receptor-subtypes in cultured cortical astroglia of the rat. Neuroscience 2009;160(4):767–83.

[187] Franke H, Schepper C, Illes P, Krügel U. Involvement of P2X and P2Y receptors in microglial activation in vivo. Purinergic Signal 2007;3(4):435–45.

[188] Fumagalli M, Brambilla R, D'Ambrosi N, Volonté C, Matteoli M, Verderio C, et al. Nucleotide-mediated calcium signaling in rat cortical astrocytes: Role of P2X and P2Y receptors. Glia 2003;43(3):218–303.

[189] Hervás C, Pérez-Sen R, Miras-Portugal MT. Coexpression of functional P2X and P2Y nucleotide receptors in single cerebellar granule cells. J Neurosci Res 2003;73(3):384–99.

[190] Kukley M, Barden JA, Steinhäuser C, Jabs R. Distribution of P2X receptors on astrocytes in juvenile rat hippocampus. Glia 2001;36(1):11–21.

[191] Lalo U, Pankratov Y, Wichert SP, Rossner MJ, North RA, Kirchhoff F, et al. $P2X_1$ and $P2X_5$ subunits form the functional P2X receptor in mouse cortical astrocytes. J Neurosci 2008;28(21):5473–80.

[192] Malin SA, Molliver DC. G_i- and G_q-coupled ADP (P2Y) receptors act in opposition to modulate nociceptive signaling and inflammatory pain behavior. Molecular pain 2010;6:21.

[193] Moore DJ, Chambers JK, Wahlin JP, Tan KB, Moore GB, Jenkins O, et al. Expression pattern of human P2Y receptor subtypes: a quantitative reverse transcription-polymerase chain reaction study. Biochim Biophys Acta 2001;1521(1–3):107–19.

[194] Nicholas RA, Watt WC, Lazarowski ER, Li Q, Harden K. Uridine nucleotide selectivity of three phospholipase C-activating P2 receptors: identification of a UDP-selective, a UTP-selective, and an ATP- and UTP-specific receptor. Mol Pharmacol 1996;50(2):224–9.

[195] Simon J, Filippov AK, Göransson S, Wong YH, Frelin C, Michel AD, et al. Characterization and channel coupling of the $P2Y_{12}$ nucleotide receptor of brain capillary endothelial cells. J Biol Chem 2002;277(35):31390–400.

[196] Takasaki J, Kamohara M, Saito T, Matsumoto M, Matsumoto S, Ohishi T, et al. Molecular cloning of the platelet $P2T_{AC}$ ADP receptor: pharmacological comparison with another ADP receptor, the $P2Y_1$ receptor. Mol Pharmacol 2001;60(3):432–9.

[197] Waldo GL, Harden TK. Agonist binding and G_q-stimulating activities of the purified human $P2Y_1$ receptor. Mol Pharmacol 2004;65(2):426–36.

[198] Xiang Z, Burnstock G. Expression of P2X receptors on rat microglial cells during early development. Glia 2005;52(2):119–26.

[199] Webb TE, Feolde E, Vigne P, Neary JT, Runberg A, Frelin C, et al. The P2Y purinoceptor in rat brain microvascular endothelial cells couple to inhibition of adenylate cyclase. Br J Pharmacol 1996;119(7):1385–92.

CHAPTER

11

Gap Junctions

OUTLINE

OVERVIEW

Electrotonic coupling between cells is accomplished through the formation of gap junctions (GJs) between cells. Composed of connexin proteins, GJs are found on multiple cell types and connexin distribution is cell type-specific. All GJs play a role in cellular communication, but astrocytic GJs are thought to play a role in K^+ and glutamate redistribution, synaptic strength regulation, and memory formation. Both human tissue studies and animal models of epilepsy have shown considerable changes in connexin expression after seizure activity. Electrophysiological studies have implicated GJs in the generation of very fast oscillations (VFOs) that precede seizures. Knockout and GJ inhibitor studies have demonstrated potential anticonvulsant effects, although these results are mixed and suffer from lack of specificity of many of the currently available GJ inhibitors. Findings implicating GJs in epilepsy as well as differences in the roles of neuronal versus glial GJs in tissue excitability are considered in this chapter.

DOI: http://dx.doi.org/10.1016/B978-0-12-802401-0.00011-9

HISTORY

In the 1940s and 1950s, it was hypothesized that cells had a surface barrier that allowed very small molecules, like inorganic ions, to pass through membrane pores or channels. This was based on permeability studies and electrical measurements of various cell types that showed that membranes were highly diffusive [1]. Subsequently, observations using electron microscopy revealed a seemingly continuous surface of membranes in various cell types. Finally, the first evidence for an electrical connection between cells came in the late 1950s when direct electrical transmission between nerve cells in both the lobster cardiac ganglion [2] and the giant motor synapses of crayfish was reported. In the 1960s, electrotonic coupling in vertebrate tissue was discovered between teleost spinal neurons [3].

Although a likely instance of electrotonic coupling was described between spinal motoneurons of the cat in the 1960s [4], it was not truly demonstrated physiologically between neurons in the mammalian central nervous system (CNS) until the early 1970s. Electrotonic coupling between mesencephalic neurons of the fifth nerve in rat [5] and between neurons in the rat lateral vestibular nucleus [6] were observed. The hypothesis that coupling between neurons could mediate the synchronization of neurons during seizure activity was proposed after MacVicar and Dudek [7] used dual intracellular recordings in rat hippocampal slices to demonstrate direct electrotonic coupling of neurons. Shortly thereafter, three independent research groups proved that chemical synaptic transmission was not necessary for the synchronization of neuronal activity in the CA1 region of the hippocampus. This was accomplished by showing that, after incubating rat hippocampal slices in a low calcium media to block chemical synaptic transmission, prolonged bursts of large population spikes in hippocampal pyramidal cells could still occur [8–13]. Over the next several decades, the role of GJs in various models of epilepsy has been examined.

STRUCTURE

Connexins are vertebrate transmembrane proteins that assemble into groups of six to form pores called connexons or hexameric hemichannels. A complete GJ is formed by two connexons, thus connecting the cytoplasm between two cells (Fig. 11.1) [14]. These assemblies can be either homomeric or heteromeric in nature. In invertebrates, six innexin protein subunits come together to form an innexon, or hemichannel. Two apposed innexons then form GJs. To date, 20 connexin genes in the mouse genome and 21 connexin genes in the human genome have been discovered [14,15]. In 2004, another family of GJ proteins homologous to invertebrate innexins, called pannexins, were cloned [16]. Pannexins were initially suspected to form GJs due to their structural similarity to connexins, however, they predominantly form hemichannels.

Kanno and Loewenstein [17] discovered that fluorescein-sodium, a 376 Da hydrophilic molecule, could traverse GJs [18]. Shortly thereafter, various tracer molecules more than 1 KDa in mass were shown to pass through GJs [19]. GJs can exchange ions, energy metabolites, neurotransmitters, and signaling molecules. The large hydrophilic pore of GJs allows

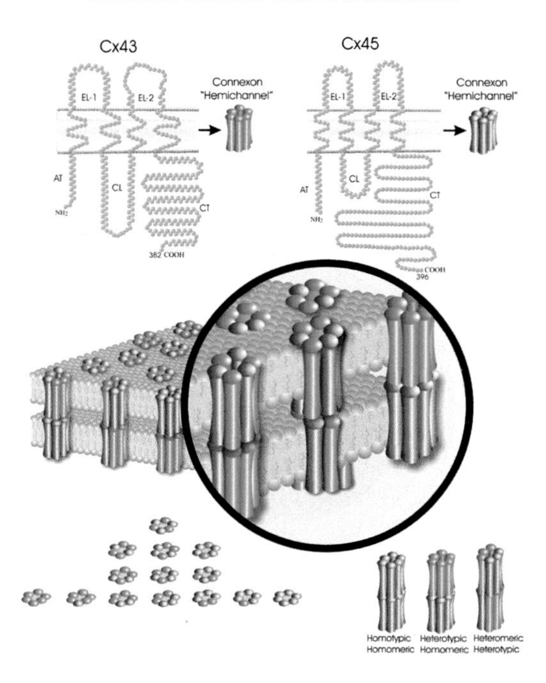

FIGURE 11.1 **Assembly of connexins into gap junctions.** Connexin 43 (Cx43) and connexin 45 (Cx45), as examples of connexin family members, typically thread through the membrane four times, with the N-terminus, C-terminus, and cytoplasmic loop exposed to the cytoplasm. Connexin arrangement in the membrane also yields two extracellular loops designated EL-1 and EL-2. Six connexins oligomerize into a connexon or hemichannel that docks in homotypic, heterotypic, and combined heterotypic/heteromeric arrangements. In total, as many as 14 different connexon arrangements can form when two members of the connexin family intermix. *Source: Reproduced with permission from Laird DW. Life cycle of connexins in health and disease. Biochem J 2006;394(Pt 3):527–43.*

chemical and metabolic coupling in addition to the well-known electrical coupling. In addition to neurotransmitter release through hemichannels and cellular communication through GJs, connexins serve a variety of functions in the brain.

CELLULAR DISTRIBUTION AND FUNCTION

Hemichannels allow the exchange of ions and small signaling molecules between the cytoplasm and extracellular environment whereas GJs provide the ability for cell-to-cell communication. Connexins exhibit a cell-specific distribution (Table 11.1) and have been studied in both human tissue and animal models of epilepsy. Several studies have demonstrated a role for GJ coupling in seizure activity, although these findings have been debated. The discrepancy, in part, could be due to the functional difference between glial and neuronal GJs; therefore, both will be discussed in this chapter.

Astrocytes are predominantly connected through the proteins connexin 30 (Cx30) and connexin 43 (Cx43), although their distribution is heterogeneous throughout the brain. For

TABLE 11.1 Cellular Distribution of Connexin Proteins

Cell Type	Connexins Expressed
Neurons	Cx30.2, Cx36, Cx45
Astrocytes	Cx26, Cx30, Cx40, Cx43, Cx47
Oligodendrocytes	Cx29, Cx32, Cx36, Cx47
Microglia	Cx43

example, Cx30 expression is highest in the cerebellum and is significantly lower in both the hippocampus and cortex; Cx43 is predominantly responsible for coupling in the hippocampus [20] and is also expressed on activated microglia [21]. Oligodendrocytes are coupled mainly through connexin 47 (Cx47) and connexin 32 (Cx32) and play a major role in myelination [22]. Griemsmann et al. [23] characterized coupled networks between astrocytes and oligodendrocytes in the murine ventrobasal thalamus, hippocampus, and cortex of mice and found large panglial networks in all three gray matter regions. They found that Cx30 is the dominant astrocytic connexin in the thalamus and functional channels formed by Cx30 and Cx32 predominantly mediate astrocyte to oligodendrocyte coupling in the thalamus. Finally, deletion of Cx30 and Cx47 resulted in the loss of panglial coupling and this could be restored with the presence of either one of the connexin alleles [23].

Astrocytic coupling is both temperature- and anisotropy-dependent [24]. GJ networks in astrocytes are known to regulate cell volume, contribute to spatial buffering of potassium, modulate glutamatergic synaptic activity and synaptic plasticity, and play a role in information processing, sensorimotor, and spatial memory tasks [25–30]. Interestingly, Pannasch et al. [26] recently discovered a new role of the astrocytic Cx30, independent of its channel function. Cx30 can regulate the extent of astrocyte invasion into glutamatergic synapses (Fig. 11.2) [31], thus determining the efficacy of glutamate clearance and excitatory synaptic strength [26]. This agrees with previous findings that astrocytic coupling is found exclusively on "classical" astrocytes expressing glutamate transporters in the hippocampus (termed "GluT" cells by Steinhäuser and colleagues) that can help modulate the strength of glutamatergic synaptic activity; coupling was largely absent from AMPA-type glutamate receptor-bearing cells ("GluR" cells) with astroglial properties [32]. Importantly, astrocytic networks can combat K^+ and glutamate hyperexcitability by removing and redistributing these excitatory molecules [27,30].

Connexin 36 (Cx36) and connexin 45 (Cx45) are the predominant proteins responsible for neuron–neuron gap junctional coupling [33,34]. In particular, Cx36 is largely expressed by interneurons [35] and appears to electrically couple hippocampal and cortical GABAergic interneurons [36]. Neuronal GJ coupling is thought to underlie synchronous neuronal activity, long-term potentiation, short-term and spatial memory formation, and motor coordination learning [35,37–40]. Importantly, electrical coupling may influence neuronal activity and has been proposed that such coupling could underlie the generation of synchronous neuronal activity [35,41]. Increasing evidence for the role of GJs in seizure initiation, epileptogenesis, and tissue synchronization has been presented in both human tissue studies and experimental models of epilepsy.

(A)

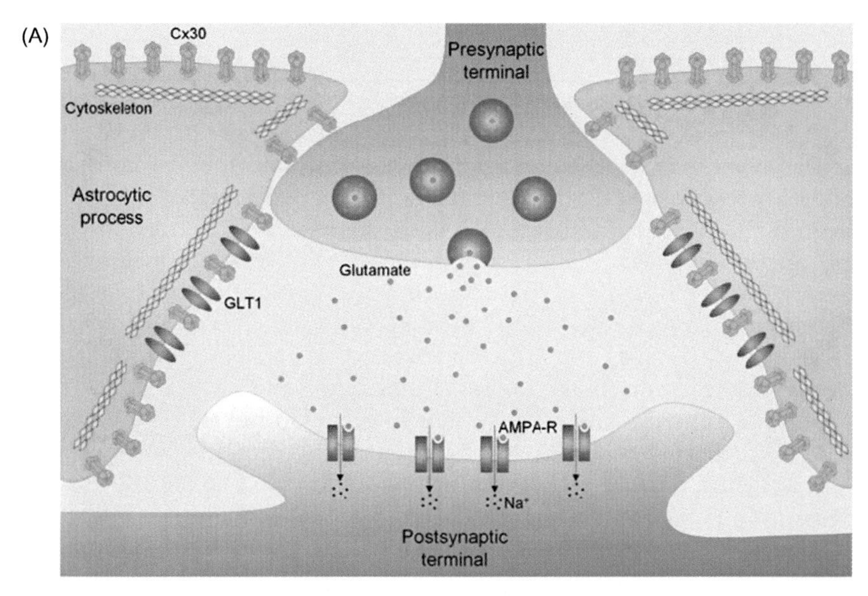

(B)

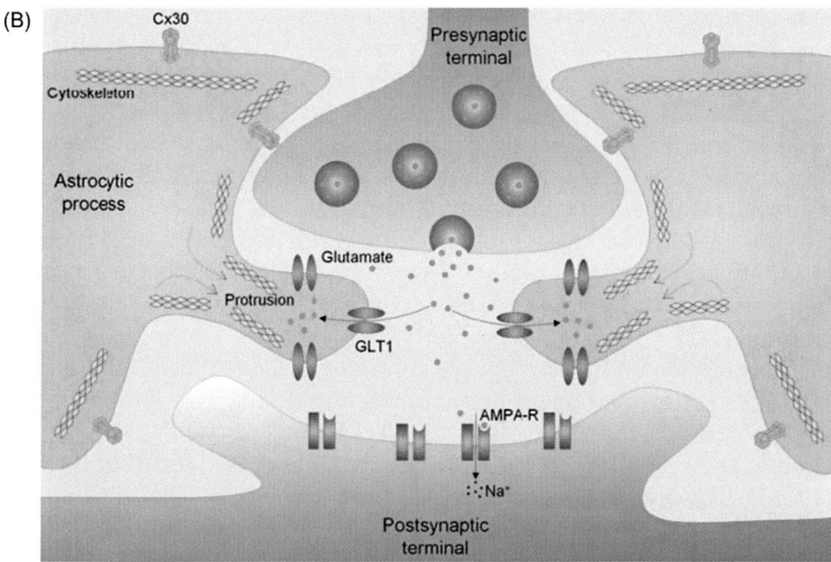

FIGURE 11.2 Schematic depicting the mechanism by which astroglial connexin 30 (Cx30) regulates the activity of AMPA receptors (AMPARs) at the hippocampal synapse. (A) The neurotransmitter glutamate released by the presynaptic terminal via exocytosis activates AMPA receptors in the postsynaptic terminal, causing excitatory synaptic currents generated by the influx of sodium (Na^+). Under steady-state conditions, astrocytic processes containing glutamate transporter 1 (GLT1) are held back from the synaptic cleft, which limits glutamate clearance and allows efficient activation of postsynaptic AMPARs. (B) Downregulation of astroglial Cx30, whose intracellular carboxy-terminal domain interacts with regulatory and structural elements of the cytoskeleton (eg, actin filaments and tubulin), promotes the extension of astrocytic processes into the synaptic cleft (protrusion). Consequently, GLT1 becomes sufficiently close to the synaptic cleft to lower synaptic glutamate levels and reduces the activity of postsynaptic AMPARs and the magnitude of excitatory synaptic transmission. *Source: Reproduced with permission from Clasadonte J, Haydon PG. Connexin 30 controls the extension of astrocytic processes into the synaptic cleft through an unconventional non-channel function. Neurosci Bull 2014;30(6):1045–8.*

HUMAN TISSUE STUDIES

The first human tissue studies to examine GJ coupling in epileptic tissue were conducted in the early 1990s. Increased Cx43 and Cx32 mRNA levels were observed in resected epileptic temporal lobe tissue for the management of intractable epilepsy compared to peritumoral temporal cortex tissue [42]. In a subset of patients with tumors associated with acute seizures, even higher levels of Cx43 mRNA were observed. In contrast, however, Elisevich et al. found no upregulation in mRNA or protein levels of Cx43 in hippocampal tissue resected from patients presenting with intractable epilepsy when compared to nonepileptic tissue from patients requiring temporal lobectomy for life-threatening situations [43]. Although there was a decline in the mean levels of Cx43 mRNA expressed predominantly in astrocytes in the epileptic tissue, there was no significant change in the corresponding protein levels. The authors point out that their findings do not rule out the possibility of a change in the dynamic state (open vs closed) of the GJs playing a role in epileptogenicity.

Lee et al. used fluorescence recovery after photobleaching (FRAP) to quantify GJ coupling in tissue from patients who underwent temporal lobectomies for intractable epilepsy. Specifically, primary astrocytes were derived ex vivo from the hippocampus, hyperexcitable parahippocampus, and the "normal" cortex. This was compared to primary astrocytes derived from the cortical margins with normal EEG activity and the surrounding hyperexcitable cortex from patients who underwent craniotomy for astrocytoma resection [44]. GJ coupling was more pronounced in astrocytes isolated from the hyperexcitable tissue, as determined by faster and more complete fluorescence recovery [44].

The most comprehensive and recent study examining the functional properties of astrocytes from human epilepsy tissue specimens was by Bedner et al. [45]. These authors studied 119 specimens from patients with mesial temporal lobe epilepsy (MTLE) with ($n = 75$) and without ($n = 44$) sclerosis. They found that in MTLE specimens with typical hippocampal sclerosis, there is a complete absence of typical "classical" astrocytes and an absence of astrocyte gap junctional coupling. In contrast, coupled astrocytes were abundant in nonsclerotic hippocampus [45].

Thus far, only three studies have used immunohistochemical techniques to analyze changes in GJ protein expression in human epileptic tissue. Two studies found that hippocampal tissue resected from patients with MTLE had elevated Cx43 immunoreactivity (IR) compared to autopsy controls [46,47]. In addition, decreased levels of Cx32 IR and preservation of Cx36 IR were observed in the epileptic tissue [47]. The third study examined protein localization of Cx43 in patients with epilepsy secondary to focal cortical dysplasia (FCD) and with cryptogenic epilepsy [48]. Patients with FCD type IA and FCD type IIA, and cryptogenic epilepsy revealed a similar distribution of Cx43 IR to that of control tissue. In tissue from patients with FCD type IIB, however, Cx43 IR clustered in large aggregates of puncta around vimentin-immunopositive and CD34-immunonegative balloon cells and hypertrophic astrocytes [48]. Interestingly, no changes in Cx43 mRNA levels were found in epileptic tissue compared to control. The authors did note, however, that the Cx43 transcript levels were relatively homogenous in control tissue but highly variable in the epileptic tissue. Investigating further, the authors found a threefold increase of Cx43 mRNA levels in a subgroup (25%) of cryptogenic epilepsy specimens compared to both the control and FCD tissue [48].

A significant disadvantage to human tissue studies is the lack of appropriate controls. As a result, many studies use tumor specimens as controls for epileptic tissue (and vice versa). Several studies, however, have shown downregulation of GJ protein in tumor tissue. For example, Cx43 expression is decreased in gliomas [49–51] and a variety of breast tumors [52]. Although Aronica et al. found decreased Cx43 IR in high-grade gliomas, they also found increased Cx43 IR in reactive astrocytes in the epileptic cortex surrounding low-grade tumors [51]. This suggests that increased Cx43 levels may contribute to tumor-related seizures. In addition, decreased GJ coupling assayed by FRAP has been observed in primary astrocyte cultures derived from glioma specimens compared to epilepsy tissue [50]. Therefore, the controls used in these studies should be taken into account when considering human tissue data. A second disadvantage of human tissue studies is that they are often a representation of the end-point of a disease. To determine changes in connexin expression and gap junctional coupling during epileptogenesis, animal models must be used.

KNOCKOUT MICE STUDIES

Neuronal Connexins

Although mice deficient in the major neuronal connexin Cx36 showed no obvious phenotypic abnormalities, Cx36 has been implicated in a number of neurological functions and behavioral tasks. For example, Cx36 knockout mice ($Cx36^{-/-}$) exhibited impaired visual transmission [53] and elevated auditory brainstem response thresholds [54]. Electrical coupling and synchronous activity among inhibitory interneurons in the neocortex [35] and neurons of the inferior olivary nucleus [55] were deficient in $Cx36^{-/-}$ mice but no significant change in coupling in the corpus callosum was observed [56]. Deletion of Cx36 led to the loss of functional synapses between GABAergic interneurons in hippocampal slices from adult mice [57]. Cx36 knockout mice exhibited reduced γ-frequency oscillations with sharp wave–burst discharges accompanied by population spikes, or "ripple" oscillations [57,58]. These burst discharges were abolished by carbenoxolone, a general GJ blocker [58]. In contrast, Maier et al. (2002) showed that hippocampal slices from mice lacking Cx36 had less frequent sharp waves and ripple oscillations. An in vivo electrophysiological analysis of the hippocampus revealed that fast-field ripple oscillations were present in both wild-type and $Cx36^{-/-}$ mice but that the power in the γ-frequency band and the magnitude of theta-phase modulation of gamma power were reduced in the knockout mice during wheel running [59].

In the striatum, interneurons provide the main inhibitory input to the principal projection medium-sized spiny neurons. Using patch clamp recordings from medium-sized spiny neurons in the striatum, Cummings et al. [60] determined that Cx36 knockout mice had reduced frequency of both excitatory and inhibitory spontaneous postsynaptic currents. Cx36 is essential for the maintenance of presynaptic inhibition, inducing the regulation of transmission from Ia muscle spindle afferents in the spinal cord [61]. This was determined by dorsal root potentials evoked by low-intensity stimulation of sensory afferents, which were reduced in amplitude and duration in Cx36 knockout mice. Additional knockout studies determined that Cx36 is also involved in long-term potentiation [39]. Recordings from behaving mice showed that spatial selectivity of hippocampal pyramidal neurons was reduced and less stable in knockout mice [38]. Mice deficient in Cx36 also exhibited slower theta oscillations, reflecting altered network activity as well as impaired short-term and spatial memory, but exhibited normal spatial reference memory [38]. In agreement with this, Zlomuzica et al. showed that $Cx36^{-/-}$ mice exhibited impaired one-trial object placement recognition [40]. In addition, Cx36 knockout mice demonstrated increased locomotion and running speed in the open-field test and exhibited more anxiety-like behavior in light-dark box [40]. More recently, coupling was significantly reduced, but not abolished in the thalamic reticular nucleus of Cx36 knockout mice [62].

Frisch et al. [37] showed that $Cx36^{-/-}$ mice had memory impairments that scale with the complexity of the stimuli presented. In contrast to the above findings, they found that the activity patterns and the exploratory- and anxiety-related responses were similar between wild-type and $Cx36^{-/-}$ mice [37]. Sensorimotor capacities and learning and memory processes were impaired in the knockout mice; Cx36-deficient mice displayed slower motor coordination learning in the rotarod test. After a retention interval of 24 hours, knockout mice showed habituation in an open-field test but failed to habituate in the more complex environment of the Y-maze. Interestingly, more pronounced memory impairment was found when Cx36 knockout mice were unable to recognize recently explored objects after short delays [37].

A few studies have examined the role of Cx36 in disease models. After ischemia, Cx36 knockout mice had fewer and shorter postischemic cortical spreading depolarization events, better function outcome, and decreased infarct size [63]. In the 4-aminopyridine (4-AP) model of epilepsy, Cx36 knockout mice exhibited attenuated epileptiform discharges [64]. $Cx36^{-/-}$ mice also had more severe seizures than wild-type mice in response to PTZ [65]. In addition, these mice showed no difference in expression levels of connexin 30.2 (Cx30.2), endothelial connexin 37 (Cx37), Cx43, Cx45, pannexin 1, pannexin 2, or $GABA_A$ receptor α1 protein levels, suggesting the increased sensitivity to PTZ was due to the lack of Cx36. In contrast, in both the low-magnesium artificial CSF (aCSF) model and aconitine (neurotoxin that increases neuronal excitability by increasing intracellular calcium levels) perfusion in low-magnesium aCSF model, neocortical slices from Cx36 knockout mice showed no difference in seizure-like events (SLE) from WT animals [66].

When examining the effect of $GABA_A$ receptor modulation on low magnesium SLE in mouse cortical slides, Voss et al. [67] found that Cx36 knockout mice had similar frequency of SLE compared to wild-type mice in response to both etomidate (augments $GABA_A$ receptors) and picrotoxin (blocks $GABA_A$ receptors). The amplitude of SLE in response to both etomidate and picrotoxin, however, was abolished in knockout mice [67]. In addition, mefloquine, a specific Cx36 blocker, did not alter the seizure-like activity in the wild-type slices, indicating that the reduction in amplitude seen in the knockout slices was likely due to preexisting compensatory changes rather than due to the lack of the interneuron GJs. These findings support the idea that Cx36 knockout mice have increased GABAergic tone as a compensation for lack of interneuron GJs and remind us to consider the possibility of compensatory mechanisms in knockout mice when looking at the literature for genetically modified animals [67].

Cx30.2 is expressed on interneurons and Cx30.2 knockout mice ($Cx30.2^{-/-}$) exhibited no differences in basal excitation and excitation-inhibition balance from wild-type mice [68].

Mice deficient in the neuronal Cx45 ($Cx45^{-/-}$) exhibited no difference from wild-type mice in motor learning, behavioral habituation to a novel environment, and object place recognition, although they did have impaired novel object recognition after short delays [69]. $Cx45^{-/-}$ mice also exhibited similar general excitability, synaptic short-time plasticity, and spontaneous high-frequency oscillations (HFOs; sharp-wave ripples) to wild-type mice. In response to bath stimulation of hippocampal slices with kainate, Cx45 mice had significantly lower γ-oscillation amplitudes in CA3, but not in CA1 subfield, and significantly larger full-width half maximum of the frequency distribution in the CA1 subfield [69].

Glial Connexins

Cx32 is highly expressed by myelinating Schwann cells in the PNS and is found on both neurons and oligodendrocytes. Mice deficient in this GJ ($Cx32^{-/-}$) protein have myelination defects, display intrinsic neuronal hyperexcitability, and have impairments in inhibitory synaptic transmission [70]. Specifically, these mice exhibited separation and splitting of myelin from the axon [71], abnormally thin myelin sheaths [72,73], unusually thick periaxonal collars, and cellular bulb formations reflecting myelin degeneration-induced Schwann cell proliferation [72]. Functionally, $Cx32^{-/-}$ mice experienced slight conductance slowing and increased latency of muscle response after distal stimulation of the sciatic nerve [72]. In neocortical slice preparations, $Cx32^{-/-}$ mice had increased membrane input resistance and enhanced intrinsic excitability in neurons [73].

A reduction in GJ coupling in oligodendrocytes can lead to myelin defects that leave the brain more vulnerable to neurological disorders. Mice deficient in Cx32 were more vulnerable to brief ischemic insults [74]. Mutations in Cx32 are associated with the X-linked form of the hereditary peripheral neuropathy Charcot-Marie-Tooth (CMT) disease [71,72], which leads to muscle weakness and atrophy, areflexia, and foot deformities. The induction of human Cx32 expression protein in Cx32 knockout mice rescued the phenotype by reducing the amount of demyelination [75].

Mice deficient in the oligodendrocyte Cx47 ($Cx47^{-/-}$) showed no obvious morphological or behavioral abnormalities, but the number of coupled cells was reduced by 80% [56]. In addition, ablation of Cx47 completely abolished coupling of oligodendrocytes to astrocytes [56]. Electron microscopic analysis revealed vacuolation of nerve fibers, particularly at sites of the optic nerve where myelination begins in $Cx47^{-/-}$ mice [76]. In addition, mutations in Cx47 are known to cause Pelizaeus-Merzbacher-like disease (PMLD), which is characterized by severe demyelination in the CNS [77]. Cx32/Cx47-double knockout mice developed severe demyelination, inflammation, and astrogliosis [22,78], all of which could be rescued by reintroducing Cx32 [22]. The deletion of both Cx32 and Cx47 in mice completely abolished coupling and caused ~10-fold increase in input resistance in the corpus callosum [56]. In addition, Cx32/Cx47-double knockout mice had an action tremor, later developed tonic seizures and sporadic convulsions, had abundant vacuolation in nerve fibers, and died within 6–10 weeks after birth [76,79].

Other connexin knockout mice revealed a variety of functional roles for connexins. Connexin 29 (Cx29) is expressed by both oligodendrocytes and myelinating Schwann cells, however, deletion of Cx29 alone did not significantly reduce the number of coupled cells in the corpus callosum in mice [56]. It is known that Cx29 is required for normal cochlear function [80], but little is known about the role of Cx29 in seizure development and spread. Connexin 26 (Cx26) is thought to be expressed in astrocytes, although this has recently been debated [81]. Knockout of Cx26 is embryonic lethal [82] and postnatal ablation of Cx26 in the ear resulted in impaired hearing [83]. Similarly, mice deficient in Cx30 also suffer from hearing impairment [84]. Cx30 knockout mice also had reduced exploratory activity in terms of rearings, but not locomotion, in the open-field and object exploration task [85]. They also showed higher open-field center avoidance and corner preference, suggesting anxiogenic behavior, but graded anxiety and rotarod performances were similar to that of wild-type [85]. Interestingly, Cx30 knockout mice experienced altered expression patterns of Cx36 and Cx32 [86].

Knockout of astrocytic Cx43 is neonatal lethal and leads to neural crest migration deficits [87–89]. Mice with astrocyte-specific ablation of Cx43 exhibited normal architecture but revealed the necessity of Cx43 in the neuron–glia interactions required for whisker-related sensory functions and plasticity [29]. These mice were insensitive to hypoxic preconditioning [90] but had increased microglial activation after acute needle stab wound in vivo [91]. Ablation of Cx43 significantly diminished the proliferation and survival of newborn cells [92]. Mice deficient in astrocytic Cx43 had oligodendrocyte–astrocyte coupling but lacked coupling to oligodendrocyte precursors [56]. In transgenic mice with deletion of the Cx43 gene in astrocytes, increased stroke volume, enhanced apoptosis, and increased inflammatory response in focal brain ischemia were observed [93]. After focal ischemia, Cx43 heterozygote mice exhibited enhanced infarct volume and apoptosis compared to wild-type mice [94,95]. Targeted postnatal inactivation of astrocytic Cx43 enhanced spreading depression [96]. Cells from Cx43 knockout mice exhibited increased reactive oxygen species (ROS)-induced astrocytic death in vitro [97].

To examine the various roles of astrocytic GJs, conditional knockout mice were created by crossing $Cx43^{-/-}$ mice with Cx30 knockout mice [25,27,30]. Similar to their wild-type counterpart, these double knockout mice ($Cx30^{-/-}$ $Cx43^{-/-}$) had astrocytes with negative resting membrane potentials, time- and voltage-independent whole-cell currents, normal astrocyte morphologies, and similar cell densities. Although $Cx30^{-/-}$ $Cx43^{-/-}$ mice exhibited no gross behavioral abnormalities, the double deficiency of Cx30 and Cx43 resulted in a complete disruption of coupling [27,30]. Despite having similar synaptic contacts as wild-type mice, double knockout mice showed increased synaptic transmission, inhibitory postsynaptic currents, whole-cell currents, AMPA/NDMA ratio, and number of functional synapses. In addition, $Cx30^{-/-}$ $Cx43^{-/-}$ mice exhibited a reduction in the rheobase and paired pulse facilitation, diminished LTP and amplified LTD at CA1 synapses [27]. Taken together, these results suggest that astrocytic connexin deficiency leads to an increase in the probability of presynaptic release, an overall increased surface density of AMPA receptors (AMPARs), and the unsilencing of synapses. Although gliotransmitter release during basal neuronal activity was unaltered in the double knockout mice, increased synaptic activity was due to enhanced GLT1 and K^+ current amplitudes and the inability of disconnected astrocytes to properly remove the elevated extracellular glutamate and potassium levels during neuronal activity. Astrocytes from $Cx30^{-/-}$ $Cx43^{-/-}$ mice were hypertrophic due to the ability to take up glutamate and potassium and the inability to redistribute them. Knockout mice exhibited prominent cell swelling and decreased extracellular space volume, thus enhancing synaptic activity and prolonging neuronal activation [27]. In agreement with these findings,

Wallraff et al. showed that $Cx30^{-/-}$ $Cx43^{-/-}$ mice exhibited slowed potassium clearance, increased K^+ accumulation during synchronized neuronal firing, and displayed a reduced threshold for the generation of epileptiform events [30].

Mice deficient in both Cx43 and Cx30 had almost no oligodendrocyte-to-astrocyte coupling [56]. In addition, these knockout mice exhibited cellular vacuolation, particularly in oligodendrocytes, reduced myelin basic protein levels in the corpus callosum and cerebellum, deficiencies in mature oligodendrocytes, impaired motor coordination and sensorimotor adaption, impaired spatial working memory, edematous astrocytes, and increased apoptosis [25]. Radial glia express Cx43 and mice lacking Cx30 and Cx43 displayed almost complete inhibition of proliferation and had a significant decline in number of radial glia-like cells and granule neurons [98]. Ablation of these GJs also reduced neurogenesis in the adult hippocampus. In a double knockout of the oligodendrocyte Cx32 and the astrocytic Cx43, animals developed white matter vacuolation without any obvious ultrasonic abnormalities in their myelin [99]. In addition, animals suffered from sensorimotor impairment seizure activity, and a progress loss of astrocytes, but not oligodendrocytes or microglia. These knockout mice suffered from early mortality around 16 weeks of age [99].

FAST OSCILLATION AND SEIZURE SYNCHRONIZATION

Very fast oscillations (VFOs) (generally defined as ~80 to >200 Hz), or ripples, have been shown to precede the onset of seizures in both human tissue [100–102] and in vitro studies [103–105]. Interestingly, network simulations of hippocampal circuitry predicted that these ripples were GJ-dependent [106] and very few axo-axonal GJs were required to produce the appropriate oscillatory activity [107]. Both direct dye coupling between CA1 pyramidal neurons and the electrophysiological detection of axo-axonal GJs confirmed this idea [108]. These VFOs were seen in rat hippocampal slices before epileptiform bursts, were abolished by carbenoxolone, and were not dependent on chemical synaptic transmission [107,109]. Similar findings have been observed in the cat neocortex, in which VFOs were observed at the onset of seizures and the general GJ inhibitor halothane blocked both the ripples and seizure activity in vivo [110,111]. The hypothesis was further strengthened by Maex and DeShutter [112] who demonstrated that VFOs could occur spontaneously during dendritic excitation, in the absence of ectopic axon spikes, and could stop and resume spontaneously, much like a seizure.

Since the conception of this theory, several in vitro studies have supported this hypothesis. Slices from human epileptic neocortex generated spontaneous interictal discharges and preseizure VFOs [113,114]. Although reducing synaptic inhibition failed to affect the occurrence of VFOs, reducing GJ conductance abolished VFOs [114]. In support of this, VFOs in human cortical tissue in vitro could be blocked by carbenoxolone [115]. Rat neocortex slices can generate VFOs when chemical synapses are blocked [113]. γ-Frequencies (30–80 Hz) in basolateral amygdala slices generated by the application of kainic acid (KA) could be blocked by GJ inhibitors carbenoxolone, octanol, and quinine [116]. Interestingly, Hu and Agmon [117] showed that the mode of coupling affects the strength of synchrony.

In vivo, the application of carbenoxolone or quinine on the entorhinal cortex of rats decreased VFOs in the hippocampus in the intracerebroventricular pilocarpine model of epilepsy [118]. Similarly, cortical application of carbenoxolone eliminated VFOs in the rat barrel cortex after multiple-whisker stimulation [119]. Taken together, these data suggest preseizure ripples are generated by electrical coupling of neurons and GJs are involved in the synchronization of VFOs.

Computational modeling has been used to expand on the idea that gap junctional coupling can influence ripple production and tissue synchronization. Traub et al. used a model to predict that VFOs could be generated between proximal axons of cerebellar Purkinje cells if electrical coupling was present [120]. After observing VFOs in patient electrocorticography recordings prior to ictal events, Cunningham et al. [121] used a computational model to demonstrate that steadily increasing GJ conductance led to a steady increase in VFO field frequency. Similar results were shown in a model that demonstrated that an axon plexus can exhibit a distinct kind of ripples and propagation became more reliable with increasing conductance of GJs [122]. Finally, Stacey et al. showed that coupling and physiological noise initiate oscillations and then recruit neighboring tissue, thus leading to network synchrony. Of note, "epileptic" VFOs were superior at recruiting neighboring tissue [123]. Altogether, these findings support the role of electric coupling through GJs in VFOs and seizure synchronization.

EXPERIMENTAL MODELS OF EPILEPSY

One major advantage of experimental models of epilepsy is the ability to examine changes in GJ expression and functionality at various stages of the disease. In addition, a variety of epilepsy and seizure models exist, allowing for the study of several different epilepsies. Both KA, a neuroexcitatory amino acid, and pilocarpine, a muscarinic receptor agonist, are commonly used to induce a model of temporal lobe epilepsy (TLE) in rodents. The drug 4-AP is a potent convulsant that acts by blocking voltage-activated K^+ channels. Picrotoxin is a $GABA_A$ receptor channel blocker with convulsant effects. Similarly, tetanus toxin blocks the release of inhibitory neurotransmitters across the synaptic cleft, thereby causing generalized muscular spasms. Finally, the kindling model of epilepsy involving repeated electrical stimulations is used to study the effects of repeated seizures on the brain. All of these models give insight into potential changes that occur in the pathology of human epilepsy. Many groups have hypothesized that increased GJ coupling may aid in the development of epilepsy and the spread of seizures, but evidence to support this has been mixed.

In the rat intraperitoneal kainic acid (IPKA) model of TLE, Sohl et al. found no change in Cx30, Cx32, and Cx43 mRNA or protein levels 30 days after KA administration. They did find, however, a 44% reduction in Cx36 mRNA with only a slight reduction in Cx36 protein levels [124]. Takahashi et al. found an increased Cx43 protein and reduced Cx30 protein levels 7 days after IPKA-induced status epilepticus [125]. Using the whole-cell patch technique, they also found increased dye coupling between astrocytes and significantly faster decay time kinetics in glutamate transporter-dependent currents in hippocampal brain slices from KA-treated rats [125].

A detailed, region-specific analysis of Cx30 mRNA and immunohistological changes following intracerebroventricular injections of KA in rats was conducted by Condorelli et al. [126]. Early and transient astrocytic Cx30 mRNA and protein IR were upregulated within

6 hours of injection in the cerebral cortex and the thalamus, including the lateral habenular nucleus. At 12–72 hours post-KA treatment, however, no difference in IR from control tissue was observed in those same brain regions. Cx30 mRNA was upregulated in the hypothalamus and in the medial amygdaloid nuclei by 6 hours. At 72 hours, Cx30 mRNA expression in the thalamic nuclei returned toward basal levels but, in the cerebral cortex, levels strongly decreased at 12 and 24 hours postinjection followed by recovery to control levels by 48 hours. In the hippocampus, laterodorsal and mediodorsal nuclei of the thalamus, and in the medial nucleus of the amygdala, Cx30 mRNA levels remained downregulated for the entire time course (6–72 hours), with partial recovery of hippocampal levels at 72 hours postinjection. A progressive decrease in Cx30 IR in the stratum oriens of CA3–CA4 was observed from 12–72 hours postinjection. Of note, Cx30 mRNA was upregulated in the CA3–CA4 pyramidal layer from 6 to 48 hours post-KA treatment and IR was intensely upregulated at 72 hours posttreatment. Interestingly, Cx30 mRNA, normally expressed exclusively on astrocytes, was found in neuronal cells undergoing cell death 6–48 hours after KA treatment [126]. Altogether, this study suggests brain region–specific regulation of Cx30 during the development of epilepsy.

Condorelli et al. [127] extended their previous study with a detailed analysis of mRNA levels of various connexins in response to intracerebroventricular injections of KA. No changes in Cx37, connexin 40 (Cx40), or Cx47 mRNA were observed after KA treatment in rats. From 6 to 48 hours after KA treatment, rats expressed Cx26 and Cx45 mRNA in CA3–CA4 hippocampal layers, localized over nuclei with pyknotic morphology [127]. Cx32 mRNA was upregulated in various regions up to 72 hours after injection. Cx36 was downregulated as early as 6 hours postinjection and was completely abolished by 48 hours post-KA treatment in CA3–CA4 regions of the hippocampus. Cx36 mRNA expression was not influenced by kainate treatment in either the CA1 subfield of the hippocampus or in the cerebral cortex. Cx43 was reduced in CA3–CA4 pyramidal layer after KA treatment but was persistently upregulated in the molecular layer, stratum radiatum, stratum oriens, and the hilus of the hippocampus [127]. These changes are potentially a response to increased neuronal activity or to cell damage.

The most recent and comprehensive evaluation of astrocytic gap junctional coupling in an animal model of TLE was performed by Bedner et al. [45]. In the intracortical kainic acid model of TLE, mice exhibited decreased astrocytic coupling 4–5 days postinjection and completely lacked coupling 3 and 6 months after status epilepticus in the sclerotic hippocampus (Fig. 11.3) [45]. In the nonsclerotic hippocampus, however, coupling remained intact (Fig. 11.3). Interestingly, decreased astrocyte coupling preceded apoptotic neuronal death and the onset of spontaneous seizures. Decreased GJ coupling also impaired K^+ clearance 4 hours postinjection. The authors found that proinflammatory cytokines induced the uncoupling of hippocampal astrocytes in vivo [45], which agreed with similar in vitro findings that proinflammatory cytokines have an inhibitory effect on astrocytic GJ coupling [128]. To test the hypothesis that inflammation may contribute to the pathophysiology of seizures, lipopolysaccharide was injected into mice. Five days postinjection, animals exhibited reduced GJ coupling and reduced Cx43 protein levels, but unchanged Cx30 protein amounts. The uncoupling effect of lipopolysaccharide could be fully prevented with the antiinflammatory and antiepileptic drug levetiracetam [45]. In addition, uncoupling was prevented in Toll-like receptor 4 knockout mice. Taken together, these important data suggest that inflammation may contribute to rapid uncoupling of astrocytes and the uncoupling of astrocytes may be involved in epileptogenesis.

Wu et al. used semiquantitive PCR, Western blotting, and immunohistochemistry to examine the progressive changes of Cx43 and Cx40 expression at 4 hours, 1 day, 1 week, and 2 months after pilocarpine-induced status epilepticus (PISE). They found that at 4 hours and 1 day post-PISE, there were no changes in Cx43 and Cx40 expression. At 1 week and 2 months post-PISE, CX43 and Cx40 mRNA, protein, and IR were increased in the hippocampus [129]. In a separate study, mRNA levels of Cx32, Cx36, and Cx43 were elevated at both the primary and mirror foci 60 minutes after cortical application of 4-AP-induced seizure activity [130,131]. In the same model, Cx26, Cx32, Cx36, and Cx43 mRNA were upregulated in the primary epileptic focus of p16 and p23 rats 1 hour after seizure induction [132]. Cx32 mRNA was upregulated at the primary focus in p10 and p14 rats, however, Cx26, Cx32, Cx36, and Cx43 were the same as control rats of the same age. In contrast, the expression of Cx43 mRNA and protein levels remained unaltered at 1, 3, and 24 hours after intraperitoneal 4-AP injections in rats. At 3 and 24 hours after seizure induction, however, a significant reduction in the ratio of phosphorylated to unphosphorylated protein was found [133]. Interestingly, pretreatment with dizocilpine maleate (MK-801), an *N*-methyl-D-aspartate (NMDA) receptor antagonist, prevented the dephosphorylation of Cx43, ameliorated seizure symptoms, and reduced astrocyte swelling. These results suggest that uncoupling of GJs may be regulated through NMDA receptors.

Jiang et al. [134] discovered that diazoxide, a selective mitochondrial K_{ATP} channel opener, enhanced astrocyte GJ coupling, while blocking them with 5-hydroxydecanoate (5-HD) decreased GJ communication in vitro. Furthermore, treatment with KA decreased both Cx30 and Cx43 expression, which could be reversed with diazoxide and replicated with 5-HD [134]. In the chronic picrotoxin kindling model of epilepsy in rats, Cx40, Cx43, and Cx45 hippocampal protein were downregulated in the seizure-free interval compared to control animals [134]. Interestingly, opening the mitochondrial K_{ATP} channel increased Cx40, Cx43, and Cx45 expression, likely increasing GJ function in the epileptic hippocampus. There was no effect on Cx36 protein [134].

Amygdala Cx43 mRNA expression was found to be decreased or unchanged relative to control tissue at 4, 9, and 10 weeks after injections of tetanus toxin (strength 3 and 9 MLD50) into the amygdala of rats [135]. Similarly, rats electrically kindled in the amygdala until they reached stage 5 seizures demonstrated reduced levels of Cx43 mRNA and IR in the basolateral amygdala that normalized with increasing number of stimulations [135]. In a separate study, no significant change in hippocampal Cx30, Cx36, or Cx43 mRNA, protein levels, or protein localization was seen after kindling in rats 2–4 weeks or 4–6 weeks after a stage 5 seizure, with the exception of decreased levels of Cx36 mRNA and protein 2–3 weeks after kindling [124]. No change in Cx32 mRNA levels in the hippocampus of kindled rats was observed at either time point [124].

In contrast to those findings, Akbarpour et al. [136] looked at Cx30 and Cx32 mRNA and protein levels at the beginning (received a single afterdischarge threshold stimulation), middle (partially kindled rats with focal seizures) and end (fully kindled rats with generalized seizures) of the amygdala kindling process. They found that Cx30 mRNA was upregulated at the beginning of the kindling process and after the acquisition of focal seizures, but was downregulated by the time animals exhibited generalized seizures [136]. Cx30 protein,

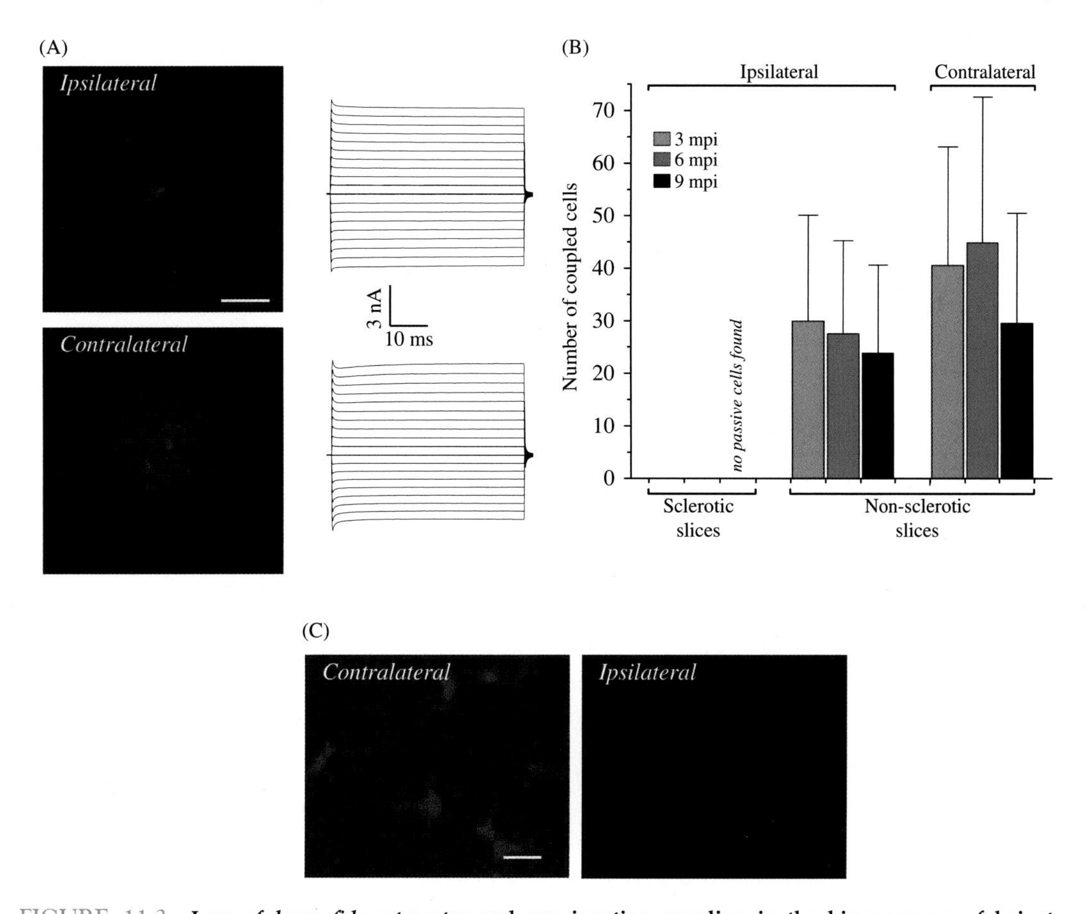

FIGURE 11.3 **Loss of *bona fide* astrocytes and gap junction coupling in the hippocampus of kainate-injected mice.** (A) Representative example showing abolished tracer coupling in sclerotic slices obtained from epileptic mice 3 months postinjection (mpi; top left). In contrast, in the contralateral hippocampus astrocytes displayed abundant gap junction coupling, resembling control conditions (bottom left). Scale bar = 100 mm. Whole-cell currents of the filled cells were elicited (left; 50 ms voltage steps ranging from −160 to +20 mV; 10 mV increments; V_{hold} = −80 mV). (B) Summary of tracer coupling experiments. The extent of intercellular biocytin diffusion was compared in sclerotic and nonsclerotic slices of the injected hemisphere, and in slices from the contralateral hippocampus. 3 and 6 mpi, no spread of biocytin could be detected in sclerotic slices (3 months: n = 12 biocytin-filled passive cells from six animals; 6 months: n = 6 filled passive cells, three animals). Astrocytes were still coupled in nonsclerotic hippocampal slices of the injected hemisphere (3 months: 30 ± 20.1 coupled cells, n = 11 slices, six animals; 6 months: 27.5 ± 17.8 coupled cells, n = 10 slices, five animals) and slices obtained from the contralateral hippocampus (3 months: 40.6 ± 22.5 coupled cells, n = 10 slices, six animals; 6 months: 44.9 ± 27.7 coupled cells, n = 11 slices, three animals). Nine months after status epilepticus, complete loss of cells with passive current pattern was observed in sclerotic segments of the hippocampus (n = 18 screened slices, six animals) whereas in nonsclerotic hippocampal slices ipsilateral to the injection, and in the contralateral hippocampus, astrocytes coupled to 23.8 ± 16.8 (n = 13 slices, five animals) and 29.6 ± 21 (n = 18 slices, five animals) cells, respectively. Gap junction coupling in nonsclerotic ipsilateral slices did not differ from the contralateral side. (C) Loss of SR101 uptake by astrocytes in the sclerotic hippocampus. Incubation with SR101 of slices from epileptic mice 9 mpi resulted in astrocytic labeling in the contralateral CA1 region. In the ipsilateral hippocampus, no labeled cells were detected (n = 3 slices, three animals). Scale bar = 15 mm. *Source: Reproduced with permission from Bedner P, Dupper A, Hüttmann K, Muller J, Herde MK, Dublin P, et al. Astrocyte uncoupling as a cause of human temporal lobe epilepsy. Brain 2015;138(Pt 5):1208–22.*

however, only increased at the beginning of the kindling process, and Cx32 exhibited no mRNA or protein changes throughout the process with the exception of decreased mRNA levels after acquisition of generalized seizures.

Recently, animal experiments have shown that exposure of the brain to albumin leads to epileptiform activity [137]. Braganza et al. used fluorescently labeled albumin injected into the mouse lateral ventricle in vivo to show that albumin was taken up by NG2 cells, astrocytes, and neurons within 1 day postinjection (dpi), with NG2 cells having the highest quantity uptake [138]. At 1 day postalbumin injection, astrocytes exhibited a significant reduction in hippocampal GJ coupling; this effect disappeared by 5 dpi. Interestingly, Cx43 mRNA expression was not altered at either time point.

Tuberous Sclerosis Complex (TSC) is a genetic disorder that causes benign tumors to form in several different organs, including the brain. It is associated with seizures and a conditional knockout of the Tsc1 gene in glia in mice ($Tsc1^{GFAP}$ cKO mice) is a commonly used animal model of this disease. In this model, $Tsc1^{GFAP}$ cKO mice expressed decreased levels of Cx43 in both the neocortex and hippocampus compared to wild-type mice, suggesting impaired GJ coupling between astrocytes [139]. In addition, hippocampal slices from $Tsc1^{GFAP}$ cKO mice have reduced potassium buffering. Both the deficient Cx43 protein expression and the impaired K^+ buffering could be reversed with an inhibitor of the mammalian target of rapamycin (mTOR), suggesting this pathway mediates these effects [139]. Using computer modeling, Amiri et al. evaluated a situation corresponding to the loss of GJ coupling. This model predicts that the uncoupling of astrocytes could contribute to neuronal hyperexcitability and the generation of seizure-like activity [140].

PHARMACOLOGICAL GJ BLOCKERS

In vitro

Numerous studies have looked at the effects of general GJ inhibitors in vitro in various models of epilepsy. Perez-Velazquez et al. [141] demonstrated that dye coupling increased in hippocampal slices made hyperexcitable by perfusion with calcium-free aCSF. Under conditions that reduced gap junctional conductance (intracellular acidification and the use of the GJ blockers halothane and octanol), epileptiform activity in CA1 was suppressed whereas conditions that increased gap junctional conductance (intracellular alkalinization) increased both the frequency and duration of field burst events [141]. Carbenoxolone reduced spontaneous burst activity in rat hippocampal slices bathed in a magnesium-free 4-AP medium [142]. In the high K^+–low Ca^{2+} perfusion fluid model of epilepsy, the GJ blockers heptanol, octanol, and carbenoxolone reduced or completely inhibited spontaneous field bursts in CA3 area of rat hippocampal slices [143]. In agreement with these findings, both low pH conditions and the GJ blockers octanol, olemide, and carbenoxolone prevented prolonged field bursts in the rat dentate granule cell layer in vitro in the low calcium–high potassium model of epilepsy [144].

In the magnesium-free model, halothane, carbenoxolone, and octanol abolished secondary epileptiform discharges (lasting several hundred milliseconds) without affecting primary discharges (typical interictal burst) in the CA3 area of rat hippocampal slices. Trimethylamine (TMA), a GJ opener, reversibly induced both secondary and tertiary (lasting for seconds) discharges [145]. Mefloquine, a Cx36 blocker, had no anticonvulsant effects in neocortical slices in either the magnesium-free aCSF [146], low-magnesium aCSF, or aconitine perfusion in low-magnesium aCSF models of SLE [66]. Carbenoxolone application resulted in a robust decoupling of SLE in the magnesium-free aCSF model [146]. Interestingly, slices from Cx36 knockout mice exhibited coupled SLE under baseline conditions and readily decoupled when exposed to etomidate, a $GABA_A$ receptor modulator known disrupt the spread of population field potential activity [146]. These results suggest that Cx36 alone cannot explain the intracortical communication. Contrary to the previous findings, the GJ blockers quinidine, quinine, and carbenoxolone increased the frequency of SLE in the low-Mg^{2+} model in rat cortical slices [147]. Mefloquine also exhibited excitatory effects in wild-type slices, but these effects were absent in slices from Cx36 knockout mice [147].

In an epilepsy model involving primary afterdischarges (PADs) after repetitive tetanization of Schaffer collaterals in rat hippocampal slices, levels of Cx26, Cx32, Cx36, and Cx43 remained unchanged; however, levels of nonphosphorylated Cx43 were decreased [148]. In this model, both carbenoxolone and acidosis reduced PADs whereas alkalinization increased PADs. Similarly, both mimetic peptides specific to Cx43 and carbenoxolone reduced FRAP, indicating reduced GJ coupling, in the stratum radiatum of the CA1 region and attenuated spontaneous seizure-like activity in rat organotypic hippocampal slice cultures; mimetic peptides, however, only exhibited these effects after longer exposure times. Only carbenoxolone could suppress epileptiform responses evoked by brief tetanic stimulations [149]. Chronic exposure to bicuculline, a $GABA_A$ receptor blocker, resulted in epileptiform discharges and increased the rate of FRAP, indicating an increase in GJ coupling, in cultured hippocampal slices [150]. Contrary to the tetanization model, bicuculline increased Cx43 and Cx32 mRNA and membrane protein levels whereas the levels of Cx26 and Cx36 remained unchanged. In this model, carbenoxolone reversibly inhibited spontaneous and evoked epileptiform discharges [150].

Sharp wave–ripple complexes (SPW-Rs) and recurrent epileptiform discharges (REDs) are both characterized by fast synchronized network oscillations. SPW-Rs are highly synchronous and potentially contribute to epileptiform activity; although the mechanism by which this occurs is unknown, it was speculated to be through GJs [151]. In hippocampal slices from adult rats, carbenoxolone blocked both SPW-Rs and REDs. The application of mefloquine did not influence SPW-Rs and had only a minor effect on REDs but significant reduced superimposed ripples [152]. Carbenoxolone decreased the frequency and cumulative duration of spontaneous ictal-like activities in the 4-AP model in thalamocortical slices from genetic absence epilepsy rats from Strasbourg (GAERS) and nonepileptic rats [153]. In neocortical slices obtained from patients with TLE and patients with FCD during the application of 4-AP, spontaneous synchronous events were reduced by carbenoxolone and octanol [154].

In vivo

Despite the accumulating evidence of the role of GJs in epileptiform activity, few studies have exampled GJ inhibition in vivo. In the rat focal seizure model in which 4-AP was applied to the cortex, carbenoxolone caused a significant decrease in seizure activity at both the primary and mirror foci when applied 60 minutes after the induction of epileptiform activity [130–132]. Pretreatment with carbenoxolone 5 minutes before a seizure, however, had no statistically significant effect on seizure activity [130]. Complimentary to these findings, the opening of GJs with trimethylamine application 60 minutes after seizure induction both lengthened seizure duration and increased seizure amplitude [131,132]. In the same 4-AP model, application of quinine, a selective blocker of Cx36, before the induction of epileptiform activity had no effect on basic cortical activity and only slightly reduced epileptogenesis [155]. Quinine applied at the active epileptic focus increased the number of seizures but the summated ictal activity was decreased due to the reduction in the duration of seizures.

Both carbenoxolone [156] and octanol [157] inhibited epileptiform activity in the penicillin model of epilepsy in rats. Similarly, carbenoxolone and the general GJ blocker meclofenamic acid both reduced seizures in focal cortical tetanus toxin model of epilepsy in rats [158]. In the 4-AP model of epilepsy in awake rats, carbenoxolone reduced the amplitude and frequency of epileptiform discharges and reduced the number and duration of epileptiform trains in the entorhinal cortex and hippocampus [159]. The duration of cortical spike-wave discharges in adult GAERS was decreased by carbenoxolone [153]. Low doses of fluorocitrate, a blocker of the citric acid cycle enzyme aconitase, disrupted astroglial metabolism and caused epileptiform EEG discharges when injected into the cerebral cortex of rats [160]. Intraperitoneal injections of octanol blocked or delayed the occurrence of epileptic discharges after cortical injections of fluorocitrate. Interestingly, halothane, typically an in vitro GJ blocker, had no effect [160].

One main limitation to pharmacological studies on GJ inhibition is the lack of specificity of many of the drugs used [161]. For example, 1 μM mefloquine, considered a specific blocker of Cx36, reduced Cx36 currents by 90% but 3 μM mefloquine has also been shown to reduce currents of the lens GJ protein Cx50 by 97% [162]. In addition, although 3 μM mefloquine had no effect on Cx43, 10 μM caused a 43% reduction in currents and 30 μM resulted in a complete block. More specific inhibitors of connexins and in vivo studies will be required to more clearly define the role of GJs in epileptogenesis.

CONCLUSION

While electrotonic coupling of cells in the CNS has been known for several decades, the precise role of GJs in the neuronal hypersynchrony of epilepsy is only beginning to be understood. Human tissue studies and animal models of epilepsy have demonstrated changes in connexin protein levels and GJ conductance. Transgenic knockout mice have individually established the functional significance of various GJ proteins while pharmacological studies have revealed powerful anticonvulsant effects of GJ inhibition. Converging evidence from in vivo, in vitro, and computational models have highlighted the role of electrical coupling in VFOs, which is thought to play a role in initiation and spread of seizure activity. Future studies should focus on the use of these tools to determine the specific roles of glial versus neuronal GJs in the pathophysiology of epilepsy.

References

[1] Loewenstein WR. Junctional intercellular communication: the cell-to-cell membrane channel. Physiol Rev 1981;61(4):829–913.
[2] Watanabe A. The interaction of electrical activity among neurons of lobster cardiac ganglion. Jpn J Physiol 1958;8(4):305–18.
[3] Bennett MV, Aljure E, Nakajima Y, Pappas GD. Electrotonic junctions between teleost spinal neurons: electrophysiology and ultrastructure. Science 1963;141:262–4.
[4] Nelson PG. Interaction between spinal motoneurons of the cat. J Neurophysiol 1966;29(2):275–87.
[5] Baker R, Llinás R. Electrotonic coupling between neurones in the rat mesencephalic nucleus. J Physiol 1971;212(1):45–63.
[6] Korn H, Sotelo C, Crepel F. Electronic coupling between neurons in the rat lateral vestibular nucleus. Exp Brain Res 1973;16(3):255–75.
[7] MacVicar BA, Dudek FE. Electrotonic coupling between pyramidal cells: a direct demonstration in rat hippocampal slices. Science 1981;213(4509):782–5.
[8] Jefferys JG, Haas HL. Synchronized bursting of CA1 hippocampal pyramidal cells in the absence of synaptic transmission. Nature 1982;300(5891):448–50.
[9] Taylor CP, Dudek FE. Synchronous neural afterdischarges in rat hippocampal slices without active chemical synapses. Science 1982;218(4574):810–2.
[10] Haas HL, Jefferys JG, Slater NT, Carpenter DO. Modulation of low calcium induced field bursts in the hippocampus by monoamines and cholinomimetics. Pflugers Arch 1984;400(1):28–33.
[11] Konnerth A, Heinemann U, Yaari Y. Slow transmission of neural activity in hippocampal area CA1 in absence of active chemical synapses. Nature 1984;307(5946):69–71.
[12] Taylor CP, Dudek FE. Synchronization without active chemical synapses during hippocampal afterdischarges. J Neurophysiol 1984;52(1):143–55.
[13] Taylor CP, Dudek FE. Excitation of hippocampal pyramidal cells by an electrical field effect. J Neurophysiol 1984;52(1):126–42.
[14] Laird DW. Life cycle of connexins in health and disease. Biochem J 2006;394(Pt 3):527–43.
[15] Willecke K, Eiberger J, Degen J, Eckardt D, Romualdi A, Güldenagel M, et al. Structural and functional diversity of connexin genes in the mouse and human genome. Biol Chem 2002;383(5):725–37.
[16] Baranova A, Ivanov D, Petrash N, Pestova A, Skoblov M, Kelmanson I, et al. The mammalian pannexin family is homologous to the invertebrate innexin gap junction proteins. Genomics 2004;83(4):706–16.
[17] Kanno Y, Loewenstein WR. Intercellular diffusion. Science 1964;143(3609):959–60.
[18] Loewenstein WR, Kanno Y. Studies on an epithelial (gland) cell junction. I. modifications of surface membrane permeability. J Cell Biol 1964;22:565–86.
[19] Kanno Y, Loewenstein WR. Cell-to-cell passage of large molecules. Nature 1966;212(5062):629–30.
[20] Gosejacob D, Dublin P, Bedner P, Hüttmann K, Zhang J, Tress O, et al. Role of astroglial connexin 30 in hippocampal gap junction coupling. Glia 2011;59(3):511–9.
[21] Eugenín EA, Eckardt D, Theis M, Willecke K, Bennett MV, Saez JC. Microglia at brain stab wounds express connexin 43 and in vitro form functional gap junctions after treatment with interferon-γ and tumor necrosis factor-α. Proc Natl Acad Sci USA 2001;98(7):4190–5.
[22] Schiza N, Sargiannidou I, Kagiava A, Karaiskos C, Nearchou M, Kleopa KA. Transgenic replacement of Cx32 in gap junction-deficient oligodendrocytes rescues the phenotype of a hypomyelinating leukodystrophy model. Hum Mol Genet 2015;24(7):2049–64.
[23] Griemsmann S, Höft SP, Bedner P, Zhang J, von Staden E, Beinhauer A, et al. Characterization of panglial gap junction networks in the thalamus, neocortex, and hippocampus reveals a unique population of glial cells. Cereb Cortex 2014;25(10):3420–33.
[24] Anders S, Minge D, Griemsmann S, Herde MK, Steinhäuser C, Henneberger C. Spatial properties of astrocyte gap junction coupling in the rat hippocampus. Phil Trans R Soc Lond Ser B Biol Sci 2014;369(1654):20130600.
[25] Lutz SE, Zhao Y, Gulinello M, Lee SC, Raine CS, Brosnan CF. Deletion of astrocyte connexins 43 and 30 leads to a dysmyelinating phenotype and hippocampal CA1 vacuolation. J Neurosci 2009;29(24):7743–52.
[26] Pannasch U, Freche D, Dallérac G, Ghézali G, Escartin C, Ezan P, et al. Connexin 30 sets synaptic strength by controlling astroglial synapse invasion. Nat Neurosci 2014;17(4):549–58.

[27] Pannasch U, Vargova L, Reingruber J, Ezan P, Holcman D, Giaume C, et al. Astroglial networks scale synaptic activity and plasticity. Proc Natl Acad Sci USA 2011;108(20):8467–72.

[28] Chever O, Pannasch U, Ezan P, Rouach N. Astroglial connexin 43 sustains glutamatergic synaptic efficacy. Phil Trans R Soc Lond Ser B Biol Sci 2014;369(1654):20130596.

[29] Han Y, Yu HX, Sun ML, Wang Y, Xi W, Yu YQ. Astrocyte-restricted disruption of connexin 43 impairs neuronal plasticity in mouse barrel cortex. Eur J Neurosci 2014;39(1):35–45.

[30] Wallraff A, Köhling R, Heinemann U, Theis M, Willecke K, Steinhäuser C. The impact of astrocytic gap junctional coupling on potassium buffering in the hippocampus. J Neurosci 2006;26(20):5438–47.

[31] Clasadonte J, Haydon PG. Connexin 30 controls the extension of astrocytic processes into the synaptic cleft through an unconventional non-channel function. Neurosci Bull 2014;30(6):1045–8.

[32] Wallraff A, Odermatt B, Willecke K, Steinhäuser C. Distinct types of astroglial cells in the hippocampus differ in gap junction coupling. Glia 2004;48(1):36–43.

[33] Rash JE, Yasumura T, Dudek FE, Nagy JI. Cell-specific expression of connexins and evidence of restricted gap junctional coupling between glial cells and between neurons. J Neurosci 2001;21(6):1983–2000.

[34] Maxeiner S, Kruger O, Schilling K, Traub O, Urschel S, Willecke K. Spatiotemporal transcription of connexin 45 during brain development results in neuronal expression in adult mice. Neuroscience 2003;119(3):689–700.

[35] Deans MR, Gibson JR, Sellitto C, Connors BW, Paul DL. Synchronous activity of inhibitory networks in neocortex requires electrical synapses containing connexin 36. Neuron 2001;31(3):477–85.

[36] Venance L, Rozov A, Blatow M, Burnashev N, Feldmeyer D, Monyer H. Connexin expression in electrically coupled postnatal rat brain neurons. Proc Natl Acad Sci USA 2000;97(18):10260–5.

[37] Frisch C, De Souza-Silva MA, Söhl G, Güldenagel M, Willecke K, Huston JP, et al. Stimulus complexity dependent memory impairment and changes in motor performance after deletion of the neuronal gap junction protein connexin 36 in mice. Behav Brain Res 2005;157(1):177–85.

[38] Allen K, Fuchs EC, Jaschonek H, Bannerman DM, Monyer H. Gap junctions between interneurons are required for normal spatial coding in the hippocampus and short-term spatial memory. J Neurosci 2011;31(17):6542–52.

[39] Wang Y, Belousov AB. Deletion of neuronal gap junction protein connexin 36 impairs hippocampal LTP. Neurosci Lett 2011;502(1):30–2.

[40] Zlomuzica A, Viggiano D, Degen J, Binder S, Ruocco LA, Sadile AG, et al. Behavioral alterations and changes in Ca^{2+}/calmodulin kinase II levels in the striatum of connexin 36 deficient mice. Behav Brain Res 2012;226(1):293–300.

[41] Traub RD, Bibbig A, Fisahn A, LeBeau FE, Whittington MA, Buhl EH. A model of γ-frequency network oscillations induced in the rat CA3 region by carbachol in vitro. Eur J Neurosci 2000;12(11):4093–106.

[42] Naus CC, Bechberger JF, Paul DL. Gap junction gene expression in human seizure disorder. Exp Neurol 1991;111(2):198–203.

[43] Elisevich K, Rempel SA, Smith BJ, Edvardsen K. Hippocampal connexin 43 expression in human complex partial seizure disorder. Exp Neurol 1997;145(1):154–64.

[44] Lee SH, Magge S, Spencer DD, Sontheimer H, Cornell-Bell AH. Human epileptic astrocytes exhibit increased gap junction coupling. Glia 1995;15(2):195–202.

[45] Bedner P, Dupper A, Hüttmann K, Müller J, Herde MK, Dublin P, et al. Astrocyte uncoupling as a cause of human temporal lobe epilepsy. Brain 2015;138(Pt 5):1208–22.

[46] Fonseca CG, Green CR, Nicholson LF. Upregulation in astrocytic connexin 43 gap junction levels may exacerbate generalized seizures in mesial temporal lobe epilepsy. Brain Res 2002;929(1):105–16.

[47] Collignon F, Wetjen NM, Cohen-Gadol AA, Cascino GD, Parisi J, Meyer FB, et al. Altered expression of connexin subtypes in mesial temporal lobe epilepsy in humans. J Neurosurg 2006;105(1):77–87.

[48] Garbelli R, Frassoni C, Condorelli DF, Trovato Salinaro A, Musso N, Medici V, et al. Expression of connexin 43 in the human epileptic and drug-resistant cerebral cortex. Neurology 2011;76(10):895–902.

[49] Huang RP, Hossain MZ, Sehgal A, Boynton AL. Reduced connexin 43 expression in high-grade human brain glioma cells. J Surg Oncol 1999;70(1):21–4.

[50] Soroceanu L, Manning Jr. TJ, Sontheimer H. Reduced expression of connexin 43 and functional gap junction coupling in human gliomas. Glia 2001;33(2):107–17.

[51] Aronica E, Gorter JA, Jansen GH, Leenstra S, Yankaya B, Troost D. Expression of connexin 43 and connexin 32 gap junction proteins in epilepsy-associated brain tumors and in the perilesional epileptic cortex. Acta Neuropathol (Berl) 2001;101(5):449–59.

[52] Laird DW, Fistouris P, Batist G, Alpert L, Huynh HT, Carystinos GD, et al. Deficiency of connexin 43 gap junctions is an independent marker for breast tumors. Cancer Res 1999;59(16):4104–10.
[53] Güldenagel M, Ammermüller J, Feigenspan A, Teubner B, Degen J, Söhl G, et al. Visual transmission deficits in mice with targeted disruption of the gap junction gene connexin 36. J Neurosci 2001;21(16):6036–44.
[54] Blakley BW, Garcia CE, da Sliva SR, Florêncio VM, Nagy JI. Elevated auditory brainstem response thresholds in mice with Connexin 36 gene ablation. Acta Otolaryngol 2015:1–5.
[55] Long MA, Deans MR, Paul DL, Connors BW. Rhythmicity without synchrony in the electrically uncoupled inferior olive. J Neurosci 2002;22(24):10898–905.
[56] Maglione M, Tress O, Haas B, Karram K, Trotter J, Willecke K, et al. Oligodendrocytes in mouse corpus callosum are coupled via gap junction channels formed by connexin 47 and connexin 32. Glia 2010;58(9):1104–17.
[57] Hormuzdi SG, Pais I, LeBeau FE, Towers SK, Rozov A, Buhl EH, et al. Impaired electrical signaling disrupts gamma frequency oscillations in connexin 36-deficient mice. Neuron 2001;31(3):487–95.
[58] Pais I, Hormuzdi SG, Monyer H, Traub RD, Wood IC, Buhl EH, et al. Sharp wave-like activity in the hippocampus in vitro in mice lacking the gap junction protein connexin 36. J Neurophysiol 2003;89(4):2046–54.
[59] Buhl DL, Harris KD, Hormuzdi SG, Monyer H, Buzsáki G. Selective impairment of hippocampal γ oscillations in connexin 36 knockout mouse in vivo. J Neurosci 2003;23(3):1013–8.
[60] Cummings DM, Yamazaki I, Cepeda C, Paul DL, Levine MS. Neuronal coupling via connexin 36 contributes to spontaneous synaptic currents of striatal medium-sized spiny neurons. J Neurosci Res 2008;86(10):2147–58.
[61] Bautista W, Nagy JI, Dai Y, McCrea DA. Requirement of neuronal connexin 36 in pathways mediating presynaptic inhibition of primary afferents in functionally mature mouse spinal cord. J Physiol 2012;590 (Pt 16):3821–39.
[62] Lee SC, Patrick SL, Richardson KA, Connors BW. Two functionally distinct networks of gap junction-coupled inhibitory neurons in the thalamic reticular nucleus. J Neurosci 2014;34(39):13170–82.
[63] Bargiotas P, Muhammad S, Rahman M, Jakob N, Trabold R, Fuchs E, et al. Connexin 36 promotes cortical spreading depolarization and ischemic brain damage. Brain Res 2012;1479:80–5.
[64] Maier N, Güldenagel M, Söhl G, Siegmund H, Willecke K, Draguhn A. Reduction of high-frequency network oscillations (ripples) and pathological network discharges in hippocampal slices from connexin 36-deficient mice. J Physiol 2002;541(Pt 2):521–8.
[65] Jacobson GM, Voss LJ, Melin SM, Mason JP, Cursons RT, Steyn-Ross DA, et al. Connexin 36 knockout mice display increased sensitivity to pentylenetetrazol-induced seizure-like behaviors. Brain Res 2010;1360:198–204.
[66] Voss LJ, Mutsaerts N, Sleigh JW. Connexin 36 gap junction blockade is ineffective at reducing seizure-like event activity in neocortical mouse slices. Epilepsy Res Treat 2010;2010:310753.
[67] Voss LJ, Melin S, Jacobson G, Sleigh JW. GABAergic compensation in connexin 36 knockout mice evident during low-magnesium seizure-like event activity. Brain Res 2010;1360:49–55.
[68] Kreuzberg MM, Deuchars J, Weiss E, Schober A, Sonntag S, Wellershaus K, et al. Expression of connexin 30.2 in interneurons of the central nervous system in the mouse. Mol Cell Neurosci 2008;37(1):119–34.
[69] Zlomuzica A, Reichinnek S, Maxeiner S, Both M, May E, Wörsdörfer P, et al. Deletion of connexin 45 in mouse neurons disrupts one-trial object recognition and alters kainate-induced γ-oscillations in the hippocampus. Physiol Behav 2010;101(2):245–53.
[70] Nemani VM, Binder DK. Emerging role of gap junctions in epilepsy. Histol Histopathol 2005;20(1):253–9.
[71] Scherer SS, Xu YT, Nelles E, Fischbeck K, Willecke K, Bone LJ. Connexin 32-null mice develop demyelinating peripheral neuropathy. Glia 1998;24(1):8–20.
[72] Anzini P, Neuberg DH, Schachner M, Nelles E, Willecke K, Zielasek J, et al. Structural abnormalities and deficient maintenance of peripheral nerve myelin in mice lacking the gap junction protein connexin 32. J Neurosci 1997;17(12):4545–51.
[73] Sutor B, Schmolke C, Teubner B, Schirmer C, Willecke K. Myelination defects and neuronal hyperexcitability in the neocortex of connexin 32-deficient mice. Cereb Cortex 2000;10(7):684–97.
[74] Oguro K, Jover T, Tanaka H, Lin Y, Kojima T, Oguro N, et al. Global ischemia-induced increases in the gap junctional proteins connexin 32 (Cx32) and Cx36 in hippocampus and enhanced vulnerability of Cx32 knockout mice. J Neurosci 2001;21(19):7534–42.
[75] Scherer SS, Xu YT, Messing A, Willecke K, Fischbeck KH, Jeng LJ. Transgenic expression of human connexin 32 in myelinating Schwann cells prevents demyelination in connexin 32-null mice. J Neurosci 2005;25(6): 1550–9.

[76] Odermatt B, Wellershaus K, Wallraff A, Seifert G, Degen J, Euwens C, et al. Connexin 47 (Cx47)-deficient mice with enhanced green fluorescent protein reporter gene reveal predominant oligodendrocytic expression of Cx47 and display vacuolized myelin in the CNS. J Neurosci 2003;23(11):4549–59.

[77] Orthmann-Murphy JL, Enriquez AD, Abrams CK, Scherer SS. Loss-of-function GJA12/Connexin 47 mutations cause Pelizaeus-Merzbacher-like disease. Mol Cell Neurosci 2007;34(4):629–41.

[78] Tress O, Maglione M, Zlomuzica A, May D, Dicke N, Degen J, et al. Pathologic and phenotypic alterations in a mouse expressing a connexin 47 missense mutation that causes Pelizaeus-Merzbacher-like disease in humans. PLoS Genet 2011;7(7):e1002146.

[79] Menichella DM, Goodenough DA, Sirkowski E, Scherer SS, Paul DL. Connexins are critical for normal myelination in the CNS. J Neurosci 2003;23(13):5963–73.

[80] Tang W, Zhang Y, Chang Q, Ahmad S, Dahlke I, Yi H, et al. Connexin 29 is highly expressed in cochlear Schwann cells, and it is required for the normal development and function of the auditory nerve of mice. J Neurosci 2006;26(7):1991–9.

[81] Tonkin RS, Mao Y, O'Carroll SJ, Nicholson LF, Green CR, Gorrie CA, et al. Gap junction proteins and their role in spinal cord injury. Front Mol Neurosci 2014;7:102.

[82] Gabriel HD, Jung D, Butzler C, Temme A, Traub O, Winterhager E, et al. Transplacental uptake of glucose is decreased in embryonic lethal connexin 26-deficient mice. J Cell Biol 1998;140(6):1453–61.

[83] Cohen-Salmon M, Ott T, Michel V, Hardelin JP, Perfettini I, Eybalin M, et al. Targeted ablation of connexin 26 in the inner ear epithelial gap junction network causes hearing impairment and cell death. Curr Biol 2002;12(13):1106–11.

[84] Teubner B, Michel V, Pesch J, Lautermann J, Cohen-Salmon M, Söhl G, et al. Connexin 30 (Gjb6)-deficiency causes severe hearing impairment and lack of endocochlear potential. Hum Mol Genet 2003;12(1):13–21.

[85] Dere E, De Souza-Silva MA, Frisch C, Teubner B, Söhl G, Willecke K, et al. Connexin 30-deficient mice show increased emotionality and decreased rearing activity in the open-field along with neurochemical changes. Eur J Neurosci 2003;18(3):629–38.

[86] Lynn BD, Tress O, May D, Willecke K, Nagy JI. Ablation of connexin 30 in transgenic mice alters expression patterns of connexin 26 and connexin 32 in glial cells and leptomeninges. Eur J Neurosci 2011;34(11):1783–93.

[87] Lo CW, Waldo KL, Kirby ML. Gap junction communication and the modulation of cardiac neural crest cells. Trends Cardiovasc Med 1999;9(3-4):63–9.

[88] Xu X, Li WE, Huang GY, Meyer R, Chen T, Luo Y, et al. Modulation of mouse neural crest cell motility by N-cadherin and connexin 43 gap junctions. J Cell Biol 2001;154(1):217–30.

[89] Fushiki S, Perez Velazquez JL, Zhang L, Bechberger JF, Carlen PL, Naus CC. Changes in neuronal migration in neocortex of connexin 43 null mutant mice. J Neuropathol Exp Neurol 2003;62(3):304–14.

[90] Lin JH, Lou N, Kang N, Takano T, Hu F, Han X, et al. A central role of connexin 43 in hypoxic preconditioning. J Neurosci 2008;28(3):681–95.

[91] Theodoric N, Bechberger JF, Naus CC, Sin WC. Role of gap junction protein connexin 43 in astrogliosis induced by brain injury. PLoS ONE 2012;7(10):e47311.

[92] Liebmann M, Stahr A, Guenther M, Witte OW, Frahm C. Astrocytic Cx43 and Cx30 differentially modulate adult neurogenesis in mice. Neurosci Lett 2013;545:40–5.

[93] Nakase T, Söhl G, Theis M, Willecke K, Naus CC. Increased apoptosis and inflammation after focal brain ischemia in mice lacking connexin 43 in astrocytes. Am J Pathol 2004;164(6):2067–75.

[94] Nakase T, Fushiki S, Naus CC. Astrocytic gap junctions composed of connexin 43 reduce apoptotic neuronal damage in cerebral ischemia. Stroke 2003;34(8):1987–93.

[95] Siushansian R, Bechberger JF, Cechetto DF, Hachinski VC, Naus CC. Connexin 43 null mutation increases infarct size after stroke. J Comp Neurol 2001;440(4):387–94.

[96] Theis M, Jauch R, Zhuo L, Speidel D, Wallraff A, Döring B, et al. Accelerated hippocampal spreading depression and enhanced locomotory activity in mice with astrocyte-directed inactivation of connexin 43. J Neurosci 2003;23(3):766–76.

[97] Le HT, Sin WC, Lozinsky S, Bechberger J, Vega JL, Guo XQ, et al. Gap junction intercellular communication mediated by connexin 43 in astrocytes is essential for their resistance to oxidative stress. J Biol Chem 2014;289(3):1345–54.

[98] Kunze A, Congreso MR, Hartmann C, Wallraff-Beck A, Hüttmann K, Bedner P, et al. Connexin expression by radial glia-like cells is required for neurogenesis in the adult dentate gyrus. Proc Natl Acad Sci USA 2009;106(27):11336–41.

[99] Magnotti LM, Goodenough DA, Paul DL. Deletion of oligodendrocyte Cx32 and astrocyte Cx43 causes white matter vacuolation, astrocyte loss and early mortality. Glia 2011;59(7):1064–74.
[100] Allen PJ, Fish DR, Smith SJ. Very high-frequency rhythmic activity during SEEG suppression in frontal lobe epilepsy. Electroencephalogr Clin Neurophysiol 1992;82(2):155–9.
[101] Fisher RS, Webber WR, Lesser RP, Arroyo S, Uematsu S. High-frequency EEG activity at the start of seizures. J Clin Neurophysiol 1992;9(3):441–8.
[102] Alarcon G, Binnie CD, Elwes RD, Polkey CE. Power spectrum and intracranial EEG patterns at seizure onset in partial epilepsy. Electroencephalogr Clin Neurophysiol 1995;94(5):326–37.
[103] Traub RD, Whittington MA, Buhl EH, LeBeau FE, Bibbig A, Boyd S, et al. A possible role for gap junctions in generation of very fast EEG oscillations preceding the onset of, and perhaps initiating, seizures. Epilepsia 2001;42(2):153–70.
[104] Khosravani H, Pinnegar CR, Mitchell JR, Bardakjian BL, Federico P, Carlen PL. Increased high-frequency oscillations precede in vitro low-Mg^{2+} seizures. Epilepsia 2005;46(8):1188–97.
[105] Chiu AW, Jahromi SS, Khosravani H, Carlen PL, Bardakjian BL. The effects of high-frequency oscillations in hippocampal electrical activities on the classification of epileptiform events using artificial neural networks. J Neural Eng 2006;3(1):9–20.
[106] Traub RD, Schmitz D, Jefferys JG, Draguhn A. High-frequency population oscillations are predicted to occur in hippocampal pyramidal neuronal networks interconnected by axoaxonal gap junctions. Neuroscience 1999;92(2):407–26.
[107] Traub RD, Michelson-Law H, Bibbig A, Buhl EH, Whittington MA. Gap junctions, fast oscillations and the initiation of seizures. Binder DK, Scharfman HE, editors. Recent advances in epilepsy research. New York, NY: Kluwer Academic/Plenum Publishers; 2004. p. 110–22.
[108] Schmitz D, Schuchmann S, Fisahn A, Draguhn A, Buhl EH, Petrasch-Parwez E, et al. Axo-axonal coupling. A novel mechanism for ultrafast neuronal communication. Neuron 2001;31(5):831–40.
[109] Draguhn A, Traub RD, Schmitz D, Jefferys JG. Electrical coupling underlies high-frequency oscillations in the hippocampus in vitro. Nature 1998;394(6689):189–92.
[110] Grenier F, Timofeev I, Steriade M. Neocortical very fast oscillations (ripples, 80–200 Hz) during seizures: intracellular correlates. J Neurophysiol 2003;89(2):841–52.
[111] Timofeev I, Steriade M. Neocortical seizures: initiation, development and cessation. Neuroscience 2004;123(2):299–336.
[112] Maex R, De Schutter E. Mechanism of spontaneous and self-sustained oscillations in networks connected through axo-axonal gap junctions. Eur J Neurosci 2007;25(11):3347–58.
[113] Traub RD, Duncan R, Russell AJ, Baldeweg T, Tu Y, Cunningham MO, et al. Spatiotemporal patterns of electrocorticographic very fast oscillations (>80 Hz) consistent with a network model based on electrical coupling between principal neurons. Epilepsia 2010;51(8):1587–97.
[114] Roopun AK, Simonotto JD, Pierce ML, Jenkins A, Nicholson C, Schofield IS, et al. A nonsynaptic mechanism underlying interictal discharges in human epileptic neocortex. Proc Natl Acad Sci USA 2010;107(1):338–43.
[115] Simon A, Traub RD, Vladimirov N, Jenkins A, Nicholson C, Whittaker RG, et al. Gap junction networks can generate both ripple-like and fast ripple-like oscillations. Eur J Neurosci 2014;39(1):46–60.
[116] Randall FE, Whittington MA, Cunningham MO. Fast oscillatory activity induced by kainate receptor activation in the rat basolateral amygdala in vitro. Eur J Neurosci 2011;33(5):914–22.
[117] Hu H, Agmon A. Properties of precise firing synchrony between synaptically coupled cortical interneurons depend on their mode of coupling. J Neurophysiol 2015;114(1):624–37.
[118] Ventura-Mejía C, Medina-Ceja L. Decreased fast ripples in the hippocampus of rats with spontaneous recurrent seizures treated with carbenoxolone and quinine. BioMed Res Int 2014;2014:282490.
[119] Kamiński J, Wróbel A, Kublik E. Gap junction blockade eliminates supralinear summation of fast (>200 Hz) oscillatory components during sensory integration in the rat barrel cortex. Brain Res Bull 2011;85(6):424–8.
[120] Traub RD, Middleton SJ, Knöpfel T, Whittington MA. Model of very fast (>75 Hz) network oscillations generated by electrical coupling between the proximal axons of cerebellar Purkinje cells. Eur J Neurosci 2008;28(8):1603–16.
[121] Cunningham MO, Roopun A, Schofield IS, Whittaker RG, Duncan R, Russell A, et al. Glissandi: transient fast electrocorticographic oscillations of steadily increasing frequency, explained by temporally increasing gap junction conductance. Epilepsia 2012;53(7):1205–14.

[122] Munro E, Börgers C. Mechanisms of very fast oscillations in networks of axons coupled by gap junctions. J Comput Neurosci 2010;28(3):539–55.

[123] Stacey WC, Krieger A, Litt B. Network recruitment to coherent oscillations in a hippocampal computer model. J Neurophysiol 2011;105(4):1464–81.

[124] Söhl G, Güldenagel M, Beck H, Teubner B, Traub O, Gutiérrez R, et al. Expression of connexin genes in hippocampus of kainate-treated and kindled rats under conditions of experimental epilepsy. Brain Res Mol Brain Res 2000;83(1-2):44–51.

[125] Takahashi DK, Vargas JR, Wilcox KS. Increased coupling and altered glutamate transport currents in astrocytes following kainic-acid-induced status epilepticus. Neurobiol Dis 2010;40(3):573–85.

[126] Condorelli DF, Mudò G, Trovato-Salinaro A, Mirone MB, Amato G, Belluardo N. Connexin 30 mRNA is upregulated in astrocytes and expressed in apoptotic neuronal cells of rat brain following kainate-induced seizures. Mol Cell Neurosci 2002;21(1):94–113.

[127] Condorelli DF, Trovato-Salinaro A, Mudò G, Mirone MB, Belluardo N. Cellular expression of connexins in the rat brain: neuronal localization, effects of kainate-induced seizures and expression in apoptotic neuronal cells. Eur J Neurosci 2003;18(7):1807–27.

[128] Même W, Calvo CF, Froger N, Ezan P, Amigou E, Koulakoff A, et al. Proinflammatory cytokines released from microglia inhibit gap junctions in astrocytes: potentiation by β-amyloid. FASEB J 2006;20(3):494–6.

[129] Wu XL, Tang YC, Lu QY, Xiao XL, Song TB, Tang FR. Astrocytic Cx43 and Cx40 in the mouse hippocampus during and after pilocarpine-induced status epilepticus. Exp Brain Res 2015;233(5):1529–39.

[130] Szente M, Gajda Z, Said Ali K, Hermesz E. Involvement of electrical coupling in the in vivo ictal epileptiform activity induced by 4-aminopyridine in the neocortex. Neuroscience 2002;115(4):1067–78.

[131] Gajda Z, Gyengési E, Hermesz E, Ali KS, Szente M. Involvement of gap junctions in the manifestation and control of the duration of seizures in rats in vivo. Epilepsia 2003;44(12):1596–600.

[132] Gajda Z, Hermesz E, Gyengési E, Szupera Z, Szente M. The functional significance of gap junction channels in the epileptogenicity and seizure susceptibility of juvenile rats. Epilepsia 2006;47(6):1009–22.

[133] Zádor Z, Weiczner R, Mihály A. Long-lasting dephosphorylation of connexin 43 in acute seizures is regulated by NMDA receptors in the rat cerebral cortex. Mol Med Rep. 2008;1(5):721–7.

[134] Jiang K, Wang J, Zhao C, Feng M, Shen Z, Yu Z, et al. Regulation of gap junctional communication by astrocytic mitochondrial K_{ATP} channels following neurotoxin administration in in vitro and in vivo models. Neuro-Signals 2011;19(2):63–74.

[135] Elisevich K, Rempel SA, Smith B, Allar N. Connexin 43 mRNA expression in two experimental models of epilepsy. Mol Chem Neuropathol 1997;32(1-3):75–88.

[136] Akbarpour B, Sayyah M, Babapour V, Mahdian R, Beheshti S, Kamyab AR. Expression of connexin 30 and connexin 32 in hippocampus of rat during epileptogenesis in a kindling model of epilepsy. Neurosci Bull 2012;28(6):729–36.

[137] Seiffert E, Dreier JP, Ivens S, Bechmann I, Tomkins O, Heinemann U, et al. Lasting blood-brain barrier disruption induces epileptic focus in the rat somatosensory cortex. J Neurosci 2004;24(36):7829–36.

[138] Braganza O, Bedner P, Hüttmann K, von Staden E, Friedman A, Seifert G, et al. Albumin is taken up by hippocampal NG2 cells and astrocytes and decreases gap junction coupling. Epilepsia 2012;53(11):1898–906.

[139] Xu L, Zeng LH, Wong M. Impaired astrocytic gap junction coupling and potassium buffering in a mouse model of tuberous sclerosis complex. Neurobiol Dis 2009;34(2):291–9.

[140] Amiri M, Bahrami F, Janahmadi M. Modified thalamocortical model: a step towards more understanding of the functional contribution of astrocytes to epilepsy. J Comput Neurosci 2012;33(2):285–99.

[141] Perez-Velazquez JL, Valiante TA, Carlen PL. Modulation of gap junctional mechanisms during calcium-free induced field burst activity: a possible role for electrotonic coupling in epileptogenesis. J Neurosci 1994;14(7):4308–17.

[142] Ross FM, Gwyn P, Spanswick D, Davies SN. Carbenoxolone depresses spontaneous epileptiform activity in the CA1 region of rat hippocampal slices. Neuroscience 2000;100(4):789–96.

[143] Margineanu DG, Klitgaard H. Can gap junction blockade preferentially inhibit neuronal hypersynchrony vs. excitability? Neuropharmacology 2001;41(3):377–83.

[144] Schweitzer JS, Wang H, Xiong ZQ, Stringer JL. pH sensitivity of non-synaptic field bursts in the dentate gyrus. J Neurophysiol 2000;84(2):927–33.

[145] Köhling R, Gladwell SJ, Bracci E, Vreugdenhil M, Jefferys JG. Prolonged epileptiform bursting induced by 0-Mg^{2+} in rat hippocampal slices depends on gap junctional coupling. Neuroscience 2001;105(3):579–87.
[146] Voss LJ, Gauffin E, Ringqvist A, Sleigh JW. Investigation into the role of gap junction modulation of intracortical connectivity in mouse neocortical brain slices. Brain Res 2014;1553:24–30.
[147] Voss LJ, Jacobson G, Sleigh JW, Steyn-Ross A, Steyn-Ross M. Excitatory effects of gap junction blockers on cerebral cortex seizure-like activity in rats and mice. Epilepsia 2009;50(8):1971–8.
[148] Jahromi SS, Wentlandt K, Piran S, Carlen PL. Anticonvulsant actions of gap junctional blockers in an in vitro seizure model. J Neurophysiol 2002;88(4):1893–902.
[149] Samoilova M, Wentlandt K, Adamchik Y, Velumian AA, Carlen PL. Connexin 43 mimetic peptides inhibit spontaneous epileptiform activity in organotypic hippocampal slice cultures. Exp Neurol 2008;210(2):762–75.
[150] Samoilova M, Li J, Pelletier MR, Wentlandt K, Adamchik Y, Naus CC, et al. Epileptiform activity in hippocampal slice cultures exposed chronically to bicuculline: increased gap junctional function and expression. J Neurochem 2003;86(3):687–99.
[151] Traub RD, Draguhn A, Whittington MA, Baldeweg T, Bibbig A, Buhl EH, et al. Axonal gap junctions between principal neurons: a novel source of network oscillations, and perhaps epileptogenesis. Rev Neurosci 2002;13(1):1–30.
[152] Behrens CJ, Ul Haq R, Liotta A, Anderson ML, Heinemann U. Nonspecific effects of the gap junction blocker mefloquine on fast hippocampal network oscillations in the adult rat in vitro. Neuroscience 2011;192:11–19.
[153] Gigout S, Louvel J, Pumain R. Effects in vitro and in vivo of a gap junction blocker on epileptiform activities in a genetic model of absence epilepsy. Epilepsy Res 2006;69(1):15–29.
[154] Gigout S, Louvel J, Kawasaki H, D'Antuono M, Armand V, Kurcewicz I, et al. Effects of gap junction blockers on human neocortical synchronization. Neurobiol Dis 2006;22(3):496–508.
[155] Gajda Z, Szupera Z, Blazsó G, Szente M. Quinine, a blocker of neuronal Cx36 channels, suppresses seizure activity in rat neocortex in vivo. Epilepsia 2005;46(10):1581–91.
[156] Bostanci MÖ, Bağirici F. Anticonvulsive effects of carbenoxolone on penicillin-induced epileptiform activity: an in vivo study. Neuropharmacology 2007;52(2):362–7.
[157] Bostanci MO, Bağirici F. The effects of octanol on penicillin induced epileptiform activity in rats: an in vivo study. Epilepsy Res 2006;71(2–3):188–94.
[158] Nilsen KE, Kelso AR, Cock HR. Antiepileptic effect of gap junction blockers in a rat model of refractory focal cortical epilepsy. Epilepsia 2006;47(7):1169–75.
[159] Medina-Ceja L, Cordero-Romero A, Morales-Villagrán A. Antiepileptic effect of carbenoxolone on seizures induced by 4-aminopyridine: a study in the rat hippocampus and entorhinal cortex. Brain Res 2008;1187:74–81.
[160] Willoughby JO, Mackenzie L, Broberg M, Thoren AE, Medvedev A, Sims NR, et al. Fluorocitrate-mediated astroglial dysfunction causes seizures. J Neurosci Res 2003;74(1):160–6.
[161] Rozental R, Srinivas M, Spray DC. How to close a gap junction channel. Efficacies and potencies of uncoupling agents. Methods Mol Biol 2001;154:447–76.
[162] Cruikshank SJ, Hopperstad M, Younger M, Connors BW, Spray DC, Srinivas M. Potent block of Cx36 and Cx50 gap junction channels by mefloquine. Proc Natl Acad Sci USA 2004;101(33):12364–9.

CHAPTER

12

Blood–Brain Barrier Disruption

OUTLINE

OVERVIEW

The blood–brain barrier (BBB) is a dynamic system that separates the peripheral blood from the neural tissue. It is composed of endothelial cells connected through gap junctional proteins and works together with various other cell types to create a unique microenvironment for proper neuronal function. BBB disruption, however, has been implicated in epilepsy. Various studies have demonstrated increased BBB permeability and albumin extravasation into the brain after prolonged seizures. Disruption of the BBB or direct application of serum albumin to the brain itself may also be sufficient to cause hyperexcitability. More recent research has focused on deciphering the mechanism of BBB dysfunction in the pathogenesis of epilepsy and several studies have suggested the involvement of the inflammatory transforming growth factor β (TGF-β) pathway. This chapter will review the current evidence for the involvement of BBB disruption in epileptogenesis and will also consider the clinical relevance of early detection of BBB damage.

Astrocytes and Epilepsy
DOI: http://dx.doi.org/10.1016/B978-0-12-802401-0.00012-0

HISTORY

The notion that the human body had an impermeable structure designed to protect the brain was first described by the London physician Humphrey Ridley (1653–1708) [1]. The first experiments to establish the existence of the BBB, however, were conducted by the German scientist and Nobel laureate, Paul Ehrlich (1854–1915) and his student Edwin Goldmann (1862–1913). In 1885, Ehrlich injected various dyes into the peritoneum of animals and noticed that almost all organs, except the brain and spinal cord, were stained [2,3]. He attributed the difference between the central nervous system (CNS) and other organs to different binding affinities to the vital dyes.

Goldmann furthered Ehrlich's work by demonstrating that intravenous injections of trypan blue into dogs and rabbits turned the whole animal blue with the exception of the brain and CNS whereas injection of trypan blue into the brain ventricular system colored only the brain tissue and choroid plexuses [4,5]. These experiments confirmed that the lack of staining was not due to the lack of affinities for the dye and led to the hypothesis that the vehicle for transport was the cerebrospinal fluid (CSF) [2].

Two English scientists, Charles Roy (1854–97) and Charles Sherrington (1857–1952) [6] noted that the brain had an "intrinsic mechanism" to maintain the brain environment "without notable interference with the blood-supply." They also determined that many lipid-soluble molecules could cross into the brain whereas many lipid-insoluble molecules could not [6]. The existence of a barrier between the blood and the brain, however, was first postulated a decade later by Max Lewandowsky (1876–1918). He discovered that intraventricular application of cholic acids or sodium ferrocyanide caused neurological symptoms whereas intravenous injections did not [7]. He thus concluded that the cerebral capillaries hindered certain compounds from entering the neuronal tissue and coined the term "blood–brain barrier" [1].

Until the 1950s, the hypothesis of a BBB was met with skepticism. It was suggested that the impermeability of the dyes were not due the presence of a brain barrier, but was instead due to their binding to albumin [1,2]. With the use of electron microscopy, several studies strengthened the hypothesis of the BBB and even described the structure and involvement of both endothelial cells and astrocytic endfeet in BBB formation [8–10]. It was not until 1967, however, that the site of the BBB was confirmed. Resse and Karnovsky [11] injected horseradish peroxidase (HRP) intravenously into mice and discovered that HRP did not pass the vascular endothelium in the cerebral cortex, thus determining the location of the barrier. Complimentary findings by Brightman demonstrated that intraventricular injections of tracer molecules are trapped between the glial and endothelial cells [12,13].

STRUCTURE AND FUNCTION

The BBB is a highly regulated, multicellular interface between the peripheral circulatory system and the CNS [14,15]. The anatomical makeup of the BBB that surrounds the capillary lumen (Fig. 12.1) [16,17] is comprised of cerebral microvascular endothelial cells connected by tight junctions that prevent paracellular diffusion of water-soluble substances [17]. Astrocytic endfoot processes interact with the endothelial cells on the abluminal membrane

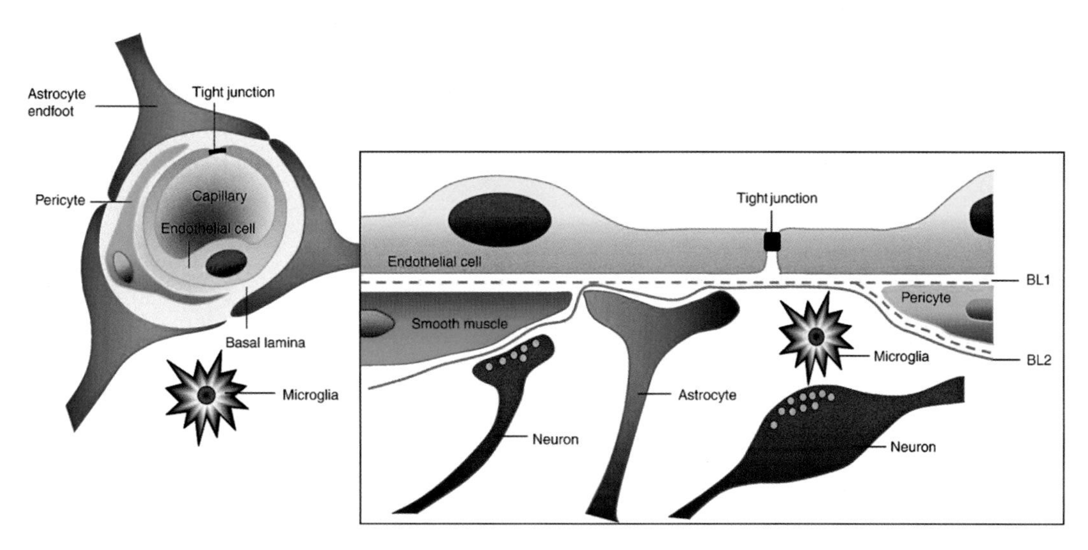

FIGURE 12.1 **Cell associations at the BBB.** The cerebral endothelial cells form tight junctions at their margins which seal the aqueous paracellular diffusional pathway between the cells. Pericytes are distributed discontinuously along the length of the cerebral capillaries and partially surround the endothelium. Both the cerebral endothelial cells and the pericytes are enclosed by, and contribute to, the local basement membrane which forms a distinct perivascular extracellular matrix (basal lamina 1, BL1), different in composition from the extracellular matrix of the glial endfeet bounding the brain parenchyma (BL2). Foot processes from astrocytes form a complex network surrounding the capillaries and this close cell association is important in induction and maintenance of barrier properties. Axonal projections from neurons onto arteriolar smooth muscle contain vasoactive neurotransmitters and peptides and regulate local cerebral blood. BBB permeability may be regulated by release of vasoactive peptides and other agents from cells associated with the endothelium. Microglia are the resident immunocompetent cells of the brain. The movement of solutes across the BBB is either passive, driven by a concentration gradient from plasma to brain, with more lipid-soluble substances entering most easily, or may be facilitated by passive or active transporters in the endothelial cell membranes. Efflux transporters in the endothelium limit the CNS penetration of a wide variety of solutes (based on Abbott et al. [16]). *Source: Reproduced with permission from Abbott NJ, Patabendige AA, Dolman DE, Yusof SR, Begley DJ. Structure and function of the blood-brain barrier. Neurobiol Dis 2010;37(1):13–25.*

(membrane of endothelial cells facing the brain) of the capillaries and play a role in maintaining BBB integrity. Pericytes are perivascular cells with elongated processes that intermittently surround the abluminal membrane of endothelial cells. More recently, pericytes have been shown to play a role in regulating cerebral blood flow [18]. Neurons may regulate local cerebral blood flow whereas microglia act as the immediate immune response within the brain. Pericytes, microglia, and neuronal projections closely associated with endothelial cells play a supportive role in the induction, maintenance, and function of the BBB [17,19,20]. The basement membrane, a thin layer of fibrous, extracellular matrix tissue, helps regulate cross talk between the various cellular components [15]. The basal lamina is a layer of extracellular matrix that, in part, makes up the basement membrane. Taken together, endothelial cells, astrocytes, pericytes, neurons, microglia, the basement membrane, and any other component of the brain parenchyma that communicates with these endothelial cells make up the neurovascular unit (NVU).

Early freeze-fracture electron microscopy studies were employed to gain insight into BBB structure [1,2,21,22]. Tight junctions, a key feature of the BBB, are arranged in a network of 6–8 anastomosing parallel strands, allowing them to be "very tight" and impermeable to most molecules [22]. In fact, the passage of molecules was restricted to a diameter of 10–15 Å [1,2]. The BBB exhibits a high transendothelial electrical resistance, which corroborated the extremely low ionic permeability of brain capillaries [23]. In addition to reducing paracellular diffusion of polar solutes, tight junctions contribute to the polarized (apical-basal) properties of endothelial cells [16]. Transmembrane proteins, including claudins and occludins, make up the junctional complexes. Claudins contribute to the high electrical resistance found throughout the endothelium whereas occludins regulate tight junction cytoskeletal activity [16,24]. Junctional adhesion molecules are involved in the formation and maintenance of tight junctions. Components of the BBB, including tight junctions, develop during embryonic stages and are well formed by birth [25–30].

In addition to the BBB, two other interfaces between the brain and epithelium exist. The blood–cerebrospinal fluid barrier is composed of epithelial cells of the choroid plexus facing the CSF. Here, epithelial cells may secrete CSF [31]. The arachnoid barrier completely seals the CNS from the rest of the body, although a relatively small amount of surface exchange between the blood and CNS occurs here. It is composed of the avascular arachnoid epithelium that underlies the dura. All three barriers play a role in modulating and regulating normal and pathological brain functions.

The BBB allows the brain to maintain a stable ionic and metabolic environment independent from the rest of the body. It has a variety of functions that help regulate and protect the brain, including responding to and passing signals between the blood and the CNS [15]. Specific ion channels and transporters found within the BBB help maintain ion balance that is idea for proper neuronal function. Because concentrations of ions such as sodium (Na^+), potassium (K^+), chloride (Cl^-), and calcium (Ca^{2+}) must be maintained within a very narrow range, this feature of the BBB is essential for neuronal health. In a similar fashion, the BBB prevents superfluous neurotransmitters, such as glutamate in the blood plasma, from entering the brain. The BBB also inhibits macromolecules and potential neurotoxins from entering the brain. For example, albumin can be damaging to nervous tissue and can cause glial scarring [17,32]. Finally, the BBB has specific transporters to ensure both the adequate supply of various essential nutrients and metabolites to the brain and the proper removal of waste products away from the brain [16,17].

BBB DISRUPTION IN EPILEPSY

Disruption of the NVU is associated with a number of neurological disorders including stroke, Alzheimer's disease, and epilepsy [33]. Prolonged seizures are associated with BBB dysfunction and a number of pathophysiological consequences including increased vascular permeability, reactive astrocytosis, leukocyte infiltration, inflammation, increased extracellular glutamate and K^+ levels, and tissue hyperexcitability. While many studies have established a role for BBB damage in epilepsy, more recent research has been geared toward understanding whether BBB disruption is a cause or a consequence of the disease.

BBB Disruption as a Result of Epileptogenesis

Both human tissue data and animal models have been used to provide convincing evidence that BBB disruption may be a common pathological feature of epilepsy. BBB tracers such as Evans Blue, horseradish peroxidase (HRP), and fluorescein have been used to determine BBB permeability after seizure activity. Measurements of serum albumin or immunoglobulin (IgG) extravasation have also been used as indicators of a leak in the BBB. Dynamic changes in BBB permeability have been observed in patients with seizures and primarily during the epileptogenic period in animal models.

Human Tissue Studies

BBB disruption has been commonly found in tissue from patients with a number of neurological disorders including epilepsy, traumatic brain injury, and brain tumors. An early study (1984) found elevated serum albumin immunoreactivity around medium-sized blood vessels in the epileptic hippocampus compared to autopsy control tissue [34]. Van Vliet et al. [35] confirmed these results by demonstrating prominent albumin extravasation throughout resected hippocampi from patients with medically intractable epilepsy. Furthermore, albumin immunoreactivity was found in both neurons and astrocytes located around blood vessels and was particularly strong in patients who died during SE, suggesting a correlation between the time of seizure and magnitude of albumin extravasation [35]. Liu et al. [36] observed higher albumin and fibrinogen immunoreactivity in neurons, glial cells, and parenchyma in hippocampi from patients with either drug-resistant or drug-sensitive epilepsy compared to control tissue. Similar findings were seen in a variety of epilepsies.

Temporal lobe epilepsy (TLE) is the most common form of epilepsy in adults. In both the entorhinal cortex and hippocampus from patients with pharmacoresistant TLE, IgG leakage was observed in the parenchyma and in degenerating neurons [37]. Pyramidal neurons and interneurons were strongly labeled for IgG whereas dentate granule cells were faintly stained. A separate study confirmed IgG leakage and accumulation in neurons, possibly due to the loss of tight junctions [38]. In addition, vessel density was higher in tissue from patients with TLE than autopsy controls and this finding was positively correlated with seizure frequency [38].

Focal cortical dysplasia (FCD) is a cortical developmental malformation often associated with medically refractory epilepsy. In patients with FCD with balloon cells, hemosiderin deposits (an indicator of BBB impairment) were rare and no substantial albumin uptake into astrocytes was observed [39]. Although this study suggests that little to no impairment of the BBB occurred in patients with FCD, more studies need to be conducted to verify this finding.

An increasingly common way to measure BBB permeability is through the use of magnetic resonance imaging (MRI). Brain MRIs revealed increased BBB permeability in the majority of patients with posttraumatic epilepsy (PTE) compared to 25–33% of patients without seizures, an observation that could be found years following the trauma [40,41]. A separate study determined that a subset of patients with postconcussion syndrome have focal cortical dysfunction in conjunction with BBB disruption [42]. Since concussions are often related to cortical seizure activity, it is possible that BBB disruption played a role in the pathophysiology of seizure development and spread.

Gangliogliomas, characterized by dysplastic neurons and neoplastic astrocytes, are the most frequent tumor entity found in young patients with drug-resistant epilepsy. Schmitz et al. [39] found strong deposits of hemosiderin in ganglioglioma tissue. Albumin deposits in astrocytes were found in all gangliogliomas, but no correlation between albumin accumulation and duration of epilepsy was found [39]. In diffuse astrocytomas, on the other hand, neither hemosiderin nor astrocytic albumin deposits were seen in lesional or perilesional regions [39]. Although hemosiderin was present in tissue from patients with renal cell carcinoma brain metastases, almost no albumin uptake into astrocytes was observed [39]. WHO grade 1 dysembryoplastic neuroepithelial tumors exhibited both intralesional hemosiderin deposits and low levels of astrocytic albumin uptake [39].

Patients with vascular malformations, such as cavernous angiomas (CAs, intracerebral vascular malformation) or arteriovenous malformations (AVMs, abnormal blood vessels connecting arteries and veins), frequently have chronic focal epilepsy. Strikingly high levels of albumin were observed in surgically removed brain parenchyma surrounding CAs and AVMs [43]. Albumin accumulation was found to occur primarily in astrocytes.

Human tissue experiments often give information about the endpoint of a disease since they are usually conducted in chronically epileptic or postmortem tissue. Bolwig et al. [44], however, conducted a unique experiment where BBB permeability was examined after inducing electroshock seizures in patients undergoing electroconvulsive therapy for depression. Radioactively labeled Na^+, Cl^-, urea, and thiourea were injected into the internal carotid artery while a continuous series of blood samples were collected from the venous catheter before, during, and after seizures. During seizures, a significant decrease in thiourea extraction and a significant increase in urea permeability-surface area products were detected; no other differences from resting measurements were observed [44]. In addition, cerebral blood flow and venous O_2-tension increased during seizures. Postictal levels were similar to resting levels [44].

Although studying postmortem human tissue has commonly been employed, some caution should be exercised when interpreting the results. In some cases, patients may have been suffering from other illnesses. For example, in the 1984 study examining albumin immunoreactivity in epileptic brain tissue, one patient suffered from cardiac malformations and died from cardiac arrest, which is known to cause severe hypoxic nerve cell damage [34]. The compounding effect of this hypoxic episode must be considered when evaluating the findings. An additional drawback to human tissue data is the inability to study a time course of the disease. For this, we must turn to studies using animal models of epilepsy.

Animal Models

Some of the earliest studies on BBB disruption after seizures showed increased permeability of BBB after pentylenetetrazol (PTZ)-induced seizures in a variety of animal models [45–48]. PTZ-induced seizures in rabbits quickly led to bilateral Evans Blue leakage into the hypothalamus (except the mammillary bodies), the preoptic area, midbrain tegmentum, and in the cerebellum (Fig. 12.2) [49]. Within a half-hour of PTZ-injections in adult rats, BBB permeability (measured by Evans Blue) was increased in the left hemisphere, right hemisphere,

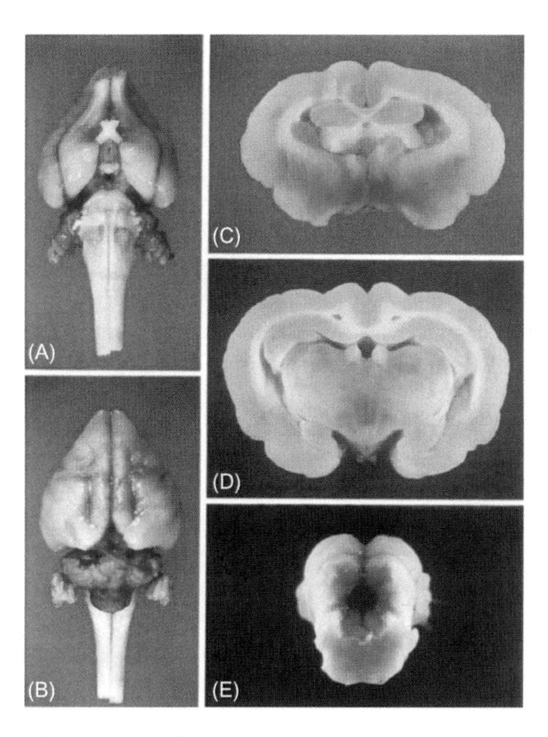

FIGURE 12.2 **Blood–brain barrier (BBB) breakdown due to PTZ-induced seizures.** (A) The brain exhibits intense leakage in hypothalamus, excluding the corpora mammillaria, in the inferior olives and in the cerebellar hemispheres. Tuberculum olfactorium and the cortex around the anterior sulcus rhinalis are also stained in this exceptionally heavily affected rabbit. (B) Dorsal view of the same brain as shown in (A). Extremely intense Evans Blue (EB) stain in the cerebellum. The fronto-dorsal cortex shows a faint blueish stain. (C) Coronal cut at the level of the chiasma opticum. Intense leakage in the preoptic area, which spreads into pallidum and claustrum. The caudate nucleus remains free of stain. (D) Coronal cut at the level of the tractus mammillothalamicus. The tracer is concentrated around the third ventricle and in the geniculate, medial and dorsal thalamic nuclei. No stain is present in the center median. (E) Distinct EB leakage in the inferior, but not superior, colliculus. *Source: Reproduced with permission from Nitsch C, Klatzo I. Regional patterns of blood-brain barrier breakdown during epileptiform seizures induced by various convulsive agents. J Neurol Sci 1983;59(3):305–22.*

brainstem, and cerebellum [50]. These rats also exhibited increased immunoreactivity of zonula occludens-1 (zo-1), the main tight junction protein, after PTZ-induced seizures [50].

Lee and Olszewski [51] demonstrated that electroshock increased BBB permeability in cats and rabbits, particularly in the thalamic and hypothalamic areas of the brain. Studies in the 1970s demonstrated that electroconvulsive shock enhanced BBB permeability [44,52–55]. Measurements of radiolabeled Na^+ and Cl^- as well as carbon labeled thiourea (^{14}C-thiourea) and glucose (^{14}C-glucose) demonstrated a decrease in glucose extraction during electroshock-induced seizures in rats; all other tracers remained unchanged [52]. Repeated electroshocks in rats resulted in Evans Blue staining or HRP extravasation throughout the brain [52,55]. This and similar electrical stimulation models of epilepsy are still employed today to study BBB permeability.

Six weeks after status epilepticus (SE) induced by electrical stimulation of the angular bundle, rats exhibited BBB leakage (measured by fluorescein) in limbic brain regions [56]. In a rat model involving tetanic stimulation of the hippocampus, albumin and Evans Blue extravasation were prominently observed in the parenchyma of the hippocampus, entorhinal cortex, piriform cortex, thalamus, and amygdala; staining was also seen in the septum and olfactory bulb, but not in the cerebellum or in the brains of control rats [35]. Extravasation was highest in the acute seizure phase (1 day after SE), was still observed in the latent phase (1 week after SE), and was nearly abolished in the chronic phase (4 months after SE). Evans Blue colocalized abundantly with a marker for neurons and, to a lesser extent, markers for astrocytes and microglia. Interestingly, a higher seizure frequency was related to a more permeable BBB [35].

A number of studies conducted in the 1980s demonstrated disrupted BBB after seizures induced by kainic acid, an ionotropic glutamate receptor agonist [57–60]. Interestingly, rats that had only mild seizures after treatment with kainic acid did not exhibit BBB disruption [60]. Rats with severe limbic seizures showed increased permeability to α-[^{14}C]aminoisobutyric acid autoradiography throughout the brain from 2 to 24 hours after kainic acid injections; BBB appeared normal from 24 hours to 7 days after kainic acid treatment, except for a small residual amount found in limbic structures [60]. Within hours of either intraperitoneal (i.p.) or systemic administration of kainic acid, rats exhibited increased permeability in the thalamus and the forebrain [58,59]. Subcutaneous injections of kainic acid in rats, however, led to only a mild increase in permeability of cerebral vessels [57]. Only about one-third of rabbits treated through the ear vein with kainic acid exhibited BBB breakdown [49]. In those rabbits, Evans Blue was found throughout the frontoparietal cortex and occasionally in the hypothalamus and thalamus, the substantia nigra, periaqueductal gray, basal forebrain, and amygdala [49].

Intraamygdala kainic acid injections into mice led to increased albumin and IgG amounts in the ipsilateral hippocampus to injection 4–24 hours after SE [61]. Levels did not differ from controls at 7 or 21 days post-SE with the exception of elevated IgG extravasation in CA3 of epileptic mice at 21 days post-SE [61]. In contrast to the early PTZ-induced seizure studies, reduced immunoreactivity for zo-1 was found at 1 hour, 8 hours, and 21 days post-SE, although no differences in Western blot protein amounts were observed [61].

After a single i.p. injection of kainic acid, strong IgG immunoreactivity was observed in adult mice 12-hours posttreatment, but it was restricted to the hippocampus [62]. By 3 days posttreatment, BBB integrity was restored. Interestingly, aged mice exhibited a low level of basal IgG staining, potentially due to age-related attenuation of BBB integrity [62]. IgG immunoreactivity was found throughout the aged mouse brain 12 hours after kainic acid treatment, particularly in circumventricular regions, increased by 24 hours, and remained visible until 7 days after treatment [62]. Contrary to these findings, however, Han et al. [63] found that i.p. administration of kainic acid in mice exhibited a dose-dependent inhibition of Evans Blue dye extravasation.

Similar to the human PTE studies, MRI has been employed in animal models to measure BBB disruption. In the intrahippocampal kainic acid model of epilepsy in rats, both MRI involving T1 mapping and postmortem microscopic analysis of fluorescein leakage determined that BBB leakage was evident in both the acute (1 day post-SE) and the chronic

(6 weeks post-SE) phase in the hippocampus, entorhinal cortex, amygdala, and piriform cortex [64,65]. BBB leakage was most evident during the acute phase in limbic brain regions. MRI performed with T2 and T1 weighted images at various time points before and after subcutaneous injections of pilocarpine (muscarinic receptor agonist) revealed BBB leakage in the thalamus 2 hours after pilocarpine treatment, which disappeared by 6 hours [66]. No leakage was observed in either the latent or chronic phase of epilepsy.

Several additional studies have shown BBB leakage in the pilocarpine model of epilepsy in rodents [37,39,66–71]. A spotty immunostaining pattern for serum proteins was observed within the first 2 hours after pilocarpine treatment in rats [72]. As early as 6 hours after pilocarpine treatment, patches of serum extravasation appeared in the substantia nigra pars reticulata, likely as a consequence of neuronal and glial cell damage; this effect persisted for several days [72]. A separate study showed an increase in serum S100β levels and Evans Blue signal when measured at the onset of SE [70]. Analogous to results seen in humans, IgG accumulation in pyramidal neurons was observed around dilated microvessels during the acute period (1–12 hours postinjection) in the rat pilocarpine model of TLE [37]. Leakage was very faint during the latent phase (2–14 days postinjection) and slightly increased during the chronic phase (21–60 days postinjection) of epilepsy.

The majority of the animal model studies of BBB dysfunction in epilepsy have been conducted in adult rodents. Young rats, however, have demonstrated increased BBB permeability in response to seizures. In both adult and p21 rats, BBB opening to [α-^{14}C]-aminoisobutyric acid occurred 90 minutes after pilocarpine-induced SE in the globus pallidus, lateral septum, and in most thalamic nuclei [69]. Adult rats also exhibited significantly increased permeability in the medial amygdala, cingulate and piriform cortices, geniculate nuclei, and anterior and ventromedial hypothalamus. In p21 rats, permeability significantly increased in the caudate nucleus, medial septum, parietal and sensorimotor cortices, hippocampal CA1, and posterior and anteromedian thalamus [69]. Although both p21 and adult rats exhibited increased BBB permeability, it was often higher in the adults.

Methylazoxymethanol acetate (MAM) is a DNA-alkylating agent that has commonly been used as a model of human developmental brain malformations. Rats treated with MAM exhibited serum albumin leakage in the malformed hippocampus. Interestingly, pilocarpine-induced seizures exacerbated BBB leakage [71], thus suggesting that seizures can lead to disruption of the BBB in the deformed brain.

Rabbits intravenously injected with the $GABA_A$ antagonist bicuculline exhibited generalized tonic-clonic convulsions which quickly resulted in Evans Blue extravasation in the pallidum and the midbrain (including substantia nigra, midbrain tegmentum, and periaqueductal gray) but not the preoptic area or the superficial layers of the superior colliculus [49]. Similar results were obtained in a separate study of bicuculline-injected rats; HRP was frequently found in the thalamus, pallidum, hippocampus, and medulla oblongata but was rare in the septum, periaqueductal gray, hypothalamus, and cerebellar cortex [73]. A more recent study has used a bicuculline arterial perfusion model in vitro with a preparation isolated from guinea pig brain that maintained physiological interactions between neuronal, glial, and vascular components [74]. Sixty minutes after pulse applications of bicuculline, cells demonstrated significant brain extravasation of FITC-albumin and the extent of BBB damage correlated to the time spent in seizure-like activity [74]. Decreased levels of zo-1 in endothelial cells were observed [74].

Methoxypyridoxine (MP) blocks GABA synthesis whereas methionine sulfoximine (MSO) inhibits the conversion of glutamate to glutamine. MP-induced seizures in rabbits resulted in Evans Blue extravasation into the temporal parts of the hippocampus, hypothalamus, and dorsal parts of thalamus [49]. MSO-induced seizures in rabbits resulted in selective Evans Blue extravasation in the mammillary bodies [49]. In both models, a seizure duration of 3–5 minutes was enough to cause BBB damage and longer seizures did not result in increased Evans Blue leakage over time.

Hyperthermia-induced seizures induced in animals share a common pathology to febrile seizures seen in humans. Rats with cortical dysplasia introduced to hyperthermia-induced seizures exhibited increased BBB permeability to Evans Blue and HRP within 30 minutes after seizure induction [75]. Pregnant rats exposed to γ radiation exhibited fluorescein extravasation in the brain after hyperthermia-induced seizures [76].

A variety of animal models have shown BBB leakage shortly after seizure induction, however, this damage may persist into the chronic phase of epilepsy. One major concern when interpreting epilepsy model data is that the chemoconvulsants themselves, rather that the seizures, may disrupt the BBB. Fortunately, the variety of epilepsy models available, including chemoconvulsants, electrical stimulation, and hyperthermia-induced seizures, may assuage this concern. Illustrating BBB disruption in a genetic epilepsy model may also be beneficial in addressing this concern.

Seizures as a Result of BBB Dysfunction

Strong evidence exists to suggest BBB dysfunction results after seizure activity. More recently, however, researchers have shifted their focus on the ability of BBB disruption itself to lead to epileptogenesis. To address this problem, a number of substances, including bile salts and mannitol, have been used to open the BBB. Direct application of albumin to the brain has also been employed to study the immediate effect of BBB leakage. Human data on BBB disruption leading to tissue excitability are limited, however, due to the rare need to damage an intact BBB in a patient. Animal models must be substituted to address this question.

Human Tissue Studies

Due to obvious ethical limitations, very few studies have been able to determine seizure susceptibility in humans after BBB disruption. A subset of patients (25%) with primary brain lymphomas who underwent a treatment involving osmotic disruption of BBB with mannitol followed by intraarterial chemotherapy immediately developed seizures [77]. When just intraarterial chemotherapy was administered in the absence of BBB disruption by mannitol, patients did not develop seizures.

Cavernous malformations, or dilated blood vessels characterized by multiple distended "caverns" of blood-vessel vasculature, have a propensity to lead to seizures in patients [78]. Morphological analysis of cavernous malformations in human patients revealed a lack of tight junctions at endothelial cells interfaces and the absence of astrocytic endfoot processes at the site of lesions [78]. Therefore, it is possible that the lack of a functional BBB may lead to red blood cells leaking into lesions and the surrounding brain, thus accounting for the tendency of patients with cavernous malformations to have seizures.

Animal Models

Early studies of BBB disruption established that bile salts could open the BBB in a concentration-dependent manner [79–81]. Seiffert et al. [80] established a model for focal disruption of the BBB in the rat by directly applying bile salts to the cortex. Exposure to bile salts resulted in activation of astrocytes and extravasation of serum albumin in the extracellular space that lasted for several days [80]. A more recent study verified that exposure to bile salts led to the formation of an Evans Blue–albumin complex as early as 1 hour after treatment [82]. MRI in vivo revealed early BBB disruption with delayed reduction in cortical volume (associated with a reduced number of neurons and increased number of astrocytes) in response to both cortical treatment with bile salts and treatment with serum albumin [83]. In vitro experiments demonstrated that epileptiform activity could be recorded in the majority of slices from animals up to several weeks after treatment with bile salts [80,81,83]. Direct perfusion of the rat cortex with either serum, denatured serum, or albumin produced similar results, therefore eliminating the possibility of a bile salt-specific effect [80,81]. Treatment with a solution containing serum levels of electrolytes, on the other hand, did not induce abnormal activity in slices [80].

Frigerio et al. [68] quantified the amount of serum albumin extravasated in the rat brain parenchyma 24 hours after pilocarpine-induced SE and injected those levels of albumin intracerebroventricularly (i.c.v.) into naïve rats. EEG analysis revealed that rats injected i.c.v. with albumin developed high-amplitude, high-frequency spiking activity within 15 minutes after injection whereas dextran-treated control rats showed EEG activity similar to preinjection baseline [68]. Three months after albumin injections, rats exhibited a significant decrease in afterdischarge threshold, suggesting long-term effects on hippocampal excitability [68]. More recent studies revealed that intraventricular administration of serum albumin in mice did not immediately result in seizures, but did induce spontaneous recurrent seizures after a latent period of 3–5 days in the majority of mice [84,85]. These studies provided direct evidence for the role of albumin extravasation in seizure susceptibility in rats.

Hyperosmotic mannitol has commonly been employed to reduce BBB integrity. Successful mannitol-induced opening of the BBB in pigs resulted in seizures [77]. EEG activity recorded from rats after mannitol infusion demonstrated paroxysmal electrical activity and, in a few instances, tonic-clonic seizures [86]. Interestingly, this activity was only observed in rats with successful opening of the BBB. In contrast, a more recent study found that mannitol infusion did not induce seizures in healthy rats [35]. Chronically epileptic rats that underwent tetanic stimulation of the hippocampus, however, had increased seizure frequency during and after 3 days of mannitol treatment [35].

When considering these data, it is important to take into account the compound used to open the BBB, the concentration, route of administration, and the species examined. In addition, assumptions on what may leak out of a damaged BBB should be limited. Kang et al. [87] demonstrated that the BBB opening to large molecules does not necessarily mean that the BBB will open to smaller ions. Future studies should focus on the role of BBB disruption in epileptogenesis and ascertain the long-term consequences relating to neuronal hyperexcitability.

Mechanisms for BBB Disruption

Theories on the mechanism of BBB disruption in epilepsy models began in the late 1970s when it was first observed that increased BBB permeability after seizures was associated with an increase in systemic blood pressure followed by vasodilation [55,88–90]. Regional differences in BBB barrier breakdown, however, did not always correlate with increased regional blood flow, suggesting other mechanisms may be at play [49]. Bolwig et al. [55] reported increased vesicular transport (pinocytosis) across the endothelial cells during seizures, noting that the vascular structure remained intact. Several studies supported this hypothesis [54,90–93] and expanded it to include regulation by neurotransmitters, such as serotonin, binding to receptors on endothelial cells of brain vessels during seizure activity [49,73,93]. A competing hypothesis was the temporary opening of endothelial tight junctions during seizure activity [73,94]. More recent studies have focused on angiogenesis, mechanistic pathways, or activation of the immune response after BBB disruption.

Rigau et al. [38] discovered BBB impairment and increased expression of vascular endothelial growth factor (VEGF) in both tissue from patients with TLE and tissue from rats shortly after pilocarpine-induced SE. A progressive increase in vascularization occurred during the latent and chronic phases in the rat pilocarpine model of TLE [38]. Although seizures are associated with robust vasodilation, spreading depolarization may be associated with vasoconstriction, increased metabolic demand, and reduced energy supplies to neurons [95,96]. This may lead to the failure of Na^+, K^+, and Ca^{2+} pumps and consequent spreading and maintenance of neuronal depolarization.

Many studies have implicated the mammalian target of rapamycin (mTOR) pathway in epileptogenesis [97–101], but only one has examined the role of rapamycin in seizure-induced BBB permeability. SE was induced in rats by electrical stimulations of the angular bundle. Rodents were then treated with either rapamycin or vehicle control once daily for 7 days, then once every other day until 6 weeks post-SE. Vehicle-treated animals displayed BBB leakage in hippocampal regions whereas rapamycin-treated rats had significantly reduced BBB permeability [99]. Rapamycin treatment also decreased the number of epileptic rats and the number of seizures per day, but did not alter seizure duration [99]. Thus, the mTOR pathway may participate in seizure-induced BBB disruption.

Strong evidence suggests that inflammation is involved in the mechanism for BBB disruption after seizures. Brain inflammation can alter synaptic signaling, plasticity, inhibitory transmission, BBB permeability, and consequently hyperexcitability [65,102]. In the i.p. pilocarpine model of epilepsy, seizures induced BBB leakage, elevated levels of vascular cell adhesion molecules, and enhanced leukocyte rolling and arrest in brain vessels in mice [67]. Damage to the BBB, however, was prevented by blocking the leukocyte-endothelial interaction. Interestingly, human data showed that leukocytes were more abundant in epileptic tissue than in controls [67], further suggesting a potential role for leukocytes in the development of epilepsy. The involvement of leukocytes in BBB damage and its contribution to seizure generation, however, is a matter of debate [103].

Several antiinflammatory treatments have been shown to reduce or prevent the breakdown of the BBB and the occurrence of seizures. Intravenous administration of interleukin-1 receptor antagonist (IL-1RA) diminished the onset of SE and BBB damage in the rat pilocarpine model [104]. Pretreatment with the antiinflammatory drug dexamethasone reduced the number of rats that developed SE, increased the latency to onset, abolished mortality, and protected against BBB damage [70]. In addition, glucocorticosteroids (dexamethasone or adrenocorticotropic hormone) reduced the number of seizures in pediatric epilepsy patients [70]. Importantly, patients with syndromes known to respond to glucocorticosteroids, including Landau-Kleffner, Lennox-Gastaut, Rasmussen encephalitis, and West, were excluded from this study. Bicuculline-induced seizure activity was associated with overexpression of interleukin-1β (IL-1β) in isolated guinea pig brains [74]. Anakinra, a selective IL-1 receptor type 1 (IL-1R1) antagonist, reduced the duration of the first seizure, prevented seizure recurrence, and resolved seizure-associated BBB breakdown [74]. Taken together, these studies imply that inflammatory mechanisms may contribute to both BBB breakdown and seizure generation.

Elevated expression of IL-1β, IL-1R1, tumor necrosis factor-α (TNF-α), and transforming growth factor β1 (TGF-β1) mRNA were observed in the amygdala, hippocampus, parietal, prefrontal, and piriform cortices 2 hours after amygdala kindling in rats [105]. TGF-β was also found to be upregulated for the first 3 weeks after electrical angular bundle stimulation-induced SE [106]. TGF-β is a cytokine that plays a role in intercellular communication and in several cell functions including the immune response, cell growth, and apoptosis. In the canonical pathway, TGF-β binds to serine-threonine kinase TGF-β receptors (TGF-βRs) which results in the phosphorylation of Smad, an intracellular mediator, by activin-like kinase (ALK). Subsequently, phosphorylated Smad (p-Smad) translocates into the nucleus and promotes transcriptome activity. Several animal studies have implicated the activation of the TGF-β pathway as a potential mechanism linking BBB disruption to the development of epilepsy [65,102,107,108].

The application of serum albumin is associated with astrocytic activation [85], uptake into astrocytes in vivo [81,82], and hypersynchronous epileptiform activity in vitro [80–83,109]. Other studies have demonstrated that in addition to astrocytes, albumin is also taken up by NG2 cells [110] and neurons [84,110], with the most prominent quantity of uptake found in NG2 cells. The use of selective inhibitors in vitro demonstrated that albumin endocytosis into astrocytes involved TGF-βR activation, caveolae-mediated endocytosis, and an additional unidentified pathway [82].

Albumin uptake may upregulate cytokines, inducing brain inflammation, and activate the TGF-βR pathway [65]. Perfusion of serum albumin into the somatosensory motor cortex of rats led to spontaneous seizures, increased levels of phosphorylated Smad2/3 2 days after treatment, and astrogliosis 7 days after treatment [82]. Losartan, an angiotensin II type 1 receptor antagonist that has been shown to inhibit TGF-β signaling, reduced the number of rats that developed epilepsy, decreased seizure duration, and prevented the increase in reactive astrocytes and p-Smad2/3 levels without affecting regional blood flow [82].

Activation of the TGF-β pathway in rat neocortical slices resulted in epileptiform activity, similar to that seen after application of albumin [109]. Co-immunoprecipitation revealed a direct interaction between albumin and TGF-βRII [109]. Assessment of Smad2/3 phosphorylation levels revealed that albumin activated this pathway in epileptogenic rats [109]. Studies on cultured cells in vitro confirmed the occurrence of Smad2/3 phosphorylation and determined that albumin activates ALK5 [82]. Taken together, these data suggest that albumin entering the brain after BBB disruption may trigger the TGF-β pathway and lead to neuronal hyperexcitability (Fig. 12.3) [82].

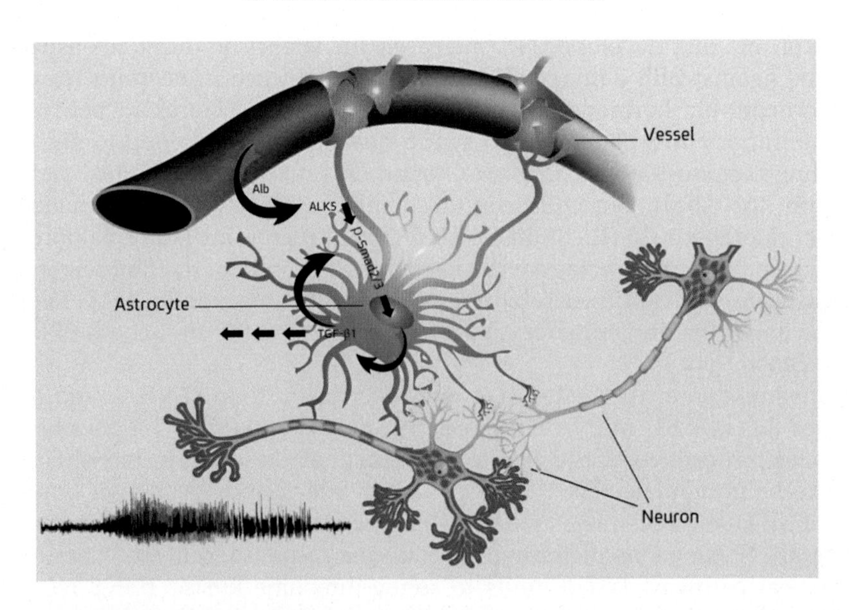

FIGURE 12.3 **Schematic diagram illustrating a dysfunctional blood–brain barrier (BBB).** With a dysfunctional BBB, the extravasation of Alb into the brain's extracellular space leads to its interaction with astrocytic TGF-β receptors, activation of TGF-β signaling, and secretion of TGF-β1, followed by neuronal hyperexcitability and unprovoked epileptic seizures. *Alb*, albumin; *ALK*, activin-like kinase; *p-Smad*, phosphorylated Smad; *TGF*, transforming growth factor. *Source: Reproduced with permission from Bar-Klein G, Cacheaux LP, Kamintsky L, Prager O, Weissberg I, Schoknecht K, et al. Losartan prevents acquired epilepsy via TGF-β signaling suppression. Ann Neurol 2014;75(6):864–75.*

Rat somatosensory cortex exposed to albumin or BBB opening using bile salts resulted in early and prominent astrocytic transcriptional changes prior to the development of epileptiform activity [111]. The observation of similar transcriptome profiles after BBB breakdown, albumin exposure, or TGF-β treatment suggested a transcriptional mechanism involving the TGF-β pathway may be responsible for BBB breakdown [109]. Transcripts coding for the astrocytic inwardly rectifying K^+ channel $K_{ir}4.1$ (Kcnj10) and the astrocytic glutamate transporters GLT1 and GLAST (SLC1A2 and SLC1A3, respectively), but not the neuronal channels or transporters, were downregulated [111]. Glutaminase and glutamine synthetase, both predominantly expressed on astrocytes, and several genes encoding gap junction proteins also exhibited reduced gene expression [111]. A NEURON model [112] simulating the consequences of reduced K^+ and glutamate uptake predicted a frequency- and *N*-methyl-D-aspartic acid (NMDA)-dependent facilitation of excitatory postsynaptic potentials (EPSPs) [111].

TGF-β-mediated uptake of albumin into astrocytes may alter the K^+ buffering capacity of astrocytes. Albumin-treated rats exhibited long-term increased brain excitability, activated astrocytes, reduced $K_{ir}4.1$ mRNA and immunolabeling [81], and increased IL-1β immunoreactivity [68]. This led to reduced K^+ buffering capacity, increased extracellular K^+ accumulation, and neuronal hyperexcitability [81]. Interestingly, TGF-β was shown to also downregulate the other K^+ inward-rectifying channel $K_{ir}2.3$ in rat primary cultured reactive astrocytes [113], although the effect this has on seizure development and spread has yet to be determined.

Gap junctional coupling is known to contribute to K^+ spatial buffering. Albumin injection into the lateral ventricle of mice caused a transient decrease in gap junction coupling [110]. In a model of photothrombic stroke, rats exhibited BBB disruption, albumin extravasation, and lowered spreading depolarization thresholds due to reduced extracellular K^+ clearance after repetitive stimulations [114]. Thus, one potential mechanism for increased seizure susceptibility after BBB disruption could be through altering astrocytic properties that lead to an imbalance in extracellular K^+ levels.

Weissberg et al. [85] quantified excitatory (presynaptic, synapsin1; postsynaptic, PSD95) and inhibitory (presynaptic, V-GAT; postsynaptic, Gephyrin) synaptic markers along dendritic lengths in primary mixed neuronal and glial cell cultures after exposure to either serum albumin or TGF-β1. Although no significant effects on inhibitory synapses were observed, both presynaptic and postsynaptic excitatory markers were increased after either albumin or TGF-β1 exposure [85]. Either treatment with SJN2511, a selective TGF-β type I receptor ALK5 inhibitor, or the absence of astrocytes prevented both the albumin- and the TGF-β-induced synaptogenesis [85]. Similar findings were seen in vivo. Mice infused i.c.v. with either albumin or TGF-β1 exhibited increased excitatory synaptic counts in the entorhinal cortex, motor cortical areas, and CA1 area of the hippocampus, but not in the dentate gyrus or CA3 areas of the hippocampus [85]. While 77% of albumin-exposed mice developed epilepsy, TGF-β1 induced spontaneous seizures in 100% of infused animals. Co-infusion with SJN2511 prevented synaptogenesis and the development of epilepsy in almost all of the infused animals [85]. Therefore, albumin extravasation may shift the balance of the neuronal network toward increased excitability in an astrocyte- and TGF-β pathway-dependent manner.

Disruption of the BBB leads to several consequences including serum extravasation into the brain, angiogenesis, activation of various inflammatory pathways, and neuronal hyperexcitability. Although activation of TGF-β may lead to increased excitability in the brain, it is likely not the only pathway involved [82]. Multiple other pathways may be involved in the development of spontaneous seizures and the exact mechanism likely will depend on the type of epilepsy or the model used. Understanding the mechanisms involved in BBB disruption-included epileptogenesis could help lead to novel identification and prevention tools for patients at risk for developing epilepsy.

BBB DISRUPTION AS A CANDIDATE BIOMARKER

Both human tissue and animal model studies have made evident that structural damage to the BBB and subsequent albumin extravasation into the brain are involved in epileptogenesis, often occurring before the onset of spontaneous seizures. Early detection of BBB disruption could serve as a potential biomarker to identify patients vulnerable to becoming epileptic. For example, a subset of patients who suffer from severe head contusion develop epilepsy months or years after injury (posttraumatic epilepsy), but no method currently exists to predict which patients are at risk.

High blood pressure has been associated with BBB disruption [55,115], but it would not be an effective biomarker because it is not specific enough to epilepsy. Assessment of serum protein levels in the CSF or examination of brain proteins in the peripheral blood may be a suitable way to detect BBB damage in patients [107]. S100β, a specific astrocyte marker, has been shown to extravasate into the peripheral circulation after the loss of BBB integrity but before the occurrence of neuronal damage. Therefore, measuring S100β levels in the blood may be a good indicator of BBB leakage [116]. Similarly, detecting serum albumin levels in the CNS through CSF sampling would be a valid clinical method to evaluate BBB dysfunction. These methods, however, are invasive, subject to producing false positives if other conditions are at play, and do not offer spatial information about BBB damage [107].

Quantitative imaging techniques may offer a noninvasive method to assess BBB damage. Potential candidates include peripheral administration of nonpermeable contrast agents followed by MRI, computerized tomography (CT), or single-photon emission CT (SPECT) [107,117]. A number of studies have already illustrated the power of MRI in detecting BBB permeability [40,41,64–66,118], often revealing a strong correlation between BBB damage and epilepsy. MRI detected BBB permeability in 82.4% of patients with PTE compared to only 25% of nonepileptic patients after trauma [41]. CT imaging detected BBB damage in about 65% of patients with various CNS disorders [115].

To be of clinical significance, a BBB biomarker needs to be reliable, quantitative, and offer high spatial resolution. An ideal diagnostic tool would be noninvasive, cost-effective, and available to the general population. Once BBB disruption is detected, patients could receive specific treatments designed to both repair BBB damage and hinder future brain dysregulation before the onset of seizures. Potential therapeutics include rapamycin, antiinflammatory agents such as IL-1RA, or TGF-β pathway blockers such as SJN2511. Advancements in imaging techniques would allow for (1) better identification and quantification of BBB breakdown; (2) determination of other CNS changes such as astrocytic reactivation or inflammation; and (3) early intervention that could prevent the development of epilepsy.

CONCLUSION

Decades of scientific research have established the occurrence of BBB dysfunction and albumin extravasation into the brain during epileptogenesis. While some studies have shown that seizures can result in BBB disruption, others have demonstrated that either BBB disruption or application of albumin alone was sufficient to cause seizures. These data suggest that BBB disruption may be an underlying cause of epileptogenesis. A dysfunctional BBB may lead to a complex cascade of events including brain inflammation and impaired K^+ and glutamate buffering. Taken together, these events may alter the excitability threshold of the brain. To prevent this, a combination of suitable detection methods and effective therapeutic interventions that reverse or prevent BBB damage should be employed.

References

[1] Liddelow SA. Fluids and barriers of the CNS: a historical viewpoint. Fluids Barriers CNS 2011;8(1):2.
[2] Ribatti D, Nico B, Crivellato E, Artico M. Development of the blood-brain barrier: a historical point of view. Anat Rec B New Anat 2006;289(1):3–8.
[3] Ehrlich P. Das Sauerstoff-Bedürfniss des Organismus: eine farbenanalytische Studie. Berlin: Hirschwald; 1885.
[4] Goldmann E. Die aussere und innere sekretion des gesunden und kranken Organismus im Licht der vitalen Farburg. Beitr Klin Chir 1909;64:192–265.
[5] Goldmann E. Vitalfärbung am Zentralnervensystem: beitrag zur Physiopathologie des plexus chorioideus der Hirnhäute. Abh Preuss Akad Wiss Physik-Math 1913;1:1–60.
[6] Roy CS, Sherrington CS. On the regulation of the blood-supply of the brain. J physiol 1890;11(1–2):85–158. 17.
[7] Lewandowsky M. Zur lehre der cerebrospinal flüssigkeit. Z Klinische Medizin 1900;40:480–94.
[8] Dempsey EW, Wislocki GB. An electron microscopic study of the blood-brain barrier in the rat, employing silver nitrate as a vital stain. J Biophys Biochem Cytol 1955;1(3):245–56.
[9] Van Breemen VL, Clemente CD. Silver deposition in the central nervous system and the hematoencephalic barrier studied with the electron microscope. J Biophys Biochem Cytol 1955;1(2):161–6.
[10] Luse SA. Electron microscopic observations of the central nervous system. J Biophys Biochem Cytol 1956;2(5):531–42.
[11] Reese TS, Karnovsky MJ. Fine structural localization of a blood-brain barrier to exogenous peroxidase. J Cell Biol 1967;34(1):207–17.
[12] Brightman MW. The distribution within the brain of ferritin injected into cerebrospinal fluid compartments. II. Parenchymal distribution. Am J Anat 1965;117(2):193–219.
[13] Brightman MW. The intracerebral movement of proteins injected into blood and cerebrospinal fluid of mice. Prog Brain Res 1968;29:19–40.
[14] Hawkins BT, Davis TP. The blood-brain barrier/neurovascular unit in health and disease. Pharmacol Rev 2005;57(2):173–85.
[15] Keaney J, Campbell M. The dynamic blood-brain barrier. FEBS J 2015;282(21):4067–79.
[16] Abbott NJ, Rönnbäck L, Hansson E. Astrocyte-endothelial interactions at the blood-brain barrier. Nat Rev Neurosci 2006;7(1):41–53.
[17] Abbott NJ, Patabendige AA, Dolman DE, Yusof SR, Begley DJ. Structure and function of the blood-brain barrier. Neurobiol Dis 2010;37(1):13–25.
[18] Hall CN, Reynell C, Gesslein B, Hamilton NB, Mishra A, Sutherland BA, et al. Capillary pericytes regulate cerebral blood flow in health and disease. Nature 2014;508(7494):55–60.
[19] Nakagawa S, Deli MA, Kawaguchi H, Shimizudani T, Shimono T, Kittel A, et al. A new blood-brain barrier model using primary rat brain endothelial cells, pericytes and astrocytes. Neurochem Int 2009;54(3–4):253–63.
[20] Shimizu F, Sano Y, Maeda T, Abe MA, Nakayama H, Takahashi R, et al. Peripheral nerve pericytes originating from the blood-nerve barrier expresses tight junctional molecules and transporters as barrier-forming cells. J Cell Physiol 2008;217(2):388–99.
[21] Nagy Z, Peters H, Hüttner I. Fracture faces of cell junctions in cerebral endothelium during normal and hyperosmotic conditions. Lab Invest 1984;50(3):313–22.
[22] Shivers RR, Betz AL, Goldstein GW. Isolated rat brain capillaries possess intact, structurally complex, interendothelial tight junctions; freeze-fracture verification of tight junction integrity. Brain Res 1984;324(2):313–22.
[23] Crone C, Olesen SP. Electrical resistance of brain microvascular endothelium. Brain Res 1982;241(1):49–55.
[24] Yu AS, McCarthy KM, Francis SA, McCormack JM, Lai J, Rogers RA, et al. Knockdown of occludin expression leads to diverse phenotypic alterations in epithelial cells. Am J Physiol Cell Physiol 2005;288(6):C1231–41.
[25] Keep RF, Ennis SR, Beer ME, Betz AL. Developmental changes in blood-brain barrier potassium permeability in the rat: relation to brain growth. J Physiol 1995;488(Pt 2):439–48.
[26] Moosa E. Brain death and organ transplantation—an Islamic opinion. S Afr Med J 1993;83(6):385–6.
[27] Olsson Y, Klatzo I, Sourander P, Steinwall O. Blood-brain barrier to albumin in embryonic new born and adult rats. Acta Neuropathologica 1968;10(2):117–22.
[28] Preston JE, al-Sarraf H, Segal MB. Permeability of the developing blood-brain barrier to ^{14}C-mannitol using the rat in situ brain perfusion technique. Brain Res Dev Brain Res 1995;87(1):69–76.

[29] Saunders NR, Knott GW, Dziegielewska KM. Barriers in the immature brain. Cell Mol Neurobiol 2000;20(1):29–40.

[30] Tauc M, Vignon X, Bouchaud C. Evidence for the effectiveness of the blood—CSF barrier in the fetal rat choroid plexus. A freeze-fracture and peroxidase diffusion study. Tissue Cell 1984;16(1):65–74.

[31] Brown PD, Davies SL, Speake T, Millar ID. Molecular mechanisms of cerebrospinal fluid production. Neuroscience 2004;129(4):957–70.

[32] Nadal A, Fuentes E, Pastor J, McNaughton PA. Plasma albumin is a potent trigger of calcium signals and DNA synthesis in astrocytes. Proc Natl Acad Sci USA 1995;92(5):1426–30.

[33] Stanimirovic DB, Friedman A. Pathophysiology of the neurovascular unit: disease cause or consequence? J Cereb Blood Flow Metab 2012;32(7):1207–21.

[34] Mihály A, Bozóky B. Immunohistochemical localization of extravasated serum albumin in the hippocampus of human subjects with partial and generalized epilepsies and epileptiform convulsions. Acta Neuropathologica 1984;65(1):25–34.

[35] van Vliet EA, da Costa Araújo S, Redeker S, van Schaik R, Aronica E, Gorter JA. Blood-brain barrier leakage may lead to progression of temporal lobe epilepsy. Brain 2007;130(Pt 2):521–34.

[36] Liu JY, Thom M, Catarino CB, Martinian L, Figarella-Branger D, Bartolomei F, et al. Neuropathology of the blood-brain barrier and pharmaco-resistance in human epilepsy. Brain 2012;135(Pt 10):3115–33.

[37] Michalak Z, Lebrun A, Di Miceli M, Rousset MC, Crespel A, Coubes P, et al. IgG leakage may contribute to neuronal dysfunction in drug-refractory epilepsies with blood-brain barrier disruption. J Neuropathol Exp Neurol 2012;71(9):826–38.

[38] Rigau V, Morin M, Rousset MC, de Bock F, Lebrun A, Coubes P, et al. Angiogenesis is associated with blood-brain barrier permeability in temporal lobe epilepsy. Brain 2007;130(Pt 7):1942–56.

[39] Schmitz AK, Grote A, Raabe A, Urbach H, Friedman A, von Lehe M, et al. Albumin storage in neoplastic astroglial elements of gangliogliomas. Seizure 2013;22(2):144–50.

[40] Tomkins O, Shelef I, Kaizerman I, Eliushin A, Afawi Z, Misk A, et al. Blood-brain barrier disruption in post-traumatic epilepsy. J Neurol Neurosurg Psychiatr 2008;79(7):774–7.

[41] Tomkins O, Feintuch A, Benifla M, Cohen A, Friedman A, Shelef I. Blood-brain barrier breakdown following traumatic brain injury: a possible role in posttraumatic epilepsy. Cardiovasc Psychiatry Neurol 2011;2011:765923.

[42] Korn A, Golan H, Melamed I, Pascual-Marqui R, Friedman A. Focal cortical dysfunction and blood-brain barrier disruption in patients with postconcussion syndrome. J Clin Neurophysiol 2005;22(1):1–9.

[43] Raabe A, Schmitz AK, Pernhorst K, Grote A, von der Brelie C, Urbach H, et al. Cliniconeuropathologic correlations show astroglial albumin storage as a common factor in epileptogenic vascular lesions. Epilepsia 2012;53(3):539–48.

[44] Bolwig TG, Hertz MM, Paulson OB, Spotoft H, Rafaelsen OJ. The permeability of the blood-brain barrier during electrically induced seizures in man. Eur J Clin Invest 1977;7(2):87–93.

[45] Bauer KF, Leonhardt H. A contribution to the pathological physiology of the blood-brain-barrier; megaphen stabilises the blood-brain-barrier. J Comp Neurol 1956;106(2):363–70.

[46] Lending M, Slobody LB, Mestern J. Effect of prolonged convulsions on the blood-cerebrospinal fluid barrier. Am J Physiol 1959;197:465–8.

[47] Barlow CF. Physiology and pathophysiology of protein permeability in the central nervous system. In: Reulen HJ, Schürmann K, editors. Springer. Berlin, Heidelberg: Steroids and Brain Edema; 1972. p. 139–146.

[48] Lorenzo AV, Shirahige I, Liang M, Barlow CF. Temporary alteration of cerebrovascular permeability to plasma protein during drug-induced seizures. Am J Physiol 1972;223(2):268–77.

[49] Nitsch C, Klatzo I. Regional patterns of blood-brain barrier breakdown during epileptiform seizures induced by various convulsive agents. J Neurol Sci 1983;59(3):305–22.

[50] Yorulmaz H, Kaptan E, Seker FB, Oztas B. Type 1 diabetes exacerbates blood-brain barrier alterations during experimental epileptic seizures in an animal model. Cell Biochem Funct 2015;33(5):285–92.

[51] Lee JC, Olszewski J. Increased cerebrovascular permeability after repeated electroshocks. Neurology 1961;11:515–9.

[52] Bolwig TG, Hertz MM, Westergaard E. Blood-brain barrier permeability to protein during epileptic seizures in the rat. Acta Neurol Scand Suppl 1977;64:226–7.

[53] Suzuki O, Takanohashi M, Yagi K. Protective effect of dexamethasone on enhancement of blood-brain barrier permeability caused by electroconvulsive shock. Arzneimittel-Forschung 1976;26(4):533–4.

[54] Westergaard E. The blood-brain barrier to horseradish peroxidase under normal and experimental conditions. Acta Neuropathol 1977;39(3):181–7.
[55] Bolwig TG, Hertz MM, Westergaard E. Acute hypertension causing blood-brain barrier breakdown during epileptic seizures. Acta Neurol Scand 1977;56(4):335–42.
[56] van Vliet EA, Holtman L, Aronica E, Schmitz LJ, Wadman WJ, Gorter JA. Atorvastatin treatment during epileptogenesis in a rat model for temporal lobe epilepsy. Epilepsia 2011;52(7):1319–30.
[57] Lassmann H, Petsche U, Kitz K, Baran H, Sperk G, Seitelberger F, et al. The role of brain edema in epileptic brain damage induced by systemic kainic acid injection. Neuroscience 1984;13(3):691–704.
[58] Ruth RE. Increased cerebrovascular permeability to protein during systemic kainic acid seizures. Epilepsia 1984;25(2):259–68.
[59] Sztriha L, Joó F, Szerdahelyi P, Lelkes Z, Adám G. Kainic acid neurotoxicity: characterization of blood-brain barrier damage. Neurosci Lett 1985;55(2):233–7.
[60] Zucker DK, Wooten GF, Lothman EW. Blood-brain barrier changes with kainic acid-induced limbic seizures. Exp Neurol 1983;79(2):422–33.
[61] Michalak Z, Sano T, Engel T, Miller-Delaney SF, Lerner-Natoli M, Henshall DC. Spatio-temporally restricted blood-brain barrier disruption after intra-amygdala kainic acid-induced status epilepticus in mice. Epilepsy Res 2013;103(2–3):167–79.
[62] Benkovic SA, O'Callaghan JP, Miller DB. Regional neuropathology following kainic acid intoxication in adult and aged C57BL/6J mice. Brain Res 2006;1070(1):215–31.
[63] Han JY, Ahn SY, Yoo JH, Nam SY, Hong JT, Oh KW. Alleviation of kainic acid-induced brain barrier dysfunction by 4-o-methylhonokiol in in vitro and in vivo models. BioMed Res Int 2015;2015:893163.
[64] van Vliet EA, Otte WM, Gorter JA, Dijkhuizen RM, Wadman WJ. Longitudinal assessment of blood-brain barrier leakage during epileptogenesis in rats. A quantitative MRI study. Neurobiol Dis 2014;63:74–84.
[65] van Vliet EA, Aronica E, Gorter JA. Blood-brain barrier dysfunction, seizures and epilepsy. Semin Cell Dev Biol 2015;38:26–34.
[66] Roch C, Leroy C, Nehlig A, Namer IJ. Magnetic resonance imaging in the study of the lithium-pilocarpine model of temporal lobe epilepsy in adult rats. Epilepsia 2002;43(4):325–35.
[67] Fabene PF, Navarro Mora G, Martinello M, Rossi B, Merigo F, Ottoboni L, et al. A role for leukocyte-endothelial adhesion mechanisms in epilepsy. Nat Med 2008;14(12):1377–83.
[68] Frigerio F, Frasca A, Weissberg I, Parrella S, Friedman A, Vezzani A, et al. Long-lasting pro-ictogenic effects induced in vivo by rat brain exposure to serum albumin in the absence of concomitant pathology. Epilepsia 2012;53(11):1887–97.
[69] Leroy C, Roch C, Koning E, Namer IJ, Nehlig A. In the lithium-pilocarpine model of epilepsy, brain lesions are not linked to changes in blood-brain barrier permeability: an autoradiographic study in adult and developing rats. Exp Neurol 2003;182(2):361–72.
[70] Marchi N, Granata T, Freri E, Ciusani E, Ragona F, Puvenna V, et al. Efficacy of anti-inflammatory therapy in a model of acute seizures and in a population of pediatric drug resistant epileptics. PloS One 2011;6(3):e18200.
[71] Marchi N, Guiso G, Caccia S, Rizzi M, Gagliardi B, Noe F, et al. Determinants of drug brain uptake in a rat model of seizure-associated malformations of cortical development. Neurobiol Dis 2006;24(3):429–42.
[72] Schmidt-Kastner R, Heim C, Sontag KH. Damage of substantia nigra pars reticulata during pilocarpine-induced status epilepticus in the rat: immunohistochemical study of neurons, astrocytes and serum-protein extravasation. Exp Brain Res 1991;86(1):125–40.
[73] Nitsch C, Goping G, Laursen H, Klatzo I. The blood-brain barrier to horseradish peroxidase at the onset of bicuculline-induced seizures in hypothalamus, pallidum, hippocampus, and other selected regions of the rabbit. Acta Neuropathol 1986;69(1–2):1–16.
[74] Librizzi L, Noé F, Vezzani A, de Curtis M, Ravizza T. Seizure-induced brain-borne inflammation sustains seizure recurrence and blood-brain barrier damage. Ann Neurol 2012;72(1):82–90.
[75] Gürses C, Orhan N, Ahishali B, Yilmaz CU, Kemikler G, Elmas I, et al. Topiramate reduces blood-brain barrier disruption and inhibits seizure activity in hyperthermia-induced seizures in rats with cortical dysplasia. Brain Res 2013;1494:91–100.
[76] Ahishali B, Kaya M, Orhan N, Arican N, Ekizoglu O, Elmas I, et al. Effects of levetiracetam on blood-brain barrier disturbances following hyperthermia-induced seizures in rats with cortical dysplasia. Life Sci 2010;87(19–22):609–19.

[77] Marchi N, Angelov L, Masaryk T, Fazio V, Granata T, Hernandez N, et al. Seizure-promoting effect of blood-brain barrier disruption. Epilepsia 2007;48(4):732–42.
[78] Clatterbuck RE, Eberhart CG, Crain BJ, Rigamonti D. Ultrastructural and immunocytochemical evidence that an incompetent blood-brain barrier is related to the pathophysiology of cavernous malformations. J Neurol Neurosurg Psychiatry 2001;71(2):188–92.
[79] Greenwood J, Adu J, Davey AJ, Abbott NJ, Bradbury MW. The effect of bile salts on the permeability and ultrastructure of the perfused, energy-depleted, rat blood-brain barrier. J Cereb Blood Flow Metab 1991;11(4):644–54.
[80] Seiffert E, Dreier JP, Ivens S, Bechmann I, Tomkins O, Heinemann U, et al. Lasting blood-brain barrier disruption induces epileptic focus in the rat somatosensory cortex. J Neurosci 2004;24(36):7829–36.
[81] Ivens S, Kaufer D, Flores LP, Bechmann I, Zumsteg D, Tomkins O, et al. TGF-β receptor-mediated albumin uptake into astrocytes is involved in neocortical epileptogenesis. Brain 2007;130(Pt 2):535–47.
[82] Bar-Klein G, Cacheaux LP, Kamintsky L, Prager O, Weissberg I, Schoknecht K, et al. Losartan prevents acquired epilepsy via TGF-β signaling suppression. Ann Neurol 2014;75(6):864–75.
[83] Tomkins O, Friedman O, Ivens S, Reiffurth C, Major S, Dreier JP, et al. Blood-brain barrier disruption results in delayed functional and structural alterations in the rat neocortex. Neurobiol Dis 2007;25(2):367–77.
[84] Weissberg I, Reichert A, Heinemann U, Friedman A. Blood-brain barrier dysfunction in epileptogenesis of the temporal lobe. Epilepsy Res Treat 2011;2011:143908.
[85] Weissberg I, Wood L, Kamintsky L, Vazquez O, Milikovsky DZ, Alexander A, et al. Albumin induces excitatory synaptogenesis through astrocytic TGF-β/ALK5 signaling in a model of acquired epilepsy following blood-brain barrier dysfunction. Neurobiol Dis 2015;78:115–25.
[86] Fieschi C, Lenzi GL, Zanette E, Orzi F, Passero S. Effects on EEG of the osmotic opening of the blood-brain barrier in rats. Life Sci 1980;27(3):239–43.
[87] Kang EJ, Major S, Jorks D, Reiffurth C, Offenhauser N, Friedman A, et al. Blood-brain barrier opening to large molecules does not imply blood-brain barrier opening to small ions. Neurobiol Dis 2013;52:204–18.
[88] Johansson B, Nilsson B. The pathophysiology of the blood-brain barrier dysfunction induced by severe hypercapnia and by epileptic brain activity. Acta Neuropathol 1977;38(2):153–8.
[89] Johansson BB. The cerebrovascular permeability to protein after bicuculline and amphetamine administration in spontaneously hypertensive rats. Evidence for increased resistance of pressure-induced blood-brain barrier dysfunction. Acta Neurol Scand 1977;56(5):397–404.
[90] Petito CK, Schaefer JA, Plum F. Ultrastructural characteristics of the brain and blood-brain barrier in experimental seizures. Brain Res 1977;127(2):251–67.
[91] Hedley-Whyte ET, Lorenzo AV, Hsu DW. Protein transport across cerebral vessels during metrazole-induced convulsions. Am J Physiol 1977;233(3):C74–85.
[92] Petito CK, Levy DE. The importance of cerebral arterioles in alterations of the blood-brain barrier. Lab Invest 1980;43(3):262–8.
[93] Westergaard E. Ultrastructural permeability properties of cerebral microvasculature under normal and experimental conditions after application of tracers. Adv Neurol 1980;28:55–74.
[94] Nagy Z, Mathieson G, Hüttner I. Blood-brain barrier opening to horseradish peroxidase in acute arterial hypertension. Acta Neuropathol 1979;48(1):45–53.
[95] Friedman A. Blood-brain barrier dysfunction, status epilepticus, seizures, and epilepsy: a puzzle of a chicken and egg? Epilepsia 2011;52(Suppl 8):19–20.
[96] Dreier JP. The role of spreading depression, spreading depolarization and spreading ischemia in neurological disease. Nat Med 2011;17(4):439–47.
[97] Ljungberg MC, Bhattacharjee MB, Lu Y, Armstrong DL, Yoshor D, Swann JW, et al. Activation of mammalian target of rapamycin in cytomegalic neurons of human cortical dysplasia. Ann Neurol 2006;60(4):420–9.
[98] Shacka JJ, Lu J, Xie ZL, Uchiyama Y, Roth KA, Zhang J. Kainic acid induces early and transient autophagic stress in mouse hippocampus. Neurosci Lett 2007;414(1):57–60.
[99] van Vliet EA, Forte G, Holtman L, den Burger JC, Sinjewel A, de Vries HE, et al. Inhibition of mammalian target of rapamycin reduces epileptogenesis and blood-brain barrier leakage but not microglia activation. Epilepsia 2012;53(7):1254–63.
[100] Zeng LH, Rensing NR, Wong M. The mammalian target of rapamycin signaling pathway mediates epileptogenesis in a model of temporal lobe epilepsy. J Neurosci 2009;29(21):6964–72.

[101] Zeng LH, Rensing NR, Wong M. Developing antiepileptogenic drugs for acquired epilepsy: targeting the mammalian target of rapamycin (mTOR) pathway. Mol Cell Pharmacol 2009;1(3):124–9.
[102] Kim SY, Buckwalter M, Soreq H, Vezzani A, Kaufer D. Blood-brain barrier dysfunction-induced inflammatory signaling in brain pathology and epileptogenesis. Epilepsia 2012;53(Suppl 6):37–44.
[103] Janigro D. Are you in or out? Leukocyte, ion, and neurotransmitter permeability across the epileptic blood-brain barrier. Epilepsia 2012;53(Suppl 1):26–34.
[104] Marchi N, Fan Q, Ghosh C, Fazio V, Bertolini F, Betto G, et al. Antagonism of peripheral inflammation reduces the severity of status epilepticus. Neurobiol Dis 2009;33(2):171–81.
[105] Plata-Salamán CR, Ilyin SE, Turrin NP, Gayle D, Flynn MC, Romanovitch AE, et al. Kindling modulates the IL-1β system, TNF-α, TGF-β1, and neuropeptide mRNAs in specific brain regions. Brain Res Mol Brain Res 2000;75(2):248–58.
[106] Aronica E, van Vliet EA, Mayboroda OA, Troost D, da Silva FH, Gorter JA. Upregulation of metabotropic glutamate receptor subtype mGluR3 and mGluR5 in reactive astrocytes in a rat model of mesial temporal lobe epilepsy. Eur J Neurosci 2000;12(7):2333–44.
[107] Friedman A, Kaufer D, Heinemann U. Blood-brain barrier breakdown-inducing astrocytic transformation: novel targets for the prevention of epilepsy. Epilepsy Res 2009;85(2–3):142–9.
[108] Heinemann U, Kaufer D, Friedman A. Blood-brain barrier dysfunction, TGF-β signaling, and astrocyte dysfunction in epilepsy. Glia 2012;60(8):1251–7.
[109] Cacheaux LP, Ivens S, David Y, Lakhter AJ, Bar-Klein G, Shapira M, et al. Transcriptome profiling reveals TGF-β signaling involvement in epileptogenesis. J Neurosci 2009;29(28):8927–35.
[110] Braganza O, Bedner P, Huttmann K, von Staden E, Friedman A, Seifert G, et al. Albumin is taken up by hippocampal NG2 cells and astrocytes and decreases gap junction coupling. Epilepsia 2012;53(11):1898–906.
[111] David Y, Cacheaux LP, Ivens S, Lapilover E, Heinemann U, Kaufer D, et al. Astrocytic dysfunction in epileptogenesis: consequence of altered potassium and glutamate homeostasis? J Neurosci 2009;29(34):10588–99.
[112] Hines ML, Carnevale NT. The NEURON simulation environment. Neural Comput 1997;9(6):1179–209.
[113] Perillan PR, Chen M, Potts EA, Simard JM. Transforming growth factor-β1 regulates $K_{ir}2.3$ inward rectifier K^+ channels via phospholipase C and protein kinase C-δ in reactive astrocytes from adult rat brain. J Biol Chem 2002;277(3):1974–80.
[114] Lapilover EG, Lippmann K, Salar S, Maslarova A, Dreier JP, Heinemann U, et al. Peri-infarct blood-brain barrier dysfunction facilitates induction of spreading depolarization associated with epileptiform discharges. Neurobiol Dis 2012;48(3):495–506.
[115] Tomkins O, Kaufer D, Korn A, Shelef I, Golan H, Reichenthal E, et al. Frequent blood-brain barrier disruption in the human cerebral cortex. Cell Mol Neurobiol 2001;21(6):675–91.
[116] Marchi N, Rasmussen P, Kapural M, Fazio V, Kight K, Mayberg MR, et al. Peripheral markers of brain damage and blood-brain barrier dysfunction. Restor Neurol Neurosci 2003;21(3–4):109–21.
[117] Volkow ND, Rosen B, Farde L. Imaging the living human brain: magnetic resonance imaging and positron emission tomography. Proc Natl Acad Sci USA 1997;94(7):2787–8.
[118] Tofts PS, Brix G, Buckley DL, Evelhoch JL, Henderson E, Knopp MV, et al. Estimating kinetic parameters from dynamic contrast-enhanced T_1-weighted MRI of a diffusible tracer: standardized quantities and symbols. J Magn Reson Imaging 1999;10(3):223–32.

C H A P T E R

13

Inflammation

OUTLINE

OVERVIEW

The immune response is an orchestra of physical barriers, distinct cell types, and inflammatory molecules that serve to protect the brain from foreign pathogens and infections. Neuroinflammation has been demonstrated in both human and experimental models of epilepsy. Cytokines, especially interleukin-1β (IL-1β), are released in a variety of epilepsies and may induce proinflammatory factors including high mobility group box 1 (HMGB1). Together, these may lead to increased excitability and seizure generation. Similarly, leukocyte infiltration after blood–brain barrier (BBB) disruption results in ion fluxes and serum protein extravasation that may induce seizure activity. Neuroinflammation can activate distinct signaling pathways, including nuclear factor-κB (NF-κB), which can trigger long-term changes in excitability; these pathways may be regulated by microRNAs (miRNAs). Therefore, limiting proinflammatory cytokine production, leukocyte infiltration, and/or

DOI: http://dx.doi.org/10.1016/B978-0-12-802401-0.00013-2

selectively altering neuroinflammation-induced gene regulation may represent therapeutic targets to prevent increased network excitability in epilepsy. The involvement of the brain immune response in epilepsy and the therapeutic potential of targeting specific inflammatory pathways will be discussed in detail in this chapter.

IMMUNE RESPONSE IN THE BRAIN

The immune response is a complex system used by the body to combat foreign pathogens. It is divided into two main types of response: the innate immune response and the adaptive immune response (Fig. 13.1). Innate immunity is the early, nonspecific inflammatory response that occurs in the brain immediately after an insult or a systemic infection [1]. It is comprised of a physical epithelial barrier to prevent initial infection (eg, the blood–brain barrier), natural killer (NK) cells, and a robust inflammatory response. NK cells are lymphocytes that rapidly detect stressed cells and destroy them. In addition, activated NK cells can produce and release cytokines, small signaling molecules that produce a specific biological response, or chemokines, a type of cytokine that can attract cells to the site of infection/inflammation. Once an inflammatory response is established, more cytokines are recruited along with phagocytes, a type of leukocyte that can recognize and engulf foreign pathogens. Phagocytes can be further subdivided into neutrophils (abundant and fast to respond), macrophages (the most versatile phagocyte), and dendritic cells (phagocytes that

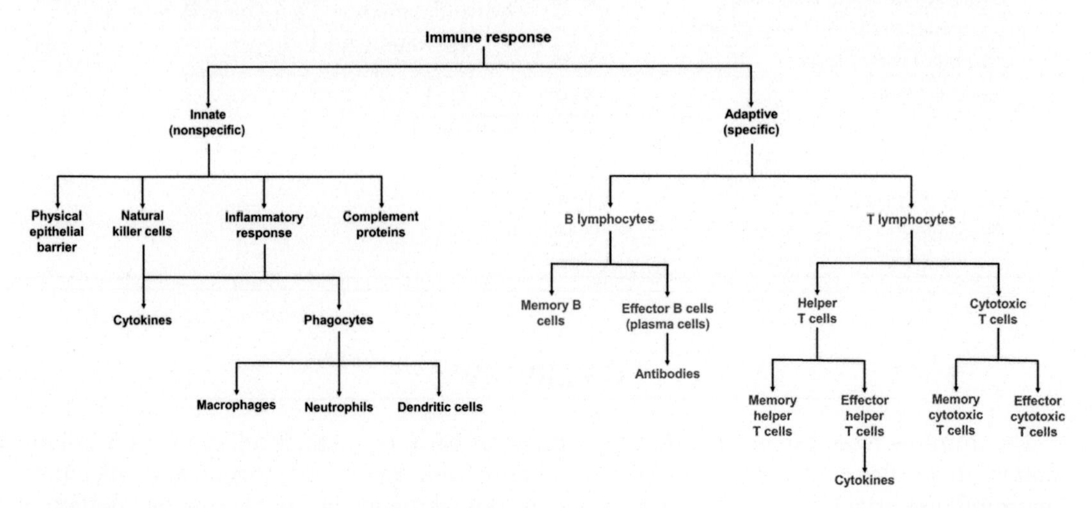

FIGURE 13.1 **The innate and adaptive immune response.** The immune response is divided into two main categories: innate and adaptive. The innate immune response is the early, nonspecific response that is comprised of a physical epithelial barrier, natural killer (NK) cells, a robust inflammatory response, and assistance by the complement cascade. The inflammatory response is carried out by cytokines and phagocytes (macrophages, neutrophils, dendritic cells). The adaptive immune response is further divided into the humoral (blue) and the cell-mediated (red) response. In the humoral response, activated B lymphocytes differentiate into memory B cells and plasma cells, or effector B cells that produce antibodies. The cell-mediated response is mediated by two types of lymphocytes, helper T cells and cytotoxic T cells. Activated helper T cells differentiate into memory helper T cells and cytokine-producing effector helper T cells. Similarly, cytotoxic T cells divide into memory and effector cytotoxic T cells. Memory cells help "remember" foreign pathogens to accelerate future immune responses to the same pathogen. Effector T cells tag cells for destruction whereas cytotoxic T cells induce apoptosis in targeted cells.

are best at activating the specific immune response). After phagocytes break down foreign pathogens, peptide fragments of the pathogen are attached to the major histocompatibility complex (MHC) type II (MHC II), this protein assembly gets transported to the cell surface of the phagocyte, and the cell is now an antigen-presenting cell. They are later recognized in the adaptive immune response.

An additional component of the innate immune system is the complement cascade. It is a type of inflammatory response that can be activated through a classical, alternative, or lectin pathway, all beginning with pattern recognition molecules [2]. In the canonical pathway, a C1 complex (composed of C1q and the serine proteases C1r and C1s) fixes antigen–antibody complexes. The complement components 2 (C2) and 4 (C4) are subsequently split, which leads to the formation of the complement component 3 (C3) convertase C4b2b. This promotes the cleavage of C3 and C5, proteins that play a role in inflammation induction. The fragments of complement proteins eventually culminate to form the membrane attack complex (MAC), a protein complex that forms a transmembrane pore allowing free diffusion of molecules into and out of the cell that can lead to target cell lysis.

The innate response may eventually progress into an adaptive immune response, a system that involves a reorganization of the immune system and remembrance of specific pathogens. Although divided into different categories, aspects of the innate immune response, including NK cells and the complement pathway, may aid in the adaptive immune response as well. The adaptive immune response occurs over a longer period of time and is mediated by activated lymphocytes recruited from the blood [3,4]. It is divided two types of response: humoral (mainly mediated by B lymphocytes) and cell mediated (involving T lymphocytes).

B lymphocytes are produced in bone marrow and are covered in membrane-bound antibodies. Once B lymphocytes bind to foreign pathogens and become activated, they start replicating and differentiating into memory B cells and effector B cells. Memory B cells are fairly permanent and "remember" the foreign pathogen to allow for an accelerated and more robust immune response in the future. Effector B cells, on the other hand, produce antibodies (effector B cells that produce antibodies are called plasma cells). These antibodies bind to the foreign pathogen and tag them for destruction by phagocytes, a process called opsonization.

T cells are lymphocytes that mature in the thymus. Helper T cells are activated by binding to MHC II and aid in the activation of B lymphocytes. Activated helper T cells also replicate and differentiate into effector helper T cells (which release cytokines) and memory helper T cells (which improve future immune responses to the same pathogen). Cytotoxic T cells, on the other hand, bind to MHC type I (MHC I), which is present on all nucleated cells. Upon binding to MHC I, cytotoxic T cells become activated and differentiate into memory and effector cytotoxic T cells. Memory cytotoxic T cells aid in future immune responses whereas effector cytotoxic T cells bind to MHC I complexes on cell surfaces and induce cell death through a variety of means, including the release of granzymes (serine proteases that induce programmed cell death) or the exocytosis of perforins to poke holes in the membrane.

T cells are often divided into CD4 expressing ($CD4^+$) or CD8 expressing ($CD8^+$) expressing cells. CD4 is a glycoprotein found on the surface of immune cells that aids in the interaction with MHC II. Therefore, most $CD4^+$ T cells are helper T cells. Similarly, CD8 is a glycoprotein that aids in the interaction with MHC I. Therefore, most $CD8^+$ T cells are cytotoxic T cells.

Is the Brain Immune Privileged?

Immune privilege is a concept that involves keeping adaptive immunity and inflammation highly controlled. The central nervous system (CNS) has been considered immune privileged since the 1920s when it was demonstrated that rat sarcoma transplanted into mouse brain parenchyma survived [5]. Over the years, these studies were furthered by a variety of experiments that showed evasion of immune recognition when bacteria, viruses, or vectors were delivered to the brain parenchyma [5]. In the 1940s, skin grafts transplanted to the brain of nonimmunized animals were successful and did not elicit an immune response [6]. Taken together, these data led to the concept of "immune privilege" in the CNS.

The idea that the brain was an immune privileged site was based on several factors, including the presence of the BBB, the lack of conventional draining lymphatics, a shortage of professional antigen-presenting cells (such as dendritic cells), low levels of MHC molecules, and the presence of many antiinflammatory factors [7]. More recently, however, the concept of an absolutely immune privileged brain has been challenged. Lymphatic vessels and antigen drainage from the brain parenchyma have been characterized [8] and dendritic cells have been found after inflammation [5]. In addition, the CNS has mechanisms for antigen uptake and transport, activated T-cell movement across the BBB, T-cell priming, and immune response cascades. In the injured brain, glial cells are activated, the BBB becomes damaged, MHC II-expressing microglia numbers increase, and inflammation (cytokine and chemokine release) is triggered. Therefore, the idea that brain immune privilege is an absolute concept has faded and been replaced with a relative model of neuroinflammation involving an intricate immune response.

NEUROINFLAMMATION AND EPILEPSY

Neuroinflammation is an integrated response to CNS injury involving several cell types, including resident CNS microglia, astrocytes, and infiltrating leukocytes. When a local inflammatory response is triggered in the brain, the activation of microglia (the resident innate immune cells) and astrocytes causes the release of a number of proinflammatory cytokines, including IL-1β, tumor necrosis factor-α (TNFα), and interleukin-6 (IL-6). In addition, damage to the BBB may allow macrophages, T cells, and B cells to enter the brain and perpetuate the immune response.

Accumulating evidence supports a role of inflammation in the pathophysiology of various types of epilepsy [1,4,9–20]. Inflammation can precede seizures and seizure-induced brain inflammation can persist long after termination of seizures [21]. Microarray analysis revealed that the immune response was the most prominently changed process during all three stages (acute, latent, and chronic) of epilepsy in the hippocampal tetanic stimulation model of epilepsy in rats [22]. Although all cell types are affected by the inflammatory response in the brain, glial cells are particularly involved.

Specific glial inflammatory pathways are chronically activated during the pathogenesis of epilepsy [23,24]. Both astrocytes and microglia were found to be reactive (astrogliosis and microgliosis, respectively) in tissue from epileptic patients [25], including the sclerotic hippocampus of patients with TLE [23,24,26–30] and in focal cortical dysplasia (FCD) [28,31,32], as well as in a number of animal models of epilepsy [24,33–36]. The release of proinflammatory molecules from these reactive cells may alter interactions between glial cells and neurons, particularly targeting astrocytes, which may lead to seizure generation and spread [10]. In addition, reactive astrocytes in the hippocampus of patients with MTLE overexpressed NFκB–p65, a transcription factor known to activate the transcription of numerous proinflammatory genes [23].

Antibody Immune Response

Rasmussen's encephalitis (RE) is a severe autoimmune form of epilepsy that is usually unihemispheric. It is a heterogeneous disease characterized by neuronal loss [37], microgliosis [37,38], chronic neuroinflammation, vacuolation, lymphatic infiltration [39], intractable seizures and progressive cognitive deterioration. Although RE primarily occurs in children, adolescent and adult cases have been reported.

Specific immunoglobulins (IgGs) were found in neurons from patients with RE, but not from patients with complex partial epilepsy [40]. In particular, circulating antibodies against the subunit 3 (GluR3) of α-amino-3-hydroxy-5-methyl-4-isoxazole-propionic acid (AMPA) receptors have been correlated to RE. Treatment of RE with plasmapheresis or IgG-selective immunoadsorption increased neurological function and reduced seizure frequency, seizure severity, and circulating GluR3 antibodies [41–43]. Therefore, the involvement of autoantibodies against GluR3 have been implicated in the pathology of RE [44–46].

A subset of rabbits immunized with either a GluR3 fusion protein or a portion of the GluR3 subunit developed seizures [47,48]. Antibodies collected from immunized rabbits or humans with RE elicited excitatory currents in cultured neurons [48]. Anti-GluR3 was primarily found on neurons in primary mixed neuronal–glial cell cultures [47]. Although it was thought that anti-GluR3 primarily destroyed neurons [47,49], it was determined that anti-GluR3 can also destroy astrocytes as well [49].

Despite the initial evidence for a role of GluR3 autoantibodies in RE, more recent evidence suggested that these antibodies were not specific to RE but, instead, were found in a variety of epilepsies [50–52]. Furthermore, others have determined that GluR3 autoantibodies were absent or infrequently found in patients with RE [37,51]. Therefore, GluR3 autoantibodies may only play a role in the pathophysiology in a subset of patients with RE.

A growing number of antibodies, including those to glutamic acid decarboxylase (GAD) or voltage-gated potassium channels, have been detected in epileptic patients [46]. Neuronal α7 nicotinic acetylcholine receptor antibodies were found in a subset of patients (2/9) with RE [53]. Two out of ten patients had autoantibodies against the presynaptic protein munc18-1, perivascular accumulation of B lymphocytes ($CD20^+$), and infiltration of plasma cells ($CD138^+$) [54]. These markers have the potential to play a role in the diagnosis and treatment of a subset of patients with RE. For further information about advances in RE and for an overview of other autoimmune epilepsies please see Refs. [55,56].

Cytokines in the Epileptic Brain

Cytokines are small signaling molecules produced by a variety of cells, including macrophages, lymphocytes, and NK cells. Interleukins (ILs), a subset of cytokines, are rapidly

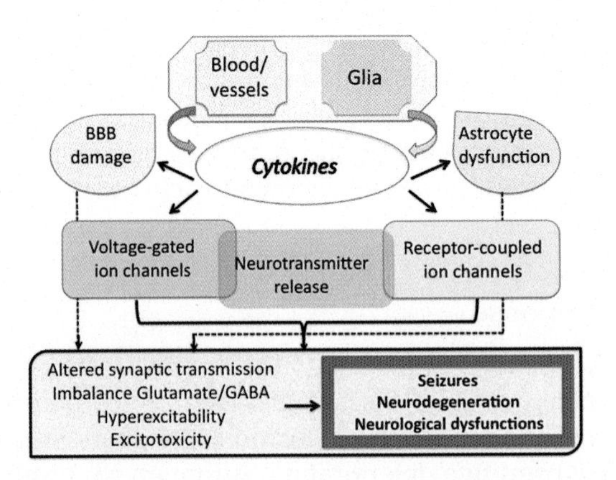

FIGURE 13.2 **Schematic representation of the cascade of pathologic events provoked by increased levels of cytokines in the nervous system.** Cytokines are synthesized and released by glia (microglia and astrocytes) or imported into nervous tissue by blood immune cells, particularly in pathologic conditions associated with a blood–brain barrier (BBB) damage. Excessive levels of cytokines can, in turn, promote both glia and BBB dysfunction, with an impact on neuronal cell excitability and viability. By activating their cognate neuronal receptors, physiological levels of cytokines can modulate voltage-gated channels (VGC) and receptor-operated ion channels (ROC), as well as presynaptic neurotransmitter release. In particular, specific interactions have been reported between IL-1β, TNF-α, IL-6, and glutamatergic or GABAergic neurotransmission. Excessive activation of cytokine receptor signaling in neurons may lead to hyperexcitability and excitotoxicity, thereby contributing to neuronal cell loss, neurological deficits and seizures. *Source: Reproduced with permission from Vezzani A, Viviani B. Neuromodulatory properties of inflammatory cytokines and their impact on neuronal excitability. Neuropharmacology 2015;96(Pt A):70–82.*

secreted from cells in response to a stimulus and bind to their target receptor on a variety of cells. IL-1α, IL-1β, IL-18, IL-12 (active heterodimer is IL-12p70), and TNFα all play a central role in regulating the immune response. IL-6, also known as B cell stimulatory factor-2, stimulates and differentiates B cells into antibody-producing cells (plasma cells). IL-8 plays a major role in chemotaxis and induction of phagocytosis whereas IL-10 inhibits the synthesis of certain cytokines, including IL-2 and TNF-α. Finally, IL-1β mediates a variety of inflammatory responses, including the induction of cyclooxygenase 2 (COX-2) synthesis, and it exerts its biological effects through the IL-1 type 1 receptor (IL-1R1). IL-1 receptor antagonist (IL-1RA) also binds to IL-1R1, therefore acting as a natural inhibitor to IL-1β. Numerous cytokines are cleaved into their active form by cysteine–aspartic proteases (caspases). For example, IL-1β is cleaved into its active form by interleukin-1 converting enzyme (ICE/caspase-1). Increased levels of cytokines can lead to a cascade of pathological events that may contribute to hyperexcitability, excitotoxicity, seizures, and neurological dysfunction (Fig. 13.2) [57].

Chemokines are a special class of cytokines that induce chemotaxis, or the movement of cells in response to a stimulus. Members of the C–C cytokine family, including CCL2 (monocyte chemoattractant protein-1, MCP-1), CCL3 (macrophage inflammatory protein 1α, MIP-1α), CCL4 (macrophage inflammatory protein-1β, MIP-1β), and CCL5 (regulated on activation, normal T cell expressed and secreted, RANTES), act as chemoattractants for a variety of cells, including monocytes, NK cells, and leukocytes. Most C–C chemokine receptors (CCR) are G-protein coupled and respond to a variety of chemokines. The CXC chemokine family are separated from the C–C family by a single amino acid ("X") but its members are also chemotactic. CXC11, for example, acts on its receptor CXCR3 and attracts activated T cells.

Inflammatory effects can be executed through a number of mechanisms. The NF-κB pathway may transcriptionally regulate inflammation and adhesion proteins. Cytokine-induced gene transcription can lead to upregulation of certain adhesion molecules, such as vascular cell adhesion molecule-1 (VCAM-1). Integrins are able to attach to the extra-cellular matrix of these cells and induce signal transduction cascades. Another mechanism of inflammation regulation is through activating transcription factor-3 (ATF-3), which acts through cAMP-responsive-element binding (CREB) protein to regulate certain inflammatory molecules. The homing cell adhesion molecule CD44 is a glycoprotein that participates in a variety of functions, including the activation, recruitment, and homing of lymphocytes. Mounting evidence has suggested a role for cytokines, particularly ILs, TNF-α, and a variety of chemokines in the pathophysiology of epilepsy.

Human Tissue Studies

Neuroinflammation has been found in human epileptic tissue. Sclerotic hippocampal tissue from patients with TLE exhibited increased levels of transforming growth factor-β (TGF-β) [58], IL-1β, IL-1R1 [24], TNF-α [59], and components of the NF-κB pathway [23,58]. While IL-1β and IL-1R1 immunoreactivity was seen in astrocytes, neurons, and microglia of sclerotic tissue, the majority of staining was found in astrocytes [24]; NF-κB staining, on the other hand, was absent from microglia [23]. Moreover, low levels of AFT-3 and IL-8 genes correlated with high seizure frequency in MTLE patients [60]. Aalbers et al. [61], however, found no difference in cytokine levels between hippocampal sclerosis (HS) and non-HS tissue.

Genetic susceptibility to TLE with HS has been linked to the occurrence of a polymorphism in the IL-1β promoter, specifically at the -511T allele. The frequency of this mutation was higher in genomic DNA extracted from peripheral blood leukocytes in TLE patients with HS than in control DNA [62]. This polymorphism was also linked to sporadic simple febrile seizures (FS) [62–65]. A polymorphism in IL-1RA has also been reported in patients with FS [65].

Patients with FS exhibited increased IL-1β production [66], specifically IL-1β allele 2 (-511) [67], as well as higher plasma IL-1RA and IL-6 levels [68]. These results, however, were controversial because IL-1β is likely to be elevated during any fever. Studies examining cytokine levels in children with febrile illness found no difference in either CSF or plasma levels of IL-1β, IL-10, or TNF-α between children with and without FS [69–72].

Tissue from patients with RE showed astrogliosis but some also showed loss in astrocyte glial fibrillary acidic protein (GFAP) and S100β immunoreactivity around blood vessels or close to the meningeal lining [37]. Activated caspase-3 reactivity, nuclear condensation, and DNA fragmentation suggested that these astrocytes were apoptotic [37]. Similarly, ICE/caspase-1 activity was increased in human tissue from TLE patients [73]. Increased TUNEL-labeled cells as well as enhanced expression of Fas and activated caspase-8 were found in tubers of patients with tuberous sclerosis complex (TSC) compared to control cortex [74]. Taken together, these data suggest that an inflammatory cascade may lead to cell death in a variety of epilepsies.

In a genome-wide study of tissue from patients with TSC, 2501 genes were differentially expressed in resected tubers compared to autopsy controls [75]. These genes include cell adhesion molecules (VCAM-1, integrins, and CD44) and inflammatory response genes (complement factors, CCL2, and various cytokines). Patients with TSC also exhibited increased protein expression of intercellular adhesion molecule-1 (ICAM-1, also called CD54), TNFα, mitogen-activated protein kinase (MAPK), and NF-κB [74]. Reactive astrocytes and activated microglia expressed IL-1β and IL-1R1 in patients with TSC [76].

Similar findings were observed in FCD. Elevated levels of both IL-1β and CCL2 immunoreactivity have been reported in FCD II tissue [32]. In fact, Ravizza et al. [77] found a positive correlation between the number of IL-1β and IL-1R1-positive neurons and seizure frequency and a negative correlation between IL-1RA-positive neurons and astrocytes and the duration of epilepsy in FCD.

Posttraumatic epilepsy (PTE) may result after an injury to the brain. A study of 256 Caucasian adults with moderate to severe TBI found that the CSF/serum IL-1β ratio was higher in patients with PTE than in patients without seizures [78]. Furthermore, the risk for developing PTE increased with the occurrence of a heterozygote (CT) genotype in the rs1143634 single nucleotide polymorphism (SNP) [78].

Human cytomegalovirus (HCMV, also known as human herpesvirus-5 or HHV-5), is a common cause of congenital nervous system infection and may result in mental retardation, deafness, motor deficits, and seizures. Primary human brain vascular pericytes infected with HCMV exhibited elevated levels of proinflammatory cytokines, including IL-8, CXCL11, and CCL5, as well as TNF-α, IL-1β, and IL-6 [79]. The inflammatory response activated by this virus likely played a role in the development of seizures.

Human tissue studies indicate that IL-1β and several other cytokines are upregulated in tissue resected from epileptic patients. Cortical protein levels of IL-1β, IL-8, IL-12p70, and CCL4 were found to be upregulated in patients with various forms of epilepsy, including FCD, encephalomalacia, RE, and MTLE [28]. One major limitation of human tissue studies, however, is that the tissue collected is representative of the endpoint of the disease. To better understand the transition from a healthy brain to an epileptic one (epileptogenesis), appropriate animal models must be used.

Animal Studies

A growing body of evidence has implicated IL-1β and its receptor IL-1R1 in epileptogenesis in a variety of epilepsy models [80–82]. A rapid increase in IL-1β immunoreactivity and mRNA levels were observed after kainic acid (KA)-[34,35,80,82–90], pilocarpine-[24,90], or bicuculline-induced status epilepticus (SE) [80,82]. IL-1β expression was often restricted to or more prominently found in astrocytes after SE. Pretreatment with IL-1β before either intrahippocampal KA-induced [80,83] or bicuculline-induced SE [81,82] significantly reduced seizure latency and increased time rodents spent in seizures [80,83]; this effect was blocked by IL-1RA [80–82]. Increased immunoreactivity [24,83] and mRNA [34,89] levels of IL-1R1, the receptor to IL-1β, were observed in both astrocytes and neurons after KA-induced SE. One study demonstrated that increased expression of the $P2X_7$ receptor was found in the hippocampus and neocortex after intraamygdala KA-induced epilepsy [91]. Antagonism of the $P2X_7$ receptor reduced seizure severity, microglia activation, neuronal injury, and IL-1β release [91].

Higher IL-1RA mRNA levels have been observed after KA-induced seizures [34,84,88,89]. Selective endogenous overexpression of IL-1RA in astrocytes under the control of the GFAP promoter resulted in increased seizure latency, reduced seizure number, and reduced seizure-related c-fos mRNA expression after bicuculline-induced excitability [81]. Similarly, intrahippocampal application of recombinant IL-1RA in mice reduced bicuculline-induced seizures and delayed seizure onset [81]. In vitro isolated guinea pig brain arterially perfused with bicuculline exhibited seizure activity that was associated with overexpression of astrocytic IL-1β [92]. Treatment with the IL-1R1 antagonist reduced the duration of the first seizure and seizure recurrence [92].

A number of other cytokines have been shown to be upregulated at various time points after KA-induced epilepsy, including IL-6, IL-10, IL-12, and TNF-α immunoreactivity [87] and IL-6, IL-10, TNF-α, and suppressor of cytokine signaling 3 (SOCS3, a cytokine-induced regulator of cytokine signaling) hippocampal mRNA levels [34,85,88,89]. Similarly, TNF-α gene and protein levels were elevated in the acute and chronic phases of pilocarpine-induced epilepsy in rats [59,90]. IL-4, IL-6, IL-10, and IL-12 were also elevated in the acute phase after pilocarpine-induced SE [90]. In contrast, young rats exhibited no changes in IL-1RA, IL-6, and TNF-α mRNA [34,89] or IL-6 and TNF-α protein [35] expression levels after intraperitoneal injections of KA compared to control animals.

Polyinosinic-polycytidylic acid (PIC), an immunostimulant structurally similar to double-stranded RNA, injected into mice during embryonic days 12–16 led to increased seizure susceptibility in postnatal day 40 offspring. Specifically, offspring exposed to PIC exhibited increased hippocampal excitability, increased seizure susceptibility, accelerated kindling rate, and diminished sociability [93]. Coadministration with antibodies to IL-6 or IL-1β abolished the effects of PIC. Administration of recombinant cytokines of both IL-6 and IL-1β together, but not individually, had similar effects as PIC [93].

Electrically induced seizures led to transiently increased levels of hippocampal IL-1β, IL-6, IL-1RA, TNF-α, and nitric oxide synthase [82,94]. All of these cytokines returned to baseline levels with the exception of IL-1β, which remained upregulated in the chronic phase [82,94]. Intracerebroventricular injections of IL-1RA reduced the total number of generalized seizures and TNF-α levels after SE [94]. In the amygdala kindling model of epilepsy, however, antiinflammatory cytokine levels (IL-1α, IL-1β, IL-6, IL-10, IL-18, CCL3, and TNFα) remained unaltered 2 and 24 hours after SE [61].

Changes in cytokine levels were also observed after intraperitoneal injections of the convulsant pentylenetetrazol (PTZ). A single dose of PTZ-induced IL-1β mRNA expression in the cerebral cortex, hypothalamus, and hippocampus within hours of treatment [86]. After repeated doses of PTZ in rats, increased mRNA and protein levels of the proinflammatory cytokines IL-1β, IL-6, and TNF-α as well as the chemokine CCL2 were increased in both the cortex and hippocampus [36].

Intrahippocampal KA-induced seizures in the rat led to BBB permeability followed by the concurrent upregulation of CCL2 and immune cell recruitment in the hippocampus [95]. Intraperitoneal injection of KA in rats led to increased levels of CCR5 and its ligands CCL3 and CCL5 in the forebrain [96,97]; reduced expression of CCR5 protected rats from KA-induced seizures, BBB damage, neuroinflammation, and neuronal cell loss [96]. Similarly, increased hippocampal levels of CCL2 and its receptor CCR2 were reported in the pilocarpine model in rats [14,98]. CCR3 and CCR2A gene and protein expressions were downregulated in the hippocampus shortly after pilocarpine-induced SE in mice [99]. CCL2 and CCL3 protein expression decreased but mRNA expression increased after pilocarpine-induced SE.

Rats with FS exhibited transiently elevated hippocampal IL-1β protein levels in astrocytes [100]. IL-1β receptor 1-deficient mice had greater seizure thresholds to experimentally induced FS [101]. Interestingly, infusion of IL-1β in wild-type, but not IL-1β receptor 1-deficient, mice reduced seizure threshold [101]. Comparable results were seen after a viral-like CNS infection in postnatal day 14 (p14) rats caused an increase in IL-1β and made these animals more susceptible to pilocarpine and PTZ-induced seizures as young adults [102]. IL-6-deficient chimeric mice resulted in decreased number of Theiler's murine encephalomyelitis viral infection-induced seizures [103].

The Tsc1-GFAP conditional knockout ($Tsc1^{GFAP}CKO$) in mice leads to seizures around 4 weeks of age. At this time, $Tsc1^{GFAP}CKO$ animals exhibited increased CCL2, IL-1β, INF-γ, CXCL10, and IL-6 mRNA levels, decreased CXCL12 mRNA levels, and increased CXCL10 protein levels [104]. In 2-week-old animals, however, only CCL2, IL-1β, and CXCL10 mRNA levels were increased [104]. IL-1β was restricted to GFAP-positive cells. Treatment with rapamycin inhibited the majority of cytokine changes in $Tsc1^{GFAP}CKO$ mice, suggesting the involvement of the mammalian target of rapamycin (mTOR) pathway in the inflammatory response [104].

Several in vitro studies have examined the involvement of IL-1β in cell excitability. The application of IL-1β concurrently with γ-aminobutyric acid (GABA) pulses to hippocampal mixed glial–neuron cultures depressed GABA-evoked currents [105]. IL-1β also decreased the mean amplitude of NMDA-induced outward currents [106] and enhanced NMDA-dependent rises in intracellular calcium [21] in hippocampal neuron cultures. Furthermore, IL-1RA antagonized the effects of IL-1β [21,105,106]. Exposure of preoptic area/anterior hypothalamus neurons in vitro to IL-1β rapidly led to decreased input resistance, hyperpolarization, and increased spontaneous inhibitory postsynaptic potentials (sIPSPs) in 26% of cells [107]. The effects involved type 1 interleukin 1 receptor (IL-1R1), the adaptor protein myeloid differentiation primary response protein (MyD88), and the second messenger ceramide; application of bicuculline abolished the hyperpolarization [107].

Adult rats with genetic absence epilepsy (GAERS—Genetic Absence Epilepsy in Rats from Strasbourg) experience generalized nonconvulsive seizures characterized by synchronous spike-and-wave discharges (SWDs). GAERS rats exhibited increased IL-1β immunoreactivity in reactive astrocytes of the somatosensory cortex [108]. Systemic administration of a specific ICE/caspase-1 blocker to inhibit IL-1β biosynthesis reduced the number and duration of SWDs [108].

Transgenic mice with GFAP promoter-driven astrocyte production of IL-6 (GFAP–IL-6 mice) exhibited enhanced sensitivity to glutamatergic (KA and NMDA) but not cholinergic (pilocarpine)-induced seizures [109]. Increased GFAP levels and astroglial hypertrophy, but not KA-induced neurodegeneration, were evident in GFAP–IL-6 mice [109]. GFAP–TNF-α transgenic mice, however, exhibited similar reactions to KA as wild-type mice [109]. In contrast, Probert et al. [110] demonstrated that overexpression of TNF-α was sufficient to trigger a neurological disorder characterized by ataxia, seizures, paresis, CNS inflammation, and white matter degeneration. Furthermore, transgenic mice with astrocytic, but not neuronal, overexpression of bioactive human transmembrane TNF-α experienced inflammation, seizures, and degeneration [111]. In a separate study, however, targeting TNF-α expression to astrocytes only led to chronic inflammatory encephalopathy [112].

Ceramide, the intracellular signaling product of the enzyme sphingomyelinase, increased intrahippocampal KA-induced seizures and this effect was blocked by preexposure to a sphingomyelinase inhibitor [83]. Furthermore, this process was dependent on phosphorylation of the NR2B subunit of NMDA receptor by the Src family of tyrosine kinases [83]. To support this, a selective NR2B-containing NMDA receptor blocker prevented the proconvulsive activity of IL-1β [21,82,83]. Taken together, these data indicate that ceramide might be a second messenger of IL-1β that exhibits its proconvulsive effects through the NMDA receptor.

High Mobility Group Box 1

HMGB1 is a nuclear DNA-binding protein that can also be released from cells to act as a chemoattractant cytokine. Extracellular HMGB1 can interact with multiple receptors, including the receptor for advanced glycation end products (RAGE), toll-like receptor (TLR) 2, and TLR4 [113]. High levels of HMGB1 have been found in clinical and experimental models of epilepsy [15]. Increased levels of both HMGB1 [114] and TLR4 [60,114] were observed in patients with epilepsy. TLR2, TLR4, and RAGE mRNA and immunoreactivity were elevated in tissue from patients with FCD, TSC, and gangliogliomas (GG) [115].

Immunoreactivity and protein levels of HMGB1 and TLR4 were rapidly increased in both the intrahippocampal KA and bicuculline injection models of epilepsy [114]. Antagonists to HMGB1 and TLR4 reduced both acute and chronic seizures. The application of HMGB1 in wild-type mice prior to KA injections led to a decrease in seizure latency and an increase in number and total time spent in seizures but had no effect in C3H/HeJ (spontaneous mutation in the Toll/IL-1 receptor domain of TLR4) mice [114].

IL-1β induced HMGB1 release in both human [115] and rat [116] cultured astrocytes. Interestingly, in human astrocyte cell culture, HMGB1 was localized in the nuclei but after IL-1β exposure, HMGB1 was also found in the cytoplasm [115]. Furthermore, HMGB1 purified from cells cultured in the presence of IL-1β, interferon-γ (IFN-γ), and TNF-α exhibited prominent proinflammatory activity; this effect was blocked by adding anti-IL-1β antibodies or IL-1RA to cell cultures [117].

Exposure of rat primary astrocytes to HMGB1 resulted in the activation of astrocytes, phosphorylation of MAPK, and increased expression of COX-2 and various chemokines (CCL2, CCL20, CCL3, CXCL1, and CXCL2), but not the inducible form of nitric oxide synthase (iNOS), CCL7, IL-6, TNF-α, or several other inflammatory molecules [118]. The HMGB1-induction of COX-2 expression was blocked by application of either an anti-RAGE antibody or a MAPK inhibitor. A complete mixture of cytokines induced COX-2 and iNOS expression in primary astrocyte culture [114].

Both IL-1β/IL-1R1 and HMGB1/TLR4 cascades are thought to be involved in the pathophysiology of epileptogenesis (Fig. 13.3) [119]. Once activated, IL-1β and HMGB1 bind to their respective receptors and activate the NF-κB pathway [120]. Activation of the NF-κB pathway can then promote chronic neuroinflammation and decreased seizure threshold. In addition, activation of IL-1R1 can lead to phosphorylation of NR2B subunits of NMDA receptors in a ceramide/Src-dependent manner [16,21,83,114,119]. This causes increased NMDA receptor function and NMDA-dependent calcium influx [21], which may also lead to hyperexcitability. Over time, this IL-1R/TLR signaling cascade may lead to chronic

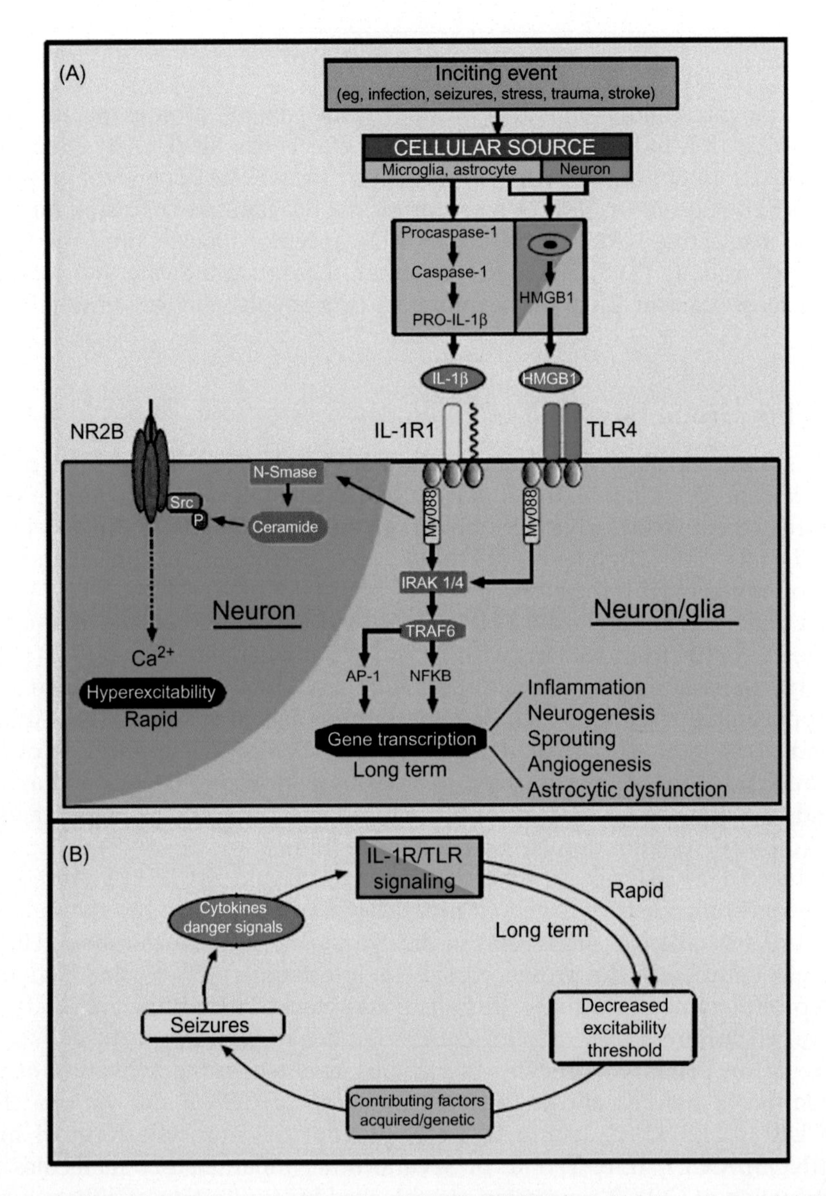

FIGURE 13.3 **Pathophysiological cascade mediated by IL-1R/TLR signaling in epilepsy.** (A) Inciting events initiated in the CNS by local injuries, or peripherally following infections, lead to activation of microglia, astrocytes, and neurons in the brain regions involved in the pathological threat. These cells release proinflammatory cytokines such as IL-1β, and danger signals such as HMGB1, thereby eliciting a cascade of inflammatory events in the target cells (ie, neurons and glia) via activation of IL-1R1 and TLR4. Signaling activation in neurons results in a rapid increase of NMDA receptor Ca^{2+} conductance via ceramide/Src mediated phosphorylation of the NR2B subunit, leading to neuronal hyperexcitability; long-term decrease in seizure threshold results from activating the transcription of genes contributing both to molecular and cellular changes involved in epileptogenesis, and to perpetuate brain inflammation. (B) The effects of brain inflammation originating by the activation of IL1R/TLR signaling contribute to the generation of individual seizures by inducing a decreased neuronal threshold of excitability. Seizure recurrence, in turn, activates further inflammation, thereby establishing a vicious cycle of events that contributes to the development of epilepsy. *Source: Reproduced with permission from Vezzani A, Maroso M, Balosso S, Sanchez MA, Bartfai T. IL-1 receptor/Toll-like receptor signaling in infection, inflammation, stress and neurodegeneration couples hyperexcitability and seizures. Brain Behav Immun 2011;25(7):1281–9.*

inflammation involving elevated cytokine levels, cytotoxic T cell–mediated apoptosis, and overall increased tissue excitability.

The Complement Pathway

The complement system consists of a number of proteins that lead to an inflammatory response including the activation of microglia, secretion of proinflammatory factors, and recruitment of macrophages. Activation of various components of the complement pathway, including C1q, C3, and C4, has been observed in reactive astrocytes, microglia, and macrophages in human tissue from epileptic patients [16,32,40,49,121–123] with temporal lobe epilepsy [26,121], TSC [75,76], and FCD [32]. Similarly, several genes and their corresponding proteins of the complement pathway were upregulated in the tetanic stimulation of the hippocampus model of epilepsy in rats [26]. Sequential infusion of individual proteins of the MAC pathway in rats leads to seizures and cytotoxicity [124]. These data strongly suggest a role for the complement pathway in the development of epilepsy.

LEUKOCYTE INFILTRATION

The migration of leukocytes is controlled by chemokines. Leukocytes may contribute to epileptogenesis both acutely through effects on BBB permeability and chronically through a variety of mechanisms including vascular alterations and release of cytotoxic enzymes [125]. Both tissue from patients with epilepsy [29,126] and from the intrahippocampal KA mouse model of epilepsy exhibited abundant levels of leukocytes [29]. Expression of adhesion molecules P-selectin and ICAM-1 on brain endothelium was rapidly induced by epileptiform activity in the in vitro isolated guinea pig brain [127]. After pilocarpine-induced seizures, mice persistently exhibited elevated levels of vascular adhesion molecules, including ICAM-1, VCAM-1, E-selectin, and P-selectin [126]. The α4 integrin chain pairs with either the β1 or β7 chain to form α4β1 or α4β7, respectively and these heterodimers play a role in T-cell migration [125]. Blocking α4 or the α4β1 ligand VCAM-1 led to a dramatic reduction of seizure frequency during the chronic phase in the pilocarpine mouse model of epilepsy [126].

It is thought that leukocyte–endothelium interactions and subsequent recruitment of leukocytes into the brain parenchyma represent a key component of the epileptic cascade (Fig. 13.4) [125]. Interactions of inflammatory molecules and extravasation of leukocytes may lead to BBB opening and subsequent leakage of serum proteins into the brain [16,18,19,128]. Serum protein (especially albumin) extravasation into the brain may contribute to a self-amplifying cycle of network hyperexcitability (see *Chapter 12: Blood–Brain Barrier Disruption*).

Phagocytes

Phagocytes, such as macrophages, neutrophils, and dendritic cells, play a major role in the immune response by recognizing and destroying foreign particles. Human tissue from patients with TLE HS [24], FCD [31], and TSC [74] exhibited increased immunoreactivity for $CD68^{+}$ macrophages [129]. Similarly, macrophages were prominently found after KA [29,87]

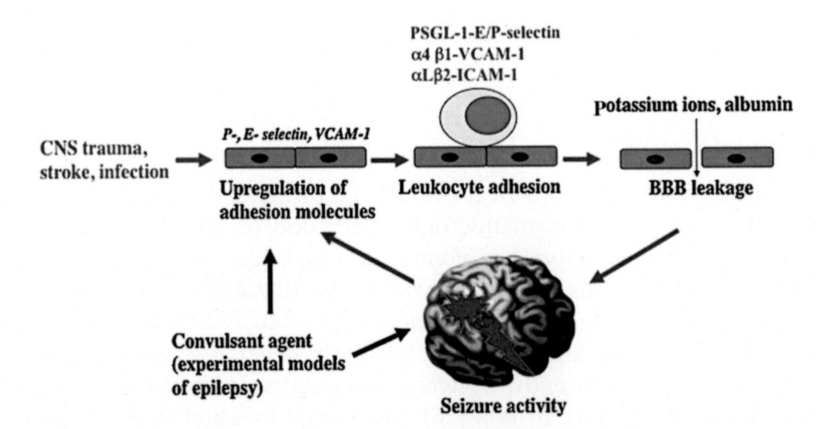

FIGURE 13.4 **The role of vascular inflammation and leukocyte trafficking in seizure generation.** Upregulation of adhesion molecules can be provoked by the convulsant agent (pilocarpine or kainic acid) or by inflammatory CNS pathologies such as stroke, infection, or traumatic brain injury. These vascular adhesion molecules mediate blood leukocyte adhesion, which compromises the endothelial barrier function resulting in leakage of plasma constituents into the brain parenchyma. Exposure of neuronal cells to plasma elements such as albumin and potassium ions may lead to neuronal electrical hyperactivity, lowering the threshold for seizures. Leukocyte adhesion, together with altered blood–brain barrier (BBB) permeability, and the seizure-dependent hyperexpression of adhesion molecules and vasodilation suggest that seizure activity and leukocyte adhesion synergize in a self-amplifying cascade of events leading to chronic epilepsy. *Source: Reproduced with permission from Fabene PF, Laudanna C, Constantin G. Leukocyte trafficking mechanisms in epilepsy. Mol Immunol 2013, 5(1):100–4.*

or pilocarpine-induced [24] epilepsy. In addition, dendritic cell immunoreactivity was elevated in FCD II tissue [32].

Recombination-activating gene 1 (RAG1) knockout mice lack mature B and T lymphocytes [130]. Intrahippocampal KA injections in RAG1 knockout mice led to early seizure onset, infiltration of neutrophils, microgliosis, elevated neurodegeneration and increased CD68 (macrophage) staining in the ipsilateral hippocampus [29]. Taken together, these data suggest a role for macrophage infiltration in epilepsy, but the sequence of events and mechanisms involved are still unclear.

T and B lymphocytes

T and B lymphocytes play a major role in the adaptive immune response. Sclerotic tissue from patients with TLE, however, lacked staining for B lymphocytes [29]. $CD3^+$ (mature) T lymphocytes, however, were found in tissue from patients with HS, TLE [24,29], FCD [32], RE [129,131], and TSC [76]. A predominance of $CD8^+$ (cytotoxic) T cells were observed in tissue from patients with TSC [76] and RE [131,132]. Interestingly, a small percentage of $CD8^+$ cells in RE were granzyme-B positive [131,132] and juxtaposed to neurons [131]. This potentially led to cell death of both neurons [131] and astrocytes [37], but not oligodendrocytes or myelin, during the pathogenesis of RE [133].

KA-induced seizures in mice led to an initial accumulation of $CD3^+$ T cells in the hippocampus [87] and blood vessels [29]. T-cell infiltration gradually increased to a maximal and persistent accumulation in the neuropil 2–4 weeks after injection. Interestingly, the majority of T cells found in the epileptic tissue were $CD8^+$, suggesting a preferential infiltration of cytotoxic T cells [29]. Similarly, in the acute, latent, and chronic phase of pilocarpine-induced epilepsy in rats, scarce NK cells, B and T lymphocytes were found associated with brain microvessels and rarely in brain parenchyma whereas control tissue exhibited none [24]. These results suggest a prominent role for cytotoxic T cell–induced apoptosis in the pathology of epilepsy.

Mutant mice lacking $CD4^+$ T cells (MHCII knockout), $CD8^+$ T cells (β2-microglobulin knockout), or both (RAG1 knockout mice) exhibited the same neurodegeneration pattern as control mice after intrahippocampal KA injections [134]. The β2-microglobulin knockout mice exhibited decreased seizure frequency over time compared to RAG1 and MCHII knockout mice. Grafting of the missing T cell population had no influence on seizure onset, but did affect macrophage infiltration. Specifically, $CD8^+$ T cells in β2-microglobulin knockout mice enhanced macrophage recruitment whereas $CD4^+$ T cells transferred into MHCII knockout and RAG1 knockout mice blocked macrophage infiltration and prevented KA-induced granule cell dispersion (Fig. 13.5) [134]. These data suggest that specific subtypes of T lymphocytes play roles in modulating epilepsy neuropathology.

Prostaglandins

Cyclooxygenases (COXs) are enzymes that convert arachidonic acid to prostaglandins and thromboxanes. Interestingly, the committed step in this pathway is initiated by two isoforms of cyclooxygenase (COX): COX-1 and COX-2 [135]. COX-1 is constitutively expressed and involved in normal tissue homeostasis whereas COX-2 is induced by inflammation [135]. In fact, the effects of IL-1β may be mediated through COX-2 and subsequent release of prostaglandin E2. A significant increase in COX-2 protein levels was observed in human patients with refractory TLE [58].

Studies using animal models of epilepsy have also implicated inflammatory induction of COX-2 in epileptogenesis. At various time points after intrahippocampal KA injection, mice exhibited an increase in hippocampal COX-2 mRNA expression [88]. Treatment with a selective COX-2 inhibitor also led to increased KA-induced seizure susceptibility in wild-type mice [136]. COX-2 knockout mice were more vulnerable to KA-induced excitotoxicity [136], exhibited increased astrogliosis and microgliosis, and had increased hippocampal gene expression of TNF-α, IL-1β, IL-6, iNOS (a major source of oxidative stress), and NF-κB compared to wild-type mice treated with KA [137].

When studying knockout mice, it is important to consider any other expression changes that may be present. COX-2 knockout mice exhibited no change in cytokine gene expression, but did show increased hippocampal and cortical mRNA expression of AMPA and KA ionotropic glutamate receptor subunits, including GluR2, GluR3, and GluR6. COX-2 knockout mice also had reduced levels of NMDA receptor gene expression [137]. Twenty-four hours after intraperitoneal injections of KA, both wild-type and COX-2 knockout mice exhibited decreased hippocampal mRNA expression of GluR subunits 1–7 and KA1–2, with the exception of GluR6 for wild-type mice and GluR7 and KA1 for COX-2 knockout mice. Thus, there are clearly many compensatory changes in glutamate receptors that occur in the absence of COX-2.

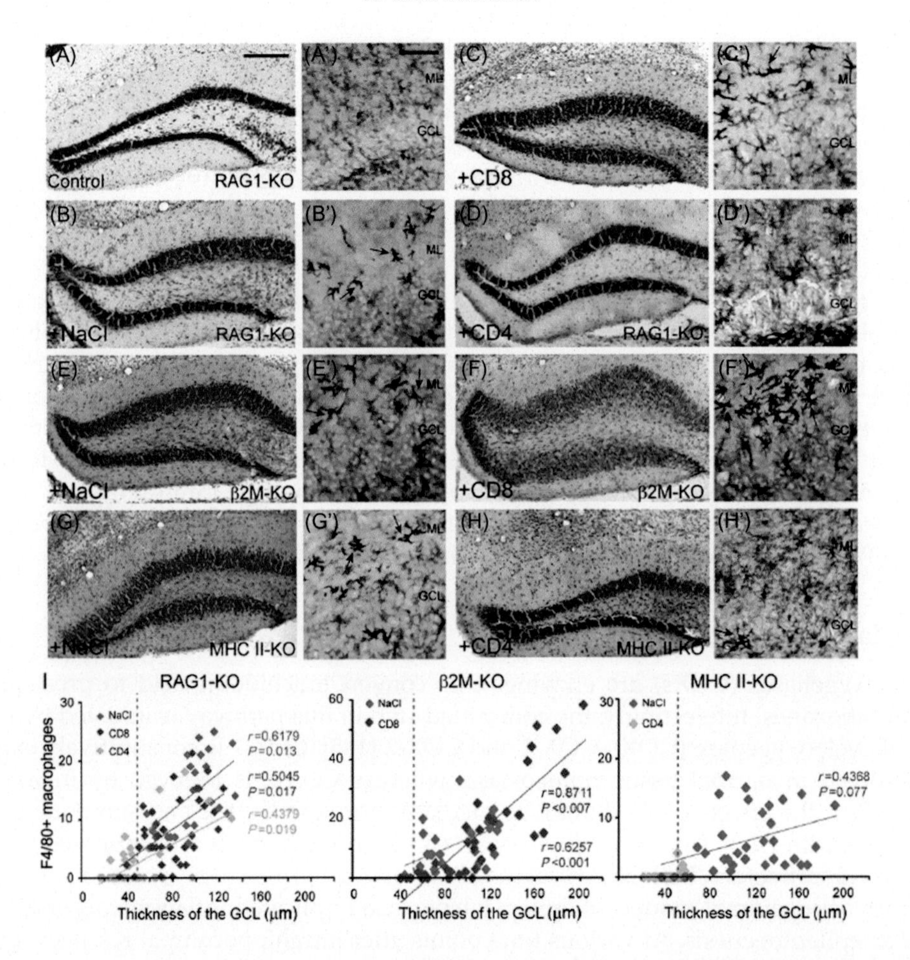

FIGURE 13.5 **Effect of lymphocyte transfer on F4/80 macrophage-like cell infiltration into the KA-lesioned dentate gyrus and on granule cell dispersion/degeneration.** (A–H) Each pair of images represents a Nissl stained section of the dentate gyrus and the pattern of F4/80 staining in an adjacent section (granule cell layer and molecular layer) from the same mouse. Arrows point to strongly stained macrophage-likes, which are distinguished from microglial cells by their shape and staining intensity. Note the absence of granule cell dispersion in the two mutants (RAG1-KO, D; MHCII-KO, H) grafted with $CD4^+$ T cells. (I) Correlation analysis between the thickness of the granule cell layer and the number of macrophage-like cells. In each section, the granule cell layer was divided into six segments, yielding six pairs of values. $N = 3–4$ mice/genotype and treatment. In each histogram, the average thickness of the granule cell layer in the contralateral dentate gyrus is indicated by the vertical dashed line. The colored lines represent the linear correlation curves, with the results of the statistical analysis indicated with the same color code. Scale bar, (A) 150 μm; (A') 50 μm. *Source: Reproduced with permission from Deprez F, Zattoni M, Mura ML, Frei K, Fritschy JM. Adoptive transfer of T lymphocytes in immunodeficient mice influences epileptogenesis and neurodegeneration in a model of temporal lobe epilepsy. Neurobiol Dis 2011;44(2):174–84.*

COX-2 is expressed in a variety of cells, including neurons, microglia, and infiltrating leukocytes. Conditional knockout (cKO) mice that lacked COX-2 in selected forebrain neurons exhibited delayed inflammation-mediated neurotoxicity in a pilocarpine model of epilepsy [138]. Furthermore, wild-type and cKO mice experienced similar seizure latency, intensity, and duration of SE [138]. This is in contrast to the effects of global COX-2 ablation.

Lipopolysaccharide-Induced Immune Response

Lipopolysaccharide (LPS) is a large molecule consisting of lipid and polysaccharide that is found in the outer membrane of gram-negative bacteria. It reliably induces a strong immune response and has been extensively used in models of both systemic and local inflammation. Direct CNS application of LPS has been investigated. Cortical application of just 0.02 μg LPS led to increased power and area of somatosensory evoked potentials; application of 0.2 μg LPS induced epileptiform discharges and, in some cases, focal seizures with motor manifestations [139]. Pretreatment with IL-1RA prevented these effects on cortical excitability [139]. Most experiments studying the involvement of the immune system in epilepsy, however, use systemic administration of LPS.

Intraperitoneal injections of LPS increased serum IL-1β and the chemokine CXCL10 levels in wild-type mice [104]. Rats injected with LPS exhibited increased IL-1β and TNF-α 6 hours after LPS treatment and hippocampal slices exhibited enhanced excitability and epileptiform activity [140]. In addition, both p7 and p14 LPS-treated rats exhibited increased seizure susceptibility in response to pilocarpine, KA, or PTZ treatment; decreased PTZ seizure threshold was blocked by concurrent intracerebroventricular administration of TNF-α antibody but not by IL-1RA [140]. Furthermore, PTZ seizure threshold was decreased 30 days after hyperthermic seizures and concomitant LPS treatment in p14 rats, suggesting that inflammation may predispose the immature brain to increased excitability in adulthood [141].

LPS injections led to more severe seizures, increased hippocampal damage, and a greater number of hypertrophied astrocytes with thickened processes in the chronic phase of the young rat pilocarpine model of epilepsy [142,143]. Similarly, p14 rats that received LPS before rapid electrical kindling of the hippocampus experienced longer seizures during the kindling process, increased afterdischarge duration (ADD), and reduced afterdischarge threshold (ADT) [142,144]. Co-application of IL-1RA with LPS blocked kindling progression but IL-1RA alone decreased ADT [144].

FS induced by intraperitoneal KA injections paired with LPS and hyperthermia in p14 rats led to a lower electrically induced ADT [145]. In a similar study, half of the rats (p14) that received intraperitoneal injections of both LPS and KA in the setting of induced hyperthermia developed febrile convulsions [146]. Increased levels of IL-1β in the hippocampus, but not the hypothalamus or cortex, were found in animals at the onset of febrile convulsions compared to those that did not seize. Three hours after SE, however, animals exhibited increased IL-1β in both the hypothalamus and hippocampus [146]. Intracerebroventricular treatment with IL-1β increased the number of rats that convulsed whereas treatment with IL-RA reduced the number of animals that convulsed [146]. Similarly, cultures of peripheral blood mononuclear cells from children with FS had increased LPS-induced IL-1 production compared to mononuclear cells from children without convulsions [147].

LPS binds to TLR4 in a variety of cells, including dendritic cells, macrophages, and B cells. This activates the NF-κB pathway and promotes the induction of a variety of proinflammatory factors including cytokines, chemokines, and inducible iNOS [148]. Although LPS decreased the PTZ-induced seizure threshold, this proconvulsive effect was reversed by pretreatment with either an NO synthase inhibitor, an antagonist of the opioid receptor, or a COX inhibitor [149]. Taken together, these data suggest that LPS acts through a variety of proinflammatory mediators to influence seizure susceptibility.

MicroRNAs in Epilepsy

MicroRNAs (miRNA) are small, noncoding endogenous RNAs that regulate gene expression by triggering degradation or translational repression of their target mRNAs. Thus, miRNAs are powerful regulators of a variety of biological processes, including the immune response. Specific miRNAs have been implicated in epilepsy and astrocyte-mediated inflammation [150–152]. Several studies have demonstrated a dysregulation of miRNAs in epileptic tissue [153–156]. Sclerotic hippocampal tissue from patients with MTLE revealed an increase in 25 miRNAs and a decrease in 5 miRNAs compared to nonsclerotic hippocampal tissue [157]. Similarly, a genome-wide miRNA profiling study of MTLE tissue revealed that 165 miRNAs were up- or downregulated with the immune response being most heavily affected [158]. In this group, miRNA-221 (miR-221) and miRNA-222 (miR-222) were found to regulate astrocytic endogenous ICAM-1 expression in tissue from patients with MTLE [158]. Large-scale reductions in miRNA expression were found in HS tissue from TLE patients and were associated with a reduction of dicer protein levels [159].

Animal models have been used extensively to study the expression profiles of miRNAs in epilepsy. After electrically induced SE in rats, hippocampal miRNA expression profiles relating to a variety of signaling pathways, including MAPK, TGF-β, and mTOR, were altered [160]. Sixty days after pilocarpine-induced SE in rats, expression of 23 of 125 detected miRNAs were altered in the hippocampus, including a significant upregulation of miRNAs miR-9, miR-23a, miR-23b, miR-126, miR-140, and miR-146a, and a significant downregulation of miR-98 and miR-352 [161]. $P2X_7$ receptor may be a target of microRNA-22 (miR-22) [162]. Inhibition of miR-22 increased $P2X_7$ receptor expression, cytokine levels, and seizure number in the intraamygdala KA model of epilepsy [162].

Proinflammatory cytokines such as TNF-α and IL-1β have been shown to induce miR-146a expression [163–171] while IL-1RA can block its expression [163]. Furthermore, increased levels of miR-146a in vitro suppressed IL 1β–induced expression of COX-2, IL-6, and other proinflammatory cytokines while knockdown of miR-146a upregulated the release of proinflammatory factors [163]. In rats, miR-146a expression was increased in latent and chronic, but not acute, phases after tetanic stimulation-induced SE [160,172].

In support of the findings from animal studies, increased levels of miR-146a were found in human sclerotic tissue from epilepsy patients [172]. Interestingly, miR-146a was confined to neurons in the healthy brain but was found in both neurons and glia in the epileptic brain [172]. High expression of miR-146a, miR-21, and miR-155 were observed in tissue from patients with epilepsy-associated glioneuronal lesions (ganglioglioma, GG), which are characterized by prominent activation of the innate immune response [163,171]. Interestingly, exposure of IL-1β to glioblastoma cells increased miR-21, miR-146a, and miR-155 expression

[171]. Tissue from both children with MTLE and rats with pilocarpine-induced epilepsy displayed upregulated miR-155 expression [59] whereas miR-21 was elevated in the electrical stimulation rat model of TLE [160]. For detailed reviews on the role of miRNAs in epilepsy, see Refs. [151,152,173].

Myeloid related protein-8 (MRP-8) is an endogenous ligand of TLR4 that induces astrocyte-related inflammation. Human astrocytoma cells exposed to MRP-8 exhibited increased miR-132, miR-146a, and miR-155 levels in a dose-dependent manner [174]. MRP-8 treatment also led to increased mRNA and protein levels of interleukin-1 receptor-associated kinase 4 (IRAK4), a kinase in TLR4 signaling that targets miR-132; upregulation of miR-132 inhibited the expression of IRAK4. In addition, downregulation of miR-132 before MRP-8 stimulation promoted the release of IL-1β and IL-6 while increased miR-132 before MRP-8 stimulation had the opposite effect [174]. In vitro astrocytes exposed to MRP-8 had dose-dependent increases in TLR4, IL-1β, IL-6, and TNF-α mRNA and protein.

ANTIINFLAMMATORY DRUGS AS A POTENTIAL TREATMENT FOR EPILEPSY

Inflammation is an attractive target for the treatment of epilepsy. In fact, some current antiepileptic drugs have already been shown to target the inflammatory pathway [175]. For example, levetiracetam, gabapentin, and phenytoin induced release of the antiinflammatory cytokine TGF-β1 in vitro [175,176]. When considering the therapeutic alteration of inflammation, however, several considerations arise. First, if an initial insult to the brain activates antiinflammatory processes, then blocking all immune signaling could potentially depress the endogenous antiinflammatory cascade [177]. Since inflammation may contribute to protection or repair of neuronal circuits, inhibition of a specific inflammatory cascade may be a better approach [1,177]. Second, the diverse action of pathways and potential cross-talk among various signaling cascades should be considered when investigating new drug treatments. Third, the profile of acute versus chronic inflammatory mediators is quite different, and only certain pathways may contribute to early versus late epileptogenesis and the chronic epileptic state. Targeting these pathways selectively would limit adverse side effects. Fourth, the issue of local (ie, at epileptogenic focus or at site of BBB breakdown) versus systemic drug/agent delivery should be considered.

The antiinflammatory agent minocycline has been investigated in several models of epilepsy. Abraham et al. [33] determined that treatment with minocycline after KA-induced SE prevented the increased microgliosis and seizure susceptibility after a second dose of KA. Minocycline also suppressed increased seizure susceptibility and IL-1β levels in a viral-like CNS infection in rats [102]. Similarly, infection of a mouse with Theiler's murine encephalomyelitis virus led to seizures in half of the mice and treatment with minocycline decreased the number of mice that developed seizures [103]. Mice that did develop seizures, however, did not differ from wild-type mice in pathological changes, including neuronal cell loss and inflammation [103].

Various drugs targeting proinflammatory cytokine production have shown promising results in the rat KA model. The thiadiazolidinone NP031112 offered neuroprotection and attenuated the production of proinflammatory cytokines [178]. Minozac, a brain-penetrant small molecule selective inhibitor of proinflammatory cytokine production, prevented astrocyte activation and the increase in proinflammatory cytokines (IL-1β, IL-6, and TNF-α) in the hippocampus [35,179]. Finally, naloxone, a drug recently shown to possess antiinflammatory effects, reduced microglia and astrocyte activation and IL-1β mRNA and protein synthesis [180]. In line with these findings, the polyphenolic diketone curcumin was shown to reduce cognitive deficits, proinflammatory cytokine and chemokine expression, and gliosis in a PTZ-kindling model of epilepsy in rats [36].

Inhibition of ICE/caspase-1 by VX-765 exhibited powerful anticonvulsant effects in the intrahippocampal KA rodent model of epilepsy in a dose-dependent manner [181,182]. Caspase-1 inhibition by either VX-765 or pralnacasan also reduced LPS + adenosine triphosphate–induced release of IL-1β in mouse organotypic hippocampal slices [182]. Anakinra, a recombinant IL-1RA, combined with VX-765 administered after either pilocarpine- or hippocampal electrical stimulation-induced SE decreased neurodegeneration and IL-1β immunoreactivity in astrocytes [183]. This treatment, however, did not modify the total duration of SE, the broad inflammatory response in SE, or seizure frequency and duration in the chronic phase of epilepsy [183].

Inhibition of the prostaglandin pathway, particularly COX-2, is an attractive candidate because COX-2 is primarily activated by inflammation. Aspirin, a nonselective COX inhibitor, reduced the frequency and duration of spontaneous recurrent seizures as well as attenuated mossy fiber sprouting, aberrant migration of granule cells, and hippocampal neuronal cell loss in the pilocarpine rat model of epilepsy [184]. Celecoxib, a more selective COX-2 inhibitor, also reduced the frequency and duration of spontaneous recurrent seizures in the rat pilocarpine model [185].

Prevention of leukocyte infiltration into the brain has also been considered as a therapeutic strategy. Blocking antibodies to either α4 integrin or VCAM-1 led to a dramatic reduction of seizure frequency during the chronic phase of the pilocarpine mouse model of epilepsy [126]. Similar results were seen when blocking αL integrin or its endothelial ligand ICAM-1; this treatment prevented convulsions and reduced leukocyte–endothelial interactions. Furthermore, pretreatment with an α4 integrin antibody completely prevented pilocarpine-induced SE in mice [126]. In line with this, inhibition of leukocyte adhesion through treatment with natalizumab, an anti-α4 integrin antibody, decreased seizures in a young adult with multiple sclerosis [186]. Inhibition of leukocyte–vascular interaction with genetic deletion of P-selectin glycoprotein ligand-1 or neutrophil depletion significantly reduced SE and spontaneous seizure occurrence [126].

Antagomirs are chemically modified antisense oligonucleotides that bind and deplete miRNA from cells. Targeting and inhibiting miRNAs has been shown to have therapeutic effects in epilepsy [187]. Antagomirs against miR-132 depleted hippocampal miR-132 levels (shown to be upregulated in epileptic tissue) and reduced seizure-induced neuronal cell death in the intraamygdala KA model of epilepsy [154]. After pilocarpine-induced SE in rats, several miRNAs are upregulated including the proapoptotic miR-34a. The use of an antagomir against miR-34a resulted in inhibition of caspase-3 and increased neuronal survival [153]. Upregulation of miR-134 has been found in both human and experimental models of epilepsy. Pretreatment with antagomirs of miR-134 reduced the number of animals that developed pilocarpine-induced SE, increased survival rate, delayed seizure onset, reduced total seizure power, and mitigated the pathological features of TLE [155,156]. An inverse correlation between IL-10 (upregulated) and miRNA-187 (miR-187, downregulated) was found in both hippocampal tissue from patients with TLE and in the pilocarpine rat model of epilepsy. Antagomirs of miR-187 increased IL-10 expression [188]. Antagomirs are a promising new therapeutic option for the treatment for epilepsy.

Imaging of Neuroinflammation

In addition to inflammation-directed drug treatment, recent research has focused on non-invasive imaging markers of inflammation. Diffusion tensor imaging (DTI) of inflammatory demyelination has been demonstrated [189] and could easily be applied to imaging glial cells in epilepsy [190]. Similarly, labeling with the isoquinoline radioligand [^{3}H]-PK-11195 combined with positron emission tomography (PET) has detected neuroinflammation and glial cell activation in patients with epilepsy [191–193]. Translocator protein 18 KDa (TSPO) overexpression was measured in patients with MTLE using ([^{11}C])-PBR28 PET imaging, suggesting this technique as a valid biomarker of neuroinflammation [194].

Quantitative T_2 measurements [195] and iron oxide contrast-enhanced MRI has been used to visualize inflammation in human patients [196]. Hippocampal T2 analysis of MRI scans of rat brains after prolonged FS demonstrated increased hippocampal T2 relaxation in 6/9 FS rats [100]. Duffy et al. [197] used MRI to image seizure-induced endothelial activation with iron oxide–conjugated antibodies to VCAM-1. They found seizure-induced expression of VCAM-1 primarily occurred in the regions most affected following SE [197]. In a mouse model of febrile status epilepticus (FSE), reduced T2 relaxation times in the amygdala hours after FSE correlated with HMGB1 translocation and predicted subsequent epileptogenesis as assessed by long-term video–EEG monitoring [198]. Based on these types of studies, advanced PET and MRI imaging techniques have the potential to become viable biomarkers of neuroinflammation and epileptogenesis.

CONCLUSION

The immune response plays a fundamental role in protecting the brain from foreign pathogens and infections. The activation of proinflammatory pathways has been demonstrated in both human and experimental models of epilepsy. In particular, IL-1β and HMGB1 seem to lead to a self-perpetuating cycle of increased excitability and seizure generation. Similarly, leukocyte adhesion to the BBB and subsequent disruption leads to the influx of ions and serum proteins, both of which may lead to seizure activity. Long-term excitability changes may be induced by signaling pathways, including NF-κB, and are regulated by miRNAs. Therefore, a combination of leukocyte infiltration, proinflammatory cytokines, and gene regulation may contribute to the spread of the inflammatory response and increased network excitability in a variety of epilepsies. While not all aspects of neuroinflammation-induced excitability are understood, the available data indicate that various proinflammatory factors and their signaling pathways may represent important novel therapeutic targets.

References

[1] Hubbard JA, Hsu MS, Fiacco TA, Binder DK. Glial cell changes in epilepsy: overview of the clinical problem and therapeutic opportunities. Neurochem Int 2013;63(7):638–51.

[2] Reis ES, Mastellos DC, Yancopoulou D, Risitano AM, Ricklin D, Lambris JD. Applying complement therapeutics to rare diseases. Clin Immunol 2015;161(2):225–40.

[3] Nguyen MD, Julien JP, Rivest S. Innate immunity: the missing link in neuroprotection and neurodegeneration? Nat Rev Neurosci 2002;3(3):216–27.

[4] Vezzani A, Balosso S, Ravizza T. Inflammation and epilepsy. Handbook of Clinical Neurology 2012;107:163–75.

[5] Galea I, Bechmann I, Perry VH. What is immune privilege (not)? Trends Immunol 2007;28(1):12–18.

[6] Medawar PB. Immunity to homologous grafted skin; the fate of skin homografts transplanted to the brain, to subcutaneous tissue, and to the anterior chamber of the eye. Br J Exp Pathol 1948;29(1):58–69.

[7] Harris MG, Hulseberg P, Ling C, Karman J, Clarkson BD, Harding JS, et al. Immune privilege of the CNS is not the consequence of limited antigen sampling. Sci Rep 2014;4:4422.

[8] Iliff JJ, Goldman SA, Nedergaard M. Implications of the discovery of brain lymphatic pathways. Lancet Neurol 2015;14(10):977–9.

[9] Aronica E, Crino PB. Inflammation in epilepsy: clinical observations. Epilepsia 2011;52(Suppl 3):26–32.

[10] Aronica E, Ravizza T, Zurolo E, Vezzani A. Astrocyte immune responses in epilepsy. Glia 2012;60(8):1258–68.

[11] Choi J, Koh S. Role of brain inflammation in epileptogenesis. Yonsei Med J 2008;49(1):1–18.

[12] Devinsky O, Vezzani A, Najjar S, De Lanerolle NC, Rogawski MA. Glia and epilepsy: excitability and inflammation. Trends Neurosci 2013;36(3):174–84.

[13] Fabene PF, Bramanti P, Constantin G. The emerging role for chemokines in epilepsy. J Neuroimmunol 2010;224(1–2):22–7.

[14] Galic MA, Riazi K, Pittman QJ. Cytokines and brain excitability. Front Neuroendocrinol 2012;33(1):116–25.

[15] Maroso M, Balosso S, Ravizza T, Liu J, Bianchi ME, Vezzani A. Interleukin-1 type 1 receptor/Toll-like receptor signalling in epilepsy: the importance of IL-1β and high-mobility group box 1. J Intern Med 2011;270(4):319–26.

[16] Vezzani A, Aronica E, Mazarati A, Pittman QJ. Epilepsy and brain inflammation. Exp Neurol 2013;244:11–21.

[17] Vezzani A, Balosso S, Ravizza T. The role of cytokines in the pathophysiology of epilepsy. Brain Behav Immun 2008;22(6):797–803.

[18] Vezzani A, French J, Bartfai T, Baram TZ. The role of inflammation in epilepsy. Nat Rev Neurol 2011;7(1):31–40.

[19] Vezzani A, Friedman A. Brain inflammation as a biomarker in epilepsy. Biomark med 2011;5(5):607–14.

[20] Xu D, Miller SD, Koh S. Immune mechanisms in epileptogenesis. Front cell neurosci 2013;7:195.

[21] Viviani B, Bartesaghi S, Gardoni F, Vezzani A, Behrens MM, Bartfai T, et al. Interleukin-1β enhances NMDA receptor-mediated intracellular calcium increase through activation of the Src family of kinases. J Neurosci 2003;23(25):8692–700.

[22] Gorter JA, van Vliet EA, Aronica E, Breit T, Rauwerda H, Lopes da Silva FH, et al. Potential new antiepileptogenic targets indicated by microarray analysis in a rat model for temporal lobe epilepsy. J Neurosci 2006;26(43):11083–110.

[23] Crespel A, Coubes P, Rousset MC, Brana C, Rougier A, Rondouin G, et al. Inflammatory reactions in human medial temporal lobe epilepsy with hippocampal sclerosis. Brain Res 2002;952(2):159–69.

[24] Ravizza T, Gagliardi B, Noé F, Boer K, Aronica E, Vezzani A. Innate and adaptive immunity during epileptogenesis and spontaneous seizures: evidence from experimental models and human temporal lobe epilepsy. Neurobiol Dis 2008;29(1):142–60.

[25] Najjar S, Pearlman D, Miller DC, Devinsky O. Refractory epilepsy associated with microglial activation. Neurologist 2011;17(5):249–54.

[26] Aronica E, Boer K, van Vliet EA, Redeker S, Baayen JC, Spliet WG, et al. Complement activation in experimental and human temporal lobe epilepsy. Neurobiol Dis 2007;26(3):497–511.

[27] Beach TG, Woodhurst WB, MacDonald DB, Jones MW. Reactive microglia in hippocampal sclerosis associated with human temporal lobe epilepsy. Neurosci Lett 1995;191(1–2):27–30.

[28] Choi J, Nordli Jr. DR, Alden TD, DiPatri Jr. A, Laux L, Kelley K, et al. Cellular injury and neuroinflammation in children with chronic intractable epilepsy. J Neuroinflammation 2009;6:38.

[29] Zattoni M, Mura ML, Deprez F, Schwendener RA, Engelhardt B, Frei K, et al. Brain infiltration of leukocytes contributes to the pathophysiology of temporal lobe epilepsy. J Neurosci 2011;31(11):4037–50.

[30] Sheng JG, Boop FA, Mrak RE, Griffin WS. Increased neuronal β-amyloid precursor protein expression in human temporal lobe epilepsy: association with interleukin-1 alpha immunoreactivity. J Neurochem 1994;63(5):1872–9.
[31] Boer K, Spliet WG, van Rijen PC, Redeker S, Troost D, Aronica E. Evidence of activated microglia in focal cortical dysplasia. J Neuroimmunol 2006;173(1–2):188–95.
[32] Iyer A, Zurolo E, Spliet WG, van Rijen PC, Baayen JC, Gorter JA, et al. Evaluation of the innate and adaptive immunity in type I and type II focal cortical dysplasias. Epilepsia 2010;51(9):1763–73.
[33] Abraham J, Fox PD, Condello C, Bartolini A, Koh S. Minocycline attenuates microglia activation and blocks the long-term epileptogenic effects of early-life seizures. Neurobiol Dis 2012;46(2):425–30.
[34] Ravizza T, Rizzi M, Perego C, Richichi C, Velísková J, Moshé SL, et al. Inflammatory response and glia activation in developing rat hippocampus after status epilepticus. Epilepsia 2005;46(Suppl 5):113–7.
[35] Somera-Molina KC, Robin B, Somera CA, Anderson C, Stine C, Koh S, et al. Glial activation links early-life seizures and long-term neurologic dysfunction: evidence using a small molecule inhibitor of proinflammatory cytokine upregulation. Epilepsia 2007;48(9):1785–800.
[36] Kaur H, Patro I, Tikoo K, Sandhir R. Curcumin attenuates inflammatory response and cognitive deficits in experimental model of chronic epilepsy. Neurochem Int 2015;89:40–50.
[37] Bauer J, Elger CE, Hans VH, Schramm J, Urbach H, Lassmann H, et al. Astrocytes are a specific immunological target in Rasmussen's encephalitis. Ann Neurol 2007;62(1):67–80.
[38] Wirenfeldt M, Clare R, Tung S, Bottini A, Mathern GW, Vinters HV. Increased activation of $Iba1^+$ microglia in pediatric epilepsy patients with Rasmussen's encephalitis compared with cortical dysplasia and tuberous sclerosis complex. Neurobiol Dis 2009;34(3):432–40.
[39] Pardo CA, Vining EP, Guo L, Skolasky RL, Carson BS, Freeman JM. The pathology of Rasmussen syndrome: stages of cortical involvement and neuropathological studies in 45 hemispherectomies. Epilepsia 2004;45(5):516–26.
[40] Whitney KD, Andrews PI, McNamara JO. Immunoglobulin G and complement immunoreactivity in the cerebral cortex of patients with Rasmussen's encephalitis. Neurology 1999;53(4):699–708.
[41] Andrews PI, Dichter MA, Berkovic SF, Newton MR, McNamara JO. Plasmapheresis in Rasmussen's encephalitis. Neurology 1996;46(1):242–6.
[42] Antozzi C, Granata T, Aurisano N, Zardini G, Confalonieri P, Airaghi G, et al. Long-term selective IgG immuno-adsorption improves Rasmussen's encephalitis. Neurology 1998;51(1):302–5.
[43] Rogers SW, Andrews PI, Gahring LC, Whisenand T, Cauley K, Crain B, et al. Autoantibodies to glutamate receptor GluR3 in Rasmussen's encephalitis. Science 1994;265(5172):648–51.
[44] Aarli JA. Epilepsy and the immune system. Arch Neurol 2000;57(12):1689–92.
[45] Moga DE, Janssen WG, Vissavajjhala P, Czelusniak SM, Moran TM, Hof PR, et al. Glutamate receptor subunit 3 (GluR3) immunoreactivity delineates a subpopulation of parvalbumin-containing interneurons in the rat hippocampus. J Comp Neurol 2003;462(1):15–28.
[46] Vincent A, Irani SR, Lang B. The growing recognition of immunotherapy-responsive seizure disorders with autoantibodies to specific neuronal proteins. Curr Opin Neurol 2010;23(2):144–50.
[47] He XP, Patel M, Whitney KD, Janumpalli S, Tenner A, McNamara JO. Glutamate receptor GluR3 antibodies and death of cortical cells. Neuron 1998;20(1):153–63.
[48] Twyman RE, Gahring LC, Spiess J, Rogers SW. Glutamate receptor antibodies activate a subset of receptors and reveal an agonist binding site. Neuron 1995;14(4):755–62.
[49] Whitney KD, McNamara JO. GluR3 autoantibodies destroy neural cells in a complement-dependent manner modulated by complement regulatory proteins. J Neurosci 2000;20(19):7307–16.
[50] Mantegazza R, Bernasconi P, Baggi F, Spreafico R, Ragona F, Antozzi C, et al. Antibodies against GluR3 peptides are not specific for Rasmussen's encephalitis but are also present in epilepsy patients with severe, early onset disease and intractable seizures. J Neuroimmunol 2002;131(1–2):179–85.
[51] Watson R, Jiang Y, Bermudez I, Houlihan L, Clover L, McKnight K, et al. Absence of antibodies to glutamate receptor type 3 (GluR3) in Rasmussen encephalitis. Neurology 2004;63(1):43–50.
[52] Wiendl H, Bien CG, Bernasconi P, Fleckenstein B, Elger CE, Dichgans J, et al. GluR3 antibodies: prevalence in focal epilepsy but no specificity for Rasmussen's encephalitis. Neurology 2001;57(8):1511–4.
[53] Watson R, Jepson JE, Bermudez I, Alexander S, Hart Y, McKnight K, et al. α7-acetylcholine receptor antibodies in two patients with Rasmussen's encephalitis. Neurology 2005;65(11):1802–4.
[54] Alvarez-Barón E, Bien CG, Schramm J, Elger CE, Becker AJ, Schoch S. Autoantibodies to Munc18, cerebral plasma cells and B-lymphocytes in Rasmussen's encephalitis. Epilepsy Res 2008;80(1):93–7.

[55] Bien CG, Bauer J. Autoimmune epilepsies. Neurotherapeutics 2014;11(2):311–8.

[56] Varadkar S, Bien CG, Kruse CA, Jensen FE, Bauer J, Pardo CA, et al. Rasmussen's encephalitis: clinical features, pathobiology, and treatment advances. Lancet Neurol 2014;13(2):195–205.

[57] Vezzani A, Viviani B. Neuromodulatory properties of inflammatory cytokines and their impact on neuronal excitability. Neuropharmacology 2015;96(Pt A):70–82.

[58] Das A, Wallace GC 4th, Holmes C, McDowell ML, Smith JA, Marshall JD, et al. Hippocampal tissue of patients with refractory temporal lobe epilepsy is associated with astrocyte activation, inflammation, and altered expression of channels and receptors. Neuroscience 2012;220:237–46.

[59] Ashhab MU, Omran A, Kong H, Gan N, He F, Peng J, et al. Expressions of tumor necrosis factor α and microRNA-155 in immature rat model of status epilepticus and children with mesial temporal lobe epilepsy. J Mol Neurosci 2013;51(3):950–8.

[60] Pernhorst K, Herms S, Hoffmann P, Cichon S, Schulz H, Sander T, et al. TLR4, ATF-3 and IL8 inflammation mediator expression correlates with seizure frequency in human epileptic brain tissue. Seizure 2013;22(8):675–8.

[61] Aalbers MW, Rijkers K, Majoie HJ, Dings JT, Schijns OE, Schipper S, et al. The influence of neuropathology on brain inflammation in human and experimental temporal lobe epilepsy. J Neuroimmunol 2014;271(1–2):36–42.

[62] Kanemoto K, Kawasaki J, Yuasa S, Kumaki T, Tomohiro O, Kaji R, et al. Increased frequency of interleukin-1β-511T allele in patients with temporal lobe epilepsy, hippocampal sclerosis, and prolonged febrile convulsion. Epilepsia 2003;44(6):796–9.

[63] Kira R, Ishizaki Y, Torisu H, Sanefuji M, Takemoto M, Sakamoto K, et al. Genetic susceptibility to febrile seizures: case-control association studies. Brain Dev 2010;32(1):57–63.

[64] Kira R, Torisu H, Takemoto M, Nomura A, Sakai Y, Sanefuji M, et al. Genetic susceptibility to simple febrile seizures: interleukin-1β promoter polymorphisms are associated with sporadic cases. Neurosci Lett 2005;384(3):239–44.

[65] Serdaroğlu G, Alpman A, Tosun A, Pehlivan S, Ozkinay F, Tekgül H, et al. Febrile seizures: interleukin 1β and interleukin-1 receptor antagonist polymorphisms. Pediatr Neurol 2009;40(2):113–6.

[66] Matsuo M, Sasaki K, Ichimaru T, Nakazato S, Hamasaki Y. Increased IL-1β production from dsRNA-stimulated leukocytes in febrile seizures. Pediatr Neurol 2006;35(2):102–6.

[67] Virta M, Hurme M, Helminen M. Increased frequency of interleukin-1β (-511) allele 2 in febrile seizures. Pediatr Neurol 2002;26(3):192–5.

[68] Virta M, Hurme M, Helminen M. Increased plasma levels of pro- and anti-inflammatory cytokines in patients with febrile seizures. Epilepsia 2002;43(8):920–3.

[69] Asano T, Ichiki K, Koizumi S, Kaizu K, Hatori T, Fujino O, et al. IL-8 in cerebrospinal fluid from children with acute encephalopathy is higher than in that from children with febrile seizure. Scand J Immunol 2010;71(6):447–51.

[70] Ichiyama T, Nishikawa M, Yoshitomi T, Hayashi T, Furukawa S. Tumor necrosis factor-α, interleukin-1β, and interleukin-6 in cerebrospinal fluid from children with prolonged febrile seizures. Comparison with acute encephalitis/encephalopathy. Neurology 1998;50(2):407–11.

[71] Lahat E, Livne M, Barr J, Katz Y. Interleukin-1β levels in serum and cerebrospinal fluid of children with febrile seizures. Pediatr Neurol 1997;17(1):34–6.

[72] Tomoum HY, Badawy NM, Mostafa AA, Harb MY. Plasma interleukin-1β levels in children with febrile seizures. J Child Neurol 2007;22(6):689–92.

[73] Henshall DC, Clark RS, Adelson PD, Chen M, Watkins SC, Simon RP. Alterations in bcl-2 and caspase gene family protein expression in human temporal lobe epilepsy. Neurology 2000;55(2):250–7.

[74] Maldonado M, Baybis M, Newman D, Kolson DL, Chen W, McKhann II G, et al. Expression of ICAM-1, TNF-α, NFκB, and MAP kinase in tubers of the tuberous sclerosis complex. Neurobiol Dis 2003;14(2):279–90.

[75] Boer K, Crino PB, Gorter JA, Nellist M, Jansen FE, Spliet WG, et al. Gene expression analysis of tuberous sclerosis complex cortical tubers reveals increased expression of adhesion and inflammatory factors. Brain Pathol 2010;20(4):704–19.

[76] Boer K, Jansen F, Nellist M, Redeker S, van den Ouweland AM, Spliet WG, et al. Inflammatory processes in cortical tubers and subependymal giant cell tumors of tuberous sclerosis complex. Epilepsy Res 2008;78(1):7–21.

[77] Ravizza T, Boer K, Redeker S, Spliet WG, van Rijen PC, Troost D, et al. The IL-1β system in epilepsy-associated malformations of cortical development. Neurobiol Dis 2006;24(1):128–43.

[78] Diamond ML, Ritter AC, Failla MD, Boles JA, Conley YP, Kochanek PM, et al. IL-1β associations with post-traumatic epilepsy development: a genetics and biomarker cohort study. Epilepsia 2015;56(7):991–1001.

[79] Alcendor DJ, Charest AM, Zhu WQ, Vigil HE, Knobel SM. Infection and upregulation of proinflammatory cytokines in human brain vascular pericytes by human cytomegalovirus. J Neuroinflammation 2012;9:95.

[80] Vezzani A, Conti M, De Luigi A, Ravizza T, Moneta D, Marchesi F, et al. Interleukin-1β immunoreactivity and microglia are enhanced in the rat hippocampus by focal kainate application: functional evidence for enhancement of electrographic seizures. J Neurosci 1999;19(12):5054–65.

[81] Vezzani A, Moneta D, Conti M, Richichi C, Ravizza T, De Luigi A, et al. Powerful anticonvulsant action of IL-1 receptor antagonist on intracerebral injection and astrocytic overexpression in mice. Proc Natl Acad Sci USA 2000;97(21):11534–9.

[82] Vezzani A, Moneta D, Richichi C, Aliprandi M, Burrows SJ, Ravizza T, et al. Functional role of inflammatory cytokines and antiinflammatory molecules in seizures and epileptogenesis. Epilepsia 2002;43(Suppl 5):30–5.

[83] Balosso S, Maroso M, Sanchez-Alavez M, Ravizza T, Frasca A, Bartfai T, et al. A novel non-transcriptional pathway mediates the proconvulsive effects of interleukin-1β. Brain 2008;131(Pt 12):3256–65.

[84] Eriksson C, Tehranian R, Iverfeldt K, Winblad B, Schultzberg M. Increased expression of mRNA encoding interleukin-1β and caspase-1, and the secreted isoform of interleukin-1 receptor antagonist in the rat brain following systemic kainic acid administration. J Neurosci Res 2000;60(2):266–79.

[85] Järvelä JT, Lopez-Picon FR, Plysjuk A, Ruohonen S, Holopainen IE. Temporal profiles of age-dependent changes in cytokine mRNA expression and glial cell activation after status epilepticus in postnatal rat hippocampus. J Neuroinflammation 2011;8:29.

[86] Minami M, Kuraishi Y, Yamaguchi T, Nakai S, Hirai Y, Satoh M. Convulsants induce interleukin-1β messenger RNA in rat brain. Biochem Biophys Res Commun 1990;171(2):832–7.

[87] Penkowa M, Florit S, Giralt M, Quintana A, Molinero A, Carrasco J, et al. Metallothionein reduces central nervous system inflammation, neurodegeneration, and cell death following kainic acid-induced epileptic seizures. J Neurosci Res 2005;79(4):522–34.

[88] Pernot F, Heinrich C, Barbier L, Peinnequin A, Carpentier P, Dhote F, et al. Inflammatory changes during epileptogenesis and spontaneous seizures in a mouse model of mesiotemporal lobe epilepsy. Epilepsia 2011;52(12):2315–25.

[89] Rizzi M, Perego C, Aliprandi M, Richichi C, Ravizza T, Colella D, et al. Glia activation and cytokine increase in rat hippocampus by kainic acid-induced status epilepticus during postnatal development. Neurobiol Dis 2003;14(3):494–503.

[90] Benson MJ, Manzanero S, Borges K. Complex alterations in microglial M1/M2 markers during the development of epilepsy in two mouse models. Epilepsia 2015;56(6):895–905.

[91] Henshall DC, Engel T. P2X purinoceptors as a link between hyperexcitability and neuroinflammation in status epilepticus. Epilepsy Behav 2015;49:8–12.

[92] Librizzi L, Noè F, Vezzani A, de Curtis M, Ravizza T. Seizure-induced brain-borne inflammation sustains seizure recurrence and blood-brain barrier damage. Ann Neurol 2012;72(1):82–90.

[93] Pineda E, Shin D, You SJ, Auvin S, Sankar R, Mazarati A. Maternal immune activation promotes hippocampal kindling epileptogenesis in mice. Ann Neurol 2013;74(1):11–19.

[94] De Simoni MG, Perego C, Ravizza T, Moneta D, Conti M, Marchesi F, et al. Inflammatory cytokines and related genes are induced in the rat hippocampus by limbic status epilepticus. Eur J Neurosci 2000;12(7):2623–33.

[95] Manley NC, Bertrand AA, Kinney KS, Hing TC, Sapolsky RM. Characterization of monocyte chemoattractant protein-1 expression following a kainate model of status epilepticus. Brain Res 2007;1182:138–43.

[96] Louboutin JP, Chekmasova A, Marusich E, Agrawal L, Strayer DS. Role of CCR5 and its ligands in the control of vascular inflammation and leukocyte recruitment required for acute excitotoxic seizure induction and neural damage. FASEB J 2011;25(2):737–53.

[97] Mennicken F, Chabot JG, Quirion R. Systemic administration of kainic acid in adult rat stimulates expression of the chemokine receptor CCR5 in the forebrain. Glia 2002;37(2):124–38.

[98] Foresti ML, Arisi GM, Katki K, Montañez A, Sanchez RM, Shapiro LA. Chemokine CCL2 and its receptor CCR2 are increased in the hippocampus following pilocarpine-induced status epilepticus. J Neuroinflammation 2009;6:40.

[99] Xu JH, Long L, Tang YC, Zhang JT, Hut HT, Tang FR. CCR3, CCR2A and macrophage inflammatory protein (MIP)-1α, monocyte chemotactic protein-1 (MCP-1) in the mouse hippocampus during and after pilocarpine-induced status epilepticus (PISE). Neuropathol Appl Neurobiol 2009;35(5):496–514.

[100] Dubé CM, Ravizza T, Hamamura M, Zha Q, Keebaugh A, Fok K, et al. Epileptogenesis provoked by prolonged experimental febrile seizures: mechanisms and biomarkers. J Neurosci 2010;30(22):7484–94.
[101] Dubé C, Vezzani A, Behrens M, Bartfai T, Baram TZ. Interleukin-1β contributes to the generation of experimental febrile seizures. Ann Neurol 2005;57(1):152–5.
[102] Galic MA, Riazi K, Henderson AK, Tsutsui S, Pittman QJ. Viral-like brain inflammation during development causes increased seizure susceptibility in adult rats. Neurobiol Dis 2009;36(2):343–51.
[103] Libbey JE, Kennett NJ, Wilcox KS, White HS, Fujinami RS. Once initiated, viral encephalitis-induced seizures are consistent no matter the treatment or lack of interleukin-6. J Neurovirol 2011;17(5):496–9.
[104] Zhang B, Zou J, Rensing NR, Yang M, Wong M. Inflammatory mechanisms contribute to the neurological manifestations of tuberous sclerosis complex. Neurobiol Dis 2015;80:70–9.
[105] Wang S, Cheng Q, Malik S, Yang J. Interleukin-1β inhibits γ-aminobutyric acid type A $GABA_A$ receptor current in cultured hippocampal neurons. J Pharmacol Exp Ther 2000;292(2):497–504.
[106] Zhang R, Yamada J, Hayashi Y, Wu Z, Koyama S, Nakanishi H. Inhibition of NMDA-induced outward currents by interleukin-1β in hippocampal neurons. Biochem Biophys Res Commun 2008;372(4):816–20.
[107] Tabarean IV, Korn H, Bartfai T. Interleukin-1β induces hyperpolarization and modulates synaptic inhibition in preoptic and anterior hypothalamic neurons. Neuroscience 2006;141(4):1685–95.
[108] Akin D, Ravizza T, Maroso M, Carcak N, Eryigit T, Vanzulli I, et al. IL-1β is induced in reactive astrocytes in the somatosensory cortex of rats with genetic absence epilepsy at the onset of spike-and-wave discharges, and contributes to their occurrence. Neurobiol Dis 2011;44(3):259–69.
[109] Samland H, Huitron-Resendiz S, Masliah E, Criado J, Henriksen SJ, Campbell IL. Profound increase in sensitivity to glutamatergic- but not cholinergic agonist-induced seizures in transgenic mice with astrocyte production of IL-6. J Neurosci Res 2003;73(2):176–87.
[110] Probert L, Akassoglou K, Kassiotis G, Pasparakis M, Alexopoulou L, Kollias G. TNF-α transgenic and knockout models of CNS inflammation and degeneration. J Neuroimmunol 1997;72(2):137–41.
[111] Akassoglou K, Probert L, Kontogeorgos G, Kollias G. Astrocyte-specific but not neuron-specific transmembrane TNF triggers inflammation and degeneration in the central nervous system of transgenic mice. J Immunol 1997;158(1):438–45.
[112] Stalder AK, Carson MJ, Pagenstecher A, Asensio VC, Kincaid C, Benedict M, et al. Late-onset chronic inflammatory encephalopathy in immune-competent and severe combined immune-deficient (SCID) mice with astrocyte-targeted expression of tumor necrosis factor. Am J Pathol 1998;153(3):767–83.
[113] Bianchi ME, Manfredi AA. High-mobility group box 1 (HMGB1) protein at the crossroads between innate and adaptive immunity. Immunol Rev 2007;220:35–46.
[114] Maroso M, Balosso S, Ravizza T, Liu J, Aronica E, Iyer AM, et al. Toll-like receptor 4 and high-mobility group box-1 are involved in ictogenesis and can be targeted to reduce seizures. Nat Med 2010;16(4):413–9.
[115] Zurolo E, Iyer A, Maroso M, Carbonell C, Anink JJ, Ravizza T, et al. Activation of Toll-like receptor, RAGE and HMGB1 signalling in malformations of cortical development. Brain 2011;134(Pt 4):1015–32.
[116] Hayakawa K, Arai K, Lo EH. Role of ERK map kinase and CRM1 in IL-1β-stimulated release of HMGB1 from cortical astrocytes. Glia 2010;58(8):1007–15.
[117] Sha Y, Zmijewski J, Xu Z, Abraham E. HMGB1 develops enhanced proinflammatory activity by binding to cytokines. J Immunol 2008;180(4):2531–7.
[118] Pedrazzi M, Patrone M, Passalacqua M, Ranzato E, Colamassaro D, Sparatore B, et al. Selective proinflammatory activation of astrocytes by high-mobility group box 1 protein signaling. J Immunol 2007;179(12):8525–32.
[119] Vezzani A, Maroso M, Balosso S, Sanchez MA, Bartfai T. IL-1 receptor/Toll-like receptor signaling in infection, inflammation, stress and neurodegeneration couples hyperexcitability and seizures. Brain Behav Immun 2011;25(7):1281–9.
[120] O'Neill LA, Bowie AG. The family of five: TIR-domain-containing adaptors in Toll-like receptor signalling. Nat Rev Immunol 2007;7(5):353–64.
[121] Aronica E, Gorter JA. Gene expression profile in temporal lobe epilepsy. Neuroscientist 2007;13(2):100–8.
[122] Başaran N, Hincal F, Kansu E, Ciğer A. Humoral and cellular immune parameters in untreated and phenytoin- or carbamazepine-treated epileptic patients. Int J Immunopharmacol 1994;16(12):1071–7.
[123] Jamali S, Bartolomei F, Robaglia-Schlupp A, Massacrier A, Peragut JC, Regis J, et al. Large-scale expression study of human mesial temporal lobe epilepsy: evidence for dysregulation of the neurotransmission and complement systems in the entorhinal cortex. Brain 2006;129(Pt 3):625–41.

[124] Xiong ZQ, Qian W, Suzuki K, McNamara JO. Formation of complement membrane attack complex in mammalian cerebral cortex evokes seizures and neurodegeneration. J Neurosci 2003;23(3):955–60.
[125] Fabene PF, Laudanna C, Constantin G. Leukocyte trafficking mechanisms in epilepsy. Mol Immunol 2013;55(1):100–4.
[126] Fabene PF, Navarro Mora G, Martinello M, Rossi B, Merigo F, Ottoboni L, et al. A role for leukocyte-endothelial adhesion mechanisms in epilepsy. Nat Med 2008;14(12):1377–83.
[127] Librizzi L, Regondi MC, Pastori C, Frigerio S, Frassoni C, de Curtis M. Expression of adhesion factors induced by epileptiform activity in the endothelium of the isolated guinea pig brain in vitro. Epilepsia 2007;48(4):743–51.
[128] Oby E, Janigro D. The blood-brain barrier and epilepsy. Epilepsia 2006;47(11):1761–74.
[129] Hildebrandt M, Amann K, Schröder R, Pieper T, Kolodziejczyk D, Holthausen H, et al. White matter angiopathy is common in pediatric patients with intractable focal epilepsies. Epilepsia 2008;49(5):804–15.
[130] Mombaerts P, Iacomini J, Johnson RS, Herrup K, Tonegawa S, Papaioannou VE. RAG-1-deficient mice have no mature B and T lymphocytes. Cell 1992;68(5):869–77.
[131] Bien CG, Bauer J, Deckwerth TL, Wiendl H, Deckert M, Wiestler OD, et al. Destruction of neurons by cytotoxic T cells: a new pathogenic mechanism in Rasmussen's encephalitis. Ann Neurol 2002;51(3):311–8.
[132] Schwab N, Bien CG, Waschbisch A, Becker A, Vince GH, Dornmair K, et al. $CD8^+$ T-cell clones dominate brain infiltrates in Rasmussen's encephalitis and persist in the periphery. Brain 2009;132(Pt 5):1236–46.
[133] Bien CG, Schramm J. Treatment of Rasmussen's encephalitis half a century after its initial description: promising prospects and a dilemma. Epilepsy Res 2009;86(2–3):101–12.
[134] Deprez F, Zattoni M, Mura ML, Frei K, Fritschy JM. Adoptive transfer of T lymphocytes in immunodeficient mice influences epileptogenesis and neurodegeneration in a model of temporal lobe epilepsy. Neurobiol Dis 2011;44(2):174–84.
[135] Rahman S, Malcoun A. Nonsteroidal antiinflammatory drugs, cyclooxygenase-2, and the kidneys. Prim Care 2014;41(4):803–21.
[136] Toscano CD, Ueda Y, Tomita YA, Vicini S, Bosetti F. Altered GABAergic neurotransmission is associated with increased kainate-induced seizure in prostaglandin-endoperoxide synthase-2 deficient mice. Brain Res Bull 2008;75(5):598–609.
[137] Caracciolo L, Barbon A, Palumbo S, Mora C, Toscano CD, Bosetti F, et al. Altered mRNA editing and expression of ionotropic glutamate receptors after kainic acid exposure in cyclooxygenase-2 deficient mice. PLoS One 2011;6(5):e19398.
[138] Serrano GE, Lelutiu N, Rojas A, Cochi S, Shaw R, Makinson CD, et al. Ablation of cyclooxygenase-2 in forebrain neurons is neuroprotective and dampens brain inflammation after status epilepticus. J Neurosci 2011;31(42):14850–60.
[139] Rodgers KM, Hutchinson MR, Northcutt A, Maier SF, Watkins LR, Barth DS. The cortical innate immune response increases local neuronal excitability leading to seizures. Brain 2009;132(Pt 9):2478–86.
[140] Galic MA, Riazi K, Heida JG, Mouihate A, Fournier NM, Spencer SJ, et al. Postnatal inflammation increases seizure susceptibility in adult rats. J Neurosci 2008;28(27):6904–13.
[141] Auvin S, Porta N, Nehlig A, Lecointe C, Vallée L, Bordet R. Inflammation in rat pups subjected to short hyperthermic seizures enhances brain long-term excitability. Epilepsy Res 2009;86(2–3):124–30.
[142] Auvin S, Mazarati A, Shin D, Sankar R. Inflammation enhances epileptogenesis in the developing rat brain. Neurobiol Dis 2010;40(1):303–10.
[143] Sankar R, Auvin S, Mazarati A, Shin D. Inflammation contributes to seizure-induced hippocampal injury in the neonatal rat brain. Acta Neurol Scand Suppl 2007;186:16–20.
[144] Auvin S, Shin D, Mazarati A, Sankar R. Inflammation induced by LPS enhances epileptogenesis in immature rat and may be partially reversed by IL-1RA. Epilepsia 2010;51(Suppl 3):34–8.
[145] Heida JG, Teskey GC, Pittman QJ. Febrile convulsions induced by the combination of lipopolysaccharide and low-dose kainic acid enhance seizure susceptibility, not epileptogenesis, in rats. Epilepsia 2005;46(12):1898–905.
[146] Heida JG, Pittman QJ. Causal links between brain cytokines and experimental febrile convulsions in the rat. Epilepsia 2005;46(12):1906–13.
[147] Helminen M, Vesikari T. Increased interleukin-1 (IL-1) production from LPS-stimulated peripheral blood monocytes in children with febrile convulsions. Acta Paediatr Scand 1990;79(8–9):810–6.
[148] Pålsson-McDermott EM, O'Neill LA. Signal transduction by the lipopolysaccharide receptor, Toll-like receptor-4. Immunology 2004;113(2):153–62.

[149] Sayyah M, Javad-Pour M, Ghazi-Khansari M. The bacterial endotoxin lipopolysaccharide enhances seizure susceptibility in mice: involvement of proinflammatory factors: nitric oxide and prostaglandins. Neuroscience 2003;122(4):1073–80.

[150] Liu H, Roy M, Tian FF. MicroRNA-based therapy: a new dimension in epilepsy treatment. Int J Neurosci 2013;123(9):617–22.

[151] Henshall DC. MicroRNA and epilepsy: profiling, functions and potential clinical applications. Curr Opin Neurol 2014;27(2):199–205.

[152] Jimenez-Mateos EM, Henshall DC. Epilepsy and microRNA. Neuroscience 2013;238:218–29.

[153] Hu K, Xie YY, Zhang C, Ouyang DS, Long HY, Sun DN, et al. MicroRNA expression profile of the hippocampus in a rat model of temporal lobe epilepsy and miR-34a-targeted neuroprotection against hippocampal neurone cell apoptosis post-status epilepticus. BMC Neurosci 2012;13:115.

[154] Jimenez-Mateos EM, Bray I, Sanz-Rodriguez A, Engel T, McKiernan RC, Mouri G, et al. miRNA Expression profile after status epilepticus and hippocampal neuroprotection by targeting miR-132. Am J Pathol 2011;179(5):2519–32.

[155] Jimenez-Mateos EM, Engel T, Merino-Serrais P, Fernaud-Espinosa I, Rodriguez-Alvarez N, Reynolds J, et al. Antagomirs targeting microRNA-134 increase hippocampal pyramidal neuron spine volume in vivo and protect against pilocarpine-induced status epilepticus. Brain Struct Funct 2015;220(4):2387–99.

[156] Jimenez-Mateos EM, Engel T, Merino-Serrais P, McKiernan RC, Tanaka K, Mouri G, et al. Silencing microRNA-134 produces neuroprotective and prolonged seizure-suppressive effects. Nat Med 2012;18(7):1087–94.

[157] Kaalund SS, Venø MT, Bak M, Møller RS, Laursen H, Madsen F, et al. Aberrant expression of miR-218 and miR-204 in human mesial temporal lobe epilepsy and hippocampal sclerosis-convergence on axonal guidance. Epilepsia 2014;55(12):2017–27.

[158] Kan AA, van Erp S, Derijck AA, de Wit M, Hessel EV, O'Duibhir E, et al. Genome-wide microRNA profiling of human temporal lobe epilepsy identifies modulators of the immune response. Cell Mol Life Sci 2012;69(18):3127–45.

[159] McKiernan RC, Jimenez-Mateos EM, Bray I, Engel T, Brennan GP, Sano T, et al. Reduced mature microRNA levels in association with dicer loss in human temporal lobe epilepsy with hippocampal sclerosis. PLoS One 2012;7(5):e35921.

[160] Gorter JA, Iyer A, White I, Colzi A, van Vliet EA, Sisodiya S, et al. Hippocampal subregion-specific microRNA expression during epileptogenesis in experimental temporal lobe epilepsy. Neurobiol Dis 2014;62:508–20.

[161] Song YJ, Tian XB, Zhang S, Zhang YX, Li X, Li D, et al. Temporal lobe epilepsy induces differential expression of hippocampal miRNAs including let-7e and miR-23a/b. Brain Res 2011;1387:134–40.

[162] Jimenez-Mateos EM, Arribas-Blazquez M, Sanz-Rodriguez A, Concannon C, Olivos-Ore LA, Reschke CR, et al. MicroRNA targeting of the $P2X_7$ purinoceptor opposes a contralateral epileptogenic focus in the hippocampus. Sci Rep 2015;5:17486.

[163] Iyer A, Zurolo E, Prabowo A, Fluiter K, Spliet WG, van Rijen PC, et al. MicroRNA-146a: a key regulator of astrocyte-mediated inflammatory response. PLoS One 2012;7(9):e44789.

[164] Lukiw WJ, Zhao Y, Cui JG. An NF-kappaB-sensitive micro RNA-146a-mediated inflammatory circuit in Alzheimer disease and in stressed human brain cells. J Biol Chem 2008;283(46):31315–22.

[165] Nakasa T, Miyaki S, Okubo A, Hashimoto M, Nishida K, Ochi M, et al. Expression of microRNA-146 in rheumatoid arthritis synovial tissue. Arthritis Rheum 2008;58(5):1284–92.

[166] Pauley KM, Chan EK. MicroRNAs and their emerging roles in immunology. Ann N Y Acad Sci 2008;1143:226–39.

[167] Pauley KM, Satoh M, Chan AL, Bubb MR, Reeves WH, Chan EK. Upregulated miR-146a expression in peripheral blood mononuclear cells from rheumatoid arthritis patients. Arthritis Res Ther 2008;10(4):R101.

[168] Sheedy FJ, O'Neill LA. Adding fuel to fire: microRNAs as a new class of mediators of inflammation. Ann Rheum Dis 2008;67(Suppl 3):iii50–5.

[169] Sonkoly E, Ståhle M, Pivarcsi A. MicroRNAs and immunity: novel players in the regulation of normal immune function and inflammation. Semin Cancer Biol 2008;18(2):131–40.

[170] Taganov KD, Boldin MP, Chang KJ, Baltimore D. NFκB-dependent induction of microRNA miR-146, an inhibitor targeted to signaling proteins of innate immune responses. Proc Natl Acad Sci USA 2006;103(33):12481–6.

[171] Prabowo AS, van Scheppingen J, Iyer AM, Anink JJ, Spliet WG, van Rijen PC, et al. Differential expression and clinical significance of three inflammation-related microRNAs in gangliogliomas. J Neuroinflammation 2015;12:97.

[172] Aronica E, Fluiter K, Iyer A, Zurolo E, Vreijling J, van Vliet EA, et al. Expression pattern of miR-146a, an inflammation-associated microRNA, in experimental and human temporal lobe epilepsy. Eur J Neurosci 2010;31(6):1100–7.
[173] Marchi N, Granata T, Janigro D. Inflammatory pathways of seizure disorders. Trends Neurosci 2014;37(2):55–65.
[174] Kong H, Yin F, He F, Omran A, Li L, Wu T, et al. The Effect of miR-132, miR-146a, and miR-155 on MRP8/TLR4-Induced Astrocyte-Related Inflammation. J Mol Neurosci 2015;57(1):28–37.
[175] Stienen MN, Haghikia A, Dambach H, Thöne J, Wiemann M, Gold R, et al. Anti-inflammatory effects of the anticonvulsant drug levetiracetam on electrophysiological properties of astroglia are mediated via TGFβ1 regulation. Br J Pharmacol 2011;162(2):491–507.
[176] Dambach H, Hinkerohe D, Prochnow N, Stienen MN, Moinfar Z, Haase CG, et al. Glia and epilepsy: experimental investigation of antiepileptic drugs in an astroglia/microglia co-culture model of inflammation. Epilepsia 2014;55(1):184–92.
[177] Dedeurwaerdere S, Friedman A, Fabene PF, Mazarati A, Murashima YL, Vezzani A, et al. Finding a better drug for epilepsy: antiinflammatory targets. Epilepsia 2012;53(7):1113–8.
[178] Luna-Medina R, Cortes-Canteli M, Sanchez-Galiano S, Morales-Garcia JA, Martinez A, Santos A, et al. NP031112, a thiadiazolidinone compound, prevents inflammation and neurodegeneration under excitotoxic conditions: potential therapeutic role in brain disorders. J Neurosci 2007;27(21):5766–76.
[179] Chrzaszcz M, Venkatesan C, Dragisic T, Watterson DM, Wainwright MS. Minozac treatment prevents increased seizure susceptibility in a mouse "two-hit" model of closed skull traumatic brain injury and electroconvulsive shock-induced seizures. J Neurotrauma 2010;27(7):1283–95.
[180] Yang L, Li F, Ge W, Mi C, Wang R, Sun R. Protective effects of naloxone in two-hit seizure model. Epilepsia 2010;51(3):344–53.
[181] Maroso M, Balosso S, Ravizza T, Iori V, Wright CI, French J, et al. Interleukin-1β biosynthesis inhibition reduces acute seizures and drug resistant chronic epileptic activity in mice. Neurotherapeutics 2011;8(2):304–15.
[182] Ravizza T, Lucas SM, Balosso S, Bernardino L, Ku G, Noè F, et al. Inactivation of caspase-1 in rodent brain: a novel anticonvulsive strategy. Epilepsia 2006;47(7):1160–8.
[183] Noè FM, Polascheck N, Frigerio F, Bankstahl M, Ravizza T, Marchini S, et al. Pharmacological blockade of IL-1β/IL-1 receptor type 1 axis during epileptogenesis provides neuroprotection in two rat models of temporal lobe epilepsy. Neurobiol Dis 2013;59:183–93.
[184] Ma L, Cui XL, Wang Y, Li XW, Yang F, Wei D, et al. Aspirin attenuates spontaneous recurrent seizures and inhibits hippocampal neuronal loss, mossy fiber sprouting and aberrant neurogenesis following pilocarpine-induced status epilepticus in rats. Brain Res 2012;1469:103–13.
[185] Jung KH, Chu K, Lee ST, Kim J, Sinn DI, Kim JM, et al. Cyclooxygenase-2 inhibitor, celecoxib, inhibits the altered hippocampal neurogenesis with attenuation of spontaneous recurrent seizures following pilocarpine-induced status epilepticus. Neurobiol Dis 2006;23(2):237–46.
[186] Sotgiu S, Murrighile MR, Constantin G. Treatment of refractory epilepsy with natalizumab in a patient with multiple sclerosis. Case report. BMC Neurol 2010;10:84.
[187] Henshall DC. Antagomirs and microRNA in status epilepticus. Epilepsia 2013;54(Suppl 6):17–19.
[188] Alsharafi WA, Xiao B, Abuhamed MM, Bi FF, Luo ZH. Correlation between IL-10 and microRNA-187 expression in epileptic rat hippocampus and patients with temporal lobe epilepsy. Front Cell Neurosci 2015;9:466.
[189] Wang Y, Wang Q, Haldar JP, Yeh FC, Xie M, Sun P, et al. Quantification of increased cellularity during inflammatory demyelination. Brain 2011;134(Pt 12):3590–601.
[190] Obenaus A. Neuroimaging biomarkers for epilepsy: advances and relevance to glial cells. Neurochem Int 2013;63(7):712–8.
[191] Banati RB. Visualising microglial activation in vivo. Glia 2002;40(2):206–17.
[192] Butler T, Ichise M, Teich AF, Gerard E, Osborne J, French J, et al. Imaging inflammation in a patient with epilepsy due to focal cortical dysplasia. J Neuroimaging 2013;23(1):129–31.
[193] Kumar A, Chugani HT, Luat A, Asano E, Sood S. Epilepsy surgery in a case of encephalitis: use of ^{11}C-PK11195 positron emission tomography. Pediatr Neurol 2008;38(6):439–42.
[194] Gershen LD, Zanotti-Fregonara P, Dustin IH, Liow JS, Hirvonen J, Kreisl WC, et al. Neuroinflammation in temporal lobe epilepsy measured using positron emission tomographic imaging of translocator protein. JAMA Neurology 2015;72(8):882–8.
[195] Bien CG, Urbach H, Deckert M, Schramm J, Wiestler OD, Lassmann H, et al. Diagnosis and staging of Rasmussen's encephalitis by serial MRI and histopathology. Neurology 2002;58(2):250–7.

[196] Stoll G, Bendszus M. Imaging of inflammation in the peripheral and central nervous system by magnetic resonance imaging. Neuroscience 2009;158(3):1151–60.

[197] Duffy BA, Choy M, Riegler J, Wells JA, Anthony DC, Scott RC, et al. Imaging seizure-induced inflammation using an antibody targeted iron oxide contrast agent. NeuroImage 2012;60(2):1149–55.

[198] Choy M, Dubé CM, Patterson K, Barnes SR, Maras P, Blood AB, et al. A novel, noninvasive, predictive epilepsy biomarker with clinical potential. J Neurosci 2014;34(26):8672–84.

CHAPTER

14

Therapeutic Targets and Future Directions

OUTLINE

OVERVIEW

Based on the wide variety of important homeostatic roles for astrocytes, their role in modulating synaptic transmission, and their key changes in response to injury, more modern conceptions of neurological diseases hold that they may have a "gliopathic" component [1–4]. From a translational standpoint, to the extent that *astrocytes* (instead of neurons) are the elements of the nervous system responsible for homeostatic function, and disease represents loss of homeostasis, therapeutic targets for disease treatment should be directed at restoring homeostasis in astrocytes [5,6]. That concept leads to many new therapeutic targets that may be putatively more astrocyte-specific. Such a strategy may not have the deleterious side effects of inhibiting normal neuronal function and synaptic transmission, such as is the case for many current antiepileptic drugs (AEDs) [7].

Astrocytes and Epilepsy
DOI: http://dx.doi.org/10.1016/B978-0-12-802401-0.00022-3

Side Effects of Current AEDs and Difficulties in Preclinical Development

Most current AEDs target neuronal voltage-gated sodium channels and calcium channels, glutamate receptors, or γ-aminobutyric acid (GABA) systems [8]. For example, Na^+ channel blockers such as phenytoin and carbamazepine reduce the frequency of neuronal action potentials, and GABA transaminase (GABAT) inhibitors, such as vigabatrin, increase GABA-mediated inhibition [8]. The mode of action of several commonly prescribed AEDs, such as valproate, is not entirely understood [8–10]. There are several drawbacks to current AEDs. First, currently used AEDs often cause some form of cognitive impairment, including memory deficiencies and mental slowing [11]. Cognitive impairments are particularly important in patients being treated with chronic AEDs. Moreover, polypharmacy has a more severe impact on cognitive function when compared to monotherapy, regardless of which type of AEDs are being used [11]. Second, about 30% of patients being treated with AEDs, even with optimal current therapy, have poor seizure control and become medically refractory. Adverse effects are frequently observed at drug doses within the recommended range [10]. Third, there is an increased risk of teratogenicity in women with epilepsy who are receiving AEDs, in particular phenobarbital and valproate [12,13]. For women taking enzyme-inducing AEDs, such as phenytoin or carbamazepine, hormonal forms of contraception are affected and the efficacy of oral contraceptive cannot be guaranteed [12], thus complicating family planning. Finally, AEDs are associated with adverse effects including mood alteration, suicidality, severe mucocutaneous reactions, hepatotoxic effects, osteoporosis, weight management difficulties, skin rash, pseudolymphoma, and many others, which often lead to treatment failure [14].

Despite the advent and introduction of many new AEDs over the past several decades, the efficacy and tolerability of the AEDs (based largely on the above mechanisms) is still not substantially improved [15]. There are many possible reasons for this including: (1) specific animal models used, in particular models such as drug efficacy against the maximum electroshock seizure (MES) test which has served as a "gatekeeper" for AED development [16] but which may not adequately model many forms of epilepsy; (2) the focus on developing *anticonvulsant* drugs rather than *antiepileptogenic* drugs which would inhibit the development of epilepsy in at-risk populations such as post-traumatic brain injury (TBI) [17]; (3) AEDs do not target underlying molecular mechanisms of disease [18]; and (4) AED mechanisms have been neuron-based and not glial-based. A main purpose of this book has been to describe the multifaceted nature of how glial-based mechanisms may be involved in epilepsy and serve as possible new therapeutic targets.

Epilepsy as a Gliopathy

Several lines of evidence have suggested that glial cells are potential therapeutic targets for the treatment of epilepsy and other central nervous system (CNS) diseases [19,20]. Glia are involved in many important physiological functions. Astrocytes play an established role in removal of glutamate at synapses and the sequestration and redistribution of K^+ during neural activity [21]. It is becoming increasingly clear that astrocytes play a direct role in seizure susceptibility and the development of epilepsy [19,20,22–27]. Stimulation of astrocytes leads to prolonged neuronal depolarization and epileptiform discharges [25]. Astrocytes release neuroactive molecules and also modulate synaptic transmission through modifications in channels, gap junctions (GJs), receptors, and transporters [19,22,24,25,28–33]. Furthermore, striking changes in astrocyte form and function occur in epilepsy. Astrocytes become reactive [23,34], lose domain organization [35], and become uncoupled [36] in epileptic tissue. The precise functional consequences are as yet unclear; however, these changes would locally or globally alter not only single-cell function but the function of the entire glial syncytium (or lack thereof after uncoupling!). These and other changes such as changes in the expression of various astrocytic enzymes, such as adenosine kinase (ADK) [37] and glutamine synthetase (GS) [38], astroglial proliferation, dysregulation of water and ion channel and glutamate transporter expression, alterations in secretion of neuroactive molecules, and increased activation of inflammatory pathways [22,23,27,34,39–42] may all contribute to hyperexcitability and epileptogenesis.

NEW ASTROCYTE-BASED THERAPEUTIC TARGETS

Consequently, many of these pathological alterations may become new astrocyte-based therapeutic targets [43,44]. In particular, if normal neuronal synaptic transmission can be preserved, judicious use of astrocyte-based therapies may not only have higher efficacy but increased tolerability (ie, fewer cognitive and other side effects from blocking synaptic transmission). "Restoring astrocyte homeostasis" would seem to be a much more logical therapeutic strategy than "blocking synaptic transmission."

Therefore let us briefly revisit some potential astrocyte-based therapeutic targets.

Potassium, Water, and Glutamate Homeostasis

Potassium (see *Chapter 7: Potassium Channels*), water (see *Chapter 8: Water Channels*), and glutamate (see *Chapter 9: Glutamate Metabolism*) form a trio of critical neuron–astrocyte synaptic systems that might be called the "trio of synaptic gliostasis" (Fig. 14.1). It is interesting to note that altering the function of one of these systems often alters another, for example, inwardly rectifying potassium channel $K_{ir}4.1^{-/-}$ mice have impaired glutamate uptake [45] and aquaporin-4 $(AQP4)^{-/-}$ mice have impaired K^+ uptake [46]. Therefore, there is functional coordination among water, potassium, and glutamate uptake. However, this may not require a direct intermolecular interaction. For example, there is no difference in expression of $K_{ir}4.1$ protein [46] or $K_{ir}4.1$ immunoreactivity [47] in $AQP4^{-/-}$ mice nor AQP4 immunoreactivity in $K_{ir}4.1^{-/-}$ mice [47]. In addition, no alterations were observed in membrane potential, barium-sensitive $K_{ir}4.1$ K^+ current, or current–voltage curves in $AQP4^{-/-}$ retinal Müller cells [48] or brain astrocytes [49]. Lack of alteration of K_{ir} channels in $AQP4^{-/-}$ mice suggests the interesting possibility that the slowed $[K^+]_o$ decay may be a secondary effect of slowed water extrusion ("deswelling") following stimulation, and this has been modeled carefully [50] but not directly demonstrated. Similarly, slowed "deswelling" could also relate to slowed seizure termination in $AQP4^{-/-}$ mice [46].

In epilepsy, changes in the synaptic gliostasis of all three have been described. For potassium, $K_{ir}4.1$-mediated K^+ uptake is diminished in epilepsy [41,42,51–54]. For water, downregulation and redistribution of AQP4 has been described in both animal and human

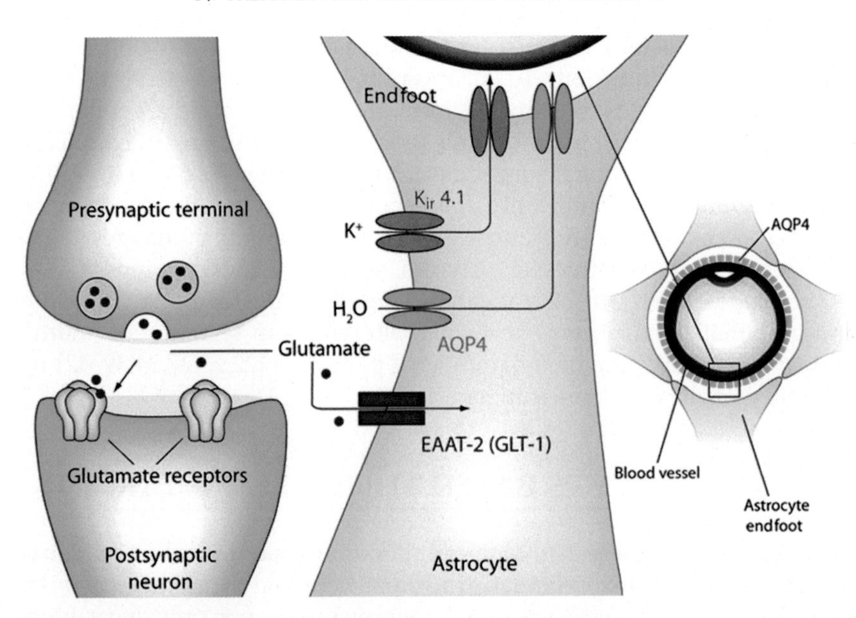

FIGURE 14.1 **Astrocyte regulation of water, potassium, and glutamate homeostasis at the tripartite synapse.** Colocalization of AQP4, $K_{ir}4.1$, and GLT1 in distinct astrocyte membrane domains (perisynaptic, perivascular) provides the basis for a critical role of astrocytes in control of water, potassium, and glutamate homeostasis ("trio of synaptic gliostasis"). *Source: Reproduced with permission from Benarroch E. Aquaporin-4, homeostasis, and neurologic disease. Neurology 2007;69:2266–8 (Figure 1).*

epilepsies [55–57]. For glutamate, both glutamate transporters (downregulation of glutamate transporter-1 (GLT1) [58]) and glutamate metabolism (downregulation of GS [59]) have been found to be altered. We have recently found dramatic reduction in expression of GLT1, which is responsible for greater than 90% of glutamate uptake at the synapse, during the latent phase of epileptogenesis in the dorsal hippocampus in the intrahippocampal kainic acid mouse model (Hubbard and Binder, unpublished observations).

All of these changes would be expected to be proepileptogenic as described in chapters "Potassium Channels," "Water Channels," and "Glutamate Metabolism." A goal for future AED development, then, would be to develop selective activators or regulators of $K_{ir}4.1$, AQP4, GLT1, and/or GS. If successful, such drugs would be potentially restorative of synaptic gliostasis. Currently, there are no drugs that selectively activate $K_{ir}4.1$ to our knowledge [60]. For AQP4, there are no selective activators despite great efforts in this area [61].

Modulation of glutamate uptake by astrocytes offers potential for decreasing excessive excitability associated with the development of epilepsy. Global deletion of GLT1 leads to intractable seizures and early postnatal lethality [62], which was confirmed recently with conditional deletion of GLT1 in astrocytes [63]. Conversely, overexpression of GLT1 in transgenic mice attenuated epileptogenesis and reduced chronic seizure frequency in a pilocarpine-induced model of epilepsy [64]. Until 2005, no pharmacological intervention was able to modulate GLT1 protein expression. Rothstein et al. [65] showed that ceftriaxone, a β-lactam antibiotic, is a potent stimulator of GLT1 transcription and glutamate uptake, acting via the nuclear factor-κB (NF-κB) signaling pathway [66], although these results are controversial [67,68]. Ceftriaxone has been shown to reduce extracellular glutamate levels [69] and have antiseizure effects [70–72]. The β-carboline alkaloid harmine can also increase GLT1 gene expression and glutamate uptake activity in vitro [73]. The β-lactamase inhibitors clavulanic acid and tazobactam decreased seizure-like activity in an invertebrate model [74], although their effects on GLT1 have not been explored. Recently, however, the effectiveness of clavulanic acid as an antiepileptic therapeutic has been debated [75,76]. Nevertheless, the approach to upregulate GLT1 expression with these related β-lactam drugs remains to be explored fully in standard animal models of epileptogenesis.

Other strategies to upregulate GLT1 include: (1) adenosine A_1 and A_{2A} receptors as well as equilibrative nucleoside transporter-1 (ENT1) can be targeted to modulate GLT1-mediated glutamate transport and release [77–84] (see *Chapter 10: Adenosine Metabolism* for details); (2) a new small molecule activator of GLT1 translation developed at Ohio State [64,85]; and (3) delivery of cells [86] or viruses [87] to express GLT1 locally at sites of pathology. Key questions that arise regarding GLT1 restoration as a therapeutic strategy include: (1) selectivity for epilepsy; (2) anatomic specificity to sites of "epileptic" GLT1 dysregulation; and (3) would GLT1 upregulation cause side effects similar to other AEDs such as sedation and cognitive problems?

Subcellular targeting and compartmentalization of $K_{ir}4.1$, AQP4, and GLT1 is another interesting opportunity for identifying new modulatory targets. Both $K_{ir}4.1$ and AQP4 demonstrate perivascular and perisynaptic targeting, and for AQP4 it is clear that the perivascular and perisynaptic pools are lost even at time points when total AQP4 protein may be elevated [56,57]. Recent studies have also shown subcellular compartmentalization of GLT1. Activity-dependent mobility of GLT1 transporters into and out of the synaptic region has recently been described [88], identifying another site and source of regulation of the efficacy of perisynaptic glutamate uptake. Biochemical studies have recently identified that GLT1 targeting to plasma membrane versus intracellular sites is regulated by sumoylation [89] and S-nitrosylation [90]. Thus, if GLT1 transporters are not only downregulated but also intracellularly dysregulated, restoration of perisynaptic GLT1 levels through intracellular targeting could be another therapeutic strategy. Further understanding of the activity regulation and intracellular (perisynaptic and perivascular) targeting and mobility of $K_{ir}4.1$, AQP4, and GLT1 will be of critical importance to establish feasibility. In this way, restoring "synaptic gliostasis" may not be as simple as upregulating mRNA or protein levels of $K_{ir}4.1$, AQP4, or GLT1, but in addition effectively manipulating plasma membrane versus intracellular fractions and perisynaptic versus extrasynaptic fractions. Furthermore, the physiological significance of homo- versus heterodimeric $K_{ir}4.1$ ($K_{ir}4.1/K_{ir}5.1$) [60] and that of different AQP4 isoforms [91] and assembly into orthogonal arrays of particles (OAPs [92,93]) remain to be explored in the context of epilepsy.

Cell Swelling and Reduction of the Extracellular Space

It has been observed for some time from studies both in vitro and in vivo that hyperosmolarity protects against seizures, whereas hypoosmolarity promotes generalized seizures [94–98]. The consequences of cell swelling and reduction of the extracellular space (ECS) include increased extracellular resistance [96,98], magnified effect of local extracellular ion and transmitter accumulation [99,100], and enhanced neuronal synchrony and excitability [95,96].

Based on the considerations above and those described in chapters "Potassium Channels," "Water Channels," and "Glutamate Metabolism," a specific role for astrocyte swelling in increasing neuronal excitability is a compelling area for future investigation (Fig. 14.2) [101]. Astrocyte susceptibility to swelling during pathological states has been suspected for quite some time, dating back to early electron microscopy (EM) studies which noted that astrocytes "are very susceptible to edema and are among the first elements of the nervous system to swell in poorly fixed preparations" [102]. Recent evidence based on real-time volume measurements using two-photon microscopy indicates that astrocytes are much more prone to swelling than neurons [103,104]. Astrocyte susceptibility to volume changes has been attributed to selective expression of AQP4 [103,105–107]. During the buildup of excitability leading up to seizure (ictal) discharges, it is speculated that elevated K^+ released from neurons during synaptic transmission is taken up into astrocytes along with water, causing astrocyte swelling and progressive reduction of the ECS. There is strong evidence in cultured astrocytes in vitro that astrocyte swelling leads to opening of volume-regulated anion channels (VRACs) to produce a regulatory volume decrease. Release of water through astrocytic VRAC is accompanied by substantial amounts of glutamate [108–111]. Among the first targets encountered by astrocytically released glutamate are extrasynaptic *N*-methyl D-aspartate receptors (NMDARs), leading to slow inward currents (SICs), and potentially interictal and ictal (seizure-like) discharges. In support of this possibility, SIC-like currents have been evoked in CA1 pyramidal neurons by cell volume changes alone in acute hippocampal slices in vitro [112]. Recent evidence indicates a very strong relationship between slight changes in osmolarity, astrocyte swelling, and the generation of NMDAR-dependent SICs [113].

The possibility for involvement of astrocyte swelling in seizure generation is even more compelling when taking into account the specific changes taking place in astrocytes during epileptogenesis. Loss and redistribution of AQP4 and $K_{ir}4.1$ away from astrocytic endfeet is expected to exacerbate astrocyte swelling by increasing water influx into astrocyte processes at synapses while decreasing efflux via endfeet into the cerebrovasculature [114]. This could prolong seizures by slowing the recovery of astrocytic volume, a possibility supported by the increased seizure duration observed in $AQP4^{-/-}$ mice [46]. Upregulated expression of group I metabotropic glutamate receptors (mGluRs) in reactive astrocytes could further enhance modulation of AQP4 [115] and exacerbate swelling and swelling-evoked release of glutamate by astrocytes. Increased release of adenosine triphosphate (ATP) due to elevated secretion of inflammatory molecules and perhaps also by ectopic expression of the NMDAR subunit NR2B by astrocytes could significantly potentiate release of glutamate from VRACs as observed in vitro [116,117]. In addition, reduced expression of GS [59] would elevate cytoplasmic concentrations of glutamate in astrocytes, providing more glutamate to be released when VRACs open. Furthermore, extrasynaptic NR2B subunit NMDAR expression increases [118], providing additional targets for astrocytically released glutamate. All of these changes would be expected to exacerbate swelling, astrocytic release of glutamate, and stimulation of extrasynaptic NMDARs, contributing to the development of epilepsy and spontaneously recurring seizures. Although the therapeutic potential of inhibition of astrocytic VRAC has not yet been tested in epilepsy, the astrocyte-specific VRAC inhibitor

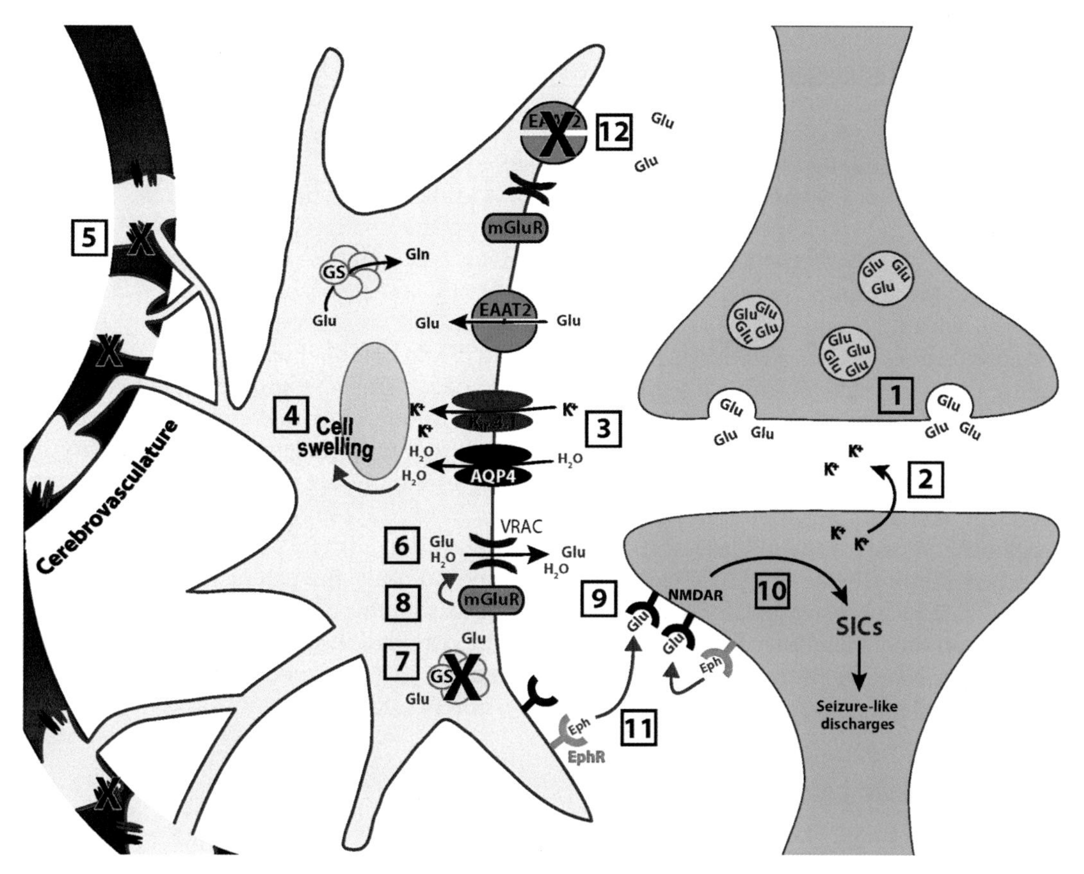

FIGURE 14.2 **Model for role of astrocyte swelling in neuronal excitability.** At an excitatory synapse the neurotransmitter glutamate (Glu) is released from vesicles within the presynaptic terminal (1). Glu binds to receptors on the postsynaptic membrane and opens ion channels (not shown) allowing the movement of ions across the membrane, including the release of potassium ions (2). Elevated levels of K^+ are taken up by perisynaptic astrocytes along with water (H_2O) predominantly by $K_{ir}4.1$ potassium channels and AQP4 channels, respectively (3). This leads to cell swelling and a reduction in ECS (4). Loss or redistribution of AQP4 and $K_{ir}4.1$ away from astrocytic endfeet (5) would exacerbate astrocyte swelling since water flux into astrocytes at synapses would be increased while efflux via endfeet into the cerebrovasculature would be decreased. Astrocyte swelling opens VRACs, which release glutamate into the ECS (6). In addition, reduced expression of GS in the epileptic condition (7) elevates cytoplasmic concentration of glutamate in astrocytes, providing more glutamate to be released. Upregulation of mGluRs in reactive astrocytes could enhance AQP4-dependent swelling and swelling-evoked release of glutamate through VRACs (8). Astrocytically released glutamate can then bind to extrasynaptic NMDARs (9), generating SICs and potentially interictal and ictal (seizure-like) discharges (10). Facilitated by the close proximity of adjacent cellular membranes during cell swelling, Eph receptor (EphR)–ephrin (Eph) ligand interactions may further enhance the stimulation of NMDARs (11). EphRs, ephrins, and certain NMDAR subunits are upregulated after neural injury on different cell types, including reactive astrocytes and neurons. Reduced excitatory amino transporter 2 (EAAT2) expression in epileptic tissue may lead to delayed clearance of glutamate from the extrasynaptic space (12). All of the above mechanisms may contribute to astrocyte control of neuronal excitability. *Source: Reproduced with permission from Hubbard JA, Hsu MS, Fiacco TA, Binder DK. Glial cell changes in epilepsy: overview of the clinical problem and therapeutic opportunities. Neurochem Int 2013;63:638–51 (Figure 1).*

DCPIB (4-(2-butyl-6,7-dichloro-2-cyclopentyl-indan-1-on-5-yl)oxobutyric acid) exhibits powerful neuroprotective effects in a rat model of ischemia [119]. Selective inhibition of astrocyte swelling, astrocyte glutamate release through VRAC, extrasynaptic NMDARs, and EphB2/ephrinB2-mediated NMDAR potentiation offer exciting avenues for the development of new strategies for the treatment of epilepsy.

A challenge facing future studies exploring specific astrocytic mechanisms of seizure generation and development of epilepsy is dissecting the relative contributions of astrocyte Ca^{2+} versus astrocyte swelling. This is difficult given that astrocyte Ca^{2+} elevations resulting from receptor activation will occur alongside glutamate and potassium uptake, water influx, and cell volume changes. This issue is complicated further by a report suggesting that ATP-induced astrocyte Ca^{2+} elevations in situ activate VRAC [120]. Furthermore, the changes taking place in reactive astrocytes during hippocampal sclerosis would be expected to affect both processes. However, tools are available to begin to differentiate between astrocyte Ca^{2+} sources and astrocyte swelling in the generation of epileptiform activity. Especially intriguing are the IP_3R2 knockout mice, in which astrocyte Ca^{2+} elevations are abolished [121]. Neuronal SICs can be readily generated in hippocampal slices from IP_3R2 knockout mice [112] suggesting that SICs occur independent of astrocyte Ca^{2+} elevations. It will be especially interesting in future studies using IP_3R2 KO mice to determine not only the extent to which astrocyte IP_3-mediated Ca^{2+} elevations play a role in the changes taking place during epileptogenesis, but also on the generation of epileptiform activity in vitro and in vivo. These mice might also be used to examine alternate Ca^{2+} sources in astrocytes that may become available over the course of epileptogenesis, such as astrocytic expression of NR2B NMDARs.

Adenosine Metabolism

Adenosine levels are elevated during seizure activity [122,123]. Adenosine exerts a powerful inhibitory effect on excitatory synaptic transmission primarily through its interaction with presynaptic A_1 adenosine receptors (A_1Rs) to suppress neurotransmitter release [124–126]. Therefore, the cycle of adenosine release and breakdown is especially important in cases of excessive excitability including epilepsy. Once released from neurons and astrocytes, ATP is rapidly converted into adenosine monophosphate (AMP) and then into adenosine by extracellular nucleotidases [127]. The reuptake of adenosine occurs through equilibrative nucleoside transporters [128], and phosphorylation by the astrocyte-specific enzyme ADK breaks down adenosine and therefore clears excess adenosine from the ECS. Minor changes in ADK activity affect the active cycle between adenosine, AMP, ADK, and 5′-nucleotidase and lead to major changes in extracellular adenosine levels [129]. Therefore, alterations in ADK are especially relevant to the generation of seizures. Increased levels of ADK are associated with seizures whereas decreased levels may lead to seizure suppression [130]. Increased ADK expression has been linked to seizure activity in both human tissue and experimental models of epilepsy [37,131–133] (see *Chapter 10: Adenosine Metabolism* for details). Seizure induction in experimental epilepsy was found to decrease extracellular adenosine concentrations through the upregulation of ADK [134]. In the kainic acid model of temporal lobe epilepsy (TLE) in mice, profound astrogliosis and increased ADK activity was observed [133]. This coincides with the findings of Aronica et al. [37], who demonstrated prolonged increases in ADK—for at least 3–4 months—in the rat hippocampus and cortex after induction of status epilepticus (SE). This increase was also detected in the hippocampus and temporal cortex of TLE patients.

Collectively, the above findings support the ADK hypothesis of epileptogenesis [131,132], including the dysregulation of ADK and its contribution to the epileptogenic cascade. Adenosine, adenosine receptor agonists, and ADK inhibitors have well-established anticonvulsant efficacy [135–138]. Intracranial injection of adenosine prevents seizures in rats [139]. In addition, the use of transgenic mice revealed that reduced forebrain ADK protects against epileptogenesis [140]. Other studies involving adenosine augmentation therapies (AAT) include a silk protein-based release system for adenosine [141] and the local release of adenosine from grafted cells [142], both of which resulted in seizure suppression. Focal adenosine delivery, such as slow-release polymers, cellular implants, gene therapy, or pump systems, has been suggested as a new pharmacological tool to treat refractory epilepsy with minimal side effects [130].

Particularly exciting is the recent finding that even a transient adenosine augmentation may have longer-lasting epigenetic effects that are antiepileptogenic [143,144]. Probably the most effective treatment would be a brain-permeant peripherally administered small molecule inhibitor of ADK. This would hopefully obviate systemic side effects seen with direct adenosine delivery. If effective in triggering long-lasting antiepileptogenesis, such a drug would ideally need to be given only during an isolated therapeutic window just after an epileptogenic stimulus. This would potentially minimize long-range side effects while maintaining efficacy.

Tumor-Associated Epilepsy

Impaired glutamate uptake and neurotoxic release of glutamate from growing gliomas have been observed in vitro [145]. It is thought that growing glial tumors actively kill surrounding neurons through the release of excessive quantities of glutamate, which may also contribute to the seizures frequently seen in conjunction with glioma [145,146]. Recent studies by Harald Sontheimer's laboratory have identified specific pathological alterations in glioma mouse models and human tissues that correlate with tumor-associated epilepsy [147]. They recently identified that a hypothesized glutamate release pathway, cystine/glutamate transporter (SXC), is active in a subset of gliomas [148]. SLC7A11/xCT, the catalytic subunit of SXC, demonstrated elevated expression in about 50% of patient tumors. Compared with tumors lacking this transporter, SLC7A11-positive tumors were associated with faster growth, peritumoral glutamate excitotoxicity, seizures, and worse survival. In a translational pilot study, use of the FDA-approved SXC inhibitor sulfasalazine in nine patients with biopsy-proven SXC expression led to inhibition of glutamate release from the tumor in vivo as assessed by magnetic resonance (MR) spectroscopy [148]. This exciting study demonstrates that phenotyping tumors for glial-associated transport molecules will lead to selective pharmacological targeting to prevent or ameliorate tumor-associated epilepsy, and gets at the pathological mechanism of glutamate release from tumor cells rather than standard AED approaches of globally suppressing synaptic transmission. Similarly, "disconnecting" invading astrocytoma cells by targeting tumor "microtubes" which connect distant membrane protrusions by GJs is a more rational approach to limit glioma invasion than standard chemo- and radiotherapeutic "cell kill" approaches which cause collateral damage on the normal brain and body [149,150].

Gap Junctions

Until recently, the literature in the field of GJs and epilepsy was confusing and contradictory, relying on studies that demonstrated up- and downregulation of astrocytic and/or neuronal connexins and positive or negative effects of GJ inhibitors in epilepsy models [151]. It is important to recognize that nearly all pharmacological GJ inhibitor drugs (eg, carbenoxolone, halothane) are "dirty" drugs with other effects, thus interpretation of the available studies is limited.

Functional coupling analysis, obtained by patch-clamping astrocytes and filling the astrocyte syncytium with dyes to quantitatively measure coupling, is clearly the gold standard and has led to recent seminal findings from Christian Steinhäuser's laboratory (see *Chapter 11: Gap Junctions* for details). In this study [36], the gap junctional connectivity of astrocytes from 119 specimens from patients with mesial temporal lobe epilepsy (MTLE) with ($n = 75$) and without ($n = 44$) sclerosis were examined. They found that in MTLE specimens with typical hippocampal sclerosis, there is a complete absence of typical "classical" astrocytes and an absence of astrocyte gap junctional coupling. In contrast, coupled astrocytes were abundant in nonsclerotic hippocampus. In the intracortical kainic acid (ICKA) model of TLE, mice exhibited decreased astrocytic coupling 4–5 days post injection and completely lacked coupling 3 and 6 months after SE in the sclerotic hippocampus [36]. In the nonsclerotic hippocampus, however, coupling remained intact. Interestingly, decreased astrocyte coupling preceded apoptotic neuronal death and the onset of spontaneous seizures. Decreased GJ coupling also impaired K^+ clearance 4 hours post injection. The authors found that proinflammatory cytokines induced the uncoupling of hippocampal astrocytes in vivo [36], which agreed with similar in vitro findings that proinflammatory cytokines have an inhibitory effect on astrocytic GJ coupling [152]. To test the hypothesis that inflammation may contribute to the pathophysiology of seizures, lipopolysaccharide was injected into mice. Five days post injection, animals exhibited reduced GJ coupling and reduced connexin 43 (Cx43) protein levels, but unchanged connexin 30 (Cx30) protein amounts. The uncoupling effect of lipopolysaccharide could be fully prevented with the anti-inflammatory and AED levetiracetam [36]. In addition, uncoupling was prevented in Toll-like receptor 4 (TLR4) knockout mice. Taken together, these important data suggest that inflammation may contribute to rapid uncoupling of astrocytes and the uncoupling of astrocytes may be involved in epileptogenesis.

Of importance for the future will be to investigate how general a mechanism uncoupling is for other forms of epilepsy. In the SE models (kainic acid, pilocarpine), there is cell death and sclerosis. Do epilepsy models without significant cell death also demonstrate uncoupling of astrocytes? What is the timing of astrocyte uncoupling and loss of K^+ homeostasis compared with the onset of epilepsy in these models? Further work in this area will be critical to establish the mechanisms and thresholds for astrocyte uncoupling in a variety of models of epilepsy. Restoration of gap junctional coupling in astrocytes, perhaps via modulation of the TLR4 pathway, represents another novel therapeutic strategy [153].

Post-Traumatic Epilepsy and Post-Stroke Epilepsy

Post-traumatic epilepsy (PTE) and post-stroke epilepsy (PSE) are good examples of injury-induced secondary epileptogenesis in which the inciting event is clear (see *Chapter 5: Neuropathology of Human Epilepsy* for details). Thus, these syndromes are in need of antiepileptogenic therapies delivered after the insult that would prevent subsequent development of epilepsy. No such therapy is currently available, and all previous antiepileptogenic trials for PTE have been ineffective [17,154–161]. Interestingly, restoration of GLT1 expression after lateral fluid percussion injury (FPI) by the β-lactam antibiotic ceftriaxone decreased gliosis and reduced cumulative post-traumatic seizure duration in rats, providing proof-of-principle for the idea of restoration of glutamate homeostasis as an antiepileptogenic strategy against PTE [162]. A recent genetics and biomarker cohort study of 256 patients with moderate to severe TBI found evidence linking interleukin-1β (IL-1β) to PTE risk, and provides rationale for testing targeted IL-1β therapies as prophylaxis against PTE [163].

One thing that PTE and PSE have in common is breakdown of the blood–brain barrier (BBB) at the time of the initial event (see *Chapter 12: Blood–Brain Barrier Disruption* for details). Here, studies of the gliovascular junction and BBB disruption-induced epileptogenesis may be therapeutically relevant [164]. The structure and roles of the gliovascular junction [165] and the role of astrocytes in BBB permeability [166] and control of microcirculation [167–170] have only recently been appreciated. Local pathological alterations in the gliovascular junction could perturb blood flow, K^+, and H_2O regulation and constitute an important mechanism in the generation of hyperexcitability.

Transient opening of the BBB is sufficient for focal epileptogenesis [171]. Extravasated albumin can be taken up by astrocytes which activates the transforming growth factor-β (TGF-β) pathway leading to focal epileptogenesis. Kaufer and colleagues [172] have worked out a mechanism by which albumin induces excitatory synaptogenesis through astrocyte TGF-β/ALK5 signaling. This mechanism provides an astrocytic basis for BBB disruption-induced epileptogenesis and suggests antiepileptogenic therapeutic approaches (TGF-β inhibition). Indeed, losartan, a TGF-β inhibitor and FDA-approved antihypertensive medication, has been found to exert antiepileptogenic effects in this model [173,174].

Another event that can occur with BBB disruption is infiltration of leukocytes into the brain. Migration of leukocytes, which are more abundant in human cortical CNS tissue of patients with epilepsy than in control tissue [175], is controlled by chemokines in physiological and pathological conditions. Leukocyte–endothelium interactions and subsequent recruitment of leukocytes in brain parenchyma represent key components of the epileptogenic cascade [176]. In a mouse model of epilepsy, Fabene et al. [175] found that seizures induced elevated expression of vascular cell adhesion molecules and enhanced leukocyte rolling and arrest in brain vessels mediated by the leukocyte mucin P-selectin glycoprotein ligand-1 (PSGL-1) and leukocyte integrins α4β1 and αLβ2. Moreover, the blockage of leukocyte–vascular adhesion attenuated BBB leakage, suggesting a pathogenic link between leukocyte–vascular interactions, BBB damage, and seizure generation [175]. Recent studies have demonstrated that lack of perforin, a downstream factor of natural killer (NK) and cytotoxic T cells, reduces BBB damage and mortality in the rat pilocarpine model of epilepsy [177]. Via astrocyte–lymphocyte interactions and expression of various adhesion molecules, astrocytes are well positioned to serve as "gatekeepers" for immune cell infiltration into the CNS [178]. Thus, adhesion molecules themselves may serve as antiepileptogenic targets to be delivered in the appropriate therapeutic time window [178].

Neuroinflammation

Neuroinflammation is an integrated response of all CNS cell types, including microglia, macroglia, neurons, and infiltrating leukocytes to an initial injury (see *Chapter 13: Inflammation* for details). Innate immunity, or the early inflammatory response triggered by an insult, occurs in the brain after systemic infection. This phenomenon can eventually progress to an adaptive immune response in which the immune system can recognize and remember specific pathogens; this is mediated by activated lymphocytes recruited from the blood [179,180]. When a local inflammatory reaction is triggered in the brain following an injury, both microglia and astrocytes become activated and release a number of proinflammatory cytokines such as IL-1β, tumor necrosis factor-α (TNF-α), and interleukin-6 (IL-6). These proinflammatory mediators can alter the properties of both glial autocrine actions and glial–neuron paracrine signaling.

Specific inflammatory pathways are chronically activated during epileptogenesis in both microglia and astrocytes [181,182]. Along with astrogliosis, microglial activation has been shown within the sclerotic hippocampus of patients with TLE [181–185] and those exhibiting focal cortical dysplasia (FCD) [186,187]. Reactive astrocytes of lesioned areas in the hippocampus of MTLE patients overexpress NF-κB-p65, a transcription factor that can activate the transcription of numerous proinflammatory genes [181]. Activated microglia and astrocytes chronically express IL-1β in the hippocampus of patients with TLE [182], which coincides with an increase in interleukin-1 receptor type 1 (IL-1R1) activation in the rat forebrain during SE [188]. The infiltration of leukocytes into sclerotic tissue was detected in both human specimens and animal models using immunohistochemistry [185]. Microglia activation and proliferation is prevalent in resected human epileptic tissue [189].

Both innate and adaptive immunity are activated in various forms of epilepsy. The complement pathway, an inflammatory response cascade that is part of innate and adaptive immunity, is overexpressed in reactive astrocytes, microglia, and macrophages in human TLE [190,191], tuberous sclerosis complex (TSC) [192,193], and FCD [187]. Alterations in BBB permeability were also associated with inflammation in TSC-associated lesions [193]. A combination of proinflammatory cytokines and the components of the complement cascade may contribute to the spread of the inflammatory response and increased network excitability in the sclerotic hippocampus of patients with TLE, in lesions of patients with TSC, and in the dysplastic tissue of FCD.

Several studies have provided evidence in support of anti-inflammatory modulation for the treatment of epilepsy. A recent study used minocycline, a known inhibitor of inflammation, to determine whether innate immunity plays a causal role in mediating the long-term epileptogenic effects of early-life seizures [194]. Mice were induced with SE at postnatal day 25, which caused an increase in microglial activation. Mice induced with a second SE 2 weeks later responded with greater microgliosis and shorter latency to seizure expression. Minocycline abolished the acute seizure-induced microglial activation and decreased seizure susceptibility, suggesting that anti-inflammatory therapy after SE may be useful in blocking the epileptogenic process and mitigating the long-term damaging effects of early-life seizures. In a different study, treatment with aspirin, a nonselective cyclooxygenase inhibitor, reduced both the frequency and duration of spontaneous and recurrent seizures following pilocarpine-induced SE in rats [195]. Moreover, aberrant migration of granule cells, mossy fiber sprouting, and hippocampal neuronal cell loss were attenuated by aspirin.

The idea to block inflammation as a treatment for epilepsy is attractive but raises at least two problems. First, several immune agents and processes are triggered in response to an initial insult, and depression of all immune signaling would also depress the endogenous anti-inflammatory agents. Second, inflammation may contribute to the repair process that protects against major neuronal circuit changes that promote the emergence of spontaneous seizures [196]. A more sensible approach may be to target a single inflammatory cascade, such as regulation of the balance between brain IL-1β and IL-1RA [197,198]. The exogenous application of IL-1β prolongs seizures in an IL-1R1-mediated manner, and intrahippocampal application of recombinant IL-1RA inhibits motor and electroencephalographic seizures induced by bicuculline in mice [198]. Inhibition of IL-1β production using selective inhibitors of interleukin-converting enzyme (ICE/caspase-1) or caspase-1 gene deletion have been shown to block seizure-induced production of IL-1β in the hippocampus of rats. Reduction in ICE/caspase-1 activity resulted in a significant decrease in seizure onset and duration [199]. A group of recent studies has indicated the importance of high-mobility group box-1 (HMGB1) release from neurons and glia and its interaction with TLR4, a key receptor of innate immunity [200]. It appears that the HMGB1–TLR4 axis is active in human epileptic tissue [200,201], providing a new target for anti-inflammatory drug therapy [202,203].

Are Cognitive Comorbidities in Epilepsy Astrocyte-Based?

Last but not least, epilepsy is much more than just seizures; nearly all epilepsy syndromes are associated with cognitive and behavioral comorbidities. Should these comorbidities be viewed now in the context of astrocyte dysregulation? Dysregulation at the level of individual gliosynaptic units and K^+/glutamate/water dyshomeostasis, and also more broadly in tissue-wide disruption of the astrocyte syncytium [36] and astrocyte domain organization [35] could conceivably lead to widespread effects on cognition, memory, and behavior.

Let us consider two examples of astrocyte-based neuromodulation gone awry that could contribute to epilepsy comorbidities. First, Boison and Aronica [204] have advanced the "adenosine hypothesis of comorbidities." Astrocyte overexpression of ADK induces tissue-wide deficiency in the "homeostatic tone" of adenosine. Boison and Aronica adduce evidence from patient samples demonstrating ADK overexpression not only in epilepsy but also in Alzheimer's disease, Parkinson's disease, and amyotrophic lateral sclerosis (ALS or Lou Gehrig's disease). A transgenic "comorbidity model" with overexpression of ADK [205] and adenosine deficiency [206] is found to cause not only seizures [140,207], but altered dopaminergic function, impairment of attention, and cognitive and sleep regulation deficits [204,208,209]. In this conception, dysregulation of astrocyte adenosine metabolism produces widespread effects not only confined to seizures and epilepsy but also to cognitive comorbidities. Thus, normalization of adenosine by augmentation therapy might effectively treat not only manifestations of the "primary" condition (eg, seizures) but also comorbidities via the same mechanism [204].

A second example is that of AQP4 dysregulation. As described earlier and in *Chapter 8: Water Channels* in detail, AQP4 exhibits downregulation and/or altered subcellular distribution in epilepsy [55–57]. What effect does deletion of the perisynaptic pool of AQP4 [99,210] have on synaptic plasticity and cognitive function? Interestingly, $AQP4^{-/-}$ mice were found to have profound and selective deficits in synaptic plasticity and memory. Specifically, $AQP4^{-/-}$ mice exhibited a selective defect in hippocampal long-term potentiation (LTP) and long-term depression (LTD) without a change in basal synaptic transmission or short-term plasticity [211]. The impairment in LTP in $AQP4^{-/-}$ mice was specific for the type of LTP that depends on the neurotrophin brain-derived neurotrophic factor (BDNF [212]), which is induced by stimulation at theta rhythm [theta-burst stimulation (TBS)-LTP], but there was no impairment in a form of LTP that is BDNF-independent, induced by high-frequency stimulation (HFS-LTP). LTD was also impaired in $AQP4^{-/-}$ mice, which was rescued by a scavenger of BDNF or blockade of Trk receptors. $AQP4^{-/-}$ mice also exhibited a cognitive defect in location-specific object memory but not Morris water maze or contextual fear conditioning. These results suggest that AQP4 channels in astrocytes may play an unanticipated role in neurotrophin-dependent plasticity and influence behavior [211]. Similar results have subsequently been obtained by other research groups [213–215]. Based on these results, downregulation of AQP4 may not only lead to increased neural excitability due to abnormalities of water and potassium homeostasis but may also lead directly to abnormalities in synaptic plasticity (both LTP and LTD). This provides a potential explanation for the way that astrocytic changes in epilepsy may contribute not only to seizures but also to cognitive comorbidities [216]. Cognitive impairment is very important because patients with TLE have many alterations in cognitive function and in particular hippocampal-dependent tasks such as spatial memory [217–220]. In addition, many other forms of plasticity are operative in the epileptic brain such as potentiation of synapses, reorganization of neuronal circuitry, and alteration in postnatal neurogenesis [221–223]. Thus, AQP4 deficiency or dysregulation could cause deficits in synaptic function and memory [224]. Restoration of AQP4 homeostasis could conceivably mitigate or reverse these deficits.

SUMMARY

An understanding of the changes taking place in glial functioning during epilepsy is gradually emerging. While animal model and human tissue studies have provided insight into glial involvement in epilepsy, both levels of investigation have certain limitations. Animal studies may not always accurately represent the disease progression as it is seen in humans; and human tissue obtained from resected specimens does not allow determination of whether observed cellular and molecular changes are a cause or a consequence of epilepsy. Future studies should focus on characterizing glial cell alterations that occur prior to spontaneous seizure onset (ie, during early epileptogenesis) in distinct models of epilepsy, as this could lead to a greater understanding of disease pathogenesis. The term "reactive gliosis" is neither accurate nor specific and should be replaced by careful morphological, biochemical, and electrophysiological studies of identified glial cell subtypes in human tissue and animal models, paying particular attention to astrocyte heterogeneity [225,226]. In addition to changes in preexisting glial cell populations, newly generated glial cells with distinct properties may migrate into the hippocampus and contribute to enhanced seizure susceptibility [227,228]. The available data likely represent only the "tip of the iceberg" in terms of the functional role of astroglial cells in epilepsy. Further study of astrocyte alterations in epilepsy should continue to unearth new molecular targets and open new avenues for the development of creative antiepileptic therapies.

In summary, we hope that this book has helped to elucidate the historical and current literature on astrocytes and epilepsy. Dramatic advances in glioscience have occurred which will potentially translate into gliotherapeutic targets not only for epilepsy but for diverse neurological diseases. Identification of gliopathic changes in multiple neurological diseases is now well advanced, and specific targeting of these changes can be contemplated that would spare normal brain and cognitive function to the greatest extent possible.

References

[1] Parpura V, Heneka MT, Montana V, Oliet SH, Schousboe A, Haydon PG, et al. Glial cells in (patho)physiology. J Neurochem 2012;121(1):4–27.
[2] Verkhratsky A, Sofroniew MV, Messing A, deLanerolle NC, Rempe D, Rodriguez JJ, et al. Neurological diseases as primary gliopathies: a reassessment of neurocentrism. ASN Neuro 2012;4:3.
[3] Chung WS, Welsh CA, Barres BA, Stevens B. Do glia drive synaptic and cognitive impairment in disease? Nat Neurosci 2015;18(11):1539–45.
[4] Barres BA. The mystery and magic of glia: a perspective on their roles in health and disease. Neuron 2008;60(3):430–40.
[5] Medina A, Watson SJ, Bunney Jr W, Myers RM, Schatzberg A, Barchas J, et al. Evidence for alterations of the glial syncytial function in major depressive disorder. J Psychiatr Res 2016;72:15–21.
[6] Koyama Y. Functional alterations of astrocytes in mental disorders: pharmacological significance as a drug target. Front Cell Neurosci 2015;9:261.
[7] Crunelli V, Carmignoto G, Steinhäuser C. Novel astrocyte targets: new avenues for the therapeutic treatment of epilepsy. Neuroscientist 2015;21(1):62–83.
[8] Rogawski MA, Löscher W. The neurobiology of antiepileptic drugs. Nat Rev Neurosci 2004;5:553–64.
[9] Kwan P, Schachter SC, Brodie JM. Drug-resistant epilepsy. New Engl J Med 2012;365:919–26.
[10] Perucca E. An introduction to antiepileptic drugs. Epilepsia 2005;46(Suppl. 4):31–7.
[11] Aldenkamp AP, De Krom M, Reijs R. Newer antiepileptic drugs and cognitive issues. Epilepsia 2003;44:21–9.
[12] Crawford P. Best practice guidelines for the management of women with epilepsy. Epilepsia 2005;46 (Suppl. 9):117–24.
[13] Wlodarczyk BJ, Palacios AM, George TM, Finnell RH. Antiepileptic drugs and pregnancy outcomes. Am J Med Genet 2012;158A(8):2071–90.
[14] Perucca P, Gilliam FG. Adverse effects of antiepileptic drugs. Lancet Neurol 2012;11(9):792–802.
[15] Galanopoulou AS, Buckmaster PS, Staley KJ, Moshé SL, Perucca E, Engel J, et al. Identification of new epilepsy treatments: issues in preclinical methodology. Epilepsia 2012;53(3):571–82.
[16] Löscher W, Schmidt D. Modern antiepileptic drug development has failed to deliver: ways out of the current dilemma. Epilepsia 2011;52(4):657–78.
[17] Temkin NR, Dikmen SS, Wilensky AJ, Keihm J, Chabal S, Winn HR. A randomized, double-blind study of phenytoin for the prevention of post-traumatic seizures. New Engl J Med 1990;323(8):497–502.
[18] Simonato M, Löscher W, Cole AJ, Dudek FE, Engel Jr. J, Kaminski RM, et al. Finding a better drug for epilepsy: preclinical screening strategies and experimental trial design. Epilepsia 2012;53(11):1860–7.
[19] Binder DK, Steinhäuser C. Functional changes in astroglial cells in epilepsy. Glia 2006;54:358–68.
[20] Friedman A, Kaufer D, Heinemann U. Blood–brain barrier breakdown-inducing astrocytic transformation: novel targets for the prevention of epilepsy. Epilepsy Res 2009;85:142–9.

[21] Ransom B, Behar T, Nedergaard M. New roles for astrocytes (stars at last). Trends Neurosci 2003;26:520–2.
[22] Binder DK, Nagelhus EA, Ottersen OP. Aquaporin-4 and epilepsy. Glia 2012;60:1203–14.
[23] Clasadonte J, Haydon PG. Astrocytes and epilepsy. In: Noebles JL, Avoli M, Rogawski MA, Olsen RW, Delgado-Escueta AV, editors. Jasper's basic mechanisms of the epilepsies (4th ed.). New York: Oxford University Press; 2012. p. 19.
[24] Hsu MS, Lee DJ, Binder DK. Potential role of the glial water channel aquaporin-4 in epilepsy. Neuron Glia Biol 2007;3(4):287–97.
[25] Tian G, Azmi H, Takano T, Xu Q, Peng W, Lin J, et al. An astrocytic basis of epilepsy. Nature Med 2005;11(9):973–81.
[26] Seifert G, Carmignoto G, Steinhäuser C. Astrocyte dysfunction in epilepsy. Brain Res Rev 2010;63:212–21.
[27] Seifert G, Schilling K, Steinhäuser C. Astrocyte dysfunction in neurological disorders: a molecular perspective. Nature Rev Neurosci 2006;7:194–206.
[28] Beenhakker MP, Huguenard JR. Astrocytes as gatekeepers of $GABA_B$ receptor function. J Neurosci 2010;30(45):15262–76.
[29] Wang F, Smith NA, Xu Q, Fujita T, Baba A, Matsuda T, et al. Astrocytes modulate neural network activity by Ca^{2+}-dependent uptake of extracellular K^+. Sci Signal 2012;5(218):ra26.
[30] Santello M, Bezzi P, Volterra A. TNFα controls glutamatergic gliotransmission in the hippocampal dentate gyrus. Neuron 2011;69(5):988–1001.
[31] Rouach N, Koulakoff A, Abudara V, Willecke K, Giaume C. Astroglial metabolic networks sustain hippocampal synaptic transmission. Science 2008;322(5907):1551–5.
[32] Volterra A, Steinhäuser C. Glial modulation of synaptic transmission in the hippocampus. Glia 2004;47:249–57.
[33] Halassa MM, Fellin T, Haydon PG. The tripartite synapse: roles for gliotransmission in health and disease. Trends Mol Med 2007;13(2):54–63.
[34] Heinemann U, Jauch GR, Schulze JK, Kivi A, Eilers A, Kovacs R, et al. Alterations of glial cell functions in temporal lobe epilepsy. Epilepsia 2000;41(Suppl. 6):S185–9.
[35] Oberheim NA, Tian GF, Han X, Peng W, Takano T, Ransom B, et al. Loss of astrocytic domain organization in the epileptic brain. J Neurosci 2008;28(13):3264–76.
[36] Bedner P, Dupper A, Hüttmann K, Muller J, Herde MK, Dublin P, et al. Astrocyte uncoupling as a cause of human temporal lobe epilepsy. Brain 2015;138(Pt 5):1208–22.
[37] Aronica E, Zurolo E, Iyer A, de Groot M, Anink J, Carbonell C, et al. Upregulation of adenosine kinase in astrocytes in experimental and human temporal lobe epilepsy. Epilepsia 2011;52(9):1645–55.
[38] Coulter DA, Eid T. Astrocytic regulation of glutamate homeostasis in epilepsy. Glia 2012;60:1215–26.
[39] Steinhäuser C, Seifert G. Glial membrane channels and receptors in epilepsy: impact for generation and spread of seizure activity. Eur J Pharmacol 2002;447:227–37.
[40] de Lanerolle NC, Lee T. New facets of the neuropathology and molecular profile of human temporal lobe epilepsy. Epilepsy Behav 2005;7:190–203.
[41] Hinterkeuser S, Schröder W, Hager G, Seifert G, Blümcke I, Elger CE, et al. Astrocytes in the hippocampus of patients with temporal lobe epilepsy display changes in potassium conductances. Eur J Neurosci 2000;12:2087–96.
[42] Kivi A, Lehmann TN, Kovács R, Eilers A, Jauch R, Meencke HJ, et al. Effects of barium on stimulus-induced rises of $[K^+]_o$ in human epileptic non-sclerotic and sclerotic hippocampal area CA1. Eur J Neurosci 2000;12:2039–48.
[43] Verkhratsky A, Steardo L, Parpura V, Montana V. Translational potential of astrocytes in brain disorders. Prog Neurobiol Sep 16, 2015. [Epub ahead of print].
[44] Kardos J, Szabo Z, Heja L. Framing neuro-glia coupling in antiepileptic drug design. J Med Chem 2015;59(3):777–87.
[45] Djukic B, Casper KB, Philpot BD, Chin LS, McCarthy KD. Conditional knock-out of $K_{ir}4.1$ leads to glial membrane depolarization, inhibition of potassium and glutamate uptake, and enhanced short-term synaptic potentiation. J Neurosci 2007;27(42):11354–65.
[46] Binder DK, Yao X, Zador Z, Sick TJ, Verkman AS. Increased seizure duration and slowed potassium kinetics in mice lacking aquaporin-4 water channels. Glia 2006;53:631–6.
[47] Hsu MS, Seldin M, Lee DJ, Seifert G, Steinhäuser C, Binder DK. Laminar-specific and developmental expression of aquaporin-4 in the mouse hippocampus. Neuroscience 2011;178:21–32.

[48] Ruiz-Ederra J, Zhang H, Verkman AS. Evidence against functional interaction between aquaporin-4 water channels and $K_{ir}4.1$ potassium channels in retinal Müller cells. J Biol Chem 2007;282(30):21866–72.

[49] Zhang H, Verkman AS. Aquaporin-4 independent $K_{ir}4.1$ K^+ channel function in brain glial cells. Mol Cell Neurosci 2008;37(1):1–10.

[50] Jin BJ, Zhang H, Binder DK, Verkman AS. Aquaporin-4-dependent K^+ and water transport modeled in brain extracellular space following neuroexcitation. J Gen Physiol 2013;141(1):119–32.

[51] Das A, Wallace GC, Holmes C, McDowell ML, Smith JA, Marshall JD, et al. Hippocampal tissue of patients with refractory temporal lobe epilepsy is associated with astrocyte activation, inflammation, and altered expression of channels and receptors. Neuroscience 2012;220:237–46.

[52] Heuser K, Eid T, Lauritzen F, Thoren AE, Vindedal GF, Tauboll E, et al. Loss of perivascular $K_{ir}4.1$ potassium channels in the sclerotic hippocampus of patients with mesial temporal lobe epilepsy. J Neuropathol Exp Neurol 2012;71(9):814–25.

[53] Bordey A, Sontheimer H. Properties of human glial cells associated with epileptic seizure foci. Epilepsy Res 1998;32:286–303.

[54] Schröder WHS, Seifert G, Schramm J, Jabs R, Wilkin GP, Steinhäuser C. Functional and molecular properties of human astrocytes in acute hippocampal slices obtained from patients with temporal lobe epilepsy. Epilepsia 2000;41(Suppl. 6):S181–4.

[55] Lee DJ, Hsu MS, Seldin MM, Arellano JL, Binder DK. Decreased expression of the glial water channel aquaporin-4 in the intrahippocampal kainic acid model of epileptogenesis. Exp Neurol 2012;235(1):246–55.

[56] Alvestad S, Hammer J, Hoddevik EH, Skare O, Sonnewald U, Amiry-Moghaddam M, et al. Mislocalization of AQP4 precedes chronic seizures in the kainate model of temporal lobe epilepsy. Epilepsy Res 2013;105(1-2):30–41.

[57] Eid T, Lee TW, Thomas MJ, Amiry-Moghaddam M, Bjørnsen LP, Spencer DD, et al. Loss of perivascular aquaporin-4 may underlie deficient water and K^+ homeostasis in the human epileptogenic hippocampus. PNAS 2005;102(4):1193–8.

[58] Proper EA, Hoogland G, Kappen SM, Jansen GH, Rensen MGA, Schrama LH, et al. Distribution of glutamate transporters in the hippocampus of patients with pharmaco-resistant temporal lobe epilepsy. Brain 2002;125:32–43.

[59] Eid T, Thomas MJ, Spencer DD, Rundén-Pran E, Lai JC, Malthankar GV, et al. Loss of glutamine synthetase in the human epileptogenic hippocampus: possible mechanisms for raised extracellular glutamate in mesial temporal lobe epilepsy. Lancet 2004;363:28–37.

[60] Hibino H, Inanobe A, Furutani K, Murakami S, Findlay I, Kurachi Y. Inwardly rectifying potassium channels: their structure, function, and physiological roles. Physiol Rev 2010;90(1):291–366.

[61] Papadopoulos MC, Verkman AS. Potential utility of aquaporin modulators for therapy of brain disorders. Prog Brain Res 2008;170:589–601.

[62] Tanaka K, Watase K, Manabe T, Yamada K, Watanabe M, Takahashi K, et al. Epilepsy and exacerbation of brain injury in mice lacking the glutamate transporter GLT1. Science 1997;276:1699–702.

[63] Petr GT, Sun Y, Frederick NM, Zhou Y, Dhamne SC, Hameed MQ, et al. Conditional deletion of the glutamate transporter GLT1 reveals that astrocytic GLT1 protects against fatal epilepsy while neuronal GLT1 contributes significantly to glutamate uptake into synaptosomes. J Neurosci 2015;35(13):5187–201.

[64] Kong Q, Takahashi K, Schulte D, Stouffer N, Lin Y, Lin CG. Increased glial glutamate transporter EAAT2 expression reduces epileptogenic processes following pilocarpine-induced status epilepticus. Neurobiol Dis 2012;47:145–54.

[65] Rothstein JD, Patel S, Regan MR, Haenggeli C, Huang YH, Bergles DE, et al. β-lactam antibiotics offer neuroprotection by increasing glutamate transporter expression. Nature 2005;433(7021):73–7.

[66] Lee S, Su Z, Emdad L, Gupta P, Sarkar D, Borjabad A, et al. Mechanism of ceftriaxone induction of excitatory amino acid transporter-2 expression and glutamate uptake in primary human astrocytes. J Biol Chem 2008;283(19):13116–23.

[67] Melzer N, Meuth SG, Torres-Salazar D, Bittner S, Zozulya AL, Weidenfeller C, et al. A β-lactam antibiotic dampens excitotoxic inflammatory CNS damage in a mouse model of multiple sclerosis. PLoS One 2008;3(9):E3149.

[68] Carbone M, Duty S, Rattray M. Riluzole elevates GLT1 activity and levels in striatal astrocytes. Neurochem Int 2012;60(1):31–8.

[69] Rasmussen BA, Baron DA, Kim JK, Unterwalk EM, Rawls SM. β-lactam antibiotic produces a sustained reduction in extracellular glutamate in the nucleus accumbens of rats. Amino Acids 2011;40:761–4.

[70] Zeng L, Bero AW, Zhang B, Holtzman DM, Wong M. Modulation of astrocyte glutamate transporters decreases seizures in a mouse model of tuberous sclerosis complex. Neurobiol Dis 2010;37(3):764–71.

[71] Jelenkovic AV, Jovanovic MD, Stanimirovic DD, Bokonjic DD, Ocic GG, Boskovic BS. Beneficial effects of ceftriaxone against pentylenetetrazole-evoked convulsions. Exper Biol Med 2008;233:1389–94.

[72] Soni N, Koushal P, Reddy BV, Deshmukh R, Kumar P. Effect of GLT1 modulator and $P2X_7$ antagonists alone and in combination in the kindling model of epilepsy in rats. Epilepsy Behav 2015;48:4–14.

[73] Li Y, Sattler R, Yang EJ, Nunes A, Ayukawa Y, Akhtar S, et al. Harmine, a natural β-carboline alkaloid, upregulates astroglial glutamate transporter expression. Neuropharmacology 2011;60:1168–75.

[74] Rawls SM, Karaca F, Madhani I, Vhojani V, Martinez RL, Abou-Gharbia M, et al. β-lactamase inhibitors display anti-seizure properties in an invertebrate assay. Neuroscience 2010;169:1800–4.

[75] Gasior M, Socala K, Nieoczym D, Wlaź P. Clavulanic acid does not affect convulsions in acute seizure tests in mice. J Neural Transm 2012;119(1):1–6.

[76] Huh Y, Ju MS, Park H, Han S, Bang Y, Ferris CF, et al. Clavulanic acid protects neurons in pharmacological models of neurodegenerative diseases. Drug Develop Res 2010;71(6):351–7.

[77] Matos M, Augusto E, Machado NJ, dos Santos-Rodrigues A, Cunha RA, Agostinho P. Astrocytic adenosine A2A receptors control the amyloid-beta peptide-induced decrease of glutamate uptake. J Alzheimers Dis 2012;31(3):555–67.

[78] Matos M, Augusto E, Santos-Rodrigues AD, Schwarzschild MA, Chen JF, Cunha RA, et al. Adenosine A_{2A} receptors modulate glutamate uptake in cultured astrocytes and gliosomes. Glia 2012;60(5):702–16.

[79] Nishizaki T. ATP- and adenosine-mediated signaling in the central nervous system: adenosine stimulates glutamate release from astrocytes via A_{2A} adenosine receptors. J Pharmacol Sci 2004;94(2):100–2.

[80] Nishizaki T, Nagai K, Nomura T, Tada H, Kanno T, Tozaki H, et al. A new neuromodulatory pathway with a glial contribution mediated via A_{2A} adenosine receptors. Glia 2002;39(2):133–47.

[81] Li XX, Nomura T, Aihara H, Nishizaki T. Adenosine enhances glial glutamate efflux via A_{2A} adenosine receptors. Life Sci 2001;68(12):1343–50.

[82] Frizzo ME, Frizzo JK, Amadio S, Rodrigues JM, Perry ML, Bernardi G, et al. Extracellular adenosine triphosphate induces glutamate transporter-1 expression in hippocampus. Hippocampus 2007;17(4):305–15.

[83] Wu J, Lee MR, Choi S, Kim T, Choi DS. ENT1 regulates ethanol-sensitive EAAT2 expression and function in astrocytes. Alcohol Clin Exp Res 2010;34(6):1110–7.

[84] Wu J, Lee MR, Kim T, Johng S, Rohrback S, Kang N, et al. Regulation of ethanol-sensitive EAAT2 expression through adenosine A_1 receptor in astrocytes. Biochem Biophys Res Commun 2011;406(1):47–52.

[85] Kong Q, Chang LC, Takahashi K, Liu Q, Schulte DA, Lai L, et al. Small-molecule activator of glutamate transporter EAAT2 translation provides neuroprotection. J Clin Invest 2014;124(3):1255–67.

[86] Li K, Javed E, Hala TJ, Sannie D, Regan KA, Maragakis NJ, et al. Transplantation of glial progenitors that overexpress glutamate transporter GLT1 preserves diaphragm function following cervical SCI. Mol Ther 2015;23(3):533–48.

[87] Falnikar A, Hala TJ, Poulsen DJ, Lepore AC. GLT1 overexpression reverses established neuropathic pain-related behavior and attenuates chronic dorsal horn neuron activation following cervical spinal cord injury. Glia 2016;64(3):396–406.

[88] Murphy-Royal C, Dupuis JP, Varela JA, Panatier A, Pinson B, Baufreton J, et al. Surface diffusion of astrocytic glutamate transporters shapes synaptic transmission. Nat Neurosci 2015;18(2):219–26.

[89] Foran E, Rosenblum L, Bogush A, Pasinelli P, Trotti D. Sumoylation of the astroglial glutamate transporter EAAT2 governs its intracellular compartmentalization. Glia 2014;62(8):1241–53.

[90] Raju K, Doulias PT, Evans P, Krizman EN, Jackson JG, Horyn O, et al. Regulation of brain glutamate metabolism by nitric oxide and S-nitrosylation. Sci Signal 2015;8(384):ra68.

[91] Smith AJ, Jin BJ, Ratelade J, Verkman AS. Aggregation state determines the localization and function of M1- and M23-aquaporin-4 in astrocytes. J Cell Biol 2014;204(4):559–73.

[92] Crane JM, Bennett JL, Verkman AS. Live cell analysis of aquaporin-4 M1–M23 interactions and regulated orthogonal array assembly in glial cells. J Biol Chem 2009;284(51):35850–60.

[93] Jin BJ, Rossi A, Verkman AS. Model of aquaporin-4 supramolecular assembly in orthogonal arrays based on heterotetrameric association of M1–M23 isoforms. Biophys J 2011;100(12):2936–45.

[94] Haglund MM, Hochman DW. Furosemide and mannitol suppression of epileptic activity in the human brain. J Neurophysiol 2005;94(2):907–18.
[95] Traynelis SF, Dingledine R. Role of extracellular space in hyperosmotic suppression of potassium-induced electrographic seizures. J Neurophysiol 1989;61(5):927–38.
[96] Andrew RD, Fagan M, Ballyk BA, Rosen AS. Seizure susceptibility and the osmotic state. Brain Res 1989;498:175–80.
[97] Maa EH, Kahle KT, Walcott BP, Spitz MC, Staley KJ. Diuretics and epilepsy: will the past and present meet? Epilepsia 2011;52(9):1559–69.
[98] Dudek FE, Obenaus A, Tasker JG. Osmolality-induced changes in extracellular volume alter epileptiform bursts independent of chemical synapses in the rat: importance of non-synaptic mechanisms in hippocampal epileptogenesis. Neurosci Lett 1990;120:267–70.
[99] Nagelhus EA, Mathiisen TM, Ottersen OP. Aquaporin-4 in the central nervous system: cellular and subcellular distribution and coexpression with $K_{ir}4.1$. Neuroscience 2004;129:905–13.
[100] Schwartzkroin PA, Baraban SC, Hochman DW. Osmolarity, ionic flux, and changes in brain excitability. Epilepsy Res 1998;32:275–85.
[101] Hubbard JA, Hsu MS, Fiacco TA, Binder DK. Glial cell changes in epilepsy: overview of the clinical problem and therapeutic opportunities. Neurochem Int 2013;63(7):638–51.
[102] Peters A, Palay SL, Webster D. The fine structure of the nervous system: neurons and their supporting cells. New York: Oxford University Press; 1991.
[103] Andrew RD, Labron MW, Boehnke SE, Carnduff L, Kirov SA. Physiological evidence that pyramidal neurons lack functional water channels. Cereb Cortex 2007;17(4):787–802.
[104] Risher WC, Andrew RD, Kirov SA. Real-time passive volume responses of astrocytes to acute osmotic and ischemic stress in cortical slices and in vivo revealed by two-photon microscopy. Glia 2009;57(2):207–21.
[105] Amiry-Moghaddam M, Williamson A, Palomba M, Eid T, de Lanerolle NC, Nagelhus EA, et al. Delayed K^+ clearance associated with aquaporin-4 mislocalization: phenotypic defects in brains of α-syntrophin-null mice. Proc Natl Acad Sci USA 2003;100(23):13615–20.
[106] Nagelhus EA, Horio Y, Inanobe A, Fujita A, Haug F, Nielsen S, et al. Immunogold evidence suggests that coupling of K^+ siphoning and water transport in rat retinal Müller cells is mediated by a coenrichment of $K_{ir}4.1$ and AQP4 in specific membrane domains. Glia 1999;26:47–54.
[107] Strohschein S, Hüttmann K, Gabriel S, Binder DK, Heinemann U, Steinhäuser C. Impact of aquaporin-4 channels on K^+ buffering and gap junction coupling in the hippocampus. Glia 2011;59:973–80.
[108] Abdullaev IF, Rudkouskaya A, Schools GP, Kimelberg HK, Mongin AA. Pharmacological comparison of swelling-activated excitatory amino acid release and Cl^- currents in cultured rat astrocytes. J Physiol 2006;572(Pt 3):677–89.
[109] Haskew-Layton RE, Rudkouskaya A, Jin Y, Feustel PJ, Kimelberg HK, Mongin AA. Two distinct modes of hypoosmotic medium-induced release of excitatory amino acids and taurine in the rat brain in vivo. PLoS One 2008;3(10):e3543.
[110] Liu HT, Tashmukhamedov BA, Inoue H, Okada Y, Sabirov RZ. Roles of two types of anion channels in glutamate release from mouse astrocytes under ischemic or osmotic stress. Glia 2006;54(5):343–57.
[111] Kimelberg HK, Macvicar BA, Sontheimer H. Anion channels in astrocytes: biophysics, pharmacology, and function. Glia 2006;54(7):747–57.
[112] Fiacco TA, Agulhon C, Taves SR, Petravicz J, Casper KB, Dong X, et al. Selective stimulation of astrocyte calcium in situ does not affect neuronal excitatory synaptic activity. Neuron 2007;54(4):611–26.
[113] Lauderdale K, Murphy T, Tung T, Davila D, Binder DK, Fiacco TA. Osmotic edema rapidly increases neuronal excitability through activation of NMDA receptor-dependent slow inward currents in juvenile and adult hippocampus. ASN Neuro 2015;7:5.
[114] Wetherington J, Serrano G, Dingledine R. Astrocytes in the epileptic brain. Neuron 2008;58(2):168–78.
[115] Gunnarson E, Zelenina M, Axehult G, Song Y, Bondar A, Krieger P, et al. Identification of a molecular target for glutamate regulation of astrocyte water permeability. Glia 2008;56:587–96.
[116] Kimelberg HK. Increased release of excitatory amino acids by the actions of ATP and peroxynitrite on volume-regulated anion channels (VRACs) in astrocytes. Neurochem Int 2004;45(4):511–9.
[117] Mongin AA, Kimelberg HK. ATP potently modulates anion channel-mediated excitatory amino acid release from cultured astrocytes. Am J Physiol Cell Physiol 2002;283(2):C569–78.

[118] Frasca A, Aalbers M, Frigerio F, Fiordaliso F, Salio M, Gobbi M, et al. Misplaced NMDA receptors in epileptogenesis contribute to excitotoxicity. Neurobiol Dis 2011;43(2):507–15.
[119] Zhang Y, Zhang H, Feustel PJ, Kimelberg HK. DCPIB, a specific inhibitor of volume regulated anion channels (VRACs), reduces infarct size in MCAo and the release of glutamate in the ischemic cortical penumbra. Exp Neurol 2008;210:514–20.
[120] Takano T, Kang J, Jaiswal JK, Simon SM, Lin JH, Yu Y, et al. Receptor-mediated glutamate release from volume sensitive channels in astrocytes. Proc Natl Acad Sci USA 2005;102(45):16466–71.
[121] Petravicz J, Fiacco TA, McCarthy KD. Loss of IP_3 receptor-dependent Ca^{2+} increases in hippocampal astrocytes does not affect baseline CA1 pyramidal neuron synaptic activity. J Neurosci 2008;28(19):4967–73.
[122] Dunwiddie TV. Adenosine and suppression of seizures. Adv Neurol 1999;79:1001–10.
[123] Winn HR, Welsh JE, Berne RM, Rubio R. Changes in brain adenosine during bicuculline-induced seizures: effect of altered arterial oxygen tensions. Trans Am Neurol Assoc 1979;104:239–41.
[124] Haas HL, Selbach O. Functions of neuronal adenosine receptors. Naunyn Schmiedebergs Arch Pharmacol 2000;362(4–5):375–81.
[125] Cunha RA. Neuroprotection by adenosine in the brain: from A_1 receptor activation to A_{2A} receptor blockade. Purinergic Signal 2005;1(2):111–34.
[126] Dunwiddie TV, Diao L. Extracellular adenosine concentrations in hippocampal brain slices and the tonic inhibitory modulation of evoked excitatory responses. J Pharmacol Exp Ther 1994;268(2):537–45.
[127] Dunwiddie TV, Diao L, Proctor WR. Adenine nucleotides undergo rapid, quantitative conversion to adenosine in the extracellular space in rat hippocampus. J Neurosci 1997;17(20):7673–82.
[128] Dunwiddie TV, Diao L. Regulation of extracellular adenosine in rat hippocampal slices is temperature dependent: role of adenosine transporters. Neuroscience 2000;95(1):81–8.
[129] Arch JRS, Newsholme EA. Activities and some properties of 5′-nucleotidase, adenosine kinase, and adenosine deaminase in tissues from vertebrates and invertebrates in relation to the control of the concentration and the physiological role of adenosine. Biochem J 1978;174:965–77.
[130] Boison D, Stewart K. Therapeutic epilepsy research: from pharmacological rationale to focal adenosine augmentaiton. Biochem Pharmacol 2009;78(12):1428–37.
[131] Boison D, Chen J, Fredholm BB. Adenosine signalling and function in glial cells. Cell Death Differ 2010;17(7):1071–82.
[132] Boison D. Adenosine dysfunction in epilepsy. Glia 2012;60:1234–43.
[133] Gouder N, Scheurer L, Fritschy J, Boison D. Overexpression of adenosine kinase in epileptic hippocampus contributes to epileptogenesis. J. Neurosci 2004;24(3):692–701.
[134] Boison D. Adenosine and epilepsy: from therapeutic rationale to new therapeutic strategies. Neuroscientist 2005;11(1):25–36.
[135] Jacobson KA, Gao ZG. Adenosine receptors as therapeutic targets. Nature Rev Drug Discov 2006;5:247–64.
[136] Spedding M, William M. Developments in purine and pyridimidine receptor-based therapeutics. Drug Develop Res 1996;39:436–41.
[137] Williams M. Development in P2 receptor targeted therapeutics. Prog Brain Res 1999;120:93–106.
[138] Williams M, Jarvis MF. Purinergic and pyrimidinergic receptors as potential drug targets. Biochem Pharmacol 2000;59:1173–85.
[139] Anschel DJ, Ortega EL, Kraus AC, Fisher RS. Focally injected adenosine prevents seizures in the rat. Exp Neurol 2004;190:544–7.
[140] Li T, Ren G, Lusardi T, Wilz A, Lan JQ, Iwasato T, et al. Adenosine kinase is a target for the prediction and prevention of epileptogenesis in mice. J Clin Invest 2008;118(2):571–82.
[141] Wilz A, Pritchard EM, Li T, Lan J, Kaplan DL, Boison D. Silk polymer based adenosine release: therapeutic potential for epilepsy. Biomaterials 2008;29:3609–16.
[142] Huber A, Padrun V, Déglon N, Aebischer P, Möhler H, Boison D. Grafts of adenosine-releasing cells suppress seizures in kindling epilepsy. Proc Natl Acad Sci USA 2001;98(13):7611–6.
[143] Boison D. Adenosinergic signaling in epilepsy. Neuropharmacology 2015;73:122–37 [Epub ahead of print].
[144] Williams-Karnesky RL, Sandau US, Lusardi TA, Lytle NK, Farrell JM, Pritchard EM, et al. Epigenetic changes induced by adenosine augmentation therapy prevent epileptogenesis. J Clin Invest 2013;123(8):3552–63.
[145] Ye Z, Sontheimer H. Glioma cells release excitotoxic concentrations of glutamate. Cancer Res 1999;59:4383–91.
[146] Buckingham SC, Campbell SL, Haas BR, Montana V, Robel S, Ogunrinu T, et al. Glutamate release by primary brain tumors induces epileptic activity. Nat Med 2011;17(10):1269–74.

[147] Robel S, Sontheimer H. Glia as drivers of abnormal neuronal activity. Nat Neurosci 2015;19(1):28–33.

[148] Robert SM, Buckingham SC, Campbell SL, Robel S, Holt KT, Ogunrinu-Babarinde T, et al. SLC7A11 expression is associated with seizures and predicts poor survival in patients with malignant glioma. Sci Transl Med 2015;7(289):289ra86.

[149] Osswald M, Jung E, Sahm F, Solecki G, Venkataramani V, Blaes J, et al. Brain tumour cells interconnect to a functional and resistant network. Nature 2015;528(7580):93–8.

[150] Binder DK, Berger MS. Proteases and the biology of glioma invasion. J Neurooncol 2002;56(2):149–58.

[151] Nemani VM, Binder DK. Emerging role of gap junctions in epilepsy. Histol Histopathol 2005;20(1):253–9.

[152] Meme W, Calvo CF, Froger N, Ezan P, Amigou E, Koulakoff A, et al. Proinflammatory cytokines released from microglia inhibit gap junctions in astrocytes: potentiation by β-amyloid. FASEB J 2006;20(3):494–6.

[153] Moore KB, O'Brien J. Connexins in neurons and glia: targets for intervention in disease and injury. Neural Regen Res 2015;10(7):1013–7.

[154] Temkin NR, Anderson GD, Winn HR, Ellenbogen RG, Britz GW, Schuster J, et al. Magnesium sulfate for neuroprotection after traumatic brain injury: a randomised controlled trial. Lancet Neurol 2007;6(1):29–38.

[155] Temkin NR, Dikmen SS, Anderson GD, Wilensky AJ, Holmes MD, Cohen W, et al. Valproate therapy for prevention of posttraumatic seizures: a randomized trial. J Neurosurg 1999;91(4):593–600.

[156] Chung MG, O'Brien NF. Prevalence of early posttraumatic seizures in children with moderate to severe traumatic brain injury despite levetiracetam prophylaxis. Pediatr Crit Care Med 2015;33:13–23.

[157] Cotton BA, Kao LS, Kozar R, Holcomb JB. Cost–utility analysis of levetiracetam and phenytoin for posttraumatic seizure prophylaxis. J Trauma 2011;71(2):375–9.

[158] Gabriel WM, Rowe AS. Long-term comparison of GOS-E scores in patients treated with phenytoin or levetiracetam for posttraumatic seizure prophylaxis after traumatic brain injury. Ann Pharmacother 2014;48(11):1440–4.

[159] Inaba K, Menaker J, Branco BC, Gooch J, Okoye OT, Herrold J, et al. A prospective multicenter comparison of levetiracetam versus phenytoin for early posttraumatic seizure prophylaxis. J Trauma Acute Care Surg 2013;74(3):766–71.

[160] Klein P, Herr D, Pearl PL, Natale J, Levine Z, Nogay C, et al. Results of phase 2 safety and feasibility study of treatment with levetiracetam for prevention of posttraumatic epilepsy. Arch Neurol 2012;69(10):1290–5.

[161] Pearl PL, McCarter R, McGavin CL, Yu Y, Sandoval F, Trzcinski S, et al. Results of phase II levetiracetam trial following acute head injury in children at risk for posttraumatic epilepsy. Epilepsia 2013;54(9):e135–7.

[162] Goodrich GS, Kabakov AY, Hameed MQ, Dhamne SC, Rosenberg PA, Rotenberg A. Ceftriaxone treatment after traumatic brain injury restores expression of the glutamate transporter, GLT1, reduces regional gliosis, and reduces post-traumatic seizures in the rat. J Neurotrauma 2013;30(16):1434–41.

[163] Diamond ML, Ritter AC, Failla MD, Boles JA, Conley YP, Kochanek PM, et al. IL-1β associations with posttraumatic epilepsy development: a genetics and biomarker cohort study. Epilepsia 2014;55(7):1109–19.

[164] Liu Z, Chopp M. Astrocytes, therapeutic targets for neuroprotection and neurorestoration in ischemic stroke. Prog Neurobiol 2015.

[165] Simard M, Arcuino G, Takano T, Liu QS, Nedergaard M. Signaling at the gliovascular interface. J Neurosci 2003;23(27):9254–62.

[166] Abbott NJ. Astrocyte–endothelial interactions and blood–brain barrier permeability. J Anat 2002;200(6):629–38.

[167] Metea MR, Newman EA. Glial cells dilate and constrict blood vessels: a mechanism of neurovascular coupling. J Neurosci 2006;26(11):2862–70.

[168] Mulligan SJ, MacVicar BA. Calcium transients in astrocyte endfeet cause cerebrovascular constrictions. Nature 2004;431(7005):195–9.

[169] Takano T, Tian GF, Peng W, Lou N, Libionka W, Han X, et al. Astrocyte-mediated control of cerebral blood flow. Nat Neurosci 2006;9(2):260–7.

[170] Zonta M, Angulo MC, Gobbo S, Rosengarten B, Hossmann KA, Pozzan T, et al. Neuron-to-astrocyte signaling is central to the dynamic control of brain microcirculation. Nat Neurosci 2003;6(1):43–50.

[171] Seiffert E, Dreier JP, Ivens S, Bechmann I, Tomkins O, Heinemann U, et al. Lasting blood–brain barrier disruption induces epileptic focus in the rat somatosensory cortex. J Neurosci 2004;24(36):7829–36.

[172] Weissberg I, Wood L, Kamintsky L, Vazquez O, Milikovsky DZ, Alexander A, et al. Albumin induces excitatory synaptogenesis through astrocytic TGF-β/ALK5 signaling in a model of acquired epilepsy following blood–brain barrier dysfunction. Neurobiol Dis 2015;78:115–25.

[173] Bar-Klein G, Cacheaux LP, Kamintsky L, Prager O, Weissberg I, Schoknecht K, et al. Losartan prevents acquired epilepsy via TGF-β signaling suppression. Ann Neurol 2014;75(6):864–75.

[174] Friedman A, Bar-Klein G, Serlin Y, Parmet Y, Heinemann U, Kaufer D. Should losartan be administered following brain injury? Expert Rev Neurother 2014;14(12):1365–75.

[175] Fabene PF, Mora GN, Martinello M, Rossi B, Merigo F, Ottoboni L, et al. A role for leukocyte–endothelial adhesion mechanism in epilepsy. Nature Med 2008;14(12):1377–83.

[176] Fabene PF, Bramanti P, Constantin G. The emerging role for chemokines in epilepsy. J Neuroimmunol 2010;224:22–7.

[177] Marchi N, Johnson AJ, Puvenna V, Johnson HL, Tierney W, Ghosh C, et al. Modulation of peripheral cytotoxic cells and ictogenesis in a model of seizures. Epilepsia 2011;52(9):1627–34.

[178] Liddelow S, Hoyer D. Astrocytes: adhesion molecules and immunomodulation. Curr Drug Targets Jan 1, 2016. [Epub ahead of print].

[179] Nguyen MD, Julien J, Rivest S. Innate immunity: the missing link in neuroprotection and neurodegeneration? Nature 2002;3:216–27.

[180] Vezzani A. Inflammation and epilepsy. Epilepsy Curr 2012;5(1):1–6.

[181] Crespel A, Coubes P, Rousset M, Brana C, Rougier A, Rondouin G, et al. Inflammatory reactions in human medial temporal lobe epilepsy with hippocampal sclerosis. Brain Res 2002;952:159–69.

[182] Ravizza T, Gagliardi B, Noé F, Boer K, Aronica E, Vezzani A. Innate and adaptive immunity during epileptogenesis and spontaneous seizures: evidence from experimental models and human temporal lobe epilepsy. Neurobiol Dis 2008;29(1):142–60.

[183] Beach TG, Woodhurst WB, MacDonald DB, Jones MW. Reactive microglia in hippocampal sclerosis associated with human temporal lobe epilepsy. Neurosci Lett 1995;191(1–2):27–30.

[184] Sheng JG, Boop FA, Mrak RE, Griffin WST. Increased neuronal β-amyloid precursor protein expression in human temporal lobe epilepsy: association with interleukin-1α immunoreactivity. J Neurochem 1994;63(5):1872–9.

[185] Zattoni M, Mura ML, Deprez F, Schwendener RA, Engelhardt B, Frei K, et al. Brain infiltration of leukocytes contributes to the pathophysiology of temporal lobe epilepsy. J Neurosci 2011;31(11):4037–50.

[186] Boer K, Spliet WGM, van Rijen PC, Redeker S, Troost D, Aronica E. Evidence of activated microglia in focal cortical dysplasia. J Neuroimmunol 2006;173:188–95.

[187] Iyer A, Zurolo E, Spliet WGM, van Rijen PC, Baayen JC, Gorter JA, et al. Evaluation of the innate and adaptive immunity in type I and type II focal cortical dysplasias. Epilepsia 2010;51(9):1763–73.

[188] Ravizza T, Vezzani A. Status epilepticus induces time-dependent neuronal and astrocytic expression of interleukin-1 receptor type 1 in the rat limbic system. Neuroscience 2006;137:301–8.

[189] Najjar S, Pearlman D, Miller DC, Devinsky O. Refractory epilepsy associated with microglial activation. Neurologist 2011;17:249–54.

[190] Aronica E, Boer K, van Vliet EA, Redeker S, Baayen JC, Spliet WGM, et al. Complement activation in experimental and human temporal lobe epilepsy. Neurobiol Dis 2007;26(3):497–511.

[191] Aronica E, Gorter JA. Gene expression profile in temporal lobe epilepsy. Neuroscientist 2007;13(2):100–8.

[192] Boer K, Jansen F, Nellist M, Redeker S, van den Ouweland AMW, Spliet WGM, et al. Inflammatory processes in cortical tubers and subependymal giant cell tumors of tuberous sclerosis complex. Epilepsy Res 2008;78:7–21.

[193] Boer K, Crino PB, Gorter JA, Nellist M, Jansen FE, Spliet WGM, et al. Gene expression analysis of tuberous sclerosis complex cortical tubers reveals increased expression of adhesion and inflammatory factors. Brain Pathol 2010;20(4):704–19.

[194] Abraham J, Fox PD, Condello C, Bartolini A, Koh S. Minocycline attenuates microglia activation and blocks the long-term epileptogenic effects of early-life seizures. Neurobiol Dis 2012;46:425–30.

[195] Ma L, Cui X, Wang Y, Li X, Yang F, Wei D, et al. Aspirin attenuates spontaneous recurrent seizures and inhibits hippocampal neuronal loss, mossy fiber sprouting and aberrant neurogenesis following pilocarpine-induced status epilepticus in rats. Brain Res 2012;1469:103–13.

[196] Dedeurwaerdere S, Friedman A, Fabene PF, Mazarati A, Murashima YL, Vezzani A, et al. Finding a better drug for epilepsy: antiinflammatory targets. Epilepsia 2012;53(7):1113–8.

[197] Vezzani A, Conti M, De Luigi A, Ravizza T, Moneta D, Marchesi F, et al. Interleukin-1β immunoreactivity and microglia are enhanced in the rat hippocampus by focal kainate application: functional evidence for enhancement of electrographic seizures. J Neurosci 1999;19(12):5054–65.

[198] Vezzani A, Moneta D, Conti M, Richichi C, Ravizza T, De Luigi A, et al. Powerful anticonvulsant action of IL-1 receptor antagonist on intracerebral injection and astrocytic overexpression in mice. Proc Natl Acad Sci USA 2000;97(21):11534–9.

[199] Ravizza T, Lucas S, Balosso S, Bernardino L, Ku G, Noé F, et al. Inactivation of caspase-1 in rodent brain: a novel anticonvulsive strategy. Epilepsia 2006;47(7):1160–8.
[200] Maroso M, Balosso S, Ravizza T, Liu J, Aronica E, Iyer AM, et al. Toll-like receptor 4 and high-mobility group box-1 are involved in ictogenesis and can be targeted to reduce seizures. Nature Med 2010;16(4):413–20.
[201] Zurolo E, Iyer A, Maroso M, Carbonell C, Anink JJ, Ravizza T, et al. Activation of Toll-like receptor, RAGE and HMGB1 signalling in malformations of cortical development. Brain 2011;134(Pt 4):1015–32.
[202] Aronica E, Ravizza T, Zurolo E, Vezzani A. Astrocyte immune responses in epilepsy. Glia 2012;60(8):1258–68.
[203] Iori V, Frigerio F, Vezzani A. Modulation of neuronal excitability by immune mediators in epilepsy. Curr Opin Pharmacol 2016;26:118–23.
[204] Boison D, Aronica E. Comorbidities in neurology: is adenosine the common link? Neuropharmacology 2015;97:18–34.
[205] Fedele DE, Gouder N, Guttinger M, Gabernet L, Scheurer L, Rulicke T, et al. Astrogliosis in epilepsy leads to overexpression of adenosine kinase, resulting in seizure aggravation. Brain 2005;128(Pt 10):2383–95.
[206] Shen HY, Lusardi TA, Williams-Karnesky RL, Lan JQ, Poulsen DJ, Boison D. Adenosine kinase determines the degree of brain injury after ischemic stroke in mice. J Cereb Blood Flow Metab 2011;31(7):1648–59.
[207] Li T, Quan Lan J, Fredholm BB, Simon RP, Boison D. Adenosine dysfunction in astrogliosis: cause for seizure generation? Neuron Glia Biol 2007;3(4):353–66.
[208] Yee BK, Singer P, Chen JF, Feldon J, Boison D. Transgenic overexpression of adenosine kinase in brain leads to multiple learning impairments and altered sensitivity to psychomimetic drugs. Eur J Neurosci 2007;26(11):3237–52.
[209] Shen HY, Singer P, Lytle N, Wei CJ, Lan JQ, Williams-Karnesky RL, et al. Adenosine augmentation ameliorates psychotic and cognitive endophenotypes of schizophrenia. J Clin Invest 2012;122(7):2567–77.
[210] Nielsen S, Nagelhus EA, Amiry-Moghaddam M, Bourque C, Agre P, Ottersen OP. Specialized membrane domains for water transport in glial cells: high-resolution immunogold cytochemistry of aquaporin-4 in rat brain. J Neurosci 1997;17(1):171–80.
[211] Skucas VA, Mathews IB, Yang J, Cheng Q, Treister A, Duffy AM, et al. Impairment of select forms of spatial memory and neurotrophin-dependent synaptic plasticity by deletion of glial aquaporin-4. J Neurosci 2011;31(17):6392–7.
[212] Binder DK, Scharfman HE. Brain-derived neurotrophic factor. Growth Factors 2004;22(3):123–31.
[213] Fan Y, Liu M, Wu X, Wang F, Ding J, Chen J, et al. Aquaporin-4 promotes memory consolidation in Morris water maze. Brain Struct Funct 2013;218(1):39–50.
[214] Li YK, Wang F, Wang W, Luo Y, Wu PF, Xiao JL, et al. Aquaporin-4 deficiency impairs synaptic plasticity and associative fear memory in the lateral amygdala: involvement of downregulation of glutamate transporter-1 expression. Neuropsychopharmacology 2012;37(8):1867–78.
[215] Yang J, Li MX, Luo Y, Chen T, Liu J, Fang P, et al. Chronic ceftriaxone treatment rescues hippocampal memory deficit in AQP4 knockout mice via activation of GLT1. Neuropharmacology 2013;75:213–22.
[216] Scharfman HE, Binder DK. Aquaporin-4 water channels and synaptic plasticity in the hippocampus. Neurochem Int 2013;63(7):702–11.
[217] Amlerova J, Laczo J, Vlcek K, Javurkova A, Andel R, Marusic P. Risk factors for spatial memory impairment in patients with temporal lobe epilepsy. Epilepsy Behav 2013;26(1):57–60.
[218] Bell B, Lin JJ, Seidenberg M, Hermann B. The neurobiology of cognitive disorders in temporal lobe epilepsy. Nat Rev Neurol 2011;7(3):154–64.
[219] Brooks-Kayal A, Bath KG, Berg A, Galanopoulou A, Holmes G, Jensen F, et al. Issues related to symptomatic and disease modifying treatments affecting cognitive and neuropsychiatric comorbidities of epilepsy. Epilepsia 2013;54(4):44–60.
[220] Chin J, Scharfman HE. Shared cognitive and behavioral impairments in epilepsy and Alzheimer's disease and potential underlying mechanisms. Epilepsy Behav 2013;26:343–51.
[221] Morimoto K, Fahnestock M, Racine RJ. Kindling and status epilepticus models of epilepsy: rewiring the brain. Prog Neurobiol 2004;73(1):1–60.
[222] Parent JM. The role of seizure-induced neurogenesis in epileptogenesis and brain repair. Epilepsy Res 2002;50(1–2):179–89.
[223] Scharfman HE. Epilepsy as an example of neural plasticity. Neuroscientist 2002;8(2):154–73.
[224] Szu JI, Binder DK. The role of astrocytic aquaporin-4 in synaptic plasticity and learning and memory. Front Integr Neurosci 2016;10:8.

[225] Degen J, Dublin P, Zhang J, Dobrowolski R, Jokwitz M, Karram K, et al. Dual reporter approaches for identification of Cre efficacy and astrocyte heterogeneity. FASEB J 2012;26(11):4576–83.

[226] Hoft S, Griemsmann S, Seifert G, Steinhäuser C. Heterogeneity in expression of functional ionotropic glutamate and GABA receptors in astrocytes across brain regions: insights from the thalamus. Philos Trans R Soc Lond B Biol Sci 2014;369(1654):20130602.

[227] Hüttmann K, Sadgrove M, Wallraff A, Hinterkeuser S, Kirchhoff F, Steinhäuser C, et al. Seizures preferentially stimulate proliferation of radial glia-like astrocytes in the adult dentate gyrus: functional and immunocytochemical analysis. Eur J Neurosci 2003;18(10):2769–78.

[228] Parent JM, von dem Bussche N, Lowenstein DH. Prolonged seizures recruit caudal subventricular zone glial progenitors into the injured hippocampus. Hippocampus 2006;16(3):321–8.

Index

Note: Page numbers followed by "*f*" and "*t*" refer to figures and tables, respectively.

D

E

F

H

K

L

M

N

O

P

Q

R

S

T